Photoelastic and Electro-Optic Properties of Crystals

Photoelastic and Electro-Optic Properties of Crystals

T. S. Narasimhamurty

Osmania University
Hyderabad, India

PLENUM PRESS • NEW YORK AND LONDON

Library of Congress Cataloging in Publication Data

Narasimhamurty, T. S.

Photoelastic and electro-optic properties of crystals.

Bibliography: p.

Includes index.

1. Crystal optics. 2. Photoelasticity. I. Title.

QD941.N37 548'.9 79-409

ISBN 0-306-31101-1

A Division of Plenum Publishing Corporation

233 Spring Street, New York, N.Y. 10013

Printed in the United States of America

Acknowledgments

The author hereby expresses his appreciation and thanks to the authors of papers and to the editors and publishers who so readily granted him their permission to include in this book the following material:

Figures

4.9	W. G. Mayer and Michigan State University. From Ref. [1534].
4.10A	S. Hirzel Verlag, Stuttgart. From L. Bergmann [110].
4.10B	M. A. Breazeale and E. A. Hiedemann, and American Institute of Physics. From Ref. [190].
4.10C	W. G. Mayer and E. A. Hiedemann, and *Acta Crystallogr.* (Denmark). From Ref. [790].
4.12	V. Chandrasekharan and *Proc. Indian Acad. Sci.* From Ref. [240].
5.6	H. E. Pettersen and Michigan State University. From Ref. [922].
5.8	K. Vedam, E. D. D. Schmidt, and W. C. Schneider, and Plenum Publishing Corp. From Ref. [1301].
5.9–5.11	*Proc. Indian Acad. Sci.* From Ref. [842].
5.12–5.14	M. Ziauddin and American Institute of Physics. From Ref. [844].
5.15	K. Veerabhadra Rao and American Institute of Physics. From Ref. [1303].
5.16	E. S. Jog and *J. Indian Inst. of Science.* From Ref. [586].
5.17	G. Jeelani and Osmania University. From Ref. [847].
5.21	G. Robertson and Institute of Physics, Bristol and London (U.K.). From Ref. [1057].
5.22	G. N. Ramachandran and V. Chandrasekharan, and *Proc. Indian Acad. Sci.* From Ref. [998].
5.23	H. F. Gates and E. A. Hiedemann, and American Institute of Physics. From Ref. [413].
5.24(a), (b)	Myron P. Hagelberg and Michigan State University. From Ref. [464].
5.24(c)	Osmania University. From Ref. [843].
5.25	K. Veerabhadra Rao and Osmania University. From Ref. [1306].
5.26	H. E. Pettersen and American Institute of Physics. From Ref. [924].
5.27	G. Alphonse and R.C.A. Review. From Ref. [32].
5.28	T. M. Smith and A. Korpel, and Institute of Electrical and Electronics Engineers Inc., New York. From Ref. [1152].
5.29, 5.30	R. W. Dixon and M.C. Cohen, and American Institute of Physics. From Ref. [294].
5.31	N. F. Borrelli and R. A. Miller, and American Institute of Physics. From Ref. [176].
5.32, 5.33	R. Adler and the Institute of Electrical and Electronics Engineers Inc., New York. From Ref. [13].
6.1	R. Srinivasan and *Zeitschrift für Physik* (West Germany). From Ref. [1175].
6.2, 6.3	A. Rahman and Osmania University. From Ref. [986].

6.4–6.6	V. G. Krishna Murty and Osmania University. From Ref. [681].
6.7	R. Ethirajan and Osmania University. From Ref. [330].
6.8	A. Rahman and K. S. Iyengar, and *Acta Crystallogr.* (Denmark). From Ref. [988].
6.9	A. Rahman and Osmania University. From Ref. [986].
8.13	American Institute of Physics. From Ref. [1419].
8.14	R. Adhav and American Institute of Physics. From Ref. [10].
8.15	American Institute of Physics. From Ref. [1417].
8.16.	S. Namba and American Institute of Physics. From Ref. [838].
8.17	American Institute of Physics. From Ref. [1417].

Tables

5.8	H. E. Pettersen and American Institute of Physics. From Ref. [925].
5.10	V. Chandrasekharan and *J. Indian Inst. Science*. From Ref. [241].
6.1	H. E. Pettersen and Michigan State University. From Ref. [922].
6.3–6.5	V. G. Krishna Murty and Osmania University. From Ref. [681].
6.6–6.8	A. Rahman and Osmania University. From Ref. [986].
8.4, 8.5	American Institute of Physics. From Ref. [1042].

Text

pp. 212–213	J. A. Mandarino and University of Michigan. From Ref. [765].

Foreword

This comprehensive treatise reviews, for the first time, all the essential work over the past 160 years on the photoelastic and the closely related linear and quadratic electro-optic effects in isotropic and crystalline materials. Emphasis is placed on the phenomenal growth of the subject during the past decade and a half with the advent of the laser, with the use of high-frequency acousto-optic and electro-optic techniques, and with the discovery of new piezoelectric materials, all of which have offered a feedback to the wide interest in these two areas of solid-state physics.

The first of these subjects, the photoelastic effect, was discovered by Sir David Brewster in 1815. He first found the effect in gels and subsequently found it in glasses and crystals. While the effect remained of academic interest for nearly a hundred years, it became of practical value when Coker and Filon applied it to measuring stresses in machine parts. With one photograph and subsequent analysis, the stress in any planar model can be determined. By taking sections of a three-dimensional model, complete three-dimensional stresses can be found. Hence this effect is widely applied in industry.

The photoelastic effect was analyzed for crystals by Pockels, who also discovered the electro-optic effect, i.e., the production of birefringence of light on the application of an electric field. Pockels produced a phenomenological theory for both of these effects for all the crystal classes. The electro-optic effect remained of purely theoretical interest for a number of years until it became desirable to produce very short light pulses. The Kerr effect in liquids had been used for this purpose for many years. This is a quadratic effect which produces a birefringence proportional to the square of the voltage. Pockels' linear electro-optic effect requires less voltage and can give a shorter light pulse than can the Kerr effect and is being considered as a modulator for obtaining very short light pulses. Hence both the photoelastic and electro-optic effects have graduated from the academic stage to the broad-application stage as acousto-optic and electro-optic modulators and deflectors of light.

The present book has the most complete description of these two effects known to the writer. It covers all the significant contributions made by the several scientists from the day of discovery up to 1976. Considerable material in the text has been collected from a score of recent Ph.D. theses that are not available to the general reader. A number of papers are by the author, T. S. Narasimhamurty.

The book also includes descriptions of ultrasonic methods for the study of the photoelastic behavior of glasses and crystals. These methods in the hands of Mueller have fully brought out the potentialities for the elasto-optic studies of amorphous and crystalline solids. The book contains a chapter on the piezoelectric effect in crystals since a knowledge of this effect is an essential prerequisite to understanding Pockels' linear electro-optic effect. A chapter is also given on the atomistic theory of photoelasticity of cubic crystals that is based mostly on Mueller's work.

Hence the book can be recommended as the most complete discussion of the two effects and related subjects known to the writer.

Columbia University Warren P. Mason

Preface

This book presents an attempt to collect between the covers of one volume some of the material, widely scattered in the literature for over 160 years, on the photoelastic and electro-optic effects in crystals. The stimulus to write this book resulted from the personal contacts that the author has had with solid-state physicists, industrial physicists, mineralogists, engineers, and graduate students. The book is intended for all such persons with varying backgrounds. Because of this broad spectrum of readers with naturally widely differing goals, some of the more elementary but fundamental ideas have been dealt with in a little more detail. The specialist, of course, can skip such topics without losing the general trend of the contents; for his benefit, the references to the existing literature on the topics are given as exhaustively as possible.

The subject of photoelasticity of crystals deals with the artificial birefringence in crystals produced by mechanical stress, and it forms an important aspect of solid-state physics. The photoelastic behavior of crystals is a fourth-rank tensor property, relating the stress tensor or strain tensor to the change in the optical-index ellipsoid.

The fundamental discovery of photoelastic birefringence in glasses and crystals was made by Sir David Brewster in 1815. This phenomenon was later observed in other solids, both amorphous and crystalline, by various investigators, notably Neumann, Mach, Wertheim, and Kerr. But it was not until 1889 that the fundamental difference in the photoelastic behavior of amorphous and crystalline solids was observed by Pockels, who evolved the phenomenological theory of photoelasticity in crystals. By developing suitable techniques, he studied the photoelastic behavior of some crystals, both cubic and noncubic, in support of his phenomenological theory. The discovery of photoelasticity made in 1815 remained just a topic of academic interest for nearly a century until it was successfully applied to structural engineering early in this century (1902) by Coker. Subsequently the interest in this subject has shifted to technical problems, and today it is the most powerful and indispensable tool in solving intricate problems in structural engineering.

After the systematic investigations by Pockels (1880–1906) there seem to have been very few contributions to this branch of crystal physics until the 1930s, when Bergmann and Fues and also Hiedemann and Hoesch successfully applied ultrasonic methods for studies on the elasto-optic behavior of glasses. Mueller's theoretical paper (1938), based on the results of Pockels (1880–1906), Bergmann and Fues (1936), and Hiedemann and Hoesch (1936) and dealing with the elasto-optic behavior of glasses and cubic crystals, has fully brought out the potentialities of ultrasonic methods for these studies. This may be considered yet another landmark in experimental techniques to study the photoelastic behavior of crystalline and amorphous solids.

Pockels' scheme for the photoelastic constants of the 32 point groups has been revised by Bhagavantam (1942), who observed from group theoretical considerations certain discrepancies in Pockels' scheme of describing the photoelastic behavior of the various classes of crystals. Some of his findings have already been confirmed experimentally, and others are awaiting such confirmation.

Almost all the experimental work done until about 1950 was confined to glasses and cubic crystals, and that too mostly for one wavelength of light radiation. The then existing methods of studying stress birefringence could not be employed for a sufficiently accurate measurement of the photoelastic dispersion in cubic and noncubic crystals and its temperature dependence; new techniques were developed to this end in subsequent years.

The photoelastic behavior of crystals plays a significant role in the Brillouin scattering of light, and the development of the laser has served as a feedback to the wide interest evinced in this area during the last 15 years.

Other interesting and closely related properties include the linear electro-optic effect (Pockels effect) and the quadratic electro-optic effect (Kerr effect) in crystals. The Pockels effect is a third-rank tensor property that can be exhibited only by noncentrosymmetric crystals, whereas the Kerr effect is a fourth-rank tensor property and is a universal effect in the sense that it can be exhibited by all crystals, both centrosymmetric and noncentrosymmetric.

Although the phenomenon of photoelasticity in plastics has rightly received its due attention in the hands of structural engineers, the photoelastic behavior of crystals, which was put on a sound phenomenological basis by Pockels as early as the 1880s, does not seem to have attained, until recently, the importance that it richly deserves. The fate of Pockels' linear electro-optic effect was similar, notwithstanding the fact that it was well established by Pockels as early as 1895. Billing's contribution in 1947 demonstrating the potentialities of ADP and KDP crystals for light modulation and as optical shutters has opened up a new field in technology, and this has spurred some more extensive investigations of the linear electro-optic effect in crystals. However, it is only with the advent of the laser in the early 1960s that there has been a phenomenal growth of activity in these two areas of crystal physics, namely photoelastic and electro-optic effects, as a result of their applications in industry. For example, the development of the laser and high-frequency acoustic techniques, along with the discovery of new piezoelectric materials with high coupling factors, has made it possible to apply the acousto-optic phenomenon to devices such as light deflectors, light modulators, and signal processors. This, in turn, has demanded a relentless search for acousto-optic materials suitable for practical use. Similarly, the Pockels effect has now almost completely replaced the hitherto indispensable quadratic electro-optic effect in polar liquids, and has found many interesting and useful applications in science and technology, such as high-voltage measurements, optical range finders, sound recording on cine films, color television, lasers, and optical elements of computer systems. Today we find that several research laboratories attached to various industries are deeply involved in the problems of light modulators and light deflectors based on acousto-optic and electro-optic effects in crystals.

Coker and Filon (1931) pointed out in the preface to their *Treatise on Photoelasticity* that: "Photoelasticity has also its value for the pure physicist, and it provides an additional means of exploring the interaction of molecules and atoms with radiation, a means to which little attention has hitherto been paid, and which should not be neglected, as it may throw much light upon conditions of matter in the solid state." It is only in the recent past that these macroscopic properties of crystals have been receiving the attention of solid-state physicists.

A large number of papers have been published on the several aspects of photoelasticity and electro-optics of crystals during the past 160 years, and yet there has been no attempt to present in one volume all the essential information contained therein.

The above circumstances have prompted the author to write the present book on the photoelastic and the closely related linear and quadratic electro-optic effects in crystals. Much of the information in Chapters 1, 3, and 4 forms a necessary background for both photoelastic and electro-optic effects in crystals.

Two parallel notations, namely Schönflies and International, are used throughout the book; this is necessary as long as books and journals employing the older Schönflies notation remain in use.

For the benefit of a wide circle of readers an extensive bibliography has been added, including some of the many important investigations to which it was not possible to refer directly in the text. The list is limited to publications dealing specifically with the photoelastic and the electro-optic effects in crystals, and it does not include references dealing wholly with photoelasticity as applied to structural engineering and the highly technically oriented and sophisticated applications of the electro-optic effects in crystals. Considerable care was taken to make the bibliography as complete as possible up to the end of 1976, and an addendum added at the proof stage of the book makes the references complete up to the end of 1979.

* * *

The author records here with immense pleasure his indebtedness to Professor S. Bhagavantam, who introduced him to the subject of crystal photoelasticity in the early 1950s at Osmania University, and who also evinced considerable interest in the manuscript while it was in preparation. The author is grateful also to Osmania University for providing him with the necessary facilities for his work.

It is with great pleasure that the author records here his deep sense of gratitude and indebtedness to Dr. Warren P. Mason for his keen personal interest in the manuscript in its final form, and also for having kindly favored the author with his foreword for this book.

The invaluable help and encouragement and the continued interest of Professor Donald J. Montgomery of Michigan State University from 1959 to the present cannot be adequately expressed. Thanks are due to Professor Sherwood K. Haynes of Michigan State University for securing the author a faculty position at MSU in 1959–60, thereby enabling him to work in the laboratories of Professor Egon A. Hiedemann, who showed immense interest in the author's work at MSU. The author's discussions with Professors Robert D. Spence, Mack A. Breazeale, and Walter G. Mayer at MSU were very helpful in the progress of his work. During that year the author had the privilege of being inspired by Professor Hans Mueller of the Massachusetts Institute of Technology. A couple of problems suggested by Professor Mueller are yet to be investigated. The author's visits to some of the American universities and industrial research laboratories during this period have enabled him to become more deeply involved in the subject.

Professor Hiedemann's interest in the author's progress with the manuscript prompted him to introduce the author to Albion College

(Michigan), and consequently, through the efforts of Professors Charles W. Ricker and Howard E. Pettersen, the author was taken on the faculty of Albion College for a year and a half (1969–1971), for which the author's grateful appreciation is due these gentlemen and Albion College. During this period the author had yet another opportunity to visit and revisit a number of American universities and industrial research laboratories, and see the phenomenal progress made since his previous visit in 1959–60, the progress being largely due to the advent of the laser, which has served as a feedback to the growth of the subject because of the innumerable applications of the acousto-optic and electro-optic phenomena to science and technology.

The period 1969–71 at Albion College was undoubtedly the most important in giving a definite shape to the manuscript. At this juncture it is most appropriate for the author to record his thanks to the National Science Foundation, USA, for enabling Albion College to offer the author a Visiting Professorship under a COSIP Grant for the year 1969–70. The author used this opportunity to have discussions with several active workers in the field, personally with some and by correspondence with several others. The author's collaborative work and helpful discussions with Professor Pettersen at Albion College helped him to finalize the first draft of the manuscript. Dr. C. W. Ricker and Albion College readily came forward to arrange secretarial assistance to get the entire handwritten manuscript typed in an orderly form. Subsequently, Professor Walter G. Cady, who had been closely following the progress of the manuscript, gave his blessings and showed his interest in the manuscript by reading the entire original version in great detail and making several valuable suggestions; the author considers this a unique privilege and gratefully acknowledges his indebtedness to Dr. Cady.

The author has incorporated several of the valuable suggestions made by Dr. Walter G. Cady and Dr. Howard E. Pettersen, both of whom have critically read the whole manuscript, and also the suggestions made by several colleagues; if in spite of this there are some changes still warranted, the author takes the full responsibility upon himself, and gladly welcomes any suggestions for improvement of the book.

The author offers his warm thanks to Plenum Publishing Corporation for the meticulous care and attention with which they have handled the production of this book, paying due regard to the minutest detail of the author's points of view; indeed, they have done an excellent job.

The author fully realizes how hard a battle his wife Shrimati Tamma Saraswati had to wage against so many odds throughout this long and arduous task of writing, rewriting, and updating the manuscript. Formal thanks cannot do any justice to her patience and cooperation in shouldering the entire responsibility of the family at home, thus leaving him free from all obligations to the family.

Physics Department
Osmania University
Hyderabad, India

Tamma Satya Narasimhamurty

Contents

Selected List of Figures

Selected List of Tables

Notation and Conventions

α_{ij}	=	Direction cosines
c_{ij}, c_{ijkl}	=	Elastic stiffness constants
d_{ij}, d_{ijk}	=	Piezoelectric constants relating strain to electric field
δ	=	Optical path retardation
$\hat{\delta}$	=	Optical phase retardation
J_0, J_1, J_m	=	Bessel's functions of order 0, 1, m
l, m, n	=	Direction cosines
MHz	=	Megahertz (megacycles per second)
p_{ij}, p_{ijkl}	=	Strain-optical (elasto-optic) constants
$P_{xx}, P_{xy}, \ldots, P_{ij}$	=	Stress tensor components in the original coordinate system
$P'_{xx}, P'_{xy}, \ldots, P'_{ij}$	=	Stress tensor components in the rotated coordinate system
q_{ij}, q_{ijkl}	=	Stress-optical (piezo-optic) constants
r_{ij}, R_{ijk}	=	Pockels' linear electro-optic constants
s_{ij}, s_{ijkl}	=	Elastic compliance constants
$\varepsilon_{xx}, \varepsilon_{xy}, \ldots, \varepsilon_{ij}$	=	Strain tensor components in the original coordinate system
$\varepsilon'_{xx}, \varepsilon'_{xy}, \ldots, \varepsilon'_{ij}$	=	Strain tensor components in the rotated coordinate system
X, Y, Z	=	Cartesian axes
X', Y', Z'	=	Rotated Cartesian axes
X_1, X_2, X_3	=	Orthogonal axes
X_1', X_2', X_3'	=	Rotated orthogonal axes
λ	=	Light wavelength
λ_a	=	Acoustic wavelength
u, v, w	=	Components of displacement of a particle in the X, Y, and Z directions, respectively, in Cartesian coordinates
n_x	=	Refractive index for light with the electric vector parallel to the X axis (propagated in the YZ plane)

n_y = Refractive index for light with the electric vector parallel to the Y axis (propagated in the ZX plane)

n_z = Refractive index for light with the electric vector parallel to the Z axis (propagated in the XY plane)

Terms such as photoelastic constants, stress-optical and strain-optical constants, piezo-optic and elasto-optic constants, and stress-birefringence and piezo-birefringence are freely used in the text with a view to acquainting the reader with all the terms occurring in the literature. The meanings of the several terms are obvious from the context.

P_{yz} is a stress component, the first subscript indicating the direction of force and the second subscript the direction of the normal to the surface on which the force acts. This component is represented by P_{23} in tensor notation and by P_4 in Voigt's single suffix notation. Similar alternative notations are employed for ε_{yz}.

Let a Cartesian coordinate system $OXYZ$ be rotated to take the position $OX'Y'Z'$. Then the direction cosines between the axes of the new system and those of the old system are represented by α_{ij} as shown in the following matrix:

	X	Y	Z
X'	α_{11}	α_{12}	α_{13}
Y'	α_{21}	α_{22}	α_{23}
Z'	α_{31}	α_{32}	α_{33}

In transforming P_{xx}, ε_{xx}, etc. from one coordinate system to another, we on occasion freely use both the Cartesian notation X, Y, Z, and its alternative form X_1, X_2, X_3, along with the tensor suffixes i, j, k, l. This should not lead to any difficulty in the interpretation, and all are in vogue in the literature.

The direction of the electric vector of a light wave represents the direction of vibration of light. The direction of polarization is perpendicular to that of the electric vector, according to the standard convention.

Throughout the book, we shall use

$$\Delta n = \Delta n_{\parallel} - \Delta n_{\perp}$$

for both photoelastic and electro-optic birefringences. This will ensure uniformity.

Further,

$$\text{the ratio of elasto-optic constants} = R = \frac{\Delta n_{\parallel}}{\Delta n_{\perp}} \text{ or } \frac{\Delta \mu_{\parallel}}{\Delta \mu_{\perp}}$$

(hkl) indicates a face

$[hkl]$ indicates a face normal

If θ is the acute angle between the face normals $[hkl]$ and $[h_1k_1l_1]$, then for a cubic crystal

$$\cos\theta = \frac{hh_1 + kk_1 + ll_1}{(h^2 + k^2 + l^2)^{1/2}(h_1^2 + k_1^2 + l_1^2)^{1/2}}$$

Occasionally, some of the symbols listed above are used with other meanings; in such cases these symbols as well as those not listed are clearly defined.

Several of the figures in the text have been taken from various research publications, including journals and theses. They have been redrawn or reproduced as they appeared in the originals to facilitate easy reference to the original texts. This has resulted in a certain amount of apparent variation from figure to figure in some cases; however, a closer study will, it is hoped, reveal the continuity of the material presented in this book.

Attention may be called here to the method of numbering the figures, each of which is labeled by a decimal system (for example, 2.1, 2.2, 2.3), with the first number corresponding to the chapter number and the second indicating the order of appearance of the figure. A similar decimal system (such as that used for the figures) is also used for tables, equations, and sections in each chapter.

Text citations of references are by numbers in square brackets. All the references are collected in alphabetical order at the end of the book, with each reference given in the following sequence: author(s), title of the article, journal, volume, page range, and year of publication. The abbreviations of journal names in the references are those listed in *Physics Abstracts–Science Abstracts Series A*, published by the Institution of Electrical Engineers (Savoy Place, London, England), catalogued in Volume 78, Part I, Author Index, pages LJ-1 to LJ-36 (January–June 1975), and in Volume 79, Part I, Author Index, pages LJ-1 to LJ-34 (January–June 1976).

Almost all the references for this century are given with their titles in English as they appeared in the *Science Abstracts–Physics*, and those for the last century in their original languages English, German, French, etc.

1

Photoelasticity of Crystals. Introduction

1.1. Discovery of the Phenomenon of Photoelasticity

The subject of photoelasticity deals with the artificial birefringence developed in a solid under the application of a mechanical stress. This phenomenon was first discovered in jellies by Sir David Brewster, who in 1815 presented a paper before the Royal Society, London, announcing the effect. He reported in 1816 the same effect in amorphous solids such as glass and in cubic crystals such as fluorspar and diamond. By 1818 he had also studied the stress-optical behavior of some uniaxial and biaxial crystals. In his earlier investigations, he placed a strip of glass between two crossed Nicols. When the strip of glass was stretched, the field of view brightened up, thereby showing that artificial birefringence was induced in the glass strip. He found that glasses and cubic crystals which are originally optically isotropic acquire the property of double refraction under a mechanical stress. Furthermore, he found that ordinary glass behaves as a negative uniaxial crystal upon the application of compressional stress, and as a positive uniaxial crystal under tension. In the case of solids which are initially birefringent, the initial birefringence is altered by the stress. Though his investigations were of a qualitative nature, he gave a correct interpretation of the stress-optical phenomenon.

1.2. Mathematical Formulation and Neumann's Constants. Pockels' Contribution

Earlier studies of the subject reveal that the artificial birefringence produced in a solid is directly proportional to the applied stress within the limits of Hooke's law. Thus, for an isotropic solid, like glass, linearly stressed along the Z axis and with observations made along the Y axis, as shown in Fig. 1.1, the birefringence produced is directly proportional to the stress applied. Thus, if δ is the optical path difference, and t the

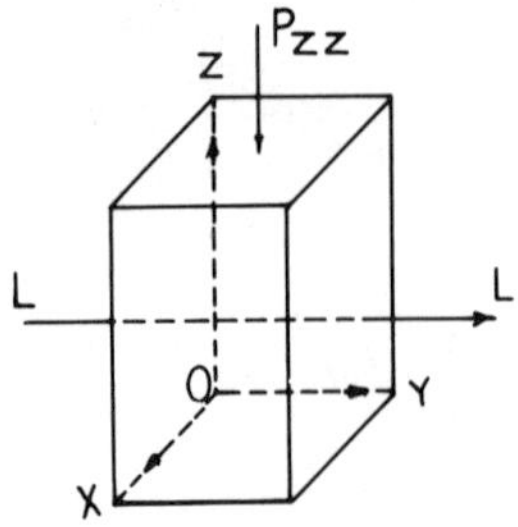

Fig. 1.1. A solid under a linear stress for stress-optical measurements (P_{zz} is the applied stress, and LL is the direction of light propagation and observation).

thickness of the glass, then $\delta \propto P_{zz}t$, or

$$\delta = C_\lambda P_{zz} t \tag{1.1}$$

where P_{zz} is the linear stress applied in the Z direction on a face normal to the Z axis, and C_λ is a constant, which may be called the relative stress-optical coefficient. The value of C_λ, also called Brewster's constant, after the celebrated scientist, depends on the material stressed and the wavelength λ of light used, as well as on the nature and direction of the stress and the direction of observation, and on temperature. The dimensions of C_λ are those of the reciprocal of stress, cm^2/dyne. When the stress is in dynes/cm^2 and the thickness of the glass in centimeters, then the value of C_λ is of the order of 10^{-13} cm^2/dyne per cm thickness. This unit of 10^{-13} cm^2/dyne per cm thickness is called a brewster. The quantity C_λ can be positive or negative and is of the order of 1–5 brewsters for ordinary glass and about 5.7 for Plexiglas.

Neumann (1841) was the first to make a quantitative study of the phenomenon. He introduced two constants p and q to describe the changes of indices of refraction for light polarized parallel and perpendicular, respectively, to the strain in isotropic solids like glasses. Equation (1.2) defines Neumann's constants p and q in terms of a typical strain ε_{zz}:

$$\begin{aligned} \Delta n_z &= n_z - n = -n^2 q \varepsilon_{zz} \\ \Delta n_x &= n_x - n = -n^2 p \varepsilon_{zz} \end{aligned} \tag{1.2}$$

Here p and q are related to the changes in the index of refraction n for light polarized parallel and perpendicular, respectively, to the strain ε_{zz}. It is the difference $(p - q)$ which is obtained for glasses by measurements using a Babinet compensator. The relation between Neumann's constants p and q and Pockels' constants p_{ij} is given in Chapter 5.

Pockels (1889) made a detailed study of the photoelastic behavior of many crystals including fluorite, rock salt, and alums, and made the important discovery that cubic crystals behave differently from isotropic materials. It was Pockels who originated the phenomenological theory of photoelasticity of crystals. His work on crystal photoelasticity formed his doctoral dissertation at Göttingen in 1889.

The phenomenon of photoelasticity, discovered by Brewster and systematically investigated by Pockels, attracted a number of workers in this field. A brief historical survey follows.

1.3. A Brief Historical Survey

1.3.1. Amorphous Solids

Sir David Brewster [192; 193] discovered the phenomenon of artificial birefringence first in jellies in 1815 and then in other transparent solids like glasses and cubic crystals in 1816. He found that when light passes through a rectangular glass plate which is stretched transversely to the direction of propagation of light, the glass plate becomes doubly refracting with the axes of polarization in the glass along and perpendicular to the direction of stress. Biot [151] showed that light was restored between crossed Nicols when a strip of glass in a state of longitudinal vibrations was introduced between them. Fresnel [391; 392] attempted a more detailed investigation of the phenomenon and separated the two rays produced by artificial double refraction. Neumann [871] gave an explanation of the observed behavior in amorphous substances, relating the birefringence with the strain developed in the solid. On the other hand, Maxwell [786] was of the opinion that the birefringence depended on the applied stress. The linear relations between the artificial birefringence and the strain and stress as found by Neumann and Maxwell, respectively, are very useful in applications to problems in engineering. Wertheim [1378; 1379] made a quantitative study of the stress-optical behavior of crystals, using an improved experimental technique, and drew several important conclusions regarding the stress-optical behavior of glasses and cubic crystals. He also made measurements on the photoelastic behavior of diamond [1379]. This phenomenon was further studied by a number of workers, among whom particular mention may be made of Mach [755], Bravais [187], and Kerr [637]. Kerr for the first time determined the stress-optical coefficients of glass using an interferometer.

Pockels [947; 949] gave a critical account of the various attempts to measure the stress-optical constants. He used a Babinet compensator, which remains until today a standard method for measuring the relative (or differential) stress-optical constants. He also used a Jamin interferometer for the determination of absolute retardation in various Jena glasses under a mechanical stress. His results are of great importance. He showed that certain heavy flint glasses containing over 80% PbO became positively uniaxial under compressional stress. He also developed the phenomenological theory which explains the photoelastic behavior of all crystals. Filon [365] studied a similar series of borosilicate glasses and concluded that the stress-optical coefficient increases with B_2O_3 content and decreases with K_2O content. The effect of chemical constitution was studied further by Adams and Williamson [7] and Savur [1085]. Twyman and Perry [1232] and Schaefer and Nassenstein [1091] are among others who studied the photoelastic behavior of glasses.

The dispersion of the photoelastic effect in glasses has attracted a number of workers. Mace de Lepinay [753; 754], Savur [1085], Pockels [954], and Filon [363–366] studied the photoelastic dispersion in glasses. Although Wertheim thought that photoelastic dispersion did not exist in glasses, Pockels [954] experimentally obtained the stress-optical constants at three different wavelengths in the visible region, employing a Babinet compensator, and showed the existence of photoelastic dispersion. The relative stress-optical constant was found to be large in the violet region compared with that in the red region. Filon [363–365] made extensive investigations on the dispersion of photoelasticity by the well-known spectroscopic method and fitted his data into an empirical formula

$$C_\lambda = C_0(1 - \lambda'/\lambda) \tag{1.3}$$

where C_λ is the relative stress-optical constant, λ is the wavelength of the light, and C_0 and λ' are constants of the medium. By Filon's method, the dispersion for a large number of wavelengths can be determined directly. Jog [585; 586] studied the photoelastic dispersion of fused silica using a Fabry–Perot interferometer in conjunction with a spectrograph, and for the first time he obtained the dispersion of the *absolute* stress-optical constants of fused silica. Narasimhamurty and Jeelani [847] have recently developed a new technique called the "interference-spectroscope" method for directly determining the *absolute* stress-optical constants of glasses. Twyman and Perry [1232] used Michelson's interferometer for determining the value of the stress-optical constants of plastics. Braybon [188] used Michelson's interferometer for the study of the dispersion of the photoelastic constants of CR 39 and Marco resin. Coker and Filon [266] in their *Treatise on Photoelasticity* have given a detailed account of the earlier investigations on several plastics, with particular reference to their applications to engineering problems.

The effect of temperature on the photoelastic behavior is another interesting aspect of the subject. Harris [470] studied the effect of temperature on the photoelastic constants of glass. Bhagavantam and Krishnamurty [131] studied the temperature variation of photoelastic constants in Perspex and polystyrene.

The first dynamical method for the determination of the ratio of Neumann's strain-optical constants was given by Bergmann and Fues [108]. Their method is based on the diffraction of light by ultrasonic waves in transparent solids when a point source of light is used. They showed that the intensity of the first-order diffraction due to a longitudinal acoustic wave is different for light polarized parallel and perpendicular to the elastic wavefront, and the ratio of these intensities leads to a measure of the ratio of the absolute strain-optical constants. They measured this ratio for different types of glasses.

Two types of light sources can be used to show the diffraction effects by ultrasonic waves traveling through a solid. Schaefer and Bergmann [1086; 1087] used a point source of light. This type of experiment produces a characteristic pattern, called the Schaefer–Bergmann pattern, in which the diffraction orders appear as spots lying on circles concentric to the central undeviated spot when an amorphous solid is excited by ultrasonic waves. Hiedemann and Hoesch [513–519] used a slit source to show the diffraction of light by plane ultrasonic waves excited in a solid. The slit is oriented parallel to the sound wavefront and the diffraction pattern is composed of a series of equally spaced images of the slit for longitudinal ultrasonic waves. The case for transverse waves is similar. In practice this line-diffraction pattern, called the Hiedemann pattern, is much more readily obtained than the Schaefer–Bergmann pattern, since the *slit* type source is brighter than the single *point* type source and since the entire ultrasonic wave energy is utilized to excite only one mode of vibration in the solid, thereby causing only one diffraction pattern. The Schaefer–Bergmann method uses forced vibrations in the solid in such a way that many modes are excited simultaneously, whereas the Hiedemann–Hoesch method utilizes the resonance technique, which obviously needs much less driving power and hence reduces the undesirable heating of the samples.

Mueller [829] discussed the potentialities of the studies on the ultrasonic-diffraction pattern in relation to the photoelastic properties of glasses and crystals. He has suggested three different experimental procedures for studying the elasto-optic behavior of solids from studies on the state of polarization of the diffraction images. Many of the conclusions from Mueller's theory have been verified experimentally by Hiedemann and Hoesch [519].

Using the method suggested by Mueller [829], various workers have studied the photoelastic behavior of several solids, both amorphous and crystalline.

Schaefer and Dransfeld [1092] and Ramavataram [1024] used the Schaefer–Bergmann pattern for photoelastic studies on glasses. The line-diffraction or Hiedemann pattern, which can be excited with greater ease and which yields results to a higher order of accuracy, has been followed by, among others, Vedam [1277; 1278], Iyengar [565], Gates and Hiedemann [413], Hagelberg [464], and Achyuthan [6] and used to study the elasto-optic behavior of a variety of glasses.

1.3.2. Cubic Crystals

After the discovery of artificial double refraction by Brewster in amorphous solids and crystals (both cubic and noncubic), the phenomenon in crystals became an interesting branch of investigation for various workers. The pioneer workers were Wertheim [1379], Bravais [187], Jannetaz [577], and Pockels [950]. Wertheim [1379] studied diamond, ammonium alum, and halite. Pockels examined halite, sylvite, potassium alum and ammonium alum, and a few noncubic crystals. The noncubic crystals will be discussed under uniaxial and biaxial crystals. Schramm [1103] studied ZnS. Herzfeld and Lee [510] verified the results on halite and sylvite. Bhagavantam [119; 121–129] and collaborators studied a large number of cubic crystals. Ramachandran [993; 996; 997] studied diamond. Ramachandran and Chandrasekharan [998] studied $NaClO_3$. Burstein and Smith [215; 216] studied a number of crystals: LiF, NaCl, KCl, KBr, KI, MgO, and TlCl. West and Makas [1382] investigated MgO, LiF, KBr, AgCl, Tl (Cl, Br), and Tl (Br, I). Grodzinski [451] and Poindexter [961; 962] carried out detailed studies on diamond. Banerjee [65], Maris [775], Giardini [418–421], Iyengar [565; 566], and Narasimhamurty [841–843; 846] are among those who made further studies on cubic crystals. Iyengar [565; 566], Bansigir and Iyengar [66; 67], and Srinivasan [1174; 1175] studied the photoelastic dispersion in cubic crystals by static methods. Rahman and Iyengar [984; 985] have studied the dispersion of the strain-optical ratios in cubic crystals and showed that the dispersion occurs predominantly in the ultraviolet region, with very little in the visible region. Jacob and

Narasimhamurty [569] studied BaF_2. Bhagavantam and Krishnamurty [132] and Krishna Rao and Krishna Murty [693–696] studied the effect of temperature on the photoelastic behavior of cubic crystals. Ultrasonic studies on the elasto-optic behavior of some cubic crystals have been made by Burstein, Smith, and Henvis [215] and by Galt [411]. These have been followed in recent times by a large number of workers including, among others, Vedam and Ramachandran [1280], Iyengar [565–567], Narasimhamurty [842; 843], Bansigir [68], Krishna Murty [681], Dixon [296], Pettersen [922–926], Rahman and Iyengar [984; 985], and Uchida [1235].

Flannery and Marburger [370] have estimated the strain-optical coefficients of polycrystalline laser window materials and showed that both the sign and magnitude of $(p_{11} - p_{12})$ may change relative to the single-crystal values.

The relation of the piezospectroscopic effect to the photoelastic effect in cubic crystals was pointed out by Kaplyanskii and Lozovskaya [626]. They studied the photoelastic effect at wavelengths close to well-defined absorption peaks in stressed crystals of CaF_2–Sm^{2+}, CaF_2–Eu^{2+}, colored LiF, and pure ZnTe. On the basis of the results of their investigations, they concluded that there is a general dependence of the photoelastic effect on the deformation-induced changes in the energy-band structure. The experimental work of Schmidt-Tiedemann [1096; 1097], Riskaer [1051], and Feldman [349] may also be mentioned in this connection; they studied the photoelastic dispersion in pure and doped Ge and Si in the spectral region (about 1.5–2.2 μm) close to the absorption peak and used the results to evaluate the deformation parameters of the energy bands. On the other hand, Riskaer and Balslev [1052] have used the known deformation parameters (of the valence band $K = 0$) in Ge to evaluate the dispersion of the photoelastic effect near the absorption edge. Recently, Feldman and Horowitz [350] studied the photoelastic dispersion in GaAs near the absorption edge.

1.3.3. Uniaxial and Biaxial Crystals

From the studies on uniaxial and biaxial crystals, Brewster concluded that a uniaxial crystal in general becomes biaxial when stressed in a direction other than its optic axis. Moigno and Soleil [821] studied the behavior of quartz under stress perpendicular to its optic axis and noticed that it became biaxial. They made the interesting observation that for positive uniaxial crystals, if $q_{12} > q_{11}$ the plane of XZ will be the plane of the optic axes in the artificially biaxial crystal, i.e., the plane of the optic axes contains the line of stress. This experimental observation can be explained using Pockels' phenomenological theory. Pockels [948; 955] studied all the photoelastic constants of quartz and calcite at one wavelength, and a few constants of Rochelle salt. West and Makas [1383] determined a few photoelastic constants of ADP (ammonium dihydrogen phosphate) and KDP (potassium dihydrogen phosphate). Later, Carpenter [232] studied ADP. But none of these workers could report the complete data. Narasimhamurty [843; 850] determined all the photoelastic constants of α-quartz and confirmed the values obtained earlier by Pockels. Eppendahl [321] was the first to investigate completely the photoelastic behavior of topaz, a biaxial crystal. Vedam [1279] studied barite. Narasimhamurty [843; 851] studied barite and Rochelle salt, and Bansigir and Iyengar [69] studied barite. Recently Narasimhamurty, Rao, and Pettersen [854] have revised completely the photoelastic constants of ADP. The photoelastic behavior of crystals —cubic and noncubic—under very large hydrostatic pressures has been studied in great detail by Vedam and collaborators [1282–1301]. Waxler and Weir [1364] studied the effect of hydrostatic pressures up to 1 kbar on the refractive indices of KBr, NaCl, LiF, diamond, MgO, quartz, and Al_2O_3. Uniaxial stresses and hydrostatic pressure were applied by Waxler and Farabaugh [1365] to determine all the photoelastic constants of ruby. Haussühl and collaborators studied calcium formate [486], α-sulfur and α-iodic acid [492], and taurin [489].

Narasimhamurty [843; 845; 848] extended the ultrasonic technique (first suggested by Mueller) to study the ratio of the strain-optical constants of uniaxial and biaxial crystals and determined the ratios of the elasto-optic constants for quartz, calcite, $NaNO_3$, barite, and Rochelle salt. Achyuthan and Breazeale [3] have applied the ultrasonic method to ADP. The Bragg-diffraction technique (being a special case of ultrasonic methods) was applied to determine the individual p_{ij} of a large number of cubic and noncubic crystals by Dixon and Cohen [294], Dixon [296], and Uchida [1239; 1241]. The Bragg-diffraction technique is of immense technological importance in developing acousto-optic modulators and deflectors [13]. An ultrasonic method of determining the absolute sign of p_{ij} has been developed by Biegelsen [138], who applied it to fused silica and cadmium molybdate. Recently Sakane and Kashiwage [1078] have developed a simple ultrasonic method for the measurement of the photoelastic constants of solids and applied their method to Vycor glass.

2

Mathematical Tools, Tensor Properties of Crystals, and Geometrical Crystallography

2.1. *Linear Transformations*

2.1.1. Coordinate Transformations

Let a point P have the coordinates x_1, x_2, x_3 in a space of three dimensions in an X system of Cartesian coordinate axes. These three coordinates of P, which may be collectively represented in an abbreviated form by x_i, are independent of one another and form a set of linearly independent variables. Let the same point P have the coordinates x_i' in a different Cartesian coordinate system called the X' system, having the same origin but with some or all of their axes not coincident with one another (Fig. 2.1). A rotation of axes and a reflection in a plane are examples of such a pair of Cartesian coordinate systems. Then x_i' and x_i are related by

$$\begin{aligned} x_1' &= \alpha_{11}x_1 + \alpha_{12}x_2 + \alpha_{13}x_3 \\ x_2' &= \alpha_{21}x_1 + \alpha_{22}x_2 + \alpha_{23}x_3 \\ x_3' &= \alpha_{31}x_3 + \alpha_{32}x_2 + \alpha_{33}x_3 \end{aligned} \tag{2.1}$$

where the coefficients α_{ij} define the direction cosines of X_i' with respect to X_i according to the scheme

	X_1	X_2	X_3
X_1'	α_{11}	α_{12}	α_{13}
X_2'	α_{21}	α_{22}	α_{23}
X_3'	α_{31}	α_{32}	α_{33}

Thus, for example, α_{11}, α_{12}, and α_{13} are the direction cosines of X_1' with respect to X_1, X_2, and X_3. Transformation of the coordinates of P from one system to another is called a linear transformation. A rotation of

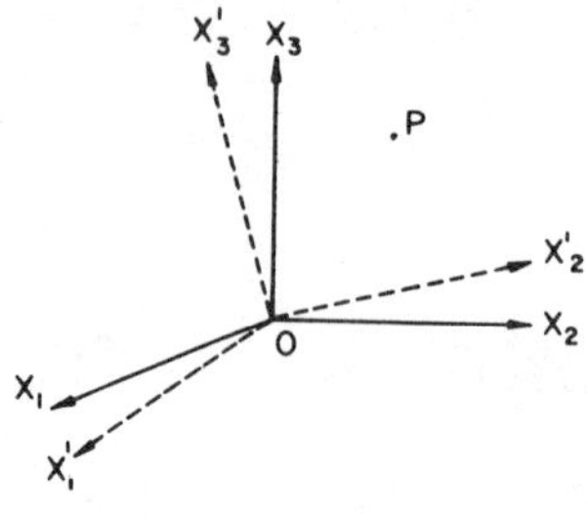

Fig. 2.1. Transformation of axes: a rotation of the axes of one coordinate system X_i' with respect to another system X_i, both having a common origin O.

axes and a reflection in a plane will cause such a linear transformation of the components of P from one system to another.

The set of equations (2.1) can also be written thus:

$$x_i' = \sum_j \alpha_{ij} x_j \qquad (i, j = 1, 2, 3) \tag{2.2}$$

Following Einstein's dummy suffix notation,* x_i' and x_j are related by

$$x_i' = \alpha_{ij} x_j \qquad (i, j = 1, 2, 3) \tag{2.3}$$

Now take the case of a point P having the coordinates x_j in a space of n dimensions constituting the X system. These n coordinates x_j of P are independent of one another and form a set of linearly independent variables. Let the same point P have the coordinates x_i' in a different coordinate system called the X' system, of the same dimensions n. Then, as in Eq. (2.3), x_i' and x_j are related thus in Einstein's convention:

$$x_i' = \alpha_{ij} x_j \qquad (i, j = 1, 2, \ldots, n) \tag{2.4}$$

Here the α_{ij} relate the coordinates of the X' system with those of the X system. The transformation of the components of P from one system to another is called a linear transformation, as mentioned earlier.

2.1.2. Orthogonality Relations

The coordinates x_i will be linearly independent if and only if the values of α_{ij} are related in such a way that

$$\alpha_{ik}\alpha_{jk} = \begin{cases} q, & j = i \\ 0, & j \neq i \end{cases} \tag{2.5}$$

If the constant $q = \pm 1$, the transformation is said to be orthogonal.

* Einstein's dummy suffix notation is explained on the following page.

It is common to use a symbol δ_{ij}, called the Kronecker delta, to specify the orthogonality condition. If we define

$$\delta_{ij} = \begin{cases} 1, & i = j \\ 0, & i \neq j \end{cases} \qquad (i, j = 1, 2, 3)$$

then the orthogonality condition can be expressed as

$$\alpha_{ik}\alpha_{jk} = \delta_{ij} \tag{2.6}$$

Equations (2.1)–(2.3) describe in different ways the same relationship between x_i' and x_j, but Eq. (2.3) is the more convenient and conventional form of Eqs. (2.1) and (2.2) in accordance with Einstein's summation convention. According to this convention, when a suffix is repeated on the same side of the equation, the summation with respect to that suffix has to be carried out over all the possible values which that suffix can take, and hence the summation sign is dropped when this convention is adopted. It should be noted that it matters little what letter is used for a repeated suffix. Thus $x_i' = \alpha_{ij}x_j = \alpha_{ik}x_k$. Here i is called the free suffix and j or k is called the dummy suffix.

The inverse transformation from the X' system to the X system can be obtained by changing the rows into columns and the columns into rows in the direction-cosine scheme given earlier in this section and interchanging x with x'. Then the law of inverse transformation is given by

$$x_i = \alpha_{ji}x_j' \qquad (i, j = 1, 2, 3) \tag{2.7}$$

A law which expresses the components of a certain quantity at a point in one coordinate system in terms of the components of the same quantity at the same point with reference to another coordinate system is called a linear transformation law.

The orthogonality relations, Eqs. (2.5) and (2.6), given earlier can now be suitably adapted to the Cartesian coordinate systems governed by the direction-cosine scheme. The direction cosines of a straight line (say, one of the axes in the X' system) with respect to the rectangular axes OX_1, OX_2, and OX_3 are given by each row in the scheme; hence the following relations hold:

$$\begin{aligned} \alpha_{11}^2 + \alpha_{12}^2 + \alpha_{13}^2 &= 1 \\ \alpha_{21}^2 + \alpha_{22}^2 + \alpha_{23}^2 &= 1 \\ \alpha_{31}^2 + \alpha_{32}^2 + \alpha_{33}^2 &= 1 \end{aligned} \tag{2.8}$$

Furthermore, since the axes in each of the systems are mutually perpendicular, we have the following additional relations among α_{ij}:

$$\begin{aligned} \alpha_{11}\alpha_{21} + \alpha_{12}\alpha_{22} + \alpha_{13}\alpha_{23} &= 0 \\ \alpha_{21}\alpha_{31} + \alpha_{22}\alpha_{32} + \alpha_{23}\alpha_{33} &= 0 \\ \alpha_{31}\alpha_{11} + \alpha_{32}\alpha_{12} + \alpha_{33}\alpha_{13} &= 0 \end{aligned} \tag{2.9}$$

Equations (2.8) and (2.9) are called orthogonality relations, which can now be written concisely as

$$\alpha_{ik}\alpha_{jk} = \begin{cases} 1 & \text{if} \quad i = j \\ 0 & \text{if} \quad i \neq j \end{cases} \tag{2.10}$$

or, in terms of the Kronecker delta, as

$$\alpha_{ik}\alpha_{jk} = \delta_{ij} \tag{2.11}$$

Similarly

$$\alpha_{ki}\alpha_{kj} = \delta_{ij} \tag{2.12}$$

Thus Eqs. (2.11) and (2.12) are the orthogonality relations as applied to the Cartesian systems of coordinates. Transformations in which the coefficients satisfy the orthogonality relations are called linear orthogonal transformations.

When two linear orthogonal transformations are successively applied, the resulting transformation from the initial to the final coordinate systems can also be shown to be linear orthogonal. Furthermore, if the transformations are described by the equations

$$x_i' = \alpha_{ij}x_j, \qquad x_i'' = \beta_{ij}x_j', \qquad x_i'' = \phi_{ij}x_j$$

then it is easily seen that

$$\phi_{ij} = \beta_{ik}\alpha_{kj} \tag{2.13}$$

Thus, the transformation matrix $[\phi_{ij}]$ is obtained as the product of transformations in terms of $[\alpha_{kj}]$ and $[\beta_{ik}]$. The result may thus be extended to the product of any number of transformations.

The distance and direction from O (the origin) to P may be described equally well in terms of either a set x_i or a set x_i'. The magnitude of this distance S is given by

$$S^2 = x_1^2 + x_2^2 + x_3^2 = x_1'^2 + x_2'^2 + x_3'^2$$

and the direction is specified in terms of angles whose cosines are x_i/S or alternatively x_i'/S. In other words, S, defined by the distance and direction of the line from O to P, is an invariant of the space. This invariant may be described by either x_i or x_i', and these sets are transformed into one another by the equation

$$x_i' = \alpha_{ij} x_j$$

where α_{ij} is the cosine of the angle between the x_i' and x_j coordinate axes.

The location of any two points relative to each other can be specified by a distance and a direction relative to a reference frame. This distance is invariant under a rotation of coordinates, and once the direction has been specified in terms of any one reference frame, it can be specified in terms of all reference frames.

2.1.3. *The Determinant of the Matrix* $[\alpha_{ij}]$ *of the Direction-Cosine Scheme*

The determinant of $[\alpha_{ij}]$ describing any linear orthogonal transformation in a three-dimensional space may be called the determinant of the transformation matrix and may be denoted by

$$\Delta = |\alpha_{ij}| = \begin{vmatrix} \alpha_{11} & \alpha_{12} & \alpha_{13} \\ \alpha_{21} & \alpha_{22} & \alpha_{23} \\ \alpha_{31} & \alpha_{32} & \alpha_{33} \end{vmatrix}$$

The product of two determinants $|\alpha_{kj}|$ and $|\beta_{ik}|$ is expressed as the determinant $|\alpha_{kj}\beta_{ik}|$. So from Eq. (2.13) we find that

$$|\phi_{ij}| = |\beta_{ik}|\,|\alpha_{kj}| \tag{2.14}$$

Now, from the orthogonality relation (2.11), we note that

$$|\alpha_{ik}\alpha_{jk}| = |\delta_{ij}| = \begin{vmatrix} 1 & 0 & 0 \\ 0 & 1 & 0 \\ 0 & 0 & 1 \end{vmatrix} = 1 \tag{2.15}$$

Note that $[\delta_{ij}]$ forms a unit matrix. But from Eq. (2.14) we note that

$$|\alpha_{kj}\alpha_{ik}| = \begin{vmatrix} \alpha_{11} & \alpha_{21} & \alpha_{31} \\ \alpha_{12} & \alpha_{22} & \alpha_{32} \\ \alpha_{13} & \alpha_{23} & \alpha_{33} \end{vmatrix} \begin{vmatrix} \alpha_{11} & \alpha_{12} & \alpha_{13} \\ \alpha_{21} & \alpha_{22} & \alpha_{23} \\ \alpha_{31} & \alpha_{32} & \alpha_{33} \end{vmatrix} \tag{2.16}$$

The two determinants on the right-hand side in Eq. (2.16) are equal* since the elements in the rows and columns of a determinant can be interchanged without altering its value. Hence putting i, j as the suffixes, as per convention, the above product is equal to

$$|\alpha_{ij}| \cdot |\alpha_{ij}| = \Delta \cdot \Delta = \Delta^2 = 1$$

Therefore $\Delta = \pm 1$. Thus for all linear orthogonal transformations (i.e., from one Cartesian system to another with the same origin), the determinant of the transformation matrix α_{ij} is equal to $+1$ or -1.

The determinant of the transformation matrix is $+1$ if the original and the transformed axes have the same sense; this is the case for all rigid-body rotations. The determinant becomes equal to -1 whenever the transformation brings about a change in the sense of the axes; such a transformation can be interpreted as the result of a rotation followed by an inversion or reflection. The converse proposition, that all the transformations which have their $\Delta = \pm 1$ retain or change the sense of the axes, follows readily.

We note that the symmetry operations which naturally transform a crystal into a geometrically identical position with itself produce linear orthogonal transformations.

2.2. Matrix Algebra

2.2.1. Introduction

Some of the physical properties of solids vary with direction, and they can be correlated with the symmetry elements of the crystalline solid. While studying such anisotropic properties of crystals, we make extensive use of transformations of the components of vectors and tensors from one coordinate system to another. The linear transformations between two coordinate systems are greatly facilitated by the use of matrix algebra. It is essentially an operational tool. Furthermore, it is very handy while interrelating two sets of tensor components, e.g., the elastic stiffness constants and compliance constants, which bear an inverse but not a reciprocal relationship. Similarly the stress-optical and strain-optical coefficients in Pockels' scheme of photoelastic constants can be interrelated in an elegant way.

* But the matrices are different.

We introduce matrix algebra very briefly just to ensure continuity in this book, since the reader is expected to be familiar with matrices.

Let X_i' and X_j be two different Cartesian systems of coordinates with a common origin O. Let α_{ij} be the scheme of direction cosines of their axes. Let **p** be a vector whose components in the X_i' system are p_i' and whose components in the X_j system are p_j. Then the components p_i' are linearly related to p_j by the equation

$$p_i' = \alpha_{ij} p_j \tag{2.17}$$

This is expressed by the matrix equation

$$[p_i'] = \mathbf{A}[p_j] \tag{2.18}$$

where α_{ij} are the elements of the matrix* **A**, and the components of **p** in the two different coordinate systems are expressed by the column matrices $[p_i']$ and $[p_j]$.

The matrix **A** can now be regarded as a rotation operator, generating a new system X' by rotating the original system X through some angle about some axis or arbitrary direction. The magnitude of **p** remains invariant under the operation of rotation though its components vary according to Eq. (2.18). Successive operations of rotations can be accomplished by the multiplication of the corresponding matrices of the rotation operators one after another.

2.2.2. *Matrix Algebra and Coordinate Transformations*

Let a point P have the coordinates x_1, x_2, x_3 and x_1', x_2', x_3' in two different Cartesian systems. Then

$$\begin{aligned} x_1' &= \alpha_{11}x_1 + \alpha_{12}x_2 + \alpha_{13}x_3 \\ x_2' &= \alpha_{21}x_1 + \alpha_{22}x_2 + \alpha_{23}x_3 \\ x_3' &= \alpha_{31}x_1 + \alpha_{32}x_2 + \alpha_{33}x_3 \end{aligned} \tag{2.19}$$

where the constants α_{ij} are the direction cosines between the axes of one system and those of another system as shown in the scheme that follows.

* Note that the matrix **A** is defined by the array or order of the elements of $[\alpha_{ij}]$.

	X_1	X_2	X_3
X_1'	α_{11}	α_{12}	α_{13}
X_2'	α_{21}	α_{22}	α_{23}
X_3'	α_{31}	α_{32}	α_{33}

This is the direction-cosine scheme for the transformation matrix.* The simultaneous linear equations (2.19) can be put in the matrix notation as

$$\begin{bmatrix} x_1' \\ x_2' \\ x_3' \end{bmatrix} = \begin{bmatrix} \alpha_{11} & \alpha_{12} & \alpha_{13} \\ \alpha_{21} & \alpha_{22} & \alpha_{23} \\ \alpha_{31} & \alpha_{32} & \alpha_{33} \end{bmatrix} \begin{bmatrix} x_1 \\ x_2 \\ x_3 \end{bmatrix} \tag{2.20}$$

In short, $[x_i'] = \mathbf{A}[x_j]$, where $\mathbf{A} = [\alpha_{ij}]$. Here $\mathbf{A}$ is a matrix and $[x_i']$ and $[x_j]$ are also matrices. Now we can define a matrix $\mathbf{A}$ in general.

2.2.3. *Some Common Types of Matrices*

A rectangular matrix $\mathbf{A}$ is an array of symbols or numbers a_{ij} arranged in rows and columns. Thus

$$\mathbf{A} = \begin{bmatrix} a_{11} & a_{12} & \cdots & a_{1j} \\ a_{21} & a_{22} & \cdots & a_{2j} \\ \vdots & \vdots & & \vdots \\ a_{k1} & a_{k2} & \cdots & a_{kj} \end{bmatrix} \tag{2.21}$$

where a_{ij} are called the elements of the matrix. The horizontal lines are called rows, and the vertical lines are called columns. The rows and columns are designated by subscripts. If the number of rows is equal to the number of columns in a matrix, then it is called a square matrix. An $n \times n$ matrix is called an n-square matrix, which is also referred to as a matrix of n dimensions. The elements $a_{11}, a_{22}, a_{33}, \ldots$ are called the diagonal elements. A matrix is called diagonal if all off-diagonal elements are zero, e.g., the matrix $\mathbf{D}$ below.

$$\mathbf{D} = \begin{bmatrix} a_{11} & 0 & 0 & \cdots & 0 \\ 0 & a_{22} & 0 & \cdots & 0 \\ 0 & 0 & a_{33} & \cdots & 0 \\ \vdots & \vdots & \vdots & & \vdots \\ 0 & 0 & 0 & \cdots & a_{nn} \end{bmatrix} \tag{2.21a}$$

* This direction-cosine scheme occurs frequently in the text and will henceforth be referred to as $[\alpha_{ij}]$ in its matrix form.

If all elements of a diagonal matrix are equal, $a_{ii} = \lambda$, the matrix is called spherical. If, in addition, $\lambda = 1$, then the matrix is called the unit matrix or identity matrix $\mathbf{I}_n$ of order n. Thus

$$\mathbf{I}_n = \begin{bmatrix} 1 & 0 & 0 & \cdots & 0 \\ 0 & 1 & 0 & \cdots & 0 \\ 0 & 0 & 1 & \cdots & 0 \\ \vdots & \vdots & \vdots & & \vdots \\ 0 & 0 & 0 & \cdots & 1 \end{bmatrix} \tag{2.21b}$$

At this stage we refer the reader, if necessary, to books on modern algebra for an exposition of some elementary topics on matrix algebra, such as addition, multiplication, and computation of the inverse of a nonsingular matrix.

2.2.4. Orthogonal Matrix

If a matrix $\mathbf{A}$ is reflected relative to its leading diagonal (that is, its rows and columns are interchanged), we obtain the transpose $\tilde{\mathbf{A}}$. The inverse of a nonsingular matrix is represented by $\mathbf{A}^{-1}$.

If $\mathbf{A}^{-1} = \tilde{\mathbf{A}}$, the matrix $\mathbf{A}$ becomes an orthogonal matrix. We may mention here that all matrices representing symmetry operations are orthogonal.*

When a column matrix $[\phi]$ is multiplied by a square or rectangular matrix $\mathbf{A}$ a linear transformation results. Thus let $\mathbf{A}[\phi] = \mathbf{B}$. We conceive of this as an operation, and sometimes call $\mathbf{A}$ an operator and $[\phi]$ an operand. Two successive operations $\mathbf{A}$ and $\mathbf{B}$ are equivalent to a single operation $\mathbf{C} = \mathbf{BA}$. The operation $\mathbf{AB}$ is not in general the same as $\mathbf{BA}$, owing to the possible lack of commutation.

2.2.5. Matrix Operators and Transformation of Tensor Components

We indicate below the use of matrix operators in the transformations of tensor components of the physical properties of crystals from one coordinate system to another. We take the simplest case of a vector property.

Let X_i and X_i' represent two systems of Cartesian coordinate axes,

* Since a matrix representing a symmetry operation is orthogonal, any transformation due to a symmetry operation is orthogonal.

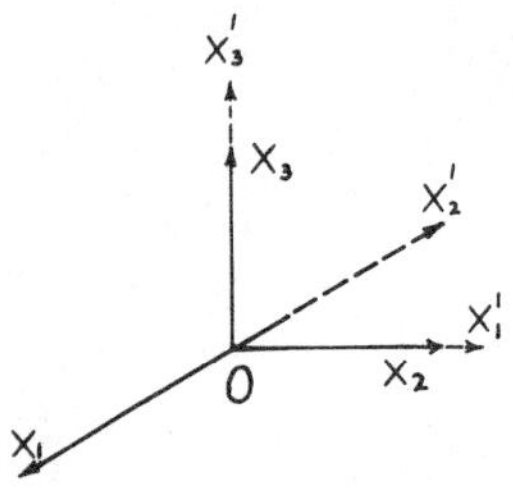

Fig. 2.2. Transformation of axes: a rotation of the coordinate system X_i through 90° about OX_3; i.e., OX_3 is a tetrad or a four-fold axis of rotation; here $OX_3' = OX_3$.

both having origin at O, and let α_{ij} be the direction cosines between the axes of one system and those of the other.

Let **p** be some vector property having components p_i in the X_i system and p_i' in the X_i' system. Then we have

$$p_i' = \alpha_{ij} p_j$$

The same relation can be expressed in matrix form thus:

$$[p_i'] = [\alpha_{ij}][p_j] \qquad \text{or} \qquad [p_i'] = \mathbf{A}[p_j] \tag{2.22}$$

Here, for example, **A** can be regarded as a rotation operator for changing the X_i system into the X_i' system. The components of **p** transform according to Eq. (2.22), although the magnitude of the vector **p** remains invariant under the transformation.

The matrix operator **A** in the particular case of rotation of the coordinate system X_i by 90° about OX_3 (Fig. 2.2) is given by

$$\mathbf{A} = \begin{bmatrix} 0 & 1 & 0 \\ -1 & 0 & 0 \\ 0 & 0 & 1 \end{bmatrix}$$

2.2.6. The Diagonalization of a Matrix

The second-rank tensor properties* of a crystal have, in general, nine components when referred to an arbitrary coordinate system as given in scheme (a) below. If, in addition, the tensor is symmetric in nature, for example, thermal expansion or electrical conductivity, there are only six independent components with reference to the general coordinate system, as given in scheme (b).

* Explained later in this chapter.

$$\begin{bmatrix} T_{11} & T_{12} & T_{13} \\ T_{21} & T_{22} & T_{23} \\ T_{31} & T_{32} & T_{33} \end{bmatrix}, \qquad \begin{bmatrix} T_{11} & T_{12} & T_{13} \\ T_{12} & T_{22} & T_{23} \\ T_{13} & T_{23} & T_{33} \end{bmatrix}, \qquad \begin{bmatrix} T'_{11} & 0 & 0 \\ 0 & T'_{22} & 0 \\ 0 & 0 & T'_{33} \end{bmatrix}$$

(a) (b) (c)

There is, however, a particular orientation of the coordinate system in which the nondiagonal components disappear. The axes of this coordinate system are called the principal axes, and the matrix representing the tensor components referred to the principal axes is given in scheme (c) above. This is called a diagonal matrix.

By introducing a similarity transformation we can deduce (1) the relationship between the components of the diagonalized matrix and those of the original matrix, and (2) the relationship between the orientation of the axes of the new coordinate system, in which the matrix of the tensor components is diagonal, and the original coordinate system.

A real symmetric matrix can be diagonalized by an orthogonal matrix, the latter being the matrix of a linear orthogonal transformation. The elements of the matrix of the linear orthogonal transformation, of course, satisfy the orthogonality relations.

2.3. *Vectors and Their Transformation Laws*

2.3.1. *Vector Components and Coordinate Transformations*

In order to locate a point quantitatively in a three-dimensional space, we introduce a Cartesian coordinate system. The location of the point is then given by three real numbers called the coordinates of the point, and they change as the frame of reference (i.e., the coordinate system) changes.

Let X and X' be two systems of reference, and let P be defined by x, y, z in X and by x', y', z' in X'. Then x', y', and z' are connected with x, y, and z thus:

$$\begin{aligned} x' &= \alpha_{11}x + \alpha_{12}y + \alpha_{13}z \\ y' &= \alpha_{21}x + \alpha_{22}y + \alpha_{23}z \\ z' &= \alpha_{31}x + \alpha_{32}y + \alpha_{33}z \end{aligned} \tag{2.23}$$

The constants α_{ij} are the direction cosines of the two coordinate systems X and X', given by the scheme $[\alpha_{ij}]$ as on page 16.

The change from one coordinate system to another is called a coordinate transformation, and the equations relating the coordinates of the same point in the two coordinate systems are called transformation equations.

Let us consider some vector field, say, an electrostatic field, in the neighborhood of a point P. The magnitude and direction of the field strength $\mathbf{E}$ at P become fully determined when we know the components of $\mathbf{E}$ with reference to some assigned coordinate system S. Let E_x, E_y, and E_z be the components of $\mathbf{E}$ at P with respect to S. Then E_x', E_y', and E_z', being components of $\mathbf{E}$ at P with reference to another system, say, S', can be computed if we know the transformation equations defining the coordinate transformation of S into S'. (These new components, E_x', E_y', and E_z', are independent of the translation of the point of origin.) The component E_x' is the sum of the projections of E_x, E_y, and E_z on the X' axis, and E_y' and E_z' are determined similarly. Thus E_i are transformed into E_j' according to the law of linear transformations.

The linear coordinate transformations from one system of rectangular coordinates to another system of rectangular coordinates are orthogonal transformations.

2.3.2. *Transformations of Coordinate Differences*

Let us start with a Cartesian coordinate system and call its three coordinates x_1, x_2, and x_3 (instead of x, y, and z). Let the coordinate differences between two points P and P' be Δx_1, Δx_2, and Δx_3. The distance S between the two points is given by the relation

$$S^2 = \sum_{i=1}^{3} (\Delta x_i)^2 \tag{2.24}$$

If we carry out a linear coordinate transformation,

$$x_i' = \sum_{k=1}^{3} \alpha_{ik} x_k \qquad (i = 1, 2, 3) \tag{2.25}$$

the new coordinate differences are

$$\Delta x_i' = \sum_{k=1}^{3} \alpha_{ik} \Delta x_k \qquad (i = 1, 2, 3) \tag{2.26}$$

Similarly, Δx_i can be expressed in terms of $\Delta x_k'$ thus:

$$\Delta x_i = \sum_{k=1}^{3} \alpha_{ki} \Delta x_k' \qquad (i = 1, 2, 3) \tag{2.27}$$

2.3.3. Transformation Law of Vectors

The transformation law of Δx_i is the general transformation law of vectors with respect to orthogonal transformations, or a vector **a** is defined as one whose components form a set of three quantities which transform like coordinate differences:

$$a_i' = \alpha_{ik} a_k \tag{2.28}$$

When the vector components are given with respect to any one Cartesian coordinate system, they can be computed with respect to every other Cartesian coordinate system provided the law of transformation between the two systems is known.

2.4. Tensor Nature of Physical Properties of Crystals and the Laws of Transformation of Cartesian Tensors

2.4.1. Concept of a Tensor Property and Some Examples of Tensor Properties

Every solid material has certain characteristic physical properties. These physical properties can be broadly divided into two groups, which must be treated in distinctly different ways. Some of the material properties are based upon qualitative concepts and hence they can be defined in terms of some specific tests: these tests facilitate the assignment of a number to the quality or property of the crystal being measured. Typical examples of such properties are hardness, brittleness, cleavage, ductility, and malleability. The numbers thus assigned will have a relative rather than an absolute meaning. There are some other properties of crystals which can be defined in absolute terms, such as density, thermal expansion, piezoelectricity, linear electro-optic effect, elastic property, photoelasticity, and Kerr electro-optic effect. These properties can be grouped together under one title, namely, anisotropic or tensor properties.

Every known physical property that is a tensor property can be associated with two measurable quantities, and, in fact, the property is defined by a relation between the two measurable quantities, which may be simple or complicated. We consider now a few typical physical properties.

2.4.1.1. Density

If we choose mass and volume as the two physical quantities, the physical property of density is associated with them. Thus density is the mass per unit volume. It may be noted that both mass and volume are scalars (also called tensors of zero rank) and their magnitudes do not depend on the directions (within the crystal) along which measurements are made. The density is thus a physical property which does not depend upon the direction in which it is measured. All crystals, irrespective of their symmetry, are therefore isotropic with respect to density.

2.4.1.2. Pyroelectricity

Certain crystals like tourmaline develop surface charges when heated. Actually, crystals exhibiting this property are spontaneously polarized at any temperature; however, the surface charge is quickly neutralized owing to atmospheric leakage, and it becomes apparent only during a transient period following a temperature change. This phenomenon is called pyroelectricity, and it is shown only by ionic crystals which lack a center of symmetry. The pyroelectricity is a consequence of an asymmetric distribution of positive and negative ions, which produces a net polarization in some one direction. The polarizability or the electric moment induced is a vector quantity (i.e., a tensor of rank one), and it is associated with temperature by a vector–scalar relation. Jewellers have noticed in window displays that tourmaline becomes dustier than other gems. As a result of temperature changes the tourmaline crystal becomes electrically charged and attracts tiny dust particles to its surface.

The reciprocal effect of pyroelectricity is also known. When an electric field $\mathbf{E}$ is applied in the direction of the pyroelectric axis of tourmaline, a heating or cooling effect results, depending on the direction of $\mathbf{E}$. For tourmaline a temperature rise of 0.001 °C takes place in a field of 1500 V/mm.

2.4.1.3. Optical Polarizability

The electric moment induced in a crystal by an external electric field (caused by incident optical radiation) and the electric field itself are the two physical quantities which are related to a physical property called the optical polarizability of a crystal. The optical polarizability may be defined as the electric (dipole) moment induced per unit volume by unit external electric field. It may be noted that the moment induced and the external electric field are vectors (first-rank tensors) and hence their magnitudes depend on the direction of measurement. Optical polarizability is thus a second-rank tensor property. But optical polarizability is related to the index of refraction. Hence the physical property of optical refraction, which is associated with the relation between the two quantities, which are first-rank tensors, will also depend on the direction of measurement within the crystal. In fact, we notice that some crystals, depending on their symmetry, are anisotropic in regard to optical refraction.

Electrical conductivity is yet another example of a property relating one vector quantity with another vector quantity, and hence it is a second-rank tensor property.

2.4.1.4. Piezoelectricity

This is a phenomenon in which some crystals, while under the influence of a mechanical stress (for example, under a compression due to a heavy weight), become electrically charged, with opposite charges appearing on opposite faces of the crystal. Under the influence of the mechanical stress, the crystal is slightly deformed, and since the crystal is made up of a large number of positive and negative ions which, in the undeformed state of the crystal, neutralize one another, it is possible that the deformation of the crystal may prevent the neutralization of the charges; if this is so, a resultant positive charge appears on one face of the crystal, and a resultant negative charge on the other face. Thus, the property of piezoelectricity relates electric (dipole) moment, a tensor of rank one, with a physical stress which is a second-rank tensor; hence piezoelectricity is a third-rank tensor property.

The *Pockels effect* (or linear electro-optic effect) is yet another example of a third-rank tensor property.

2.4.1.5. Elasticity

The elastic property of a crystal is a relation between stress and strain, both of which are examples of a second-rank tensor and thus sensitive to direction. Hence the elastic stiffness constants and the elastic compliance constants are examples of fourth-rank tensor properties.

Photoelasticity is yet another fourth-rank tensor property.

The *quadratic electro-optic effect* (also called the *Kerr effect*) comes under this category of fourth-rank tensor properties.

We have described above a few of the important physical properties, in order of increasing complexity. Every one of the above properties could be associated with the interaction of two tensors of appropriate rank. Each of the properties is influenced by the symmetry of the crystal class under study, thus governing the extent of anisotropy of the physical property for the given crystal class. (See also pages 128–133.)

The tensor method of treating the macroscopic properties of single crystals very clearly brings out the allowed and forbidden effects. However, one cannot make any predictions as to whether a particular crystal of any symmetry shows a strong or weak effect on considerations of symmetry alone. Technological applications require crystals possessing optimum effects. In this book we discuss the conditions which determine the classes of crystals capable of showing any desired effect and also the experimental methods by which such an effect can be detected. It should be noted that all macroscopic properties caused by interaction on a microscopic level should fit into a general framework, governed entirely by the crystal symmetry and the nature of the applied interaction.

Tensor properties, which may also be called tensors, are classified according to their rank. A tensor of rank n is a quantity which has 3^n components with reference to a system of Cartesian coordinates (that is, in a three-dimensional Euclidean space); these components vary from one set of Cartesian coordinates to another, and the law of transformation depends upon the rank of the tensor. In fact, it is usual to define the rank of a tensor from the law of transformation which it obeys whenever a transformation of the Cartesian coordinates occurs.

2.4.2. Transformation Law of Cartesian Tensors

We now deduce the law of transformation of Cartesian tensors. It is pertinent to reiterate that the tensor (or the tensor property) itself is a quantity which exists independently of the system of coordinate axes in which its components are represented. These components, however, change whenever there is a change of the Cartesian system according to a particular law of transformation, but the tensor itself does not change under such a transformation of the coordinate axes. The rank of a Cartesian tensor depends only on the number of its components and the way these components transform whenever there is a change of the Cartesian axes.

2.4.2.1. Tensors of Rank Zero (Also Called Scalars)

The simplest tensor property of a crystal is density. The density of a crystal is independent of the direction in which it is measured; hence it is a scalar, which is termed a tensor of rank zero. When a transformation of coordinates occurs from one Cartesian system to another, the property of density remains unaltered. If the scalar property is represented by T and T' in the old and new systems respectively, we have the transformation law expressed by the relation

$$T' = T \tag{2.29}$$

2.4.2.2. *Tensors of Rank One (Also Called Vectors)*

A typical physical property that is a first-rank tensor is pyroelectricity, discussed earlier in this chapter. Here the change in the induced electric moment $\Delta\mathbf{P}$ is proportional to the rise of temperature ΔT of the crystal. Hence

$$\Delta\mathbf{P} \propto \Delta T \qquad \text{or} \qquad \Delta\mathbf{P} = \mathbf{p}\,\Delta T$$

where **p** is defined as the induced electric moment per 1°C rise, and it is this **p** that represents the tensor property of rank one. Let **p** have components p_x, p_y, and p_z along the OX, OY, and OZ axes of the original coordinate system and p_x', p_y', and p_z' along the OX', OY', and OZ' axes of the transformed (new) coordinate system (Fig. 2.3), with

	X	Y	Z
X'	α_{11}	α_{12}	α_{13}
Y'	α_{21}	α_{22}	α_{23}
Z'	α_{31}	α_{32}	α_{33}

It is of great convenience, in discussing the tensor properties, to represent p_x, p_y, and p_z by p_1, p_2, and p_3; similarly to represent p_x', p_y', and p_z' by p_1', p_2', and p_3'. We can treat p_i' as a vector quantity having components along the OX, OY, and OZ axes (i.e., the OX_1, OX_2, and OX_3 axes) of the original coordinate system. Hence we have

$$p_i' = \text{sum of the three components in the old coordinate system}$$

Thus

$$\begin{aligned} p_1' &= \alpha_{11}p_1 + \alpha_{12}p_2 + \alpha_{13}p_3 \\ p_2' &= \alpha_{21}p_1 + \alpha_{22}p_2 + \alpha_{23}p_3 \\ p_3' &= \alpha_{31}p_1 + \alpha_{32}p_2 + \alpha_{33}p_3 \end{aligned} \tag{2.30}$$

$$\therefore p_i' = \alpha_{ij}p_j \quad \text{(new in terms of old)}$$

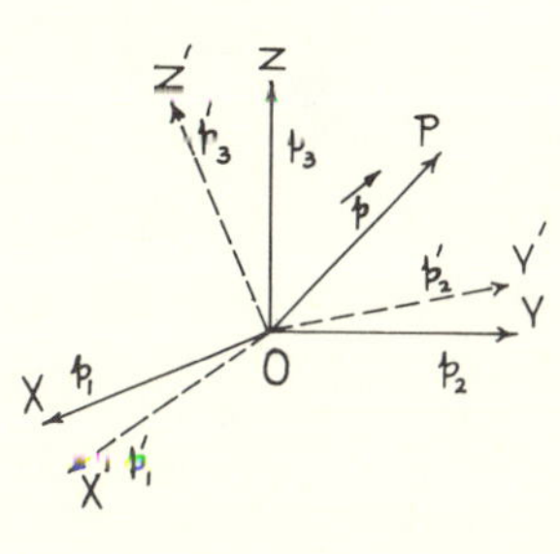

Fig. 2.3. Transformation of the components of a vector **p** from the X system to the X' system.

Similarly,

$$
\begin{aligned}
p_1 &= \alpha_{11}p_1' + \alpha_{21}p_2' + \alpha_{31}p_3' \\
p_2 &= \alpha_{12}p_1' + \alpha_{22}p_2' + \alpha_{32}p_3' \\
p_3 &= \alpha_{13}p_1' + \alpha_{23}p_2' + \alpha_{33}p_3' \\
\therefore p_i &= \alpha_{ji}p_j' \quad \text{(old in terms of new)}
\end{aligned}
\tag{2.31}
$$

Note that $\alpha_{ij} \neq \alpha_{ji}$ in general.

In the transformation of the first-rank tensor property, we have put p_i as a tensor component instead of T_i. Hence, using the symbol T_i for the first-rank tensor property, the law of transformation is

$$
\underset{\text{(new in terms of old)}}{T_i' = \alpha_{ij}T_j} \qquad \underset{\text{(old in terms of new)}}{T_i = \alpha_{ji}T_j'} \tag{2.32}
$$

Note also the positions of the dummy indexes j in the two cases. In the first case the dummy indexes j are nearest, while in the second case, they are separated by the free index i.

2.4.2.3. Tensors of Rank Two

We now examine in detail the electrical conductivity of a crystal and then define a second-rank tensor. Let an electric field **E** (parallel to OP) be applied to a crystal in an arbitrary direction. Then the electric current **i** through the crystal may not be parallel to **E**. Let **i** be in some direction parallel to OQ (Fig. 2.4).

We know that **E**, being an electric field, is a vector; let its components be E_1, E_2, and E_3 along OX_1, OX_2, and OX_3, respectively.

Let us treat the current density **i** also as a vector, its components being i_1, i_2, and i_3 along OX_1, OX_2, and OX_3, respectively. Next we assume, as in the case of the simple Ohm's law, that

$$
\begin{aligned}
i_1 &= \sigma_{11}E_1 + \sigma_{12}E_2 + \sigma_{13}E_3 \\
i_2 &= \sigma_{21}E_1 + \sigma_{22}E_2 + \sigma_{23}E_3 \\
i_3 &= \sigma_{31}E_1 + \sigma_{32}E_2 + \sigma_{33}E_3
\end{aligned}
\tag{2.33}
$$

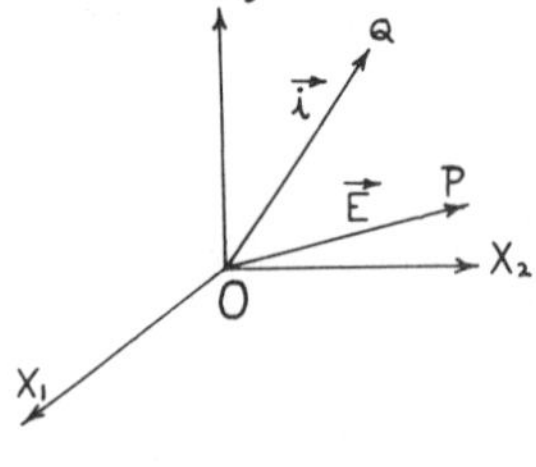

Fig. 2.4. The relation between two vectors, the electric current **i** and the electric field **E**, in an anisotropic crystal.

Thus there are nine components of σ_{ij} in general for a crystal, instead of one as in the case of an isotropic solid. This quantity σ_{ij}, which defines the electrical conductivity, is a tensor property of second rank, relating vector **i** and vector **E**.

How does σ_{ij} transform from one system of rectangular coordinates to another?

We now replace σ_{ij} by T_{ij}, the letter T representing in general a tensor property; the number of suffixes indicates the rank of the tensor. Similarly, the current i is replaced by the letter p, and E by q. Then Eq. (2.33) changes to

$$p_i = T_{ij}q_j \tag{2.34}$$

when the property T_{ij} is described in terms of the original Cartesian coordinate system. When it is referred to a new coordinate system, we have

$$p_k' = T'_{kl}q_l' \tag{2.35}$$

We know the direct relation between p_k' and p_i and also between q_l' and q_j. We can now establish a relation between p_k' and q_l' through p_i and q_j thus:

$$p_k' \rightarrow p_i \rightarrow q_j \rightarrow q_l'$$

Then

$p_k' = \alpha_{ki}p_i$ (this is due to the law of transformation from one coordinate system to another)

$p_i = T_{ij}q_j$ (this establishes the tensor nature of T_{ij} relating p_i and q_j in the same coordinate system)

$q_j = \alpha_{lj}q_l'$ (this is again due to the law of transformation from one coordinate system to another)

Hence

$$\begin{aligned} p_k' &= \alpha_{ki}p_i \\ &= \alpha_{ki}T_{ij}q_j \\ &= \alpha_{ki}T_{ij}\alpha_{lj}q_l' \end{aligned}$$

But $p_k' = T'_{kl}q_l'$, so

$$T'_{kl} = \alpha_{ki}\alpha_{lj}T_{ij} \tag{2.36}$$

Equation (2.36) establishes the law of transformation of a second-rank tensor. It should be clearly noted that the α_{ij} describe the direction cosines among the axes of the new and old coordinate systems. The α_{ij} have nothing to do with the nature of the physical property under consideration. Furthermore, we have the orthogonal relations among the several α_{ij} [see Eqs. (2.11) and (2.12)]. The T_{kl} describe the tensor nature of the physical property in the same coordinate system. Thus the T_{kl} relate two physical quantities, say, electrical intensity and current density, in the same coordinate system, while the α_{ij} relate two different coordinate systems and have nothing to do with the tensor property under consideration.

To express T_{ij} in terms of T'_{kl} we proceed thus:

$$p_i \rightarrow p_k' \rightarrow q_l' \rightarrow q_j$$

Now

$$p_i = \alpha_{ki} p_k', \qquad p_k' = T'_{kl} q_l', \qquad q_l' = \alpha_{lj} q_j$$

$$\begin{aligned} \therefore p_i &= \alpha_{ki} p_k' \\ &= \alpha_{ki} T'_{kl} q_l' \\ &= \alpha_{ki} T'_{kl} \alpha_{lj} q_j \end{aligned}$$

But $p_i = T_{ij} q_j$. Hence

$$T_{ij} = \alpha_{ki} \alpha_{lj} T'_{kl} \tag{2.37}$$

2.4.2.4. Tensors of Rank Higher than Two

Proceeding along the lines described above, we can show that, for a third-rank tensor,

$$\begin{aligned} T'_{ijk} &= \alpha_{il} \alpha_{jm} \alpha_{kn} T_{lmn} \\ T_{ijk} &= \alpha_{li} \alpha_{mj} \alpha_{nk} T'_{lmn} \end{aligned} \tag{2.38}$$

and for a fourth-rank tensor,

$$\begin{aligned} T'_{ijkl} &= \alpha_{im} \alpha_{jn} \alpha_{ko} \alpha_{lp} T_{mnop} \\ T_{ijkl} &= \alpha_{mi} \alpha_{nj} \alpha_{ok} \alpha_{pl} T'_{mnop} \end{aligned} \tag{2.39}$$

We thus see that the law of transformation of the tensor property of any rank is closely related to that of the coordinates themselves and makes use of the same basic quantities, namely the direction cosines α_{ij}.

2.4.3. Physical Properties and Crystal Symmetry

The index of refraction of NaCl crystal is 1.542; calcite has two indices of refraction, $n_o = 1.658$ and $n_e = 1.486$; barite has three indices of refraction, $n_x = 1.648$, $n_y = 1.637$, and $n_z = 1.636$. Now the important question arises as to why NaCl should have only one index of refraction, calcite two, designated by n_o and n_e, and barite three, designated by n_x, n_y, and n_z; we may also ask how these three crystals behave in regard to their elastic properties. In fact, we are interested in knowing in a very general way the complexity of anisotropy of every physical property exhibited by various crystals. Before we answer the above, there are two aspects of the problem which we should understand: (1) the tensor nature of the physical property and (2) the structural symmetry of the crystal under consideration. We have given in an earlier section some typical physical properties and their tensor nature. We shall now proceed somewhat in detail to examine the structural symmetry and then the tensor properties, and the influence of the crystal symmetry on each of the tensor properties. It should be noted that crystal symmetry eliminates some of the components while establishing mathematical relations among the remaining components of each rank. Some of these relations are obvious without detailed calculation, while others require a complete tensor calculation of the effect of symmetry on the tensor components.

2.5. *Crystal Symmetry and Geometrical Crystallography. The 32 Point Groups*

2.5.1. *The 32 Crystallographic Point Groups: Their Symmetry Elements and Some Examples of Crystals*

All the crystalline solids are broadly divided into seven systems, and these seven systems are subdivided into 32 crystal classes, also called point groups [782; 552].* Table 2.5a lists these crystal classes or point groups in International and Schönflies notation along with the elements of symmetry possessed by each crystal class. It may be noted that each class has at least the minimum elements of symmetry specified for the crystal system, but it lacks some of the symmetry elements of the higher classes; the classes themselves have been arranged in order of increasing symmetry. The importance of the generating elements of symmetry listed in column 5 will be brought out in Chapters 3, 7, and 8.

Table 2.1 gives a detailed geometric description of the various elements of symmetry and some common examples of crystals of the 32 classes.

* For the IEEE convention of crystal axes see [782] and [552].

Table 2.1. The 32 Crystallographic Point Groups, Their Symmetry Elements, and Some Examples of Crystals in Each Class

Class symbol		Symmetry operations of the point group	Examples
Schönflies	International		
C_1	1	None[a]	Strontium ditartrate tetrahydrate, $SrH_2(C_4H_4O_6)_2 \cdot 4H_2O$; calcium thiosulfate
$C_i = S_2$	$\bar{1}$	A center	$CuSO_4 \cdot 5H_2O$
C_2	2	Diad $\parallel y$	$Li_2SO_4 \cdot H_2O$; tartaric acid
$C_s = C_{1h} = S_1$	m	Plane $\perp y$ (= inversion diad)	KNO_2; sodium silicate
C_{2h}	$2/m$	Diad $\parallel y$; plane $\perp y$; a center	$KClO_3$; mica; gypsum
C_{2v}	$mm2$	Diad, two planes containing it	Calamine; picric acid; resorcinol
$D_2 = V$	222	3 diads $\perp$ to each other	Rochelle salt; $MgSO_4 \cdot 7H_2O$
$D_{2h} = V_h$	mmm	3 planes intersecting in 3 diads; a center	$BaSO_4$; $PbSO_4$; K_2SO_4; α-sulfur
C_3	3	1 triad	$NaIO_4 \cdot 3H_2O$
$C_{3i} = S_6$	$\bar{3}$	Inversion triad (= a rotation triad plus a center)	Na_2SO_3; Li_2MoO_4; Li_2WO_4
C_{3v}	$3m$	Triad along which three vertical planes intersect	$KBrO_3$; proustite; tourmaline
D_{3d}	$\bar{3}m$	Inversion triad along which 3 vertical planes intersect (= 3 vertical planes intersecting in a triad, 3 horizontal diads $\perp$ planes of symmetry, and a center)	Calcite; $NaNO_3$; Corundum
D_3	32	Triad $\perp$ 3 diads	Cinnabar; α-quartz
C_4	4	1 tetrad	Barium antimonyl tartrate; metaldehyde; $KSr_2Nb_5O_{15}$
S_4	$\bar{4}$	1 inversion tetrad	$Ca_2Al_2SiO_7$
C_{4h}	$4/m$	Tetrad, a plane $\perp$ to tetrad, and a center	$NaIO_4$ (anhydrous); $CaWO_4$

Table 2.1 (*continued*)

Class symbol		Symmetry operations of the point group	Examples
Schönflies	International		
C_{4v}	$4mm$	2 pairs of planes at right angles intersecting in a single tetrad	Diaboleite; pentaerythritol
$V_d = D_{2d} = S_{4v}$	$\bar{4}2m$	2 planes at right angles intersecting in an inversion tetrad, 2 diads at 45° to the planes; diads are x and y axes	Urea; ADP; KDP
D_4	422	Tetrad axis ⊥ two pairs of diad axes	$NiSO_4 \cdot 6H_2O$; methyl ammonium iodide
D_{4h}	$4/mmm$	Tetrad axis at the intersection of 2 pairs of planes, two pairs of horizontal diads normal to the planes, a plane ⊥ to tetrad, and a center	Rutile; zircon; mercurous chloride
C_6	6	Hexad	$LiKSO_4$; CHI_3 (iodoform)
$S_3 = C_{3h}$	$\bar{6}$ $(=3/m)$	Inversion hexad (= triad ⊥ plane of symmetry)	
C_{6h}	$6/m$	Hexad ⊥ a plane, and a center	Apatite
C_{6v}	$6mm$	Hexad at the intersection of 2 sets of 3 planes of symmetry	ZnO (zincite); ZnS (wurtzite); CdS (Greenockite)
D_{3h}	$\bar{6}m2$	Inversion hexad at the intersection of 3 vertical planes (= a triad ⊥ a plane, 3 planes intersecting in the triad, 3 diads lying in these planes ⊥ to triad axis)	$BaTiSi_3O_9$ (benitoite)
D_6	622	Hexad normal to 6 diad axes	β-quartz; $LiIO_3$
D_{6h}	$6/mmm$	Hexad at intersection of 2 sets of 3 vertical planes of symmetry; 2 sets of 3 diads ⊥ to these planes, a plane of symmetry ⊥ hexad, and a center	Beryl

continued overleaf

Table 2.1 (continued)

Class symbol		Symmetry operations of the point group	Examples
Schönflies	International		
T	23	3 mutually ⊥ diads and 4 triads	$NaClO_3$; $NaBrO_3$
T_h	$m3$	3 diads at the intersections of 3 mutually ⊥ planes, 4 triads, and a center	$Ba(NO_3)_2$; $Ca(NO_3)_2$
T_d	$\bar{4}3m$	3 diads (actually these are inversion tetrads), 4 triads, and 6 planes of symmetry	ZnS (sphalerite); CuCl
O	432	3 tetrads, 4 triads, and 6 diads	
O_h	$m3m$	3 tetrads, 4 triads, 6 diads, 3 cubic planes, 6 dodecahedral planes, and a center	NaCl (halite); MgO (periclase); CaF_2 (fluorite)

[a] The identity element E or the trivial element of symmetry, describing a rotation through 360°, being common to all the classes, is omitted here; but it is an important symmetry element in group theoretical calculations of the tensor components of crystal properties. In the Schönflies notation $S_n = (C_n \cdot \sigma_h)$.

2.5.2. *Some Symmetry Operations and Their Representation by Symbols*

The elements of symmetry of a crystal are due to a set of geometric operations which leave the unit cell of the crystal in a state indistinguishable from the original one. These operations consist of (1) rotations about certain axes, (2) reflection through planes, and (3) reflection through points (inversion centers). Some of these operations are defined in Table 2.2.

Some combinations of symmetry elements necessarily imply the existence of others. For example, *mm* also implies the presence of a two-fold axis of symmetry. Most of the crystal classes, therefore, possess symmetry elements beyond those which entitle them to belong to a particular class.

The symbol *mmm* may imply that there are three mutually perpendicular planes of symmetry. The symbol could also imply that there are three planes of symmetry intersecting along a line all having dihedral angles of 60° with one another.

Table 2.2. *Some Symmetry Operations and Their Representation by Symbols*[a]

Symmetry operation	Short name	Symbol	
		Schönflies	International
Rotation about an axis through an angle 360°/n	n-fold axis of symmetry	C_n	n
Reflection through a plane	Plane of symmetry	$C_s = C_{1h}$	m
Reflection through a point	Center of symmetry (inversion center)	C_i	$\bar{1}$
Rotation about an axis through 360°/n along with a center of symmetry	n-fold axis with a center of symmetry	C_{ni}	
Rotation about an axis through 360°/n with a point of symmetry situated on the axis (this is a composite symmetry operation)	n-fold rotation–inversion axis		$\bar{n}$
Rotation about an axis through 360°/n with a plane of reflection normal to the axis	n-fold axis of symmetry normal to a plane of symmetry	C_{nh}	n/m
Rotation about an axis through 360°/n with a plane of reflection normal to the axis (this is a composite symmetry operation)	n-fold rotation–reflection axis	S_n	
Rotation about an axis through 360°/n contained by n planes of symmetry	n-fold axis of symmetry contained by n planes of symmetry	C_{nv}	nm

[a] A plane of reflection is also represented by σ. In this book a rotation–inversion axis is represented by $\tilde{S}_n$ in order to describe the complete elements of symmetry of a point group. See also pages 38 and 39 for further information on symmetry operations and their symbols.

2.5.3. *The 32 Crystallographic Point Groups in the Schönflies Notation. Geometric Derivation*

We now give a geometric method of deriving the 32 point groups in the Schönflies notation, shown in Table 2.3.

1. The simplest symmetry operation in a solid is an axis of rotation, and in order to fill space completely, the axis can be of 1-, 2-, 3-, 4-, and 6-fold rotation. Hence we have the five crystal classes C_n with n-fold axis in the first row, viz., C_1 C_2 C_3 C_4 C_6. All these classes are monoaxial, in

Table 2.3. *The 32 Crystallographic Point Groups in the Schönflies Notation. Geometric Derivation*[a,b]

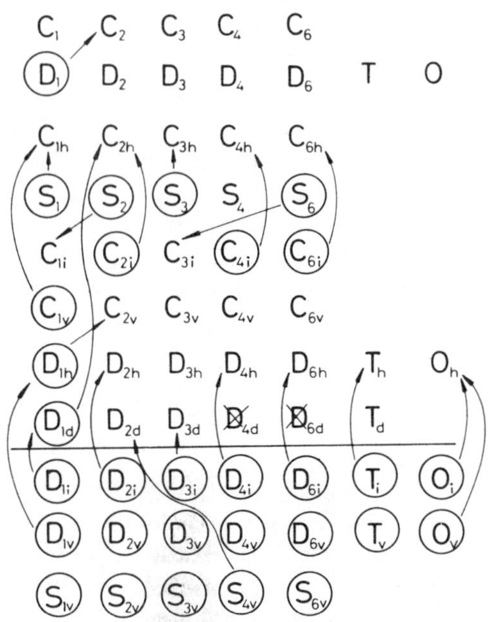

[a] In this notation the composite symmetry element $S_n = (C_n \cdot \sigma_h)$.

[b] The arrows join equivalent classes, i.e., classes with the same symmetry; in each case the arrow points to the preferred (standard) notation.

the sense that each one of them has only one axis of rotation. The classes C_3, C_4, and C_6, however, are uniaxial as well.

2. Starting from the first row of the C_n class, we can develop other classes thus: To each of these axes can be added n two-fold axes which are secondary and perpendicular to the C_n axis. The symbol for them is D_n. These five classes are called the dihedral classes. Tetrahedral T and octahedral O are two more classes having several sets of axes. This row then accounts for seven classes.

3. In the third row, a plane of reflection is added perpendicular to the C_n axis of the first row. The new symmetry is designated as C_{nh}.

4. In the fourth row, all the C_n axes become axes of rotation–reflection, also called alternating axes. They are designated as S_1 S_2 S_3 S_4 S_6. However, all S_n except S_4 become equal to other crystal classes (as shown by the arrows in Table 2.3), so that ultimately only S_4 appears among the

32 point groups in the Schönflies notation: $S_n = (\sigma_h \cdot C_n)$. This is a composite symmetry.

5. In the fifth row we have a center of symmetry in addition to the C_n axis of the first row. Hence all the five classes in this row are monoaxial (not necessarily uniaxial), having a center of inversion.

6. In the sixth row we have the classes represented by C_{nv} obtained by adding *n* planes of symmetry which intersect along the C_n axis of the first row.

Two more rows, namely the seventh and eighth, are derived from the second row; thus we have the following:

7. The seventh row is obtained from D_n by adding a plane of symmetry normal to the principal axis (n) and then adding *n* planes of symmetry each containing the principal and one secondary axes.

8. The eighth row D_{nd} is characterized by the fact that it contains the symmetry D_n with planes of symmetry containing the C_n axis and bisecting the angles between the secondary axes. Thus the symbol *d* represents diagonal planes, i.e., vertical planes between the two-fold axes of the dihedral groups.

Out of the several classes thus obtained, there are only 32 possible independent classes, and all others are either equivalent to some of the 32 classes, or crystallographically not possible. Some of the equivalent classes are shown by arrows in Table 2.3, while some others which are crystallographically untenable are crossed out. For example, in the eighth row, D_{4d} and D_{6d} are ruled out from the possible crystal classes, since they demand the existence of crystallographic axes of 8- and 12-fold rotation; this is due to the fact that in the case of classes D_{nd} the principal axis is not just an *n*-fold axis but rather a 2*n*-fold rotation–reflection axis.

Further addition of any symmetry element to any of the point groups in the above eight rows does not yield any new point groups. To substantiate this statement we proceed thus:

9. The ninth row is obtained by adding a center of inversion to D_n in order to get D_{ni} classes.

10. To D_n are added *n* vertical planes (i.e., planes containing the principal axis) in order to get the D_{nv} of the tenth row.

11. To S_n of the fourth row are added *n* vertical planes (i.e., planes containing the principal axis) in order to get S_{nv} of the eleventh row.

The equivalence of some of the new classes with some of the classes occurring in the first eight rows can be seen from the following remarks:

No new groups would result if a vertical plane were added to the D_n, T, and O classes, because $D_{nv} = D_{nh}$, $T_v = T_h$, and $O_v = O_h$. Hence the tenth row is completely overlapped by the seventh row.

Addition of an inversion center to D_n gives nothing new when n is even. When n is odd, D_{3i}, for example, is equal to D_{3d}.

The class C_{1h} is equal to C_{1v} because the one-fold axis can be taken normal to the mirror to give C_{1h}, and parallel to the mirror to give C_{1v}.

Addition of a vertical plane to S_n results in S_{nv}, but no new groups result, e.g., $S_{3v} = D_{3d}$ and $S_{4v} = D_{2d}$. Similar arguments can be extended to all the remaining classes of the ninth, tenth, and eleventh rows. See, for example, page 92 of [339].

Some special features of the crystal classes are listed below.

1. All the 11 classes (C_n D_n T O) have no plane of symmetry and no center of symmetry. They have only axial symmetry.

2. In all these 11 classes, two different types of the same species may exist, one type being a mirror image of the other. They are called enantiomorphous. Crystals of these classes offer a good example of Neumann's principle, since certain directed physical properties have different signs for the two types.

3. Of all these 11 classes, 10 are piezoelectric. Only class O is not piezoelectric. Although class O has no center of symmetry, nevertheless it has other symmetry elements that combine to exclude the piezoelectric property. Hence it cannot show the Pockels effect (see Chapter 8 for further comments).

4. All the 11 classes are included among the 15 optically active classes.

5. The tetrahedral class is T. There are three orthogonal (i.e., mutually perpendicular) 2-fold axes and four 3-fold axes; T_h is symmetry T with a plane of symmetry perpendicular to each of the 2-fold axes; T_d is symmetry T with six planes of symmetry, each containing two of the 3-fold axes.

6. The octohedral class is O, containing three orthogonal 4-fold axes, six 2-fold axes, and four 3-fold axes; O_h is symmetry O with planes of symmetry of both T_d and T_h.

2.6. Symmetry Operations and Their Transformation Matrices

When a crystal is subjected to a symmetry operation, each atom of the crystal (or each lattice point) is moved from its original position; however,

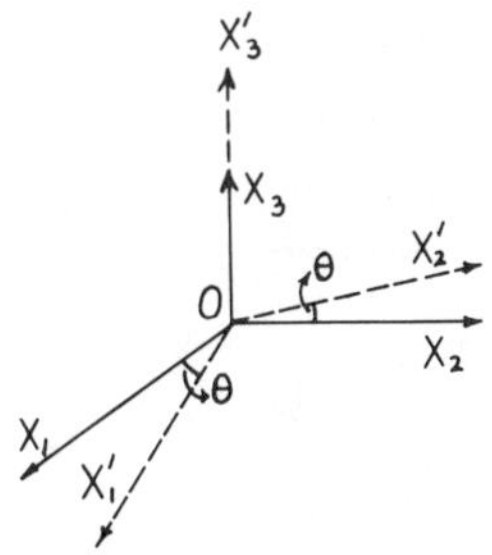

Fig. 2.5. Transformation of axes: a rotation of the X_i system through an angle θ about the X_3 axis; here $OX_3' = OX_3$.

after completion of the symmetry operation, the new positions of the atoms will be indistinguishable from the original positions, that is, the two positions are identical or self-congruent. Instead of performing the symmetry operation on the crystal itself, we might choose some orthogonal system of coordinates X_1, X_2, and X_3 and locate every lattice point of the crystal in this coordinate system. Then the same symmetry operation performed on the coordinate system transforms it into a new coordinate system, but the lattice points in the two sets of coordinates become identical and hence indistinguishable.

The transformation of the original coordinate system into the new coordinate system can be performed by matrix multiplication, the matrix being the one corresponding to the relevant or chosen symmetry operation. For example, a rotation of the coordinate system about the X_3 axis through an angle θ (Fig. 2.5) gives rise to the transformation matrix

$$\begin{bmatrix} \cos\theta & \sin\theta & 0 \\ -\sin\theta & \cos\theta & 0 \\ 0 & 0 & 1 \end{bmatrix}$$

Table 2.4 presents a set of transformation matrices corresponding to some typical symmetry operations. It may be noted that symmetry operations such as a three-fold axis of rotation require that the crystal come into itself or be congruent by a rotation of either 120° or 240°. Hence two different transformations can be expressed, though only one is given in Table 2.4. Furthermore, symmetry operations such as $2/m$ correspond to a transformation matrix equivalent to the product of two different matrices. In this particular case, $2/m$ would have a transformation matrix which is the product of the transformation matrix corresponding to a rotation through 180° and the matrix corresponding to reflection through a plane normal to the diad or two-fold axis.

Table 2.4. Some Symmetry Operations and Their Transformation Matrices

Symmetry element		Detailed description of the symmetry operation represented by the symmetry element	Corresponding transformation matrix
E	1	Identity	$\begin{bmatrix} 1 & 0 & 0 \\ 0 & 1 & 0 \\ 0 & 0 & 1 \end{bmatrix}$
i	$\bar{1}$	Center of symmetry (origin being the center of symmetry)	$\begin{bmatrix} -1 & 0 & 0 \\ 0 & -1 & 0 \\ 0 & 0 & -1 \end{bmatrix}$
C_2	2	Two-fold axis of rotation about OX_3 axis	$\begin{bmatrix} -1 & 0 & 0 \\ 0 & -1 & 0 \\ 0 & 0 & 1 \end{bmatrix}$
C_2'	2	Two-fold axis of rotation about OX_1 axis	$\begin{bmatrix} 1 & 0 & 0 \\ 0 & -1 & 0 \\ 0 & 0 & -1 \end{bmatrix}$
σ_h	m	Reflection plane perpendicular to OX_3 axis	$\begin{bmatrix} 1 & 0 & 0 \\ 0 & 1 & 0 \\ 0 & 0 & -1 \end{bmatrix}$
σ_v	m	Reflection plane perpendicular to OX_1 axis	$\begin{bmatrix} -1 & 0 & 0 \\ 0 & 1 & 0 \\ 0 & 0 & 1 \end{bmatrix}$
C_3	3	Three-fold axis of symmetry about OX_3 axis (rotation through 120°)	$\begin{bmatrix} -1/2 & 3^{1/2}/2 & 0 \\ -3^{1/2}/2 & -1/2 & 0 \\ 0 & 0 & 1 \end{bmatrix}$
C_3'	3	Three-fold axis of symmetry about a body diagonal of a cubic crystal	$\begin{bmatrix} 0 & 1 & 0 \\ 0 & 0 & 1 \\ 1 & 0 & 0 \end{bmatrix}$
C_4	4	Four-fold axis of symmetry about OX_3 axis (rotation through 90°)	$\begin{bmatrix} 0 & 1 & 0 \\ -1 & 0 & 0 \\ 0 & 0 & 1 \end{bmatrix}$
C_4'	4	Four-fold axis of symmetry about OX_1 axis (rotation through 90°)	$\begin{bmatrix} 1 & 0 & 0 \\ 0 & 0 & 1 \\ 0 & -1 & 0 \end{bmatrix}$
$\tilde{S}_4$	$\bar{4}$	Four-fold axis of symmetry about OX_3 axis, along with a center of symmetry (this is a composite symmetry)—inversion tetrad	$\begin{bmatrix} 0 & -1 & 0 \\ 1 & 0 & 0 \\ 0 & 0 & -1 \end{bmatrix}$
C_6	6	Six-fold axis of symmetry about OX_3 axis (rotation through 60°)	$\begin{bmatrix} 1/2 & 3^{1/2}/2 & 0 \\ -3^{1/2}/2 & 1/2 & 0 \\ 0 & 0 & 1 \end{bmatrix}$

2.7. Symmetry Elements of the 32 Point Groups

We now proceed to enumerate the complete set of symmetry elements of the 32 point groups.

2.7.1. Symmetry Elements of the 32 Point Groups

Table 2.5a gives the complete set of symmetry elements of the 32 crystallographic point groups. The orientation of the various symmetry elements in Tables 2.5a and 2.5b can be correlated with those described in detail in Table 2.1 or by referring to the stereograms given in [931; 885; 1396].

For the purpose of computing by group theoretical methods the total number of nonvanishing independent tensor components of any crystal class, enumeration of the complete set of elements of symmetry of the crystal class is required, which is given in detail in Table 2.5a. Some of the more common symmetry elements are defined in detail in Table 2.4.

For a clear understanding of those elements not defined in Table 2.4, one should consider all the symmetry operations noted in column 5 of Table 2.5a, and also examine the consequential elements that accompany them.

For example, for class 3 (C_3) in Table 2.5a under column 5 this has only one triad as the symmetry element parallel to the principal axis OX_3. But this generates the symmetry elements C_3, $(C_3)^2 = (C_3)^{-1}$, and $(C_3)^3 = E$. For the purpose of calculating the tensor properties by group theoretical methods, $C_3{}^+$ and $C_3{}^{-1}$ belong to the same class; hence they are grouped together, thereby giving $E\,2C_3$ in column 6 of Table 2.5a.

Similarly, for class 6 (C_6), we get $E\,C_2\,2C_3\,2C_6$ as the complete set of symmetry elements for the purpose of group theoretical calculations of the tensor properties.

Invariably for all the 32 point groups the principal axis of rotation is parallel to the OX_3 axis of the Cartesian coordinate system, except for the monoclinic point groups for which the principal diad axis is parallel to OX_2 (see Reference [552]). An axis of rotation different from the principal axis is distinguished by C'. Thus in class 23 (T), the diad axis C_2 is parallel to the principal axis and $C_3{}'$ is a secondary triad axis. Yet another secondary axis, if it exists, is indicated by C'' as in the case of the point group 422 (D_4) which has $E\,2C_4\,C_2\,2C_2{}'\,2C_2{}''$.

The symbols $\sigma_v{}'$ and $\sigma_v{}''$ represent two differently oriented reflection planes which intersect in the principal axis. For example, for the class

Table 2.5a. *The 32 Point Groups and Their Symmetry Elements (as Operators Forming the Elements of a Finite Group)*[a]

Crystal system	No.	Class symbol: International	Class symbol: Schönflies	Generating elements	Complete set of elements of symmetry of the crystallographic point group
Triclinic	1	1	C_1	E	E
	2	$\bar{1}$	$S_2 = C_i$	i	$E\, i$
Monoclinic	3	2	$C_2 = D_1$	C_2	$E\, C_2$
	4	$m = \bar{2}$	$S_1 = C_s = C_{1h}$	σ_h	$E\, \sigma_h$
	5	$2/m$	$C_{2h} = C_{2i}$	$(C_2\, \sigma_h)$ or $(C_2\, i)$	$E\, C_2\, \sigma_h\, i$
Orthorhombic	6	222	$D_2 = V$	$C_2\, C_2'$	$E\, C_2\, C_2'\, C_2''$
	7	$mm2$	C_{2v}	$(C_2\, \sigma_v')$	$E\, C_2\, \sigma_v'\, \sigma_v''$
	8	mmm	$D_{2h} = V_h$	$(C_2\, C_2'\, i)$ or $(C_2\, C_2'\, \sigma_h)$	$E\, C_2\, C_2'\, C_2''\, i\, \sigma_h\, \sigma_v'\, \sigma_v''$
Tetragonal	9	4	C_4	C_4	$E\, 2C_4\, C_2$
	10	$\bar{4}$	S_4	S_4	$E\, 2S_4\, C_2$
	11	$4/m$	C_{4h}	$(C_4\, \sigma_h)$ or $(C_4\, i)$	$E\, 2C_4\, C_2\, i\, 2S_4\, \sigma_h$
	12	422	D_4	$C_4\, C_2'$	$E\, 2C_4\, C_2\, 2C_2'\, 2C_2''$
	13	$4mm$	C_{4v}	$C_4\, \sigma_v'$	$E\, 2C_4\, C_2\, 2\sigma_v'\, 2\sigma_v''$
	14	$\bar{4}2m$	$D_{2d} = V_d$	$(S_4\, C_2')$ or $(S_4\, \sigma_v')$	$E\, C_2\, C_2'\, C_2''\, 2S_4\, \sigma_v'\, \sigma_v''$
	15	$4/mmm$	D_{4h}	$C_4\, C_2'\, \sigma_h$	$E\, 2C_4\, C_2\, 2C_2'\, 2C_2''\, i\, 2S_4\, \sigma_h\, 2\sigma_v'\, 2\sigma_v''$

Trigonal	16	3	C_3	C_3	$E\,2C_3$
	17	$\bar{3}$	$S_6 = C_{3i}$	$C_3\,i$	$E\,2C_3\,i\,2S_6$
	18	32	D_3	$C_3\,C_2'$	$E\,2C_3\,3C_2'$
	19	$3m$	C_{3v}	$C_3\,\sigma_v$	$E\,2C_3\,3\sigma_v$
	20	$\bar{3}m$	D_{3d}	$C_3\,C_2'\,i$	$E\,2C_3\,3C_2'\,i\,2S_6\,3\sigma_v$
Hexagonal	21	6	C_6	C_6 or $(C_3\,C_2)$	$E\,2C_6\,2C_3\,C_2$
	22	$\bar{6} = 3/m$	$C_{3h} = S_3$	$C_3\,\sigma_h$	$E\,2C_3\,\sigma_h\,2S_3$
	23	$6/m$	C_{6h}	$(C_6\,\sigma_h)$ or $(C_6\,i)$ or $(C_3\,C_2\,\sigma_h)$ or $(C_3\,C_2\,i)$	$E\,2C_6\,2C_3\,C_2\,i\,2S_3\,2S_6\,\sigma_h$
	24	622	D_6	$(C_6\,C_2')$ or $(C_3\,C_2\,C_2')$	$E\,2C_6\,2C_3\,C_2\,3C_2'\,3C_2''$
	25	$6mm$	C_{6v}	$(C_6\,\sigma_v')$ or $(C_3\,C_2\,\sigma_v')$	$E\,2C_6\,2C_3\,C_2\,3\sigma_v'\,3\sigma_v''$
	26	$\bar{6}m2 = 3/mm$	D_{3h}	$(C_3\,C_2'\,\sigma_h)$ or $(C_3\,C_2'\,\sigma_v)$ or $(C_3\,\sigma_h\,\sigma_v)$	$E\,2C_3\,3C_2'\,\sigma_h\,2S_3\,3\sigma_v$
	27	$6/mmm$	D_{6h}	$(C_6\,C_2'\,\sigma_h)$ or $(C_6\,C_2'\,i)$ or $(C_6\,\sigma_v'\,i)$ or $(C_3\,C_2\,C_2'\,i)$	$E\,2C_6\,2C_3\,C_2\,3C_2'\,3C_2''\,i\,2S_3\,2S_6\,\sigma_h\,3\sigma_v'\,3\sigma_v''$
Cubic	28	23	T	$C_2\,C_3'$	$E\,3C_2\,8C_3'$
	29	$m3$	T_h	$(C_2\,C_3'\,i)$ or $(C_2\,C_3'\,\sigma_h)$	$E\,3C_2\,8C_3'\,i\,8S_6\,3\sigma_h$
	30	432	O	$C_4\,C_3'$	$E\,3C_2\,8C_3'\,6C_2''\,6C_4$
	31	$\bar{4}3m$	T_d	$S_4\,C_3'$	$E\,3C_2\,8C_3'\,6\sigma_d\,6S_4$
	32	$m3m$	O_h	$(C_4\,C_3'\,\sigma_h)$ or $(C_4\,C_3'\,i)$	$E\,3C_2\,8C_3'\,6C_2''\,6C_4\,i\,8S_6\,3\sigma_h\,6\sigma_d\,6S_4$

[a] In the last column, the composite symmetry element S_n is equal to $(C_n \cdot \sigma_h)$, following Schönflies. See further comments on pages 42 and 43.

Table 2.5b. *Appendix to Table 2.5a*[a]

Crystal system	No.	Class symbol: International	Class symbol: Schönflies (modified)	Generating elements	Complete set of elements of symmetry of the crystallographic point group
Tetragonal	10	$\bar{4}$	$\tilde{S}_4$	$\tilde{S}_4$	$E\ 2\tilde{S}_4\ C_2$
	11	$4/m$	C_{4h}	$(C_4\ \sigma_h)$ or $(C_4\ i)$	$E\ 2C_4\ C_2\ i\ 2\tilde{S}_4\ \sigma_h$
	14	$\bar{4}2m$	$D_{2d} = V_d$	$(\tilde{S}_4\ C_2')$ or $(\tilde{S}_4\ \sigma_v')$	$E\ C_2\ C_2'\ C_2''\ 2\tilde{S}_4\ \sigma_v'\ \sigma_v''$
	15	$4/mmm$	D_{4h}	$C_4\ C_2'\ \sigma_h$	$E\ 2C_4\ C_2\ 2C_2'\ 2C_2''\ i\ 2\tilde{S}_4\ \sigma_h\ 2\sigma_v'\ 2\sigma_v''$
Trigonal	17	$\bar{3}$	$C_{3i} = \tilde{S}_3$	$(C_3\ i)$ or $\tilde{S}_3$	$E\ 2C_3\ i\ 2\tilde{S}_3$
	20	$\bar{3}m$	D_{3d}	$C_3\ C_2'\ i$	$E\ 2C_3\ 3C_2'\ i\ 2\tilde{S}_3\ 3\sigma_v$
Hexagonal	22	$\bar{6} = 3/m$	$C_{3h} = \tilde{S}_6$	$(C_3\ \sigma_h)$ or $\tilde{S}_6$	$E\ 2C_3\ \sigma_h\ 2\tilde{S}_6$
	23	$6/m$	C_{6h}	$(C_6\ \sigma_h)$ or $(C_6\ i)$ or $(C_3\ C_2\ \sigma_h)$ or $(C_3\ C_2\ i)$	$E\ 2C_6\ 2C_3\ C_2\ i\ 2\tilde{S}_3\ 2\tilde{S}_6\ \sigma_h$
	26	$\bar{6}m2 = 3/mm$	D_{3h}	$(C_3\ C_2'\ \sigma_h)$ or $(C_3\ C_2'\ \sigma_v)$ or $(C_3\ \sigma_h\ \sigma_v)$	$E\ 2C_3\ 3C_2'\ \sigma_h\ 2\tilde{S}_6\ 3\sigma_v$
	27	$6/mmm$	D_{6h}	$(C_6\ C_2'\ \sigma_h)$ or $(C_6\ C_2'\ i)$ or $(C_6\ \sigma_v'\ i)$ or $(C_3\ C_2\ C_2'\ i)$	$E\ 2C_6\ 2C_3\ C_2\ 3C_2'\ 3C_2''\ i\ 2\tilde{S}_3\ 2\tilde{S}_6\ \sigma_h\ 3\sigma_v'\ 3\sigma_v''$
Cubic	29	$m3$	T_h	$(C_2\ C_3'\ i)$ or $(C_2\ C_3'\ \sigma_h)$	$E\ 8C_3'\ 3C_2\ i\ 8\tilde{S}_3\ 3\sigma_h$
	31	$\bar{4}3m$	T_d	$\tilde{S}_4\ C_3'$	$E\ 8C_3'\ 3C_2\ 6\sigma_d\ 6\tilde{S}_4$
	32	$m3m$	O_h	$(C_4\ C_3'\ \sigma_h)$ or $(C_4\ C_3'\ i)$	$E\ 8C_3'\ 3C_2\ 6C_2''\ 6C_4\ i\ 8\tilde{S}_3\ 3\sigma_h\ 6\sigma_d\ 6\tilde{S}_4$

[a] We note that the matrix of transformation of the symmetry element $\tilde{S}_6$ is equal to the inverse (and also the transpose) of that of the symmetry element S_3. Similarly, $\tilde{S}_4 = (S_4)^{-1}$ and $\tilde{S}_3 = (S_6)^{-1}$. Hence in the final analysis of the tensor properties of crystals, no distinction can be made between the two sets of tensor components arrived at on the basis of the two notations. See also page 48 of [573] and page 195 of [588] for the generating elements in the International notation, and also page 82 of [339] and pages 29, 31, 34, and 286 of [183].

$6mm$ (C_{6v}) they intersect in the principal axis. For the class $m3m$ (O_h), there are six reflection planes respectively parallel to each pair of parallel face diagonals of opposite cube faces. These planes are called dodecahedral planes. All of them do not intersect in any single axis, but fall under the same class in group theory, since each pair of dodecahedral planes intersects in a cube axis.

2.7.2. Comments on the 32 Crystallographic Point Groups and Their Symmetry Elements as Listed in Tables 2.3 and 2.5a

In crystal physics, according to the IEEE Standards on Piezoelectric Crystals, 1949, the composite symmetry element $\bar{n}$ in the International notation is defined as $(C_n \cdot i)$ instead of $(S_n = C_n \cdot \sigma_h)$ in the Schönflies notation. In the comments below, the composite symmetry $\bar{n}$ will be referred to as $\tilde{S}_n$ in order to distinguish it from S_n in the Schönflies notation.

If we use the definition $\tilde{S}_n = (C_n \cdot i)$, then the point groups in the fourth row of Table 2.3 become $\tilde{S}_1$, $\tilde{S}_2$, $\tilde{S}_3$, $\tilde{S}_4$, and $\tilde{S}_6$, and these are equal to the C_{nh} or C_{ni} as indicated below:

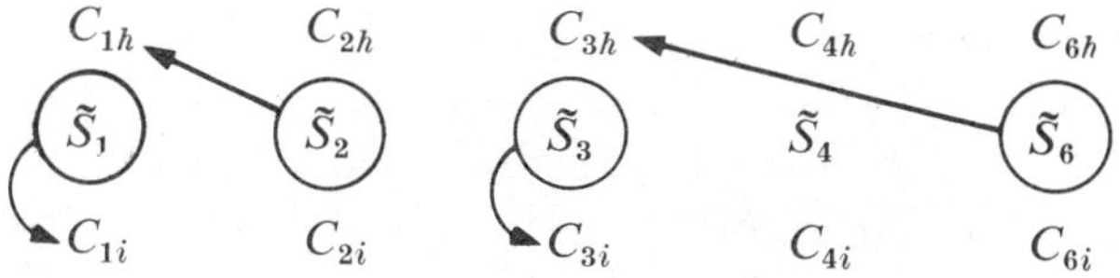

This scheme also leaves $\tilde{S}_4$ unrelated to others.

When this change is introduced in Table 2.5a, we find that the enumeration of the complete set of symmetry elements of the crystallographic point groups involving $\tilde{S}_n$ in the Schönflies notation changes as in Table 2.5b. The generating elements are also altered suitably.

The enumeration of the elements of symmetry in the last column is consistent with the International notation and hence is preferred when discussing the crystallographic point groups and their tensor properties by classical, tensor, and group theoretical methods. Accordingly, we shall follow this notation for the computation of the total number of non vanishing independent constants of a variety of tensor properties for some chosen crystal classes by group theoretical methods in Chapter 3, and also in deriving the individual nonvanishing independent R_{ijk} of Pockels' (linear electro-optic) constants by tensor methods in Chapter 8.

Incidentally, we may point out that Kaminow has given this latter notation in his recent book ([622], page 6). We may further state that as early as 1930 Wigner [1389] suggested the alternative definition $S_n = (C_n \cdot i)$ for the alternating axis while describing the point groups (see also page 57 of [412] and the footnote on page 79 of [1391]).

2.8. Neumann's Principle and Effect of Crystal Symmetry on Physical Properties

Every physical property of a crystal capable of being represented as a tensor property of appropriate rank is governed by Neumann's principle; this principle states that the components of a tensor representing a property will remain invariant under a transformation of coordinates governed by a symmetry operation valid for the crystal symmetry under consideration; that is, the crystal symmetry imposes restrictions such that the tensor components will not change under any such transformations. For example, if a cubic crystal is subjected to an electric field in a given direction, the electric moment induced is the same in magnitude and direction, being independent of the direction of the incident field.

Another way of stating Neumann's principle is that the symmetry elements of any physical property of a crystal must include all the symmetry elements of the point group of the crystal. This implies that any given physical property may possess a higher symmetry than that possessed by the crystal, but it cannot be of a lower symmetry than that of the crystal. The two statements given above can be combined together thus: Every physical property of a crystal must possess at least the symmetry of the point group of the crystal. We elaborate Neumann's principle, taking some concrete examples. Optical refraction and elasticity, for example, are inherently centrosymmetric, which is a consequence of the physical nature of the two properties, and they remain centrosymmetric irrespective of whether the crystal possesses a center of symmetry or not. Even if the crystal lacks a center, Neumann's principle demands a superposition of all the symmetry elements of the point group of the crystal on a center of inversion already characteristic of the physical property, so far as that property of the crystal is concerned. Thus the property exhibits a higher order of symmetry than that possessed by the point group of the crystal.

The nth-rank tensor property of the most anisotropic crystal has 3^n components. This number decreases depending on the nature of the tensor property. There are two aspects to consider in bringing about the reduction in the number of components.

The first aspect relates to the property itself and the symmetry conditions that govern the property. A property like thermal expansion, for which $T_{ij} = T_{ji}$, is a second-rank symmetric tensor property. Hence, the maximum number of independent constants is 6 instead of 9 (which is 3^2). Another property is the optical index of refraction. Here the index of refraction is governed by the optical index ellipsoid (Fresnel's index ellipsoid). The most general case is that of a triaxial ellipsoid, which is centrosymmetric, requiring six parameters to describe it completely. Hence, the maximum number of independent constants for the index of refraction is 6 (3 for the axial lengths and 3 for their orientations). Similarly, elastic stiffness constants are 21 for the triclinic crystal instead of the 81 expected because the elastic property is a fourth-rank tensor property; this reduction in number is due, first, to the fact that the elasticity tensor relates two second-rank symmetric tensors, and second, because $c_{ijkl} = c_{klij}$, as explained in Chapter 4. Similarly, the stress-optical and strain-optical constants are 36 each.

The second aspect concerns the symmetry conditions of the crystal. This is demanded by Neumann's principle that no physical property associated with any crystal can be of a lower symmetry than that of the crystal itself. If any, the property can be of a higher symmetry; for example, density is the same in all directions even for a triclinic crystal. Thus, a further reduction in the number of independent constants characteristic of each physical property is brought about by the symmetry operations governing the point group of the crystal. This reduction can be accomplished by classical methods, tensor methods, or group theoretical methods.

Putting the two conditions together, we find that the symmetry of a property associated with a crystal is higher than the higher of the two—the property and the point group.

3

Pockels' Phenomenological Theory of Photoelasticity of Crystals

3.1. Introduction

The refractive index n, one of the most important optical properties of all transparent media, does not transform like a tensor. However, the dielectric tensor K_{ij} or its reciprocal, the impermeability tensor $(1/K)_{ij}$, does transform like a second-rank tensor.

When a second-rank tensor T_{ij} of a crystal is symmetric, then we note that $T_{ij} = T_{ji}$. It is possible to make a geometric representation of any second-rank symmetric tensor, as will be shown now. Multiplying T_{ij} by the product of the two coordinate vectors x_i, x_j, we have

$$T_{ij}x_ix_j = 1 \qquad (i, j = 1, 2, 3) \tag{3.1a}$$

which can be expanded in the form

$$T_{11}x_1^2 + T_{22}x_2^2 + T_{33}x_3^2 + 2T_{23}x_2x_3 + 2T_{31}x_3x_1 + 2T_{12}x_1x_2 = 1 \tag{3.1b}$$

Equation (3.1b) is the equation of a second-degree surface, also called a quadric, whose center is the origin of the coordinate system.

It is easily seen that the quadric surface described by Eq. (3.1b) obeys the same law of transformation that governs the transformation of a second-rank tensor T_{ij} from one coordinate system to another, and for this reason the quadric surface is called the representation quadric for the tensor T_{ij}.

The quadric surface described by Eq. (3.1a) or (3.1b) is characterized by its possessing three mutually perpendicular directions such that, when referred to them, Eq. (3.1b) reduces to the simple form

$$T'_{11}x_1'^2 + T'_{22}x_2'^2 + T'_{33}x_3'^2 = 1 \tag{3.1c}$$

In the quadric described by Eq. (3.1c) T'_{11}, T'_{22}, and T'_{33} represent the three principal components of the second-rank symmetric tensor T_{ij}.

We are particularly interested in Eq. (3.1c) when the three principal tensor components T'_{11}, T'_{22}, and T'_{33} are all positive, in which case the representation quadric described by Eq. (3.1c) becomes a triaxial ellipsoid, whose semiaxes are, respectively, equal to $1/(T'_{11})^{1/2}$, $1/(T'_{22})^{1/2}$, and $1/(T'_{33})^{1/2}$.

If now the impermeability tensor $(1/K)_{ij}$ is denoted by B_{ij}, the following equation describes a triaxial ellipsoidal surface:

$$\sum_{i,j} B_{ij}x_i x_j = 1 \qquad (i, j = 1, 2, 3) \tag{3.2}$$

This ellipsoid, more explicitly called the optical index ellipsoid, or indicatrix, plays an important role in the phenomenological descriptions of the optical behavior of single crystals. The photoelastic effect and the electro-optic effects relate the optical parameters B_{ij} to the externally applied mechanical stress or electric field on the crystal. Since $K = n^2$ for an isotropic medium, we have $B_{ij} = (1/K)_{ij} = 1/n_{ij}^2$. It is relevant at this juncture to recall the interpretation of the geometry of the indicatrix or the optical index ellipsoid.

In order to explain the phenomenon of double refraction in crystalline solids in general, Fresnel developed the well-known elastic solid theory, according to which, to any given crystal, there corresponds, for a given wavelength of incident light radiation and at a given temperature, a certain ellipsoid of elasticity with its axes uniquely oriented with respect to the crystal lattice. This led him to the discovery of laws governing the propagation of light in crystalline media. Further information on Fresnel's deduction of the equation of the ellipsoid of elasticity and its properties regarding the propagation of light can be obtained from advanced textbooks on physical optics, for example, [1397].

It is more useful, for the purpose of this chapter, to express the optical properties of crystals in terms of the refractive index ellipsoid, whose semiaxes are proportional to the three principal refractive indices. The equation of the refractive index ellipsoid is given by

$$x^2/n_x^2 + y^2/n_y^2 + z^2/n_z^2 = 1 \tag{3.3}$$

For plane waves propagated in the direction of any radius vector, the two refractive indices are given by the principal axes of the ellipse in which the ellipsoid is intersected by a plane perpendicular to this radius vector. The directions of these principal axes are those of the corresponding electric vectors, i.e., the vibration directions.

The equation of this refractive index ellipsoid or optical index ellipsoid can therefore be written in the form

$$B_{11}x^2 + B_{22}y^2 + B_{33}z^2 = 1 \tag{3.4}$$

where the three optical parameters B_{ij} are equal to the reciprocals of the squares of the respective refractive indices. Equation (3.4) thus describes both the indicatrix and the representation quadric for the tensor B_{ij}.

The number of optical parameters required to fix the lengths and orientations of the axes of the optical index ellipsoid decreases as the crystal symmetry increases. These optical parameters are called the polarization constants. In the most general case, namely, for triclinic crystals, their number is six (three for the axial lengths, and three for their orientation). Monoclinic crystals require four parameters, and orthorhombic crystals, three. These three systems comprise the biaxial crystals, for which the velocities *a*, *b*, and *c* and hence the polarization constants B_{ij} are all different. For uniaxial crystals (comprising trigonal, hexagonal, and tetragonal systems) the number of polarization constants is two, and for anaxial crystals (cubic system) only one polarization constant is sufficient to describe completely the optical index ellipsoid.

It is customary to use an orthogonal axial system in describing the tensor properties of crystals, including the effects of external influences, say, mechanical stress or electric field, on the optical properties of crystals. For all crystal systems except triclinic and monoclinic, Eq. (3.4) can be used as it stands; for these two crystal systems, however, one has to use an equation for the optical index ellipsoid in terms of six parameters.

3.2. *Phenomenological Theory, Stress-Optical and Strain-Optical Constants in Four- and Two-Suffix Notations; q_{ij} and p_{ij} Matrices for the 32 Crystallographic Point Groups*

3.2.1. *The Assumptions Forming the Basis of Pockels' Theory*

The choice of the *X*, *Y*, *Z* Cartesian axes, coinciding with the crystallographic axes of all crystal systems, except triclinic and monoclinic, is now followed to present the phenomenological theory as developed by Pockels. Pockels' theory is based on the following two assumptions:

(1) In a homogeneously deformed solid the effect of deformation is only to alter the optical parameters of the optical index ellipsoid.

(2) When the strain is within elastic limits, the change of an optical parameter (polarization constant) of the solid due to deformation can be expressed as a homogeneous linear function of all the nine stress components $P_{xx}, P_{yy}, P_{zz}, P_{yz}, P_{zx}, P_{xy}, P_{zy}, P_{xz}, P_{yx}$, or the nine strain components $\varepsilon_{xx}, \varepsilon_{yy}, \varepsilon_{zz}, \varepsilon_{yz}, \varepsilon_{zx}, \varepsilon_{xy}, \varepsilon_{zy}, \varepsilon_{xz}, \varepsilon_{yx}$. The stress and strain tensors are discussed in Chapter 4 on elasticity.

The first assumption implies that the effect of the elastic deformation leads only to a change in the polarization constants. The second assumption is a reformulation of the experimentally observed Brewster's law so as to make it applicable to crystalline solids. Henceforth, $P_{xx}, P_{yz}, \ldots$, will be changed to $P_{11}, P_{23}, \ldots$, and similarly for $\varepsilon_{xx}, \varepsilon_{yz}, \ldots$.

3.2.2. Mathematical Formulation of Photoelasticity in Terms of q_{ijkl} and p_{ijkl}

In its most general form, Eq. (3.4) describing the optical index ellipsoid for an unstressed crystal, referred to an arbitrary right-handed rectangular coordinate system (OX_1, OX_2, OX_3), can be put thus:

$$B^0_{11}x_1^2 + B^0_{22}x_2^2 + B^0_{33}x_3^2 + B^0_{23}x_2x_3 + B^0_{31}x_3x_1 + B^0_{12}x_1x_2 + B^0_{32}x_3x_2 + B^0_{13}x_1x_3 + B^0_{21}x_2x_1 = 1 \tag{3.5a}$$

The superscript 0 designates the undeformed state of the crystal, and B^0_{ij} are the polarization constants. Here $B^0_{11} = 1/(n^0_{11})^2$, $B^0_{23} = 1/(n^0_{23})^2, \ldots$, and n^0_{ij} are the refractive indices.

The equation for the stressed crystal becomes

$$B_{11}x_1^2 + B_{22}x_2^2 + B_{33}x_3^2 + B_{23}x_2x_3 + B_{31}x_3x_1 + B_{12}x_1x_2 + B_{32}x_3x_2 + B_{13}x_1x_3 + B_{21}x_2x_1 = 1 \tag{3.5b}$$

Pockels assumes that the changes in the B_{ij} are linear functions of all the nine stress components P_{kl}. Thus

$$B_{ij} - B^0_{ij} = -q_{ijkl}P_{kl} \tag{3.6a}$$

where i, j, k, and l take the values 1, 2, 3. The q_{ijkl} are called the stress-optical or piezo-optic coefficients. Following Pockels, we consider here an extensional stress as negative for the purpose of defining q_{ijkl}.

Equation (3.6a) when expanded yields Eq. (3.6b), showing that there are 81 q_{ijkl} for the most anisotropic crystal.

$$\begin{bmatrix} B_{11}-B_{11}^0 \\ B_{22}-B_{22}^0 \\ B_{33}-B_{33}^0 \\ B_{23}-B_{23}^0 \\ B_{31}-B_{31}^0 \\ B_{12}-B_{12}^0 \\ B_{32}-B_{32}^0 \\ B_{13}-B_{13}^0 \\ B_{21}-B_{21}^0 \end{bmatrix} = - \begin{bmatrix} q_{1111} & q_{1122} & q_{1133} & q_{1123} & q_{1131} & q_{1112} & q_{1132} & q_{1113} & q_{1121} \\ q_{2211} & q_{2222} & q_{2233} & q_{2223} & q_{2231} & q_{2212} & q_{2232} & q_{2213} & q_{2221} \\ q_{3311} & q_{3322} & q_{3333} & q_{3323} & q_{3331} & q_{3312} & q_{3332} & q_{3313} & q_{3321} \\ q_{2311} & q_{2322} & q_{2333} & q_{2323} & q_{2331} & q_{2312} & q_{2332} & q_{2313} & q_{2321} \\ q_{3111} & q_{3122} & q_{3133} & q_{3123} & q_{3131} & q_{3112} & q_{3132} & q_{3113} & q_{3121} \\ q_{1211} & q_{1222} & q_{1233} & q_{1223} & q_{1231} & q_{1212} & q_{1232} & q_{1213} & q_{1221} \\ q_{3211} & q_{3222} & q_{3233} & q_{3223} & q_{3231} & q_{3212} & q_{3232} & q_{3213} & q_{3221} \\ q_{1311} & q_{1322} & q_{1333} & q_{1323} & q_{1331} & q_{1312} & q_{1332} & q_{1313} & q_{1321} \\ q_{2111} & q_{2122} & q_{2133} & q_{2123} & q_{2131} & q_{2112} & q_{2132} & q_{2113} & q_{2121} \end{bmatrix} \begin{bmatrix} P_{11} \\ P_{22} \\ P_{33} \\ P_{23} \\ P_{31} \\ P_{12} \\ P_{32} \\ P_{13} \\ P_{21} \end{bmatrix} \tag{3.6b}$$

As mentioned in Chapter 2, a repeated index on the same side of Eq. (3.6a) implies a summation over all values 1, 2, 3. The q_{ijkl} relate the components of a second-rank tensor ΔB_{ij} with those of another second-rank tensor P_{kl}. Hence the q_{ijkl} form the components of a fourth-rank tensor and are expressed in units of cm²/dyne. Since B_{ij} and P_{kl} are symmetric tensors of second rank, the q_{ijkl} are not all independent. Hence relations of the type $q_{ijkl} = q_{jikl}$ and $q_{ijkl} = q_{ijlk}$ exist, and the number of independent coefficients q_{ijkl} is reduced from 81 to 36 (see page 80).

Since an elastic solid acted upon by a stress undergoes a physical deformation, Pockels assumes that the changes in the B_{ij} are also linear functions of all the nine components of the strain tensor describing the deformation. Thus, starting with Eqs. (3.5a) and (3.5b), we put

$$B_{ij} - B_{ij}^0 = p_{ijkl}\varepsilon_{kl} \tag{3.7}$$

The p_{ijkl} are called the strain-optical or elasto-optic constants. Here, again following Pockels, extensional strains are considered positive. Equation (3.7) can also be expanded into a form similar to Eq. (3.6b) by putting p_{ijkl} in place of q_{ijkl} and ε_{kl} in place of P_{kl}, showing thereby that there are 81 p_{ijkl} for the most anisotropic crystal.

The p_{ijkl} relate the components of a second-rank tensor ΔB_{ij} with those of another second-rank tensor ε_{kl}; hence the p_{ijkl} form the components of a fourth-rank tensor. Since B_{ij} and ε_{kl} are symmetric tensors of second rank, the p_{ijkl} are not all independent. As in the case of q_{ijkl}, here also we have relations of the type $p_{ijkl} = p_{jikl}$ and $p_{ijkl} = p_{ijlk}$ and hence the number of independent constants is reduced from 81 to 36 even for the most anisotropic crystal.

Furthermore, the p_{ijkl} and q_{ijkl} are related through the elastic compliance constants s_{ijkl} and the elastic stiffness constants c_{ijkl} as follows:

$$q_{ijkl} = p_{ijmn}s_{mnkl}, \qquad p_{ijmn} = q_{ijkl}c_{klmn}$$

In the above discussion, the mathematical formulation of the photoelastic behavior of crystals has been presented as a fourth-rank tensor property, and hence the q_{ijkl} and p_{ijkl} are introduced in the tensor or four-suffix notation.

3.2.3. Mathematical Formulation of Photoelasticity in Terms of q_{ij} and p_{ij}

An alternative notation which is in common use in the literature describes the stress-optical and strain-optical constants in the matrix or two-suffix notation, as follows:

The quantities B_{ij}^0 in Eq. (3.5a) are the components of a symmetric tensor of second rank, similar to the familiar stress and elastic-strain tensors and to the dielectric susceptibility tensor. Hence they are only six in number.

Thus, in the most general form, Eq. (3.5a) describing the optical index ellipsoid for an unstrained crystal can be written:

$$B_{11}^0x^2 + B_{22}^0y^2 + B_{33}^0z^2 + 2B_{23}^0yz + 2B_{31}^0zx + 2B_{12}^0xy = 1 \qquad (3.8a)$$

When the crystal is deformed, Eq. (3.8a) changes to

$$B_{11}x^2 + B_{22}y^2 + B_{33}z^2 + 2B_{23}yz + 2B_{31}zx + 2B_{12}xy = 1 \qquad (3.8b)$$

Using the single-suffix notation, let $B_k = B_{ij}$; accordingly $B_1{}^0 = B_{11}^0$, $\ldots$, $B_4{}^0 = B_{23}^0$, $\ldots$, and $B_1 = B_{11}$, $\ldots$, $B_4 = B_{23}$. In terms of the strain components ε_j (written in the single-suffix notation)

$$\Delta B_i = B_i - B_i{}^0 = \sum_{i,j} p_{ij}\varepsilon_j \qquad (i, j = 1\text{–}6) \qquad (3.9a)$$

and in terms of stress components P_j (also in the single-suffix notation),

$$\Delta B_i = B_i - B_i{}^0 = \sum_{i,j} -q_{ij}P_j \qquad (i, j = 1\text{–}6) \qquad (3.9b)$$

There are thus 36 p_{ij} and 36 q_{ij} for the most anisotropic crystal class. Here p_{ij} are called the strain-optical or elasto-optic constants, while q_{ij} are called the stress optical or piezo-optic constants. The constants p_{ij} and q_{ij} together are termed photoelastic constants. The constants p_{ij} are dimensionless, while the q_{ij} have the dimensions of an elastic compliance. It is obvious from the definition of q_{ij} and p_{ij} that both of them are tensor components of the fourth rank, since both P_{kl} and ε_{kl}, and B_{ij}, are components of second-rank symmetric tensors.

Equation (3.9a) when written in full takes the form

$$
\begin{aligned}
B_1 - B_1^0 &= p_{11}\varepsilon_1 + p_{12}\varepsilon_2 + p_{13}\varepsilon_3 + p_{14}\varepsilon_4 + p_{15}\varepsilon_5 + p_{16}\varepsilon_6 \\
B_2 - B_2^0 &= p_{21}\varepsilon_1 + p_{22}\varepsilon_2 + p_{23}\varepsilon_3 + p_{24}\varepsilon_4 + p_{25}\varepsilon_5 + p_{26}\varepsilon_6 \\
B_3 - B_3^0 &= p_{31}\varepsilon_1 + p_{32}\varepsilon_2 + p_{33}\varepsilon_3 + p_{34}\varepsilon_4 + p_{35}\varepsilon_5 + p_{36}\varepsilon_6 \\
B_4 - B_4^0 &= p_{41}\varepsilon_1 + p_{42}\varepsilon_2 + p_{43}\varepsilon_3 + p_{44}\varepsilon_4 + p_{45}\varepsilon_5 + p_{46}\varepsilon_6 \\
B_5 - B_5^0 &= p_{51}\varepsilon_1 + p_{52}\varepsilon_2 + p_{53}\varepsilon_3 + p_{54}\varepsilon_4 + p_{55}\varepsilon_5 + p_{56}\varepsilon_6 \\
B_6 - B_6^0 &= p_{61}\varepsilon_1 + p_{62}\varepsilon_2 + p_{63}\varepsilon_3 + p_{64}\varepsilon_4 + p_{65}\varepsilon_5 + p_{66}\varepsilon_6
\end{aligned} \tag{3.9c}
$$

Similarly, in terms of stress components, Eq. (3.9b) takes the form

$$
\begin{aligned}
B_1 - B_1^0 &= -(q_{11}P_1 + q_{12}P_2 + q_{13}P_3 + q_{14}P_4 + q_{15}P_5 + q_{16}P_6) \\
B_2 - B_2^0 &= -(q_{21}P_1 + q_{22}P_2 + q_{23}P_3 + q_{24}P_4 + q_{25}P_5 + q_{26}P_6) \\
B_3 - B_3^0 &= -(q_{31}P_1 + q_{32}P_2 + q_{33}P_3 + q_{34}P_4 + q_{35}P_5 + q_{36}P_6) \\
B_4 - B_4^0 &= -(q_{41}P_1 + q_{42}P_2 + q_{43}P_3 + q_{44}P_4 + q_{45}P_5 + q_{46}P_6) \\
B_5 - B_5^0 &= -(q_{51}P_1 + q_{52}P_2 + q_{53}P_3 + q_{54}P_4 + q_{55}P_5 + q_{56}P_6) \\
B_6 - B_6^0 &= -(q_{61}P_1 + q_{62}P_2 + q_{63}P_3 + q_{64}P_4 + q_{65}P_5 + q_{66}P_6)
\end{aligned} \tag{3.9d}
$$

Equations (3.9c) and (3.9d) can be put in matrix form as in Eq. (3.6b). The negative sign* in Eqs. (3.6b) and (3.9d) arises from Pockels' convention that positive stress produces negative strain. Here the stress is taken as positive when compressional, and negative when extensional. The strain, however, is considered positive for an extension, and negative for a compression.

* The new convention is: Tensile stress causing a positive strain is positive. It may be noted that the negative signs on the right-hand sides of Eqs. (3.10a) and (3.10b) arise out of the fact that P_{ij} and P_{kl} and also P_i and P_l are considered positive by Pockels for a compression. However, in the theory of elasticity discussed on pages 142, 143, and 145, they are considered positive for a tensile stress.

Considering a compressional stress P_{xx} as positive, $-P_{xx}$ describes a tensile stress, and q_{11} is therefore the coefficient (in both magnitude and sign) of an extensional stress. If ε_{xx} is the consequent elongation strain caused by $-P_{xx}$, then p_{11} is the coefficient of ε_{xx}; here ε_{xx} is a positive strain. Hence a positive stress P_{xx} produces a negative strain $-\varepsilon_{xx}$.

The stress and strain components are related by the following equations:

$$P_{ij} = -c_{ijkl}\varepsilon_{kl}, \qquad \varepsilon_{ij} = -s_{ijkl}P_{kl} \tag{3.10a}$$

where the c_{ijkl} are called the stiffness coefficients, and s_{ijkl} are called the compliance coefficients of the crystal. It is obvious that both c_{ijkl} and s_{ijkl} are components* of fourth-rank tensors. In the two-suffix notation c_{ijkl} are shown as c_{mn}, and s_{ijkl} as s_{op}, where all the four suffixes m, n, o, p take the values 1 to 6. Thus in the two-suffix notation,

$$P_i = -c_{ij}\varepsilon_j, \qquad \varepsilon_j = -s_{jl}P_l \tag{3.10b}$$

(i, j, k, and l take the values 1 to 6).

The two sets of constants c_{ij} and s_{kl} are related to each other by the following equations:

$$\sum_{j=1}^{6} c_{ij}s_{ij} = 1 \quad \text{and} \quad \sum_{j=1}^{6} c_{ij}s_{kj} = 0 \quad \text{when} \quad i \neq k \tag{3.11}$$

Since P_i and ε_k are components of second-rank symmetric tensors, the components c_{ij} (and s_{kl}) are components of fourth-rank tensors and they have, in general, 36 components each. In the classical theory of elasticity these two fourth-rank tensors are symmetric in i and j; hence†

$$c_{ij} = c_{ji} \quad \text{and} \quad s_{ij} = s_{ji} \tag{3.12}$$

and there are therefore only 21 independent constants of each type for the most anisotropic crystals, namely, triclinic.

The recent theory of elasticity due to Raman [1014] and Laval [719] does not require the stress and strain tensors to be symmetric for a general deformation; consequently P_i and ε_i have nine components each, and the tensors c_{ij} and s_{ij} should have 81 components in general. But since both of the constants are symmetric in the suffixes representing stress and strain (thus making $c_{ij} = c_{ji}$ and $s_{kl} = s_{lk}$), there are only 45 independent components. The consequence of the Raman–Laval theory demands 54 photoelastic constants of q_{ijkl} (as also of p_{ijkl}) for the triclinic crystal. The bearing of their theory on the photoelastic behavior of crystals will not be considered any further in this book. However, the reader is referred to the more recent

* The two sets of coefficients c_{ijkl} and s_{ijkl} are also called the elastic constants and elastic moduli, respectively.

† See Chapter 4.

investigations of Nelson and Lax [866–868] and of Vacher and Boyer [1246–1248] on the photoelastic behavior obtained from studies on the Brillouin scattering in crystals.

From Eqs. (3.9c) and (3.9d) it is seen that Pockels' fundamental equations are relations between the components of two second-rank tensors, namely, the polarization constants B_{ij} and either the components ε_{kl} of elastic strain or those of elastic stress P_{kl}. The matrices of p_{ij} and q_{ij} (which are fourth-rank tensors) for the various classes of crystals follow closely the matrices for the elastic constants except in two essential respects: (1) It is not in general true that $p_{mn} = p_{nm}$, whereas in elasticity $c_{ij} = c_{ji}$ and $s_{kl} = s_{lk}$ as deduced from thermodynamic considerations which are not applicable to the photoelastic phenomenon. Consequently the maximum number of independent photoelastic constants (which exist for the triclinic system) is 36, while for elasticity it is only 21; the additional 15 photoelastic constants appear below the leading diagonal of the matrix, and their subscripts are reversed with respect to those above the diagonal of the matrix in the two-suffix notation. (2) The 32 point groups are divided into nine groups by Pockels on the basis of their photoelastic behavior, like the nine groups for elasticity, whereas Bhagavantam has shown that they should be divided into 11 groups for photoelasticity, unlike the nine groups for elasticity. This aspect will be discussed in greater detail in a later section of this chapter.

The first three equations in (3.9c) and (3.9d) express a change in the principal velocities, which can be measured by observations on light propagated parallel to the three axes, respectively, by the usual methods for measuring small changes in the refractive index. For example, if the applied stress is P_1 ($= P_{11}$ or P_{xx}), then the product $q_{11}P_1$ may be interpreted as expressing the change in the refractive index n_1 caused by P_1 in accordance with the relation

$$(B_{11} - B_{11}^0) \quad \text{or} \quad (B_1 - B_1{}^0) = (1/n_1)^2 - (1/n_1{}^0)^2 = -q_{11}P_1$$

where n_1 and $n_1{}^0$ are the values of this particular principal refractive index with and without application of stress. The last three equations in (3.9c) and (3.9d) represent a rotation of the optical ellipsoid around its axes. This rotation is determined by the traditional ellipsometric methods; however, such methods are more difficult, and the computation (therefrom) of the coefficients p_{ij} and q_{ij} more involved.

In the two sets of equations (3.9a) and (3.9b), P_{ij} and ε_{ij} are related through c_{ijkl} and s_{ijkl} for any crystal and hence p_{ijkl} and q_{ijkl} are also interrelated through c_{ijkl} and s_{ijkl}. It can be shown that in the two-suffix notation

$$p_{ij} = \sum_{k=1}^{6} q_{ik}c_{kj}, \qquad q_{ij} = \sum_{k=1}^{6} p_{ik}s_{kj}, \qquad (i, j = 1\text{–}6) \qquad (3.13)$$

where c_{kj} and s_{kj} are stiffness constants and compliance constants, respectively. The various p_{ij} and q_{ij} are experimentally determinable, the methods for which are given in Chapter 5. It may be mentioned, however, that the piezo-optic constants q_{ij} can be determined more directly by experiment, although the elasto-optic constants p_{ij} are more significant from the theoretical point of view. Hence it is usual to determine the q_{ij} experimentally and make use of these q_{ij} and the values of the elastic constants of the crystal to obtain the values of the p_{ij}. It is for this reason that an accurate determination of the c_{ij} and s_{ij} is very important for quantitative measurements of p_{ij} and q_{ij}; obviously the order of accuracy in the determination of the elastic constants and the values assumed for c_{ij} and s_{ij} in calculating p_{ij} from the experimentally determined values of q_{ij} plays an important role in the final calculated values of p_{ij}.

In fact, even for the experimental determination of q_{ij} from the measurements on the relative path retardation, one has to use the elastic constants since one has to allow for the change in thickness of the crystal under a known load.

Furthermore, oftentimes the photoelastic constants of a crystal are determined, for reasons explained in detail in Chapter 5, by combining some equations involving q_{ij} and some more equations involving p_{ij}. In order to obtain the numerical values of the individual p_{ij} and q_{ij} involved, one must assume the values of the elastic constants.

Thus a study of the elastic behavior of a crystal forms a necessary prerequisite to determining the values of the individual photoelastic constants p_{ij} and q_{ij}.

The relationship between q_{ijkl} and q_{mn} is discussed on pages 77–80 in the section dealing with the derivation of the nonvanishing independent photoelastic constants by tensor methods.

Before concluding this section,* it is pertinent to note that Pockels tacitly assumes that the symmetry of the crystal is not appreciably altered

* Analogous to the effect of a mechanical stress on the optical index ellipsoid in a crystal, an electric field also both deforms and rotates the optical index ellipsoid; but whereas piezo-optic effects are universal, the linear electro-optic effects are possible only in piezoelectric crystals. The quadratic electro-optic effects are also universal. The electro-optic effects are discussed in Chapter 8.

by a uniaxial stress. From more recent observations on cubic crystals, it has become apparent that this assumption is not strictly valid. The explanation, as pointed out by Mueller [831], is that the deformation lowers the symmetry, thus increasing the number of nonvanishing photoelastic constants p_{ij} and q_{ij}; the magnitude of these deformation-induced coefficients is not constant but is proportional to the stress. For effects of this nature, Mueller suggests the term "morphic." It follows therefore that the photoelastic effect, which according to Eqs. (3.9c) and (3.9d) should be strictly linear, requires (at least for systems of higher symmetry) correction terms involving the square of the stress. The treatment in this book ignores this morphic (or second-order) effect. However, the reader is referred to the studies on the photoelastic behavior of crystals—cubic and noncubic—under very large hydrostatic pressures by Vedam and collaborators [1282–1301].

3.2.4. *Crystal Symmetry and the Number of Photoelastic Constants*

It has been indicated in the previous section that the relations

$$c_{ij} = c_{ji} \qquad \text{and} \qquad s_{ij} = s_{ji}$$

which hold in elasticity could be derived from thermodynamic considerations. But such considerations are not applicable to the photoelastic phenomenon; hence p_{ij} need not be equal to p_{ji} unless it is warranted by symmetry considerations applicable to the particular crystal. Consequently the number of p_{ij} (as also q_{ij}) is 36 for the most asymmetric triclinic crystal. This number would, however, become less for crystals possessing various symmetry elements. This reduction in the number of the photoelastic constants arises due to the fact that all expressions involving the photoelastic constants should be invariant when each of the symmetry operations is applied. Hence it is possible to derive relationships among some of the 36 photoelastic constants [573], and thus the number of independent constants is reduced.

Pockels classified the 32 point groups into nine photoelastic groups, depending upon the number and nature of the surviving nonvanishing and independent photoelastic constants [947; 949]. Bhagavantam [117] showed that the 32 point groups should be classified into 11 photoelastic groups instead of Pockels' nine groups. Bhagavantam's 11 groups are the same as the so-called Laue-symmetry groups and are what one would obtain if an additional symmetry, namely a center of symmetry (inversion), were introduced.

Bhagavantam [117] has developed group theoretical methods to arrive

Table 3.1. Photoelastic Constants: Discrepancies between Pockels' and Bhagavantam's Schemes

Group number	Classes	Number of nonvanishing constants		Remarks	
		Pockels' scheme	Bhagavantam's scheme	Non-equivalence	New constants
10	T, T_h	3	4	q_{13}, q_{12}	—
8	C_{3h}, C_6, C_{6h}	6	8	—	q_{62}, q_{45}
6	C_4, S_4, C_{4h}	9	10	—	q_{45}
4	C_3, S_6	11	12	—	q_{62}

at the number of surviving constants. Subsequently his results were confirmed by Bhagavantam and Suryanarayana [119]. The discrepancies between the two schemes as given by Pockels and by Bhagavantam arise in two ways. First, some of the constants which according to Pockels are equal have been shown to be unequal; second, some others which vanish in Pockels' scheme do exist in Bhagavantam's scheme. The discrepancies between the two schemes are listed in Table 3.1.

The number of the surviving independent constants and the actual matrices of the surviving photoelastic constants belonging to the 11 photoelastic groups as classified by Bhagavantam are given in Table 3.2. It is necessary to point out at this stage that the phenomenological theory of the photoelastic effect as developed by Pockels cannot give any indication of the magnitude of the photoelastic constants in a crystal of any point group. The atomistic theory of the photoelastic behavior of cubic crystals due to Mueller is briefly indicated in Chapter 6.

3.3. Derivation of the Nonvanishing and Independent Photoelastic Constants for the Various Crystal Classes by Different Methods

The number of nonvanishing and independent constants for each of the crystallographic point groups can be derived in different ways. Three different methods are described below:

a. Classical method
b. Tensor method
c. Group theoretical method

Table 3.2. Matrices of Pockels' Photoelastic Constants as Revised by Bhagavantam

Group 1. Triclinic system, C_1, $C_i(1, \bar{1})$ (36 constants)

p_{11}	p_{12}	p_{13}	p_{14}	p_{15}	p_{16}	q_{11}	q_{12}	q_{13}	q_{14}	q_{15}	q_{16}
p_{21}	p_{22}	p_{23}	p_{24}	p_{25}	p_{26}	q_{21}	q_{22}	q_{23}	q_{24}	q_{25}	q_{26}
p_{31}	p_{32}	p_{33}	p_{34}	p_{35}	p_{36}	q_{31}	q_{32}	q_{33}	q_{34}	q_{35}	q_{36}
p_{41}	p_{42}	p_{43}	p_{44}	p_{45}	p_{46}	q_{41}	q_{42}	q_{43}	q_{44}	q_{45}	q_{46}
p_{51}	p_{52}	p_{53}	p_{54}	p_{55}	p_{56}	q_{51}	q_{52}	q_{53}	q_{54}	q_{55}	q_{56}
p_{61}	p_{62}	p_{63}	p_{64}	p_{65}	p_{66}	q_{61}	q_{62}	q_{63}	q_{64}	q_{65}	q_{66}

Group 2. Monoclinic system, C_s, C_2, $C_{2h}(m, 2, 2/m)$ (20 constants); diad $\parallel X_3$

p_{11}	p_{12}	p_{13}	0	0	p_{16}	q_{11}	q_{12}	q_{13}	0	0	q_{16}
p_{21}	p_{22}	p_{23}	0	0	p_{26}	q_{21}	q_{22}	q_{23}	0	0	q_{26}
p_{31}	p_{32}	p_{33}	0	0	p_{36}	q_{31}	q_{32}	q_{33}	0	0	q_{36}
0	0	0	p_{44}	p_{45}	0	0	0	0	q_{44}	q_{45}	0
0	0	0	p_{54}	p_{55}	0	0	0	0	q_{54}	q_{55}	0
p_{61}	p_{62}	p_{63}	0	0	p_{66}	q_{61}	q_{62}	q_{63}	0	0	q_{66}

Group 2. Monoclinic system, C_s, C_2, $C_{2h}(m, 2, 2/m)$ (20 constants); diad $\parallel X_2$ according to IEEE Standards on Piezoelectric Crystals (1949)

p_{11}	p_{12}	p_{13}	0	p_{15}	0	q_{11}	q_{12}	q_{13}	0	q_{15}	0
p_{21}	p_{22}	p_{23}	0	p_{25}	0	q_{21}	q_{22}	q_{23}	0	q_{25}	0
p_{31}	p_{32}	p_{33}	0	p_{35}	0	q_{31}	q_{32}	q_{33}	0	q_{35}	0
0	0	0	p_{44}	0	p_{46}	0	0	0	q_{44}	0	q_{46}
p_{51}	p_{52}	p_{53}	0	p_{55}	0	q_{51}	q_{52}	q_{53}	0	q_{55}	0
0	0	0	p_{64}	0	p_{66}	0	0	0	q_{64}	0	q_{66}

continued overleaf

Table 3.2 *(continued)*

Group 3. Orthorhombic system, D_2, C_{2v}, D_{2h} (222, *mm*2, *mmm*) (12 constants)

p_{11}	p_{12}	p_{13}	0	0	0	q_{11}	q_{12}	q_{13}	0	0	0
p_{21}	p_{22}	p_{23}	0	0	0	q_{21}	q_{22}	q_{23}	0	0	0
p_{31}	p_{32}	p_{33}	0	0	0	q_{31}	q_{32}	q_{33}	0	0	0
0	0	0	p_{44}	0	0	0	0	0	q_{44}	0	0
0	0	0	0	p_{55}	0	0	0	0	0	q_{55}	0
0	0	0	0	0	p_{66}	0	0	0	0	0	q_{66}

Group 4. Trigonal system, C_3, C_{3i} (3, $\bar{3}$) (12 constants)

p_{11}	p_{12}	p_{13}	p_{14}	$-p_{25}$	$p_{62}{}^a$	q_{11}	q_{12}	q_{13}	q_{14}	$-q_{25}$	$2q_{62}{}^a$
p_{12}	p_{11}	p_{13}	$-p_{14}$	p_{25}	$-p_{62}{}^a$	q_{12}	q_{11}	q_{13}	$-q_{14}$	q_{25}	$-2q_{62}{}^a$
p_{31}	p_{31}	p_{33}	0	0	0	q_{31}	q_{31}	q_{33}	0	0	0
p_{41}	$-p_{41}$	0	p_{44}	p_{45}	p_{52}	q_{41}	$-q_{41}$	0	q_{44}	q_{45}	$2q_{52}$
$-p_{52}$	p_{52}	0	$-p_{45}$	p_{44}	p_{41}	$-q_{52}$	q_{52}	0	$-q_{45}$	q_{44}	$2q_{41}$
$-p_{62}{}^a$	$p_{62}{}^a$	0	p_{25}	p_{14}	$\frac{1}{2}(p_{11}-p_{12})$	$-q_{62}{}^a$	$q_{62}{}^a$	0	q_{25}	q_{14}	$(q_{11}-q_{12})$

Group 5. Trigonal system, C_{3v}, D_3, D_{3d}(3*m*, 32, $\bar{3}m$) (8 constants)

p_{11}	p_{12}	p_{13}	p_{14}	0	0	q_{11}	q_{12}	q_{13}	q_{14}	0	0
p_{12}	p_{11}	p_{13}	$-p_{14}$	0	0	q_{12}	q_{11}	q_{13}	$-q_{14}$	0	0

p_{31}	p_{31}	p_{33}	0	0	0	q_{31}	q_{31}	q_{33}	0	0	0
p_{41}	$-p_{41}$	0	p_{44}	0	0	q_{41}	$-q_{41}$	0	q_{44}	0	0
0	0	0	0	p_{44}	p_{41}	0	0	0	0	q_{44}	$2q_{41}$
0	0	0	0	p_{14}	$\frac{1}{2}(p_{11}-p_{12})$	0	0	0	0	q_{14}	$(q_{11}-q_{12})$

Group 6. Tetragonal system, C_4, S_4, $C_{4h}(4, \bar{4}, 4/m)$ (10 constants)

p_{11}	p_{12}	p_{13}	0	0	p_{16}	q_{11}	q_{12}	q_{13}	0	0	q_{16}
p_{12}	p_{11}	p_{13}	0	0	$-p_{16}$	q_{12}	q_{11}	q_{13}	0	0	$-q_{16}$
p_{31}	p_{31}	p_{33}	0	0	0	q_{31}	q_{31}	q_{33}	0	0	0
0	0	0	p_{44}	p_{45}[a]	0	0	0	0	q_{44}	q_{45}[a]	0
0	0	0	$-p_{45}$[a]	p_{44}	0	0	0	0	$-q_{45}$[a]	q_{44}	0
p_{61}	$-p_{61}$	0	0	0	p_{66}	q_{61}	$-q_{61}$	0	0	0	q_{66}

Group 7. Tetragonal system, D_{2d}, C_{4v}, D_4, $D_{4h}(\bar{4}2m, 4mm, 422, 4/mmm)$ (7 constants)

p_{11}	p_{12}	p_{13}	0	0	0	q_{11}	q_{12}	q_{13}	0	0	0
p_{12}	p_{11}	p_{13}	0	0	0	q_{12}	q_{11}	q_{13}	0	0	0
p_{31}	p_{31}	p_{33}	0	0	0	q_{31}	q_{31}	q_{33}	0	0	0
0	0	0	p_{44}	0	0	0	0	0	q_{44}	0	0
0	0	0	0	p_{44}	0	0	0	0	0	q_{44}	0
0	0	0	0	0	p_{66}	0	0	0	0	0	q_{66}

continued overleaf

Table 3.2 *(continued)*

Group 8. Hexagonal system, C_{3h}, C_6, $C_{6h}(\bar{6}, 6, 6/m)$ (8 constants)

p_{11}	p_{12}	p_{13}	0	0	$-p_{61}{}^{a}$	q_{11}	q_{12}	q_{13}	0	0	$-2q_{61}{}^{a}$
p_{12}	p_{11}	p_{13}	0	0	$p_{61}{}^{a}$	q_{12}	q_{11}	q_{13}	0	0	$2q_{61}{}^{a}$
p_{31}	p_{31}	p_{33}	0	0	0	q_{31}	q_{31}	q_{33}	0	0	0
0	0	0	p_{44}	$p_{45}{}^{a}$	0	0	0	0	q_{44}	$q_{45}{}^{a}$	0
0	0	0	$-p_{45}{}^{a}$	p_{44}	0	0	0	0	$-q_{45}{}^{a}$	q_{44}	0
$p_{61}{}^{a}$	$-p_{61}{}^{a}$	0	0	0	$\frac{1}{2}(p_{11}-p_{12})$	$q_{61}{}^{a}$	$-q_{61}{}^{a}$	0	0	0	$(q_{11}-q_{12})$

Group 9. Hexagonal system, D_{3h}, C_{6v}, D_6, $D_{6h}(\bar{6}, 2m, 6mm, 622, 6/mmm)$ (6 constants)

p_{11}	p_{12}	p_{13}	0	0	0	q_{11}	q_{12}	q_{13}	0	0	0
p_{12}	p_{11}	p_{13}	0	0	0	q_{12}	q_{11}	q_{13}	0	0	0
p_{31}	p_{31}	p_{33}	0	0	0	q_{31}	q_{31}	q_{33}	0	0	0
0	0	0	p_{44}	0	0	0	0	0	q_{44}	0	0
0	0	0	0	p_{44}	0	0	0	0	0	q_{44}	0
0	0	0	0	0	$\frac{1}{2}(p_{11}-p_{12})$	0	0	0	0	0	$(q_{11}-q_{12})$

Group 10. Cubic system, T, $T_h(23, m3)$ (4 constants)

p_{11}	$p_{12}{}^{b}$	$p_{13}{}^{b}$	0	0	0	q_{11}	$q_{12}{}^{b}$	$q_{13}{}^{b}$	0	0	0
$p_{13}{}^{b}$	p_{11}	$p_{12}{}^{b}$	0	0	0	$q_{13}{}^{b}$	q_{11}	$q_{12}{}^{b}$	0	0	0

$p_{12}{}^{b}$	$p_{13}{}^{b}$	p_{11}	0	0	0	$q_{12}{}^{b}$	$q_{13}{}^{b}$	q_{11}	0	0	0
0	0	0	p_{44}	0	0	0	0	0	q_{44}	0	0
0	0	0	0	p_{44}	0	0	0	0	0	q_{44}	0
0	0	0	0	0	p_{44}	0	0	0	0	0	q_{44}

Group 11. Cubic system, T_d, O, $O_h(\bar{4}3m, 43, m3m)$ (3 constants)

p_{11}	p_{12}	p_{12}	0	0	0	q_{11}	q_{12}	q_{12}	0	0	0
p_{12}	p_{11}	p_{12}	0	0	0	q_{12}	q_{11}	q_{12}	0	0	0
p_{12}	p_{12}	p_{11}	0	0	0	q_{12}	q_{12}	q_{11}	0	0	0
0	0	0	p_{44}	0	0	0	0	0	q_{44}	0	0
0	0	0	0	p_{44}	0	0	0	0	0	q_{44}	0
0	0	0	0	0	p_{44}	0	0	0	0	0	q_{44}

Group 12. Isotropic solids (2 constants)

p_{11}	p_{12}	p_{12}	0	0	0	q_{11}	q_{12}	q_{12}	0	0	0
p_{12}	p_{11}	p_{12}	0	0	0	q_{12}	q_{11}	q_{12}	0	0	0
p_{12}	p_{12}	p_{11}	0	0	0	q_{12}	q_{12}	q_{11}	0	0	0
0	0	0	$\frac{1}{2}(p_{11}-p_{12})$	0	0	0	0	0	$(q_{11}-q_{12})$	0	0
0	0	0	0	$\frac{1}{2}(p_{11}-p_{12})$	0	0	0	0	0	$(q_{11}-q_{12})$	0
0	0	0	0	0	$\frac{1}{2}(p_{11}-p_{12})$	0	0	0	0	0	$(q_{11}-q_{12})$

[a] These constants are nonvanishing in Bhagavantam's scheme, and zero in Pockels' scheme.
[b] These constants are unequal in Bhagavantam's scheme, and equal in Pockels' scheme.

3.3.1. Classical Method

The constant q_{ij} in the following discussion is given in the two-suffix notation.

3.3.1.1. The 36 q_{ij} for the Most Anisotropic Crystal

Let OX, OY, OZ be the principal axes of the optical index ellipsoid for the unstressed crystal. In its most general form the equation describing the optical index ellipsoid for an unstrained crystal can be put (as on page 52) thus:

$$B_{11}^0 x^2 + B_{22}^0 y^2 + B_{33}^0 z^2 + 2B_{23}^0 yz + 2B_{31}^0 zx + 2B_{12}^0 xy = 1$$

Under the influence of an applied stress P the modified optical index ellipsoid of the crystal need not have the axes of coordinates for its principal axes. Let the equation of the deformed ellipsoid be (as on page 52)

$$B_{11} x^2 + B_{22} y^2 + B_{33} z^2 + 2B_{23} yz + 2B_{31} zx + 2B_{12} xy = 1$$

Pockels assumes that the differences between the coefficients of the deformed and undeformed optical index ellipsoids are small quantities of the first order, being linear in the stresses.

Thus, putting B_{ij} and P_{kl} in the two-suffix notation, and q_{ijkl} also in the two-suffix notation, Eqs. (3.9d) take the form

$$\begin{aligned} B_{11} - B_{11}^0 &= -[q_{11}P_{xx} + q_{12}P_{yy} + q_{13}P_{zz} + q_{14}P_{yz} + q_{15}P_{zx} + q_{16}P_{xy}] \\ &\vdots \\ B_{23} - B_{23}^0 &= -[q_{41}P_{xx} + q_{42}P_{yy} + q_{43}P_{zz} + q_{44}P_{yz} + q_{45}P_{zx} + q_{46}P_{xy}] \\ &\vdots \end{aligned} \tag{3.14}$$

Thus there are 36 q_{ij}, which for the most unsymmetrical crystal (of the triclinic system) cannot be reduced. The strain–energy relationship which exists for the elastic constants of a crystal does not have an analogue in the photoelastic behavior of crystals.

3.3.1.2. Influence of Some Symmetry Elements on the Number of q_{ij}

For crystals possessing various elements of symmetry, however, there will be a reduction in the number of constants.

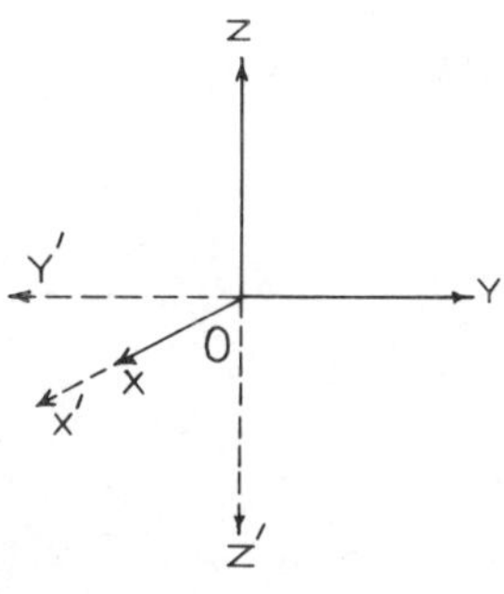

Fig. 3.1. Transformation of axes: OX is a diad or a two-fold axis of rotation; here $OX' = OX$.

Three typical cases of symmetry elements, namely, axis of rotation, center, and plane of reflection, will now be considered with regard to their influence on the number of nonvanishing independent constants.

a. Diad (Axis of 180° Rotation), or Digonal Axis, or a Two-fold Axis. Let the diad axis be the OX axis (Fig. 3.1). Then the coordinate system is rotated through 180° about the OX axis. Consequently

$$X' = X, \qquad Y' = -Y, \qquad \text{and} \qquad Z' = -Z$$

From the second-rank tensor nature of P_{ij}, weh ave, for this particular transformation,

$$\begin{aligned} P'_{x'x'} &= P_{xx}, & P'_{y'y'} &= P_{(-y)(-y)} = P_{yy} \\ P'_{z'z'} &= P_{(-z)(-z)} = P_{zz}, & P'_{y'z'} &= P_{(-y)(-z)} = P_{yz} \\ P'_{z'x'} &= P_{(-z)x} = -P_{zx}, & P'_{x'y'} &= P_{x(-y)} = -P_{xy} \end{aligned}$$

Henceforth $P'_{x'x'}$ will be represented by P'_{xx}, and similarly $P'_{y'z'}$ by P'_{yz}, etc. (see page xxvii of Notation and Conventions).

From the above set of equalities, it is clear that only P_{xy} and P_{zx} are reversed while all other stress components remain unaltered.*

Also if in the equation of the deformed optical index ellipsoid the signs of OY and OZ are reversed to yield OY' and OZ', it is obvious that B_{11}, B_{22}, B_{33}, and B_{23} are unaltered, and B_{31} and B_{12} are reversed in sign because the equation must retain its original form. The same conclusion can also be arrived at on account of the simple fact that both P_{ij} and B_{ij} are sym-

* A linear stress when reversed changes a compression into tension. A shearing stress when reversed causes a twist, say, in the clockwise rotation to change into a twist in the anticlockwise direction.

metric tensors of second rank. So whatever happens to P_{ij} also happens to B_{ij}.

Thus, if B_{ij} and B^0_{ij} refer to the original position of the crystal, and A_{ij} and A^0_{ij} refer to the transformed position of the crystal (the transformed position being brought about by the symmetry operation on the crystal), then the following relations hold:

$$A_{11} - A^0_{11} = B_{11} - B^0_{11}, \qquad A_{23} - A^0_{23} = B_{23} - B^0_{23}$$
$$A_{22} - A^0_{22} = B_{22} - B^0_{22}, \qquad A_{31} - A^0_{31} = -(B_{31} - B^0_{31})$$
$$A_{33} - A^0_{33} = B_{33} - B^0_{33}, \qquad A_{12} - A^0_{12} = -(B_{12} - B^0_{12})$$

Now from Eq. (3.14) take

$$B_{11} - B^0_{11} = -[q_{11}P_{xx} + q_{12}P_{yy} + q_{13}P_{zz} + q_{14}P_{yz} + q_{15}P_{zx} + q_{16}P_{xy}]$$

In the transformed position

$$A_{11} - A^0_{11} = -[q_{11}P'_{xx} + q_{12}P'_{yy} + q_{13}P'_{zz} + q_{14}P'_{yz} + q_{15}P'_{zx} + q_{16}P'_{xy}]$$

By the symmetry operation we have

$$A_{11} - A^0_{11} = B_{11} - B^0_{11}$$

Therefore

$$\begin{aligned}-[q_{11}P_{xx} &+ q_{12}P_{yy} + q_{13}P_{zz} + q_{14}P_{yz} + q_{15}P_{zx} + q_{16}P_{xy}]\\ &= B_{11} - B^0_{11} = A_{11} - A^0_{11}\\ &= -[q_{11}P'_{xx} + q_{12}P'_{yy} + q_{13}P'_{zz} + q_{14}P'_{yz} + q_{15}P'_{zx} + q_{16}P'_{xy}]\end{aligned}$$

In the above equation we have $P_{xx} = P'_{xx}$, $P_{yy} = P'_{yy}$, $P_{zz} = P'_{zz}$, $P_{yz} = P'_{yz}$, $P_{zx} = -P'_{zx}$, $P_{xy} = -P'_{xy}$. Hence the equation reduces to

$$\begin{aligned}-q_{15}P_{zx} - q_{16}P_{xy} &= -q_{15}P'_{zx} - q_{16}P'_{xy}\\ &= -q_{15}(-P_{zx}) - q_{16}(-P_{xy})\end{aligned}$$

Therefore, equating like coefficients, we obtain $-q_{15} = +q_{15}$ and $-q_{16} = +q_{16}$, or $q_{15} = 0$ and $q_{16} = 0$.

Similarly, taking $(B_{22} - B^0_{22}) = (A_{22} - A^0_{22})$, we get $q_{25} = 0$, $q_{26} = 0$; taking $(B_{33} - B^0_{33}) = (A_{33} - A^0_{33})$, we get $q_{35} = 0$; $q_{36} = 0$; and taking $(B_{23} - B^0_{23}) = (A_{23} - A^0_{23})$, we get $q_{45} = 0$; $q_{46} = 0$.

Take $(B_{31} - B_{31}^0) = -(A_{31} - A_{31}^0)$. Hence

$$-[q_{51}P_{xx} + q_{52}P_{yy} + q_{53}P_{zz} + q_{54}P_{yz} + q_{55}P_{zx} + q_{56}P_{xy}]$$
$$= -\{-[q_{51}P'_{xx} + q_{52}P'_{yy} + q_{53}P'_{zz} + q_{54}P'_{yz} + q_{55}P'_{zx} + q_{56}P'_{xy}]\}$$

Here again, since $P_{xx} = P'_{xx}$, $P_{yy} = P'_{yy}$, $P_{zz} = P'_{zz}$, $P_{yz} = P'_{yz}$, $P_{zx} = -P'_{zx}$, and $P_{xy} = -P'_{xy}$, we obtain $q_{51} = 0$, $q_{52} = 0$, $q_{53} = 0$, $q_{54} = 0$, $q_{55}P_{zx} = -q_{55}P'_{zx}$. But $P_{zx} = -P'_{zx}$. Therefore $q_{55} = q_{55}$, i.e., q_{55} survives; similarly, q_{56} survives.

Take $+(B_{12} - B_{12}^0) = -(A_{12} - A_{12}^0)$. As in the case above, we get $q_{61} = 0$, $q_{62} = 0$, $q_{63} = 0$, $q_{64} = 0$; and q_{65} and q_{66} survive. Thus 16 constants vanish.

The above results, in all the cases put together, can be expressed thus: $q_{ij} = 0$ when one of the i, j is either 5 or 6, the other being 1, 2, 3, or 4.

b. Center of Symmetry. Owing to the existence of a center of symmetry (Fig. 3.2), $X \to X'$, $Y \to Y'$, and $Z \to Z'$. Thus $X' = -X$, $Y' = -Y$, and $Z' = -Z$. Hence $P'_{xx} = P_{(-x)(-x)} = P_{xx}$; similarly, $P'_{yy} = P_{yy}$, $P'_{zz} = P_{zz}$, $P'_{yz} = P_{yz}$, $P'_{zx} = P_{zx}$, and $P'_{xy} = P_{xy}$. Further, $A_{11}^0 = B_{11}^0$, $A_{22}^0 = B_{22}^0$, $A_{33}^0 = B_{33}^0$, $A_{23}^0 = B_{23}^0$, $A_{31}^0 = B_{31}^0$, $A_{12}^0 = B_{12}^0$. Thus $(A_{11} - A_{11}^0) = (B_{11} - B_{11}^0)$ and $(A_{23} - A_{23}^0) = (B_{23} - B_{23}^0)$.

Now taking $(A_{11} - A_{11}^0) = (B_{11} - B_{11}^0)$, we get

$$-[q_{11}P_{xx} + q_{12}P_{yy} + q_{13}P_{zz} + q_{14}P_{yz} + q_{15}P_{zx} + q_{16}P_{xy}]$$
$$= -[q_{11}P'_{xx} + q_{12}P'_{yy} + q_{13}P'_{zz} + q_{14}P'_{yz} + q_{15}P'_{zx} + q_{16}P'_{xy}]$$

Since $P_{xx} = P'_{xx}, \ldots,$ and $P_{yz} = P'_{yz}, \ldots,$ all the six constants q_{11}, q_{12}, q_{13}, q_{14}, q_{15}, and q_{16} survive.

Similarly, taking the other five equalities of $(A_{ij} - A_{ij}^0) = (B_{ij} - B_{ij}^0)$, we find that all the remaining 30 q_{ij} survive. The resulting influence of a center of symmetry, therefore, on the existence or otherwise of q_{ij} is nil.

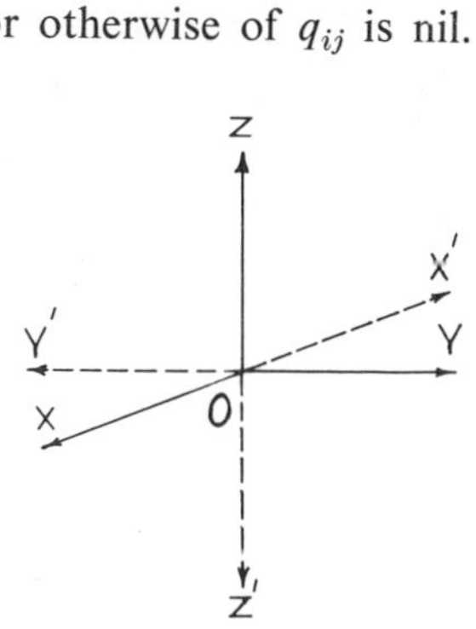

Fig. 3.2. Transformation of axes: the origin O is a center of symmetry or point of inversion.

The important conclusion that can be drawn from the above discussion is that crystal classes C_1 and C_i of the triclinic system, which differ only in the existence of a center of symmetry for the latter, have the same matrix $[q_{ij}]$ for both.

One can as well conclude from Neumann's principle (page 44) that the photoelastic property is centrosymmetric because it is related by two measurable quantities, the stress P_{ij} and the optical parameters B_{ij}, both of which are themselves centrosymmetric. Similarly, the elastic property of crystals is centrosymmetric because it deals with stress and strain, both being centrosymmetric.

c. Plane of Symmetry (*Reflection Plane*). Let the reflection plane be perpendicular to the OZ (or X_3) axis (Fig. 3.3).

Then $X' = X$, $Y' = Y$, and $Z' = -Z$. Hence

$$P'_{xx} = P_{xx}, \quad P'_{yy} = P_{yy}, \quad P'_{zz} = P_{zz}, \quad P'_{yz} = -P_{yz}, \quad P'_{zx} = -P_{zx}, \quad P'_{xy} = P_{xy}$$

Thus P_{xx}, P_{yy}, P_{zz} and P_{xy} are unaltered while P_{yz} and P_{zx} reverse. Also

$$A^0_{11} = B^0_{11}, \quad A^0_{22} = B^0_{22}, \quad A^0_{33} = B^0_{33}, \quad A^0_{23} = -B^0_{23}, \quad A^0_{31} = -B^0_{31}, \quad A^0_{12} = B^0_{1}$$

$$\begin{aligned} A_{11} - A^0_{11} &= B_{11} - B^0_{11}, \qquad (A_{23} - A^0_{23}) = -(B_{23} - B^0_{23}) \\ A_{22} - A^0_{22} &= B_{22} - B^0_{22}, \qquad (A_{31} - A^0_{31}) = -(B_{31} - B^0_{31}) \\ A_{33} - A^0_{33} &= B_{33} - B^0_{33}, \qquad (A_{12} - A^0_{12}) = +(B_{12} - B^0_{12}) \end{aligned}$$

Now taking $(A_{11} - A^0_{11}) = (B_{11} - B^0_{11})$, we get

$$\begin{aligned} &-[q_{11}P'_{xx} + q_{12}P'_{yy} + q_{13}P'_{zz} + q_{14}P'_{yz} + q_{15}P'_{zx} + q_{16}P'_{xy}] \\ &\quad = -[q_{11}P_{xx} + q_{12}P_{yy} + q_{13}P_{zz} + q_{14}P_{yz} + q_{15}P_{zx} + q_{16}P_{xy}] \end{aligned}$$

Therefore $q_{11}, q_{12}, q_{13}, q_{16}$ exist; $q_{14} = 0$, $q_{15} = 0$.

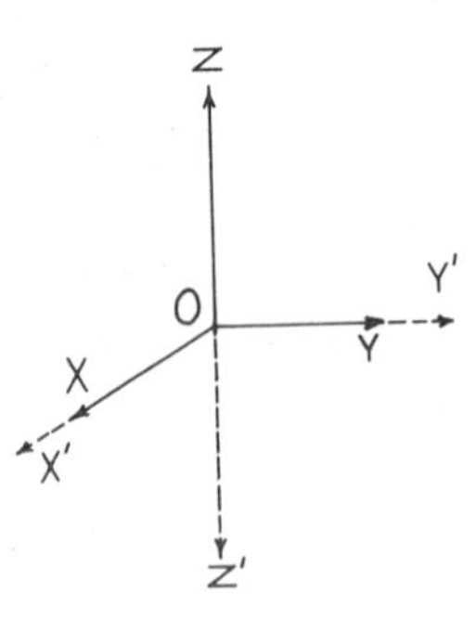

Fig. 3.3. Transformation of axes: the plane XOY is a reflection plane.

Taking $(A_{22} - A_{22}^0) = (B_{22} - B_{22}^0)$, we get $q_{24} = 0$, $q_{25} = 0$; taking $(A_{33} - A_{33}^0) = (B_{33} - B_{33}^0)$, we get $q_{34} = 0$, $q_{35} = 0$; and taking $(A_{12} - A_{12}^0) = (B_{12} - B_{12}^0)$, we get $q_{64} = 0$, $q_{65} = 0$.

Taking $(A_{23} - A_{23}^0) = -(B_{23} - B_{23}^0)$, we get

$$-[q_{41}P'_{xx} + q_{42}P'_{yy} + q_{43}P'_{zz} + q_{44}P'_{yz} + q_{45}P'_{zx} + q_{46}P'_{xy}]$$
$$= -\{-[q_{41}P_{xx} + q_{42}P_{yy} + q_{43}P_{zz} + q_{44}P_{yz} + q_{45}P_{zx} + q_{46}P_{xy}]\}$$

Therefore $q_{41} = 0$, $q_{42} = 0$, $q_{43} = 0$, $q_{46} = 0$; q_{44} and q_{45} survive. Similarly, q_{51}, q_{52}, q_{53}, and q_{56} vanish; q_{54} and q_{55} survive. Thus, in all, 16 constants vanish, namely,

$$\begin{matrix} q_{14} & q_{24} & q_{34} & q_{64} & \quad & q_{41} & q_{42} & q_{43} & q_{46} \\ q_{15} & q_{25} & q_{35} & q_{65} & \quad & q_{51} & q_{52} & q_{53} & q_{56} \end{matrix}$$

Hence $q_{ij} = 0$, where $i = 1, 2, 3$, or 6; $j = 4$ or 5; and $q_{ij} = 0$, where $i = 4$ or 5; $j = 1, 2, 3$, or 6.

Summary. Proceeding along the lines given above, we obtain the results listed in Table 3.3. Having thus seen the influence of the three important symmetry elements on the matrix of photoelastic constants q_{ij}, we shall now proceed to examine in detail the surviving constants for some of the 32 point groups.

3.3.1.3. Derivation of the Surviving q_{ij} for Some Crystal Classes

Triclinic System

1. Class C_1 (1) (Symmetry Element E). Identity or rotation by 360° is the only symmetry element possessed by this class. Hence all the $36 q_{ij}$ exist.

2. Class C_i ($\bar{1}$) (Symmetry Elements E i). Besides the identity, this has an inversion or center of symmetry. But we have already seen in the previous section that a center of symmetry does not cause any change or reduction in the matrix of the q_{ij}. Hence, here too, all the 36 q_{ij} survive.

The two classes C_1 and C_i, having 36 q_{ij}, are grouped together in their photoelastic group. They belong to Group 1.

Monoclinic System

3. Class C_2 (2) (Symmetry Elements E C_2). It is customary to designate the OY axis as the principal axis. It has been shown in the previous section

Table 3.3. The q_{ij} That Vanish for Some Symmetry Elements

Symmetry element	q_{ij} that vanish															
OX diad	q_{15}	q_{25}	q_{35}	q_{45}	q_{16}	q_{26}	q_{36}	q_{46}	q_{51}	q_{52}	q_{53}	q_{54}	q_{61}	q_{62}	q_{63}	q_{64}
OY diad	q_{14}	q_{24}	q_{34}	q_{54}	q_{16}	q_{26}	q_{36}	q_{56}	q_{41}	q_{42}	q_{43}	q_{45}	q_{61}	q_{62}	q_{63}	q_{65}
OZ diad	q_{14}	q_{24}	q_{34}	q_{64}	q_{15}	q_{25}	q_{35}	q_{65}	q_{41}	q_{42}	q_{43}	q_{46}	q_{51}	q_{52}	q_{53}	q_{56}
XY reflection plane	q_{14}	q_{24}	q_{34}	q_{64}	q_{15}	q_{25}	q_{35}	q_{65}	q_{41}	q_{42}	q_{43}	q_{46}	q_{51}	q_{52}	q_{53}	q_{56}
YZ reflection plane	q_{15}	q_{25}	q_{35}	q_{45}	q_{16}	q_{26}	q_{36}	q_{46}	q_{51}	q_{52}	q_{53}	q_{54}	q_{61}	q_{62}	q_{63}	q_{64}
ZX reflection plane	q_{16}	q_{26}	q_{36}	q_{56}	q_{14}	q_{24}	q_{34}	q_{54}	q_{61}	q_{62}	q_{63}	q_{65}	q_{41}	q_{42}	q_{43}	q_{45}

that the presence of a diad axis parallel to OY results in the vanishing of 16 constants, namely,

$$\begin{array}{cccccccc} q_{14} & q_{24} & q_{34} & q_{54} & q_{41} & q_{42} & q_{43} & q_{45} \\ q_{16} & q_{26} & q_{36} & q_{56} & q_{61} & q_{62} & q_{63} & q_{65} \end{array}$$

4. Class $C_s\,(= C_{1h})$ (m) (Symmetry Elements E, σ_h). As already indicated, the principal axis is OY. Hence the plane of reflection is perpendicular to the OY axis, or the plane OXZ is the mirror plane. It has been shown in the previous section that, when the plane XZ is the mirror plane, the constants that vanish are again 16. In fact, it has already been shown that these 16 vanishing constants are the same as those for a crystal with a diad axis parallel to OY. The vanishing constants are

$$\begin{array}{cccccccc} q_{14} & q_{24} & q_{34} & q_{54} & q_{41} & q_{42} & q_{43} & q_{45} \\ q_{16} & q_{26} & q_{36} & q_{56} & q_{61} & q_{62} & q_{63} & q_{65} \end{array}$$

5. Class C_{2h} $(2/m)$ (Symmetry Elements $E\ C_2\ i\ \sigma_h$). This class may be considered as having combined in itself all the elements of symmetry possessed by both classes $C_s\,(= C_{1h})$ and C_2. Hence all the constants that vanish for those two classes together vanish for this class also. Furthermore, the inversion i, appearing as an extra element, is actually a consequential element— a diad and a mirror perpendicular to it generate an inversion. Thus no additional constants disappear. This can also be understood from the fact that in a previous section the inversion symmetry has been shown to have no influence on the number of photoelastic constants. Hence the same constants that survive for the crystal classes C_s and C_2 survive here also. Thus there are all the 20 constants that survive for the previous two classes of the monoclinic system.

The three classes C_2, C_s, and C_{2h} form one group for the purpose of discussing the photoelastic effect in crystals.

Orthorhombic System

6. Class C_{2v} $(mm2)$ (Symmetry Elements $E\ C_2\ \sigma_v'\ \sigma_v''$). Here the diad axis C_2 is parallel to the OX_3, the principal axis; σ_v' is a mirror plane perpendicular to the OX_1 axis and parallel to OX_2X_3; σ_v'' is a mirror plane perpendicular to the OX_2 axis and parallel to OX_1X_3.

Based on the earlier discussion, the vanishing constants can be written straightaway as follows (reproduced from Table 3.3):

a. Due to $\sigma_v' \perp OX_1$: The vanishing constants are

$$q_{15} \quad q_{25} \quad q_{35} \quad q_{45} \qquad q_{51} \quad q_{52} \quad q_{53} \quad q_{54}$$
$$q_{16} \quad q_{26} \quad q_{36} \quad q_{46} \qquad q_{61} \quad q_{62} \quad q_{63} \quad q_{64}$$

b. Due to $\sigma_v'' \perp OX_2$: The vanishing constants are

$$q_{14} \quad q_{24} \quad q_{34} \quad q_{54} \qquad q_{41} \quad q_{42} \quad q_{43} \quad q_{45}$$
$$q_{16} \quad q_{26} \quad q_{36} \quad q_{56} \qquad q_{61} \quad q_{62} \quad q_{63} \quad q_{65}$$

c. Due to $C_2 \parallel OX_3$: The vanishing constants are

$$q_{14} \quad q_{24} \quad q_{34} \quad q_{64} \qquad q_{41} \quad q_{42} \quad q_{43} \quad q_{46}$$
$$q_{15} \quad q_{25} \quad q_{35} \quad q_{65} \qquad q_{51} \quad q_{52} \quad q_{53} \quad q_{56}$$

Altogether 24 constants vanish. Hence the 12 surviving constants are

$$q_{11} \quad q_{12} \quad q_{13} \qquad q_{31} \quad q_{32} \quad q_{33}$$
$$q_{21} \quad q_{22} \quad q_{23} \qquad q_{44} \quad q_{55} \quad q_{66}$$

7. Class D_2 (222) (Symmetry Elements E C_2 C_2' C_2''). We list the vanishing constants.

a. Due to $C_2' \parallel X_1$: The vanishing constants are

$$q_{15} \quad q_{25} \quad q_{35} \quad q_{45} \qquad q_{51} \quad q_{52} \quad q_{53} \quad q_{54}$$
$$q_{16} \quad q_{26} \quad q_{36} \quad q_{46} \qquad q_{61} \quad q_{62} \quad q_{63} \quad q_{64}$$

b. Due to $C_2'' \parallel X_2$: The vanishing constants are

$$q_{14} \quad q_{24} \quad q_{34} \quad q_{54} \qquad q_{41} \quad q_{42} \quad q_{43} \quad q_{45}$$
$$q_{16} \quad q_{26} \quad q_{36} \quad q_{56} \qquad q_{61} \quad q_{62} \quad q_{63} \quad q_{65}$$

c. Due to $C_2 \parallel OX_3$: The vanishing constants are

$$q_{14} \quad q_{24} \quad q_{34} \quad q_{64} \qquad q_{41} \quad q_{42} \quad q_{43} \quad q_{46}$$
$$q_{15} \quad q_{25} \quad q_{35} \quad q_{65} \qquad q_{51} \quad q_{52} \quad q_{53} \quad q_{56}$$

Hence the number of vanishing constants is 24 and they are the same as those for the previous class C_{2v}. Hence the surviving constants are as follows:

$$q_{11} \quad q_{12} \quad q_{13} \qquad q_{31} \quad q_{32} \quad q_{33}$$
$$q_{21} \quad q_{22} \quad q_{23} \qquad q_{44} \quad q_{55} \quad q_{66}$$

8. Class D_{2h} (*mmm*) (Symmetry Elements $E\ C_2\ C_2'\ C_2''\ i\ \sigma_h\ \sigma_v'\ \sigma_v''$). Here E and i do not cause any further reduction (see discussion earlier); $C_2\ C_2'\ C_2''$ cause the vanishing of 24 constants (as discussed earlier); $\sigma_h\ \sigma_v'\ \sigma_v''$ cause the vanishing of 24 constants which are the same as above.

Hence the surviving constants q_{ij} are the same as for the two previous classes in the orthorhombic system, namely,

$$\begin{array}{ccccccc} q_{11} & q_{12} & q_{13} & \quad & q_{31} & q_{32} & q_{33} \\ q_{21} & q_{22} & q_{23} & \quad & q_{44} & q_{55} & q_{66} \end{array}$$

Tetragonal System

9. Class C_4 (4) (Symmetry Elements $E\ 2C_4\ C_2$). This class has a tetragonal axis parallel to OZ. But the tetragonal axis of symmetry includes a digonal axis in the same direction. We have seen above that the existence of a digonal axis parallel to OZ eliminates 16 q_{ij} (which are given in the earlier section). Furthermore, the presence of the tetragonal axis may be expected to reduce a few more and to equalize a few others. To investigate this, we proceed as follows: Since OZ is a tetrad axis, let $OX \rightarrow OY$ and $OY \rightarrow -OX$ (Fig. 3.4). Then $OX' = OY$, $OY' = -OX$, $OZ' = OZ$; $P'_{xx} = P_{yy}$, $P'_{yy} = P_{xx}$, $P'_{zz} = P_{zz}$, $P'_{yz} = -P_{xz} = -P_{zx}$, $P'_{zx} = P_{zy}$, $P'_{xy} = -P_{xy}$. Also $A^0_{11} = B^0_{22}$, $A^0_{22} = B^0_{11}$, $A^0_{33} = B^0_{33}$, $A^0_{23} = -B^0_{13}$, $A^0_{31} = B^0_{32}$, $A^0_{12} = -B^0_{21} = -B^0_{12}$. Here the primed parameters refer to the transformed coordinate system.

Therefore Pockels' equations are reduced to the following: Taking $(A_{11} - A^0_{11}) = (B_{22} - B^0_{22})$, we have

$$q_{11}P'_{xx} + q_{12}P'_{yy} + q_{13}P'_{zz} = q_{21}P_{xx} + q_{22}P_{yy} + q_{23}P_{zz}$$

$$\therefore q_{11} = q_{22}, \qquad q_{12} = q_{21}, \qquad q_{13} = q_{23}$$

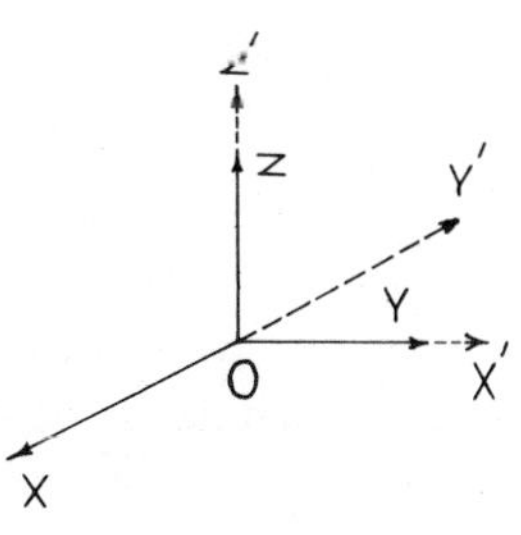

Fig. 3.4. Transformation of axes: the axis OZ is a tetrad or a four-fold axis of rotation; here $OZ' = OZ$.

Taking $(A_{33} - A^0_{33}) = (B_{33} - B^0_{33})$, we have

$$q_{31}P'_{xx} + q_{32}P'_{yy} + q_{33}P'_{zz} = q_{31}P_{xx} + q_{32}P_{yy} + q_{33}P_{zz}$$
$$\therefore q_{31} = q_{32}, \qquad q_{32} = q_{31}, \qquad q_{33} = q_{33}$$

Taking $(A_{22} - A^0_{22}) = (B_{11} - B^0_{11})$, we have

$$q_{21}P'_{xx} + q_{22}P'_{yy} + q_{23}P'_{zz} = q_{11}P_{xx} + q_{12}P_{yy} + q_{13}P_{zz}$$
$$\therefore q_{21} = q_{12}, \qquad q_{22} = q_{11}, \qquad q_{23} = q_{13}$$

We take $(A_{23} - A^0_{23}) = -(B_{13} - B^0_{13}) = -(B_{31} - B^0_{31})$. In fact, $B^0_{23} = 0 = B^0_{31}$, i.e., $-q_{44}P'_{yz} = -(-q_{55}P_{zx})$ and $P'_{yz} = -P_{zx}$. Therefore $q_{44} = q_{55}$. Taking $(A_{31} - A^0_{31}) = (B_{32} - B^0_{32})$, we have

$$-q_{55}P'_{zx} = -q_{44}P_{yz} \qquad \text{and} \qquad P'_{zx} = P_{yz}$$

Therefore $q_{55} = q_{44}$. Taking $(A_{12} - A^0_{12}) = -(B_{12} - B^0_{12})$, we have

$$q_{66}P'_{xy} = -q_{66}P_{xy} \qquad \text{and} \qquad P'_{xy} = -P_{xy}$$

Therefore $q_{66} = q_{66}$. Hence q_{66} survives.

Thus we have out of the surviving 12 constants,

$$q_{11} = q_{22}, \qquad q_{12} = q_{21}, \qquad q_{13} = q_{23}, \qquad q_{31} = q_{32}, \qquad q_{44} = q_{55}$$
$$q_{66} = q_{66}, \qquad q_{22} = q_{11}, \qquad q_{21} = q_{12}, \qquad q_{23} = q_{13}, \qquad q_{55} = q_{44}$$

Five q_{ij} are equal to five others of the 12 surviving constants. Hence only seven constants survive for class C_4 of the tetragonal system. They are

$$q_{11} \quad q_{12} \quad q_{13} \quad q_{31} \quad q_{33} \quad q_{44} \quad q_{66}$$

Isotropic Solids

10. An isotropic solid should exhibit the symmetries of all the point groups. We take, in particular, the combined effect of a six-fold symmetry over a four-fold symmetry. For all hexagonal classes, we observe from their q_{ij} matrices that $q_{66} = (q_{11} - q_{12})$. For all cubic classes, $q_{44} = q_{55} = q_{66}$. Combining these conditions, we find that for an isotropic solid

$$q_{44} = q_{55} = q_{66} = q_{11} - q_{12}$$

Hence for an isotropic solid there are only two constants q_{11} and q_{12} when compared with the three constants q_{11}, q_{12}, and q_{44} of a cubic crystal of the highest symmetry class $m3m$ (O_h).

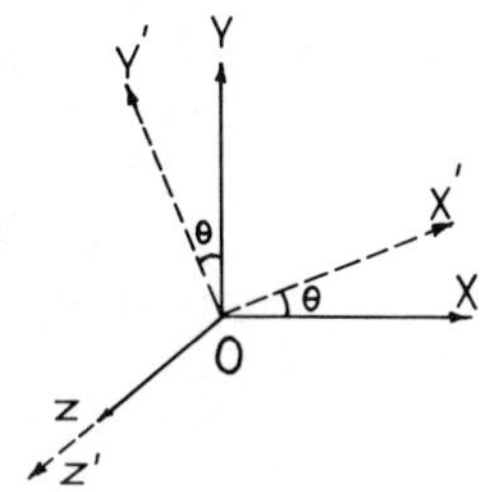

Fig. 3.5. Transformation of axes: a rotation of the X system through an angle θ about the Z axis; here $OZ' = OZ$.

Alternatively, we can arrive at the above result from the principle that an isotropic solid is a special case of a cubic crystal with the additional condition that here the tensor properties are invariant under a transformation of rotation about any axis through any angle θ.

We now study the effect of rotation of a cubic crystal of class $m3m$ (O_h) about its Z axis. The transformed axes are shown in Fig. 3.5, and the transformation matrix is given by

	X	Y	Z
X'	$\cos\theta$	$\sin\theta$	0
Y'	$-\sin\theta$	$\cos\theta$	0
Z'	0	0	1

A general stress P has the following six stress components in the original coordinate system:

$$P_{xx} \quad P_{yy} \quad P_{zz} \quad P_{yz} \quad P_{zx} \quad P_{xy}$$

and the same stress P has six components in the new coordinate system:

$$P'_{x'x'} \qquad P'_{y'y'} \qquad P'_{z'z'} \qquad P'_{y'z'} \qquad P'_{z'x'} \qquad P'_{x'y'}$$

Note that in this section $P'_{x'x'}$ stands for P'_{xx}, $P'_{y'z'}$ for P'_{yz}, etc. Further, $\alpha_{x'x}$ represents the direction cosine between the X' and X axes; $\alpha_{x'y}$ has a similar meaning.

Since $P'_{x'x'}$ is a second-rank symmetric tensor we note that

$$P'_{ij} = \alpha_{ik}\alpha_{jl}P_{kl} \qquad \text{and} \qquad P_{kl} = \alpha_{ik}\alpha_{jl}P'_{ij}$$

Hence we obtain $P'_{x'x'}, \ldots, P'_{y'z'}, \ldots$ in terms of $P_{xx}, \ldots, P_{yz}, \ldots$ from the above relation thus:

$$\begin{aligned} P'_{x'x'} &= \alpha_{x'k}\alpha_{x'l}P_{kl} = \alpha_{x'l}\{\alpha_{x'x}P_{xl} + \alpha_{x'y}P_{yl}\} \\ &= \alpha_{x'x}(\alpha_{x'x}P_{xx} + \alpha_{x'y}P_{xy}) + \alpha_{x'y}(\alpha_{x'x}P_{yx} + \alpha_{x'y}P_{yy}) \\ &= \alpha^2_{x'x}P_{xx} + 2\alpha_{x'x}\alpha_{x'y}P_{xy} + \alpha^2_{x'y}P_{yy} \\ &= \cos^2\theta P_{xx} + \sin^2\theta P_{yy} + \sin 2\theta P_{xy} \\ P'_{y'y'} &= \sin^2\theta P_{xx} + \cos^2\theta P_{yy} - \sin 2\theta P_{xy} \\ P'_{z'z'} &= P_{zz} \\ P'_{y'z'} &= \cos\theta P_{yz} - \sin\theta P_{zx} \\ P'_{z'x'} &= \sin\theta P_{yz} + \cos\theta P_{zx} \\ P'_{x'y'} &= \cos 2\theta P_{xy} + \tfrac{1}{2}\sin 2\theta(P_{yy} - P_{xx}) \end{aligned} \tag{3.15}$$

Since the solid is isotropic, q_{11}, q_{12}, and q_{44} above should exist as for a cubic crystal of class $m3m$ (O_h), and, furthermore, these must be independent of the direction of stress and of observation. Thus for the new coordinate system,

$$\begin{aligned} A_{11} - A^0_{11} &= q_{11}P'_{x'x'} + q_{12}P'_{y'y'} + q_{12}P'_{z'z'} \\ &= q_{11}(\cos^2\theta P_{xx} + \sin^2\theta P_{yy} + \sin 2\theta P_{xy}) \\ &\quad + q_{12}(\sin^2\theta P_{xx} + \cos^2\theta P_{yy} - \sin 2\theta P_{xy}) + q_{12}P_{zz} \end{aligned} \tag{3.16}$$

But $A_{11} = B_{11}\cos^2\theta + B_{22}\sin^2\theta + B_{12}\sin 2\theta$ and $A^0_{11} = B^0_{11}$. Therefore for the left-hand side of Eq. (3.16) we have

$$\begin{aligned} &(B_{11}\cos^2\theta + \sin^2\theta B_{22} + \sin 2\theta B_{12}) - B^0_{11} \\ &\quad = \cos^2\theta(B_{11} - B^0_{11}) + \sin^2\theta(B_{22} - B^0_{22}) + \sin 2\theta B_{12} \\ &\quad = \cos^2\theta(q_{11}P_{xx} + q_{12}P_{yy} + q_{12}P_{zz}) \\ &\qquad + \sin^2\theta(q_{12}P_{xx} + q_{11}P_{yy} + q_{12}P_{zz}) + \sin 2\theta q_{44}P_{xy} \\ &\quad = \cos^2\theta[(q_{11} - q_{12})P_{xx} + q_{12}(P_{xx} + P_{yy} + P_{zz})] \\ &\qquad + \sin^2\theta[(q_{11}-q_{12})P_{yy} + q_{12}(P_{xx}+P_{yy}+P_{zz}) + \sin 2\theta q_{44}P_{xy}] \end{aligned} \tag{3.17}$$

From Eqs. (3.16) and (3.17),

$$\begin{aligned} &(q_{11} - q_{12})(\cos^2\theta P_{xx} + \sin^2\theta P_{yy}) + q_{12}(P_{xx} + P_{yy} + P_{zz}) + \sin 2\theta q_{44}P_{xy} \\ &\quad = q_{11}\cos^2\theta P_{xx} + q_{12}\sin^2\theta P_{xx} + q_{11}\sin^2\theta P_{yy} + q_{12}\cos^2\theta P_{yy} \\ &\qquad + \sin 2\theta P_{xy}(q_{11} - q_{12}) + q_{12}P_{zz} \end{aligned}$$

which on simplification yields

$$q_{44} = (q_{11} - q_{12})$$

3.3.2. Tensor Method

3.3.2.1. Introduction

The photoelastic phenomenon describes the effect of a mechanical stress on the optical index ellipsoid of the crystal. Mechanical stress is a second-rank symmetric tensor, having in general nine components (which can be put in terms of six components). The optical index ellipsoid can also be described by a second-rank symmetric tensor of nine optical parameters (which can also be put in terms of six components). Under the influence of a stress tensor, changes are brought about in the optical tensor. Pockels assumes that the changes in the optical-tensor components are linear functions of all the components of the stress tensor, as given in Eqs. (3.9d), which are equivalent to Eqs. (3.6b). In the earlier equations the stress components and the *B*'s are given in one-suffix notation; hence the coefficients q_{ij} (Pockels' stress-optic coefficients) are in two-suffix notation. This does not mean that q_{ij} are a second-rank tensor components.

The relationship between q_{ijkl} and q_{mn} can be deduced as shown below:

From Eqs. (3.6b) we have

$$B_{11} - B_{11}^0 = -[q_{1111}P_{11} + q_{1122}P_{22} + q_{1133}P_{33} + q_{1123}P_{23} + q_{1131}P_{31} + q_{1112}P_{12} + q_{1132}P_{32} + q_{1113}P_{13} + q_{1121}P_{21}] \quad (3.18a)$$

Also, from Eq. (3.9d) we have

$$B_{11} - B_{11}^0 = B_1 - B_1{}^0 = -[q_{11}P_1 + q_{12}P_2 + q_{13}P_3 + q_{14}P_4 + q_{15}P_5 + q_{16}P_6] \quad (3.18b)$$

But as shown on page 138, P_{ij} and P_k are related thus:

$$P_{11} = P_1, \quad P_{22} = P_2, \quad P_{33} = P_3$$
$$P_{23} = P_{32} = P_4, \quad P_{31} = P_{13} = P_5, \quad P_{12} = P_{21} = P_6$$

Therefore, comparing the coefficients of P_{ij} and P_k in Eqs. (3.18a) and (3.18b) we obtain

$$q_{1111} = q_{11}, \quad q_{1122} = q_{12}, \quad q_{1133} = q_{13}$$
$$(q_{1123} + q_{1132}) = q_{14}, \therefore q_{1123} = \tfrac{1}{2}q_{14}$$
$$(q_{1131} + q_{1113}) = q_{15}, \therefore q_{1131} = \tfrac{1}{2}q_{15}$$
$$(q_{1112} + q_{1121}) = q_{16}, \therefore q_{1112} = \tfrac{1}{2}q_{16}$$

Similarly, we observe that

$$B_{23} - B_{23}^0 = -[q_{2311}P_{11} + q_{2322}P_{22} + q_{2333}P_{33} + q_{2323}P_{23} + q_{2331}P_{31} + q_{2312}P_{12} + q_{2332}P_{32} + q_{2313}P_{13} + q_{2321}P_{21}] \quad (3.18c)$$

and

$$B_{23} - B_{23}^0 = B_4 - B_4^0 = -[q_{41}P_1 + q_{42}P_2 + q_{43}P_3 + q_{44}P_4 + q_{45}P_5 + q_{46}P_6] \quad (3.18d)$$

Therefore

$$q_{2311} = q_{41}, \quad q_{2322} = q_{42}, \quad q_{2333} = q_{43}$$
$$(q_{2323} + q_{2332}) = q_{44}, \therefore q_{2323} = \tfrac{1}{2}q_{44}$$
$$(q_{2331} + q_{2313}) = q_{45}, \therefore q_{2331} = \tfrac{1}{2}q_{45}$$
$$(q_{2312} + q_{2321}) = q_{46}, \therefore q_{2312} = \tfrac{1}{2}q_{46}$$

Thus, on examining all the q_{ijkl} and the corresponding q_{mn}, we note that

$$\begin{aligned} q_{ijkl} &= q_{mn}, \text{ where } m = 1, 2, \ldots, \text{ or } 6, \; n = 1, 2, \text{ or } 3 \\ 2q_{ijkl} &= q_{mn}, \text{ where } m = 1, 2, \ldots, \text{ or } 6, \; n = 4, 5, \text{ or } 6 \end{aligned} \quad (3.18e)$$

We can proceed on similar lines to relate p_{ijkl} with p_{mn}. Here, from Eq. (3.7) we find

$$B_{11} - B_{11}^0 = [p_{1111}\varepsilon_{11} + p_{1122}\varepsilon_{22} + p_{1133}\varepsilon_{33} + p_{1123}\varepsilon_{23} + p_{1131}\varepsilon_{31} + p_{1112}\varepsilon_{12} + p_{1132}\varepsilon_{32} + p_{1113}\varepsilon_{13} + p_{1121}\varepsilon_{21}] \quad (3.19a)$$

From Eqs. (3.9c) we find

$$B_{11} - B_{11}^0 = B_1 - B_1^0 = [p_{11}\varepsilon_1 + p_{12}\varepsilon_2 + p_{13}\varepsilon_3 + p_{14}\varepsilon_4 + p_{15}\varepsilon_5 + p_{16}\varepsilon_6] \quad (3.19b)$$

Now, as shown on pages 141–142, ε_{ij} and ε_k are related thus:

$$\varepsilon_{11} = \varepsilon_1, \quad \varepsilon_{22} = \varepsilon_2, \quad \varepsilon_{33} = \varepsilon_3$$
$$\varepsilon_{23} = \varepsilon_{32} = \tfrac{1}{2}\varepsilon_4, \quad \varepsilon_{31} = \varepsilon_{13} = \tfrac{1}{2}\varepsilon_5, \quad \varepsilon_{12} = \varepsilon_{21} = \tfrac{1}{2}\varepsilon_6$$

Therefore, comparing the coefficients of ε_{ij} and ε_k in Eqs. (3.19a) and (3.19b), we obtain

$$p_{1111} = p_{11}, \quad p_{1122} = p_{12}, \quad p_{1133} = p_{13}$$
$$(p_{1123} + p_{1132})\tfrac{1}{2}\varepsilon_4 = p_{14}\varepsilon_4, \therefore p_{1123} = p_{14}$$
$$(p_{1131} + p_{1113})\tfrac{1}{2}\varepsilon_5 = p_{15}\varepsilon_5, \therefore p_{1131} = p_{15}$$
$$(p_{1112} + p_{1121})\tfrac{1}{2}\varepsilon_6 = p_{16}\varepsilon_6, \therefore p_{1112} = p_{16}$$

Similarly, we note that

$$B_{23} - B_{23}^0 = [p_{2311}\varepsilon_{11} + p_{2322}\varepsilon_{22} + p_{2333}\varepsilon_{33} + p_{2323}\varepsilon_{23} + p_{2331}\varepsilon_{31} + p_{2312}\varepsilon_{12} + p_{2332}\varepsilon_{32} + p_{2313}\varepsilon_{13} + p_{2321}\varepsilon_{21}] \tag{3.19c}$$

and

$$B_{23} - B_{23}^0 = B_4 - B_4{}^0 = [p_{41}\varepsilon_1 + p_{42}\varepsilon_2 + p_{43}\varepsilon_3 + p_{44}\varepsilon_4 + p_{45}\varepsilon_5 + p_{46}\varepsilon_6] \tag{3.19d}$$

Therefore, comparing the coefficients of ε_{ij} and ε_k, we obtain

$$p_{2311} = p_{41}, \qquad p_{2322} = p_{42}, \qquad p_{2333} = p_{43}$$
$$(p_{2323} + p_{2332})\tfrac{1}{2}\varepsilon_4 = p_{44}\varepsilon_4, \;\therefore\; p_{2323} = p_{44}$$
$$(p_{2331} + p_{2313})\tfrac{1}{2}\varepsilon_5 = p_{45}\varepsilon_5, \;\therefore\; p_{2331} = p_{45}$$
$$(p_{2312} + p_{2321})\tfrac{1}{2}\varepsilon_6 = p_{46}\varepsilon_6, \;\therefore\; p_{2312} = p_{46}$$

Thus, on examining all the p_{ijkl} and the corresponding p_{mn}, we note that

$$p_{ijkl} = p_{mn} \text{ for } m = 1\text{–}6,\ n = 1\text{–}3$$

and

$$p_{ijkl} = p_{mn} \text{ for } m = 1\text{–}6,\ n = 4\text{–}6$$

Thus

$$p_{ijkl} = p_{mn} \qquad \text{for } m = 1\text{–}6 \text{ and } n = 1\text{–}6, \text{ and hence for all values of } m \text{ and } n. \tag{3.19e}$$

Thus it is important to note the difference between the matrices of q_{mn} and p_{mn}.

It can be shown that $p_{mn} = q_{mj}c_{jn}$ and $q_{mn} = p_{mj}s_{jn}$ [see Eq. (3.13)].

In the two-suffix notation, we are concerned with only 36 constants of q_{ij} or p_{ij} to study their survival and equivalence. But in the four-suffix notation we will have to deal with 81 constants of q_{ijkl}. Here the problem can be simplified by taking advantage of the fact that B_{ij} and P_{kl} are second-rank symmetric tensors, and hence we have the following general relationship:

$$(B_{ij} - B_{ij}^0) = -q_{ijkl}P_{kl} \qquad \text{and} \qquad (B_{ji} - B_{ji}^0) = -q_{jilk}P_{lk}$$
$$(B_{ij} - B_{ij}^0) = -q_{ijlk}P_{lk} \qquad \text{and} \qquad (B_{ji} - B_{ji}^0) = -q_{jikl}P_{kl}$$

Furthermore $(B_{ij} - B_{ij}^0) = (B_{ji} - B_{ji}^0)$ and $P_{kl} = P_{lk}$, which are components of second-rank symmetric tensors. Hence all the four equations are equivalent. Therefore

$$q_{ijkl} = q_{ijlk} = q_{jilk} = q_{jikl} \tag{3.20}$$

Table 3.4. Relations between q_{ijkl} and q_{mn}

$q_{1111} = q_{11}$	$q_{2311} = q_{3211} = q_{41}$
$q_{1122} = q_{12}$	$q_{2322} = q_{3222} = q_{42}$
$q_{1133} = q_{13}$	$q_{2333} = q_{3233} = q_{43}$
$q_{1123} = q_{1132} = \frac{1}{2}q_{14}$	$q_{2323} = q_{3223} = q_{2332} = q_{3232} = \frac{1}{2}q_{44}$
$q_{1131} = q_{1113} = \frac{1}{2}q_{15}$	$q_{2331} = q_{3231} = q_{2313} = q_{3213} = \frac{1}{2}q_{45}$
$q_{1112} = q_{1121} = \frac{1}{2}q_{16}$	$q_{2312} = q_{3212} = q_{2321} = q_{3221} = \frac{1}{2}q_{46}$
$q_{2211} = q_{21}$	$q_{3111} = q_{1311} = q_{51}$
$q_{2222} = q_{22}$	$q_{3122} = q_{1322} = q_{52}$
$q_{2233} = q_{23}$	$q_{3133} = q_{1333} = q_{53}$
$q_{2223} = q_{2232} = \frac{1}{2}q_{24}$	$q_{3123} = q_{1323} = q_{3132} = q_{1332} = \frac{1}{2}q_{54}$
$q_{2231} = q_{2213} = \frac{1}{2}q_{25}$	$q_{3131} = q_{1331} = q_{3113} = q_{1313} = \frac{1}{2}q_{55}$
$q_{2212} = q_{2221} = \frac{1}{2}q_{26}$	$q_{3112} = q_{1312} = q_{3121} = q_{1321} = \frac{1}{2}q_{56}$
$q_{3311} = q_{31}$	$q_{1211} = q_{2111} = q_{61}$
$q_{3322} = q_{32}$	$q_{1222} = q_{2122} = q_{62}$
$q_{3333} = q_{33}$	$q_{1233} = q_{2133} = q_{63}$
$q_{3323} = q_{3332} = \frac{1}{2}q_{34}$	$q_{1223} = q_{2123} = q_{1232} = q_{2132} = \frac{1}{2}q_{64}$
$q_{3331} = q_{3313} = \frac{1}{2}q_{35}$	$q_{1231} = q_{2131} = q_{1213} = q_{2113} = \frac{1}{2}q_{65}$
$q_{3312} = q_{3321} = \frac{1}{2}q_{36}$	$q_{1212} = q_{2112} = q_{1221} = q_{2121} = \frac{1}{2}q_{66}$

Using Eq. (3.20), the $81 q_{ijkl}$ constants can be put in the form of 36 sest each of which contains one, two, or four constants q_{ijkl}. Table 3.4 lists all 36 sets and their equivalents in the two-suffix notation.

The 81 stress-optic constants q_{ijkl}, which are components of a fourth-rank tensor, have thus been reduced to 36 independent components. Now, to determine the total number of surviving constants which describe the photoelastic behavior of any particular point group, we will have to study the influence of each of the generating elements of the point group on the 36 constants (which form the leading members of the 36 sets of constants given in Table 3.4).

Before applying the method to any particular point group, we shall first of all study the effect of some important elements of symmetry, viz. (a) axis, (b) plane, and (c) center of symmetry.

The tensor method of deducing whether a particular tensor component vanishes can be described in broad outline form as follows. Any symmetry which a crystalline material possesses will reduce the number of constants because of the inherent equivalence among combinations of *ijkl.* For example, an axis of two-fold symmetry will result in a certain $q_o = q_t$ or $q_o = -q_t$. Here q_o is the original constant and q_t is the transformed constant. By symmetry, since $q_o = q_t$, if as a result of symmetry operation (i.e., transformation of the axes) q_o is still equal to q_t, then q_o remains as a nonvanishing component. On the other hand, if the transformation brings about $q_o = -q_t$, then $q_t = q_o = -q_t$. So $q_t = 0$ and thus $q_o = 0$. Thus the influence of each of the generating elements of the crystal under examination must be checked and the resulting nonvanishing constants enumerated.

Equivalence among the 81 constants (which are combinations of *ijkl*, where *i*, *j*, *k*, and *l* are 1, 2, or 3) as a result of symmetry transformation can be evaluated by applying the tensor transformation rule thus:

$$q'_{ijkl} = \alpha_{im}\alpha_{jn}\alpha_{ko}\alpha_{lp}q_{mnop}$$

where α_{im}, α_{jn}, etc., are the respective direction cosines between the two sets of coordinate axes before and after symmetry transformation, respectively, in the case of Cartesian tensors.

3.3.2.2. *Influence of Some Symmetry Elements on the* q_{ijkl}

We now consider the effect of each symmetry element.

a. Axis of Symmetry. Let the axis chosen be a diad axis $\parallel OZ$. Then the coordinate system is rotated through 180° about this OZ axis (Fig. 3.6). Here $X' = -X$, $Y' = -Y$, and $Z' = Z$. The direction-cosine scheme is given by

	X	Y	Z
X'	-1	0	0
Y'	0	-1	0
Z'	0	0	$+1$

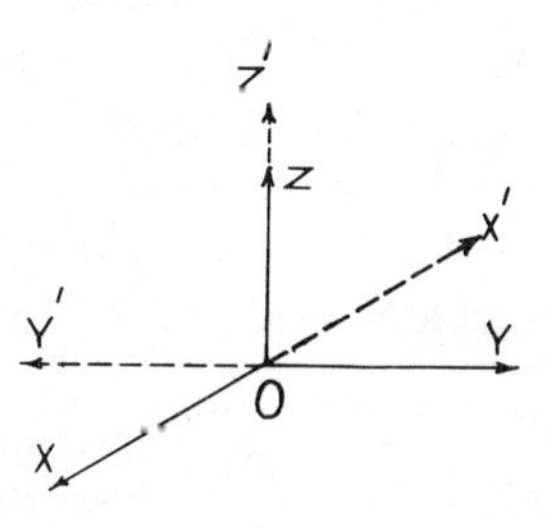

Fig. 3.6. Transformation of axes: OZ is a diad or a two-fold axis of rotation; here $OZ' = OZ$.

Now, to ascertain if a particular q_{ijkl}, say q_{1123}, vanishes, we express q'_{1123} in terms of q_{mnop} and expand the relation in full:

$$q'_{ijkl} = \alpha_{im}\alpha_{jn}\alpha_{ko}\alpha_{lp}q_{mnop}$$

where m, n, o, p take values 1, 2, 3. Substituting the values of α_{im}, etc., from the above direction-cosine matrix we obtain*

$$\begin{aligned}
q'_{1123} &= \alpha_{1m}\alpha_{1n}\alpha_{2o}\alpha_{3p}q_{mnop} \\
&= \alpha_{11}[\alpha_{1n}\alpha_{2o}\alpha_{3p}]q_{1nop} + \cancel{\alpha_{12}}[\alpha_{1n}\alpha_{2o}\alpha_{3p}]q_{2nop} + \cancel{\alpha_{13}}[\alpha_{1n}\alpha_{2o}\alpha_{3p}]q_{3nop} \\
&= \alpha_{11}[\alpha_{11}\{\alpha_{2o}\alpha_{3p}\}]q_{11op} + \alpha_{11}[\cancel{\alpha_{12}}\{\alpha_{2o}\alpha_{3p}\}]q_{12op} \\
&\quad + \alpha_{11}[\cancel{\alpha_{13}}\{\alpha_{2o}\alpha_{3p}\}]q_{13op} \\
&= \alpha_{11}[\alpha_{11}\{\cancel{\alpha_{21}}(\alpha_{3p})\}]q_{111p} + \alpha_{11}[\alpha_{11}\{\alpha_{22}(\alpha_{3p})\}]q_{112p} \\
&\quad + \alpha_{11}[\alpha_{11}\{\cancel{\alpha_{23}}(\alpha_{3p})\}]q_{113p} \\
&= \alpha_{11}[\alpha_{11}\{\alpha_{22}(\cancel{\alpha_{31}})\}]q_{1121} + \alpha_{11}[\alpha_{11}\{\alpha_{22}(\cancel{\alpha_{32}})\}]q_{1122} \\
&\quad + \alpha_{11}[\alpha_{11}\{\alpha_{22}(\alpha_{33})\}]q_{1123} \\
&= \alpha_{11}\alpha_{11}\alpha_{22}\alpha_{33}q_{1123} = (-1)(-1)(-1)(+1)q_{1123} = -q_{1123}
\end{aligned}$$

But, by symmetry operation, q'_{1123} must be the same as q_{1123}. Thus $q_{1123} = q'_{1123} = -q_{1123}$. So $q_{1123} = 0$ or $q_{14} = 0$, a result consistent with what we obtained earlier by the classical method (see page 70).

We are now in a position to draw the general conclusion for this symmetry element thus: All q_{ijkl} whose coefficients $\alpha_{im}\alpha_{jn}\alpha_{ko}\alpha_{lp}$ involve α_{11} and α_{22} an odd number of times, together or separately, vanish, because q_{ijkl} is equal to $-q_{ijkl}$; and if α_{11} and α_{22} are involved an even number of times, separately or together, then q_{ijkl} is retained.

It is now easy to see that the following $16 q_{ijkl}$ (and also their equivalents in Table 3.3) vanish:

q_{1123}	q_{1131}	q_{2223}	q_{2231}	q_{3323}	q_{3331}	q_{2311}	q_{2322}
q_{2333}	q_{2312}	q_{3111}	q_{3122}	q_{3133}	q_{3112}	q_{1223}	q_{1231}

The tediousness in the tensor method lies in the fact that each of the 36 q_{ijkl} (the maximum number for the most anisotropic crystal) must be examined individually.

b. Plane of Symmetry. Let the plane chosen be perpendicular to the principal axis OZ. Here $X' = X$, $Y' = Y$, and $Z' = -Z$ after transfor-

* The term $[\alpha_{1n}\alpha_{2o}\alpha_{3p}]$ is shown separately to indicate that it is α_{1m} that is expanded first, and similarly for other terms.

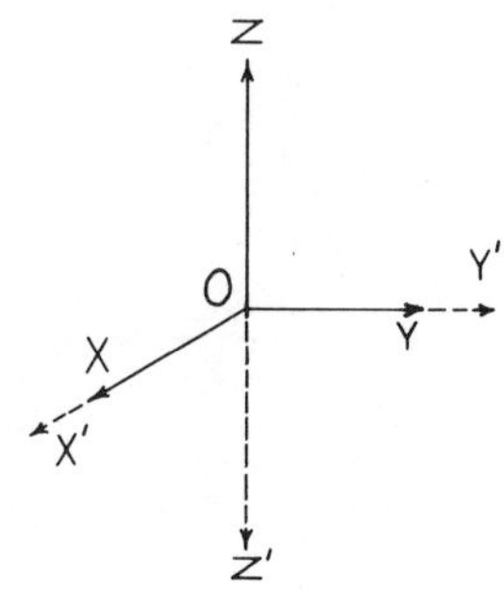

Fig. 3.7. Transformation of axes: the plane XOY is a reflection plane.

mation (Fig. 3.7). The direction-cosine scheme is

	X	Y	Z
X'	1	0	0
Y'	0	1	0
Z'	0	0	−1

We take the case of q_{1123}. Here

$$\begin{aligned} q'_{1123} &= \alpha_{1m}\alpha_{1n}\alpha_{2o}\alpha_{3p}q_{mnop} \\ &= \alpha_{11}[\alpha_{1n}\alpha_{2o}\alpha_{3p}]q_{1nop} + \cancel{\alpha_{12}}[\alpha_{1n}\alpha_{2o}\alpha_{3p}]q_{2nop} + \cancel{\alpha_{13}}[\alpha_{1n}\alpha_{2o}\alpha_{3p}]q_{3nop} \\ &= \alpha_{11}\alpha_{11}[\alpha_{2o}\alpha_{3p}]q_{11op} + \alpha_{11}\cancel{\alpha_{12}}[\alpha_{2o}\alpha_{3p}]q_{12op} + \alpha_{11}\cancel{\alpha_{13}}[\alpha_{2o}\alpha_{3p}]q_{13op} \\ &= \alpha_{11}\alpha_{11}\cancel{\alpha_{21}}[\alpha_{3p}]q_{111p} + \alpha_{11}\alpha_{11}\alpha_{22}[\alpha_{3p}]q_{112p} + \alpha_{11}\alpha_{11}\cancel{\alpha_{23}}[\alpha_{3p}]q_{113p} \\ &= \alpha_{11}\alpha_{11}\alpha_{22}\cancel{\alpha_{31}}q_{1121} + \alpha_{11}\alpha_{11}\cancel{\alpha_{22}}\alpha_{32}q_{1122} + \alpha_{11}\alpha_{11}\alpha_{22}\alpha_{33}q_{1123} \\ &= \alpha_{11}\alpha_{11}\alpha_{22}\alpha_{33}q_{1123} = (1)(1)(1)(-1)q_{1123} = -q_{1123} \end{aligned}$$

i.e., $q_{1123} = q'_{1123} = -q_{1123}$. So q_{1123} vanishes. Here again we can summarize thus: All q_{ijkl} which involve α_{33} an odd number of times vanish for this symmetry element. Thus the 16 vanishing q_{ijkl} are

$$\begin{matrix} q_{1123} & q_{1131} & q_{2223} & q_{2231} & q_{3323} & q_{3331} & q_{2311} & q_{2322} \\ q_{2333} & q_{2312} & q_{3111} & q_{3122} & q_{3133} & q_{3112} & q_{1223} & q_{1231} \end{matrix}$$

c. Center of Symmetry. Let O be the center of symmetry (Fig. 3.8). The direction-cosine scheme is

	X	Y	Z
X'	−1	0	0
Y'	0	−1	0
Z'	0	0	−1

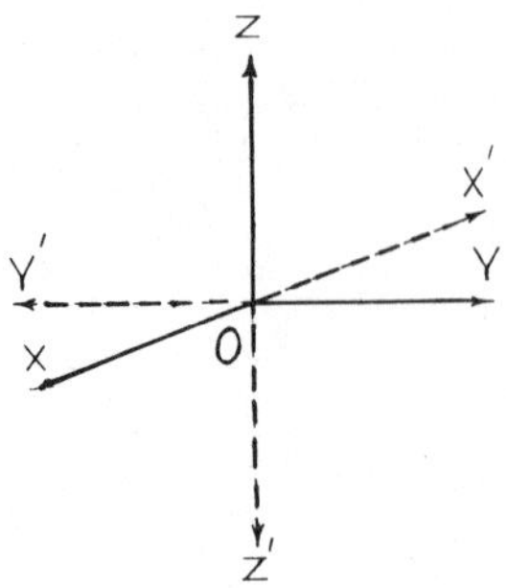

Fig. 3.8. Transformation of axes: the origin O is a center of symmetry or point of inversion.

We now examine q_{2331} (chosen arbitrarily). Then

$$\begin{aligned} q'_{2331} &= \alpha_{2m}\alpha_{3n}\alpha_{3o}\alpha_{1p}q_{mnop} \\ &= \cancel{\alpha_{21}}[\alpha_{3n}\alpha_{3o}\alpha_{1p}]q_{1nop} + \alpha_{22}[\alpha_{3n}\alpha_{3o}\alpha_{1p}]q_{2nop} + \cancel{\alpha_{23}}[\alpha_{3n}\alpha_{3o}\alpha_{1p}]q_{3nop} \\ &= \alpha_{22}\cancel{\alpha_{31}}[\alpha_{3o}\alpha_{1p}]q_{21op} + \alpha_{22}\cancel{\alpha_{32}}[\alpha_{3o}\alpha_{1p}]q_{22op} + \alpha_{22}\alpha_{33}[\alpha_{3o}\alpha_{1p}]q_{23op} \\ &= \alpha_{22}\alpha_{33}\cancel{\alpha_{31}}[\alpha_{1p}]q_{231p} + \alpha_{22}\alpha_{33}\cancel{\alpha_{32}}[\alpha_{1p}]q_{232p} + \alpha_{22}\alpha_{33}\alpha_{33}[\alpha_{1p}]q_{233p} \\ &= \alpha_{22}\alpha_{33}\alpha_{33}\alpha_{11}q_{2331} + \alpha_{22}\alpha_{33}\alpha_{33}\cancel{\alpha_{12}}q_{2332} + \alpha_{22}\alpha_{33}\alpha_{33}\cancel{\alpha_{13}}q_{2333} \\ &= (-1)(-1)(-1)(-1)q_{2331} = q_{2331} \end{aligned}$$

Hence q_{2331} survives.

Alternatively, a very simple and compact way of deducing the above result can be expressed thus: To examine the tensor component q_{ijkl}, we take q'_{ijkl} and express it in terms of q_{mnop}. Thus

$$q'_{ijkl} = \alpha_{im}\alpha_{jn}\alpha_{ko}\alpha_{lp}q_{mnop}$$

From the direction-cosine scheme for the center of symmetry, we note that unless $i = m$, $j = n$, $k = o$, and $l = p$, the α's are zero. This imposes a condition for the suffixes m, n, o, p. Thus m, n, o, and p are respectively equal to i, j, k, and l. Consequently $q_{mnop} = q_{ijkl}$. Thus $q'_{ijkl} = q_{ijkl}$ by transformation of the tensor components, and by symmetry, of course, $q'_{ijkl} = q_{ijkl}$; hence q_{ijkl} exists. Since no sequence of the suffixes i, j, k, l is imposed, all the 36 q_{ijkl} can exist.

By either of the arguments, we find that the center of symmetry does not reduce the number of constants.

In conclusion, we note that photoelasticity is a centrosymmetric property (like elasticity, thermal expansion, etc.) and does not distinguish between crystals with or without center. This means that even in those classes of crystals where a polar axis is present the coefficients of this physical property along the polar axis are the same in either direction. Piezoelectricity, for example, is a noncentrosymmetric property.

Above we have seen the influence of a diad, a plane of reflection, and a center of inversion on the stress-optical tensor coefficients q_{ijkl} of a crystal in general. Proceeding along similar lines, we can deduce the non-vanishing q_{ijkl} of every one of the 32 point groups starting with the class 1 (C_1) and introducing symmetry elements one after another until we reach the class $m3m$ (O_h) of the cubic system.

3.3.2.3. *Derivation of the Surviving q_{ijkl} for Some Crystal Classes*

We shall next examine in detail the surviving q_{ijkl} for some of the 32 crystallographic point groups.

Triclinic System

1. Class 1 (C_1): Symmetry Element E; Generating Element E. Identity or rotation by 360° is the only element of symmetry possessed by this class. The two coordinate systems, before and after the symmetry operation, are shown in Fig. 3.9, and their direction cosines are given by

	X	Y	Z
X'	1	0	0
Y'	0	1	0
Z'	0	0	1

According to the law of transformation of a fourth-rank tensor component

$$q'_{ijkl} = \alpha_{im}\alpha_{jn}\alpha_{ko}\alpha_{lp}q_{mnop}$$

From the direction-cosine scheme given in Fig. 3.9, we observe that when

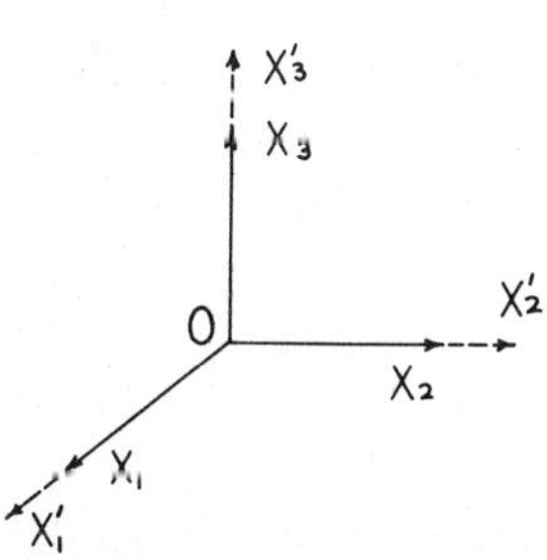

Fig. 3.9. Transformation of axes due to the symmetry element identity E or a rotation by 360° about any axis.

$i = m$, $j = n$, $k = o$, and $l = p$, in each case $\alpha = 1$, and all other α's are equal to zero. Therefore

$$q'_{ijkl} = \alpha_{ii}\alpha_{jj}\alpha_{kk}\alpha_{ll}q_{ijkl} = q_{ijkl}$$

From Neumann's principle too, $q'_{ijkl} = q_{ijkl}$, since the change in coordinate system resulting from any symmetry operation must leave the physical property unaltered. Hence the q_{ijkl} survive. Furthermore, since we have not imposed any restriction on the sequence of i, j, k, and l, we conclude that all the 36 q_{ijkl} exist for the crystals belonging to this class.

2. Class $\bar{1}$ (C_i); Symmetry Elements E i; Generating Element i. This class has an inversion center in addition to E of class 1 (C_1). But we have already discussed the influence of i and concluded that the presence of i does not cause any change or reduction in the q_{ijkl} matrix. Hence for this class also all the 36 independent q_{ijkl} survive.

Monoclinic System

3. Class 2 (C_2); Symmetry Elements E C_2; Generating Element C_2. Here if OZ is the principal diad axis, we have seen on page 82 that the following 16 q_{ijkl} (and also their equivalents) vanish:

$$\begin{array}{cccccccc} q_{1123} & q_{1131} & q_{2223} & q_{2231} & q_{3323} & q_{3331} & q_{2311} & q_{2322} \\ q_{2333} & q_{2312} & q_{3111} & q_{3122} & q_{3133} & q_{3112} & q_{1223} & q_{1231} \end{array}$$

If, according to the standard orientation, the OY is taken as the principal diad axis, it can easily be seen that the following 16 q_{ijkl} (and also their equivalents) vanish:

$$\begin{array}{cccccccc} q_{1123} & q_{2223} & q_{3323} & q_{3123} & q_{1112} & q_{2212} & q_{3312} & q_{3112} \\ q_{2311} & q_{2322} & q_{2333} & q_{2331} & q_{1211} & q_{1222} & q_{1233} & q_{1231} \end{array}$$

4. Class m (C_{1h}); Symmetry Elements E σ_h. If OZ is taken as the principal diad axis, the plane of reflection is the XOY plane. This case has already been discussed on page 83, and it has been shown that the 16 q_{ijkl} that vanish are

$$\begin{array}{cccccccc} q_{1123} & q_{1131} & q_{2223} & q_{2231} & q_{3323} & q_{3331} & q_{2311} & q_{2322} \\ q_{2333} & q_{2312} & q_{3111} & q_{3122} & q_{3133} & q_{3112} & q_{1223} & q_{1231} \end{array}$$

However, if OY is taken as the diad axis according to the standard orientation and the reflection plane is perpendicular to OY, the 16 q_{ijkl} that vanish are

$$q_{1123} \quad q_{2223} \quad q_{3323} \quad q_{3123} \quad q_{1112} \quad q_{2212} \quad q_{3312} \quad q_{3112}$$
$$q_{2311} \quad q_{2322} \quad q_{2333} \quad q_{2331} \quad q_{1211} \quad q_{1222} \quad q_{1233} \quad q_{1231}$$

5. Class $2/m$ (C_{2h}); Symmetry Elements $E\ C_2\ \sigma_h\ i$; Generating Elements $C_2\ \sigma_h$. With OY as the diad axis, as in the standard orientation, the following 16 q_{ijkl} vanish:

$$q_{1123} \quad q_{2223} \quad q_{3323} \quad q_{3123} \quad q_{1112} \quad q_{2212} \quad q_{3312} \quad q_{3112}$$
$$q_{2311} \quad q_{2322} \quad q_{2333} \quad q_{2331} \quad q_{1211} \quad q_{1222} \quad q_{1233} \quad q_{1231}$$

With a reflection plane perpendicular to OY, we have seen in the earlier case of class m (C_{1h}) that the same constants vanish. Hence there is no reduction or change in the number or nature of the nonvanishing tensor components.

Orthorhombic System

6. Class 222 (D_2); Symmetry Elements $E\ C_2\ C_2'\ C_2''$; Generating Elements $C_2\ C_2'$. Here three mutually perpendicular diad axes exist; they are respectively parallel to the OZ, OX, and OY crystallographic axes. While discussing the case of class 2 (C_2) we have already seen that with OZ as the diad axis 16 q_{ijkl} vanish, and with OY as the diad axis 16 q_{ijkl} vanish. However, by comparing the two sets of the vanishing q_{ijkl}, it will be observed that 8 q_{ijkl} of each set are common; thus with OZ and OY as the diad axes simultaneously 24 q_{ijkl} vanish, and the 12 surviving q_{ijkl} are

$$q_{1111} \quad q_{1122} \quad q_{1133} \quad q_{2211} \quad q_{2222} \quad q_{2233}$$
$$q_{3311} \quad q_{3322} \quad q_{3333} \quad q_{2323} \quad q_{3131} \quad q_{1212}$$

The superposition of OX as the diad axis on the above two diads will not alter the 12 surviving q_{ijkl}. This is consistent with the fact that, in deducing the nonvanishing independent components of any tensor property of a crystal, it will be quite sufficient if we study the influence of only the generating elements of the crystal, and for the class 222 (D_2), any two of the three mutually perpendicular diads form the generating elements. The other two classes $mm2$ (C_{2v}) and mmm (D_{2h}) can also be shown to have the same 12 q_{ijkl} as the surviving tensor components.

Trigonal System

7. Class 3 (C_3); Symmetry Elements $E\ 2C_3$; Generating Element C_3. Here the deduction is more involved, although the method of approach is the same. Hence we shall examine this case in detail. It is also interesting because it has the lowest symmetry of the 10 crystal classes for which Bhagavantam's scheme is at variance with that of Pockels.

The orientation of the new and old coordinate systems is shown in Fig. 3.10, and the matrix of their direction cosines is given by

$$\begin{bmatrix} -1/2 & (3)^{1/2}/2 & 0 \\ -(3)^{1/2}/2 & -1/2 & 0 \\ 0 & 0 & 1 \end{bmatrix}$$

Now let us consider q_{1111}. As shown earlier,

$$\begin{aligned} q'_{1111} &= \alpha_{1m}\alpha_{1n}\alpha_{1o}\alpha_{1p}q_{mnop} \\ &= \frac{1}{16}q_{1111} - \frac{2(3)^{1/2}}{16}q_{1112} + \frac{3}{16}q_{1122} - \frac{2(3)^{1/2}}{16}q_{1211} + \frac{12}{16}q_{1221} \\ &\quad - \frac{6(3)^{1/2}}{16}q_{1222} + \frac{3}{16}q_{2211} - \frac{6(3)^{1/2}}{16}q_{2212} + \frac{9}{16}q_{2222} \end{aligned}$$

(where m, n, o, p take values 1, 2, 3). But q'_{1111} should be equal to q_{1111} due to a symmetry operation. Hence, we have, by transposition,

$$\begin{aligned} 15q_{1111} &+ 2(3)^{1/2}q_{1112} - 3q_{1122} + 2(3)^{1/2}q_{1211} - 12q_{1221} + 6(3)^{1/2}q_{1222} \\ &- 3q_{2211} + 6(3)^{1/2}q_{2212} - 9q_{2222} = 0 \end{aligned} \tag{3.21}$$

Thus we see that q'_{1111} is related not only to q_{1111} but also to eight more q_{ijkl}. Examining each of them one after another and putting together all

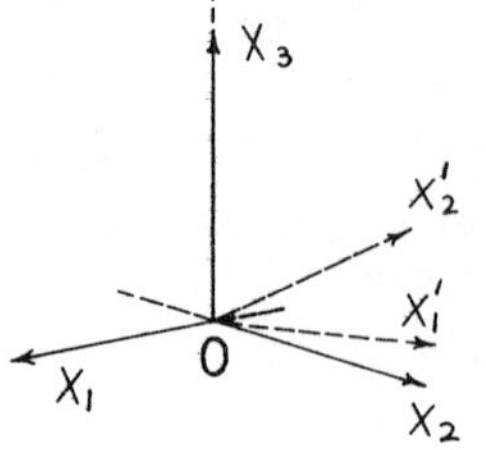

Fig. 3.10. Transformation of axes: OX_3 is a triad or a three-fold axis of rotation; here $OX_3' = OX_3$.

the nine equations, we have

$$15q_{1111} - 3q_{1122} + 2(3)^{1/2}q_{1112} - 3q_{2211} - 9q_{2222} + 6(3)^{1/2}q_{2212} + 2(3)^{1/2}q_{1211} + 6(3)^{1/2}q_{1222} - 12q_{1221} = 0 \tag{3.21}$$

$$-3q_{1111} + 15q_{1122} - 2(3)^{1/2}q_{1112} - 9q_{2211} - 3q_{2222} - 6(3)^{1/2}q_{2212} + 6(3)^{1/2}q_{1211} + 2(3)^{1/2}q_{1222} + 12q_{1221} = 0 \tag{3.22}$$

$$(3)^{1/2}q_{1111} - (3)^{1/2}q_{1122} - 18q_{1112} + 3(3)^{1/2}q_{2211} - 3(3)^{1/2}q_{2222} - 6q_{2221} - 6q_{1211} + 6q_{1222} + 4(3)^{1/2}q_{1212} = 0 \tag{3.23}$$

$$3q_{1111} + 9q_{1122} - 6(3)^{1/2}q_{1112} - 15q_{2211} + 3q_{2222} - 2(3)^{1/2}q_{2212} + 2(3)^{1/2}q_{1211} + 6(3)^{1/2}q_{1222} - 12q_{1212} = 0 \tag{3.24}$$

$$-9q_{1111} - 3q_{1122} - 6(3)^{1/2}q_{1112} - 3q_{2211} + 15q_{2222} - 2(3)^{1/2}q_{2212} - 6(3)^{1/2}q_{1211} - 2(3)^{1/2}q_{2122} - 12q_{1212} = 0 \tag{3.25}$$

$$3(3)^{1/2}q_{1111} - 3(3)^{1/2}q_{1122} - 6q_{1112} + (3)^{1/2}q_{2211} - (3)^{1/2}q_{2222} - 18q_{2221} + 6q_{1211} - 6q_{1222} - 4(3)^{1/2}q_{1212} = 0 \tag{3.26}$$

$$(3)^{1/2}q_{1111} + 3(3)^{1/2}q_{1122} - 6q_{1112} - (3)^{1/2}q_{2211} - 3(3)^{1/2}q_{2222} + 6q_{2221} - 18q_{1211} - 6q_{1222} + 4(3)^{1/2}q_{1212} = 0 \tag{3.27}$$

$$3(3)^{1/2}q_{1111} + (3)^{1/2}q_{1122} + 6q_{1112} - 3(3)^{1/2}q_{2211} - (3)^{1/2}q_{2222} - 6q_{2221} - 6q_{1211} - 18q_{1222} - 4(3)^{1/2}q_{1212} = 0 \tag{3.28}$$

$$3q_{1111} - 3q_{1122} - 2(3)^{1/2}q_{1112} - 3q_{2211} + 3q_{2222} + 2(3)^{1/2}q_{2212} - 2(3)^{1/2}q_{1211} + 2(3)^{1/2}q_{1222} - 12q_{1212} = 0 \tag{3.29}$$

Adding Eqs. (3.21) and (3.22), we have

$$12(q_{1111} - q_{2222}) + 12(q_{1122} - q_{2211}) + 8(3)^{1/2}(q_{1211} + q_{1222}) = 0 \tag{3.30}$$

Subtracting Eq. (3.23) from (3.27), we have

$$12(q_{2221} + q_{1112}) + 4(3)^{1/2}(q_{1122} - q_{2211}) - 12(q_{1211} + q_{1222}) = 0 \tag{3.31}$$

Subtracting Eq. (3.24) from Eq. (3.21), we have

$$12(q_{1111} - q_{2222}) + 8(3)^{1/2}(q_{1112} + q_{2212}) - 12(q_{1122} - q_{2211}) = 0 \tag{3.32}$$

Adding Eqs. (3.27) and (3.28), we have

$$4(3)^{1/2}(q_{1111} - q_{2222}) + 4(3)^{1/2}(q_{1122} - q_{2211}) - 24(q_{1211} + q_{1222}) = 0 \tag{3.33}$$

From Eqs. (3.30)–(3.33) we obtain

$$(q_{1111} - q_{2222}) = 0, \qquad (q_{1122} - q_{2211}) = 0$$
$$(q_{1112} + q_{2212}) = 0, \qquad (q_{1211} + q_{1222}) = 0 \tag{3.34}$$

$$\therefore q_{1111} = q_{2222}, \quad q_{1122} = q_{2211}, \quad q_{1112} = -q_{2212}, \quad q_{1211} = -q_{1222} \tag{3.35}$$

Substituting the q_{ijkl} equalities from Eq. (3.35) in Eq. (3.28) and Eq. (3.29), respectively, we have

$$\{2(3)^{1/2}(q_{1111} - q_{1122}) - 4(3)^{1/2}q_{1212}\} + 12(q_{1112} - q_{1222}) = 0 \tag{3.36}$$

$$\{6(q_{1111} - q_{1122}) - 12q_{1212}\} - 4(3)^{1/2}(q_{1112} - q_{1222}) = 0 \tag{3.37}$$

From Eqs. (3.36) and (3.37) we obtain

$$q_{1111} - q_{1122} = 2q_{1212}, \qquad q_{1112} = q_{1222} \tag{3.38}$$

We have thus established interrelationships among the nine q_{ijkl} appearing in Eqs. (3.21) to (3.29), and, furthermore, we note that all the nine q_{ijkl} survive.

Let us start with q'_{1123}. Then we obtain

$$q'_{1123} = -\frac{(3)^{1/2}}{8}q_{1113} - \frac{1}{8}q_{1123} + \frac{3}{8}q_{1213} + \frac{(3)^{1/2}}{8}q_{1223} + \frac{3}{8}q_{2113} + \frac{(3)^{1/2}}{8}q_{2123} - \frac{3(3)^{1/2}}{8}q_{2213} - \frac{3}{8}q_{2223}$$

$$\therefore -9q_{1123} - (3)^{1/2}q_{1113} - 3q_{2223} - 3(3)^{1/2}q_{2213} + 2(3)^{1/2}q_{1223} + 6q_{1213} = 0 \tag{3.39}$$

Thus q_{1123} involves five more q_{ijkl} in Eq. (3.39). We list below all the six simultaneous equations involving q_{1123} and the other five q_{ijkl}:

$$-9q_{1123} - (3)^{1/2}q_{1113} - 3q_{2223} - 3(3)^{1/2}q_{2213} + 2(3)^{1/2}q_{1223} + 6q_{1213} = 0 \tag{3.40}$$

$$(3)^{1/2}q_{1123} - 9q_{1113} + 3(3)^{1/2}q_{2223} - 3q_{2213} - 6q_{1223} + 2(3)^{1/2}q_{1213} = 0 \tag{3.41}$$

$$-3q_{1123} - 3(3)^{1/2}q_{1113} - 9q_{2223} - (3)^{1/2}q_{2213} - 2(3)^{1/2}q_{1223} - 6q_{1213} = 0 \tag{3.42}$$

$$3(3)^{1/2}q_{1123} - 3q_{1113} + (3)^{1/2}q_{2223} - 9q_{2213} + 6q_{1223} - 2(3)^{1/2}q_{1213} = 0 \tag{3.43}$$

$$-(3)^{1/2}q_{1123} - 3q_{1113} + (3)^{1/2}q_{2223} + 3q_{2213} - 6q_{1223} + 2(3)^{1/2}q_{1213} = 0 \tag{3.44}$$

$$3q_{1123} - (3)^{1/2}q_{1113} - 3q_{2223} + (3)^{1/2}q_{2213} - 2(3)^{1/2}q_{1223} - 6q_{1213} = 0 \tag{3.45}$$

Adding Eqs. (3.41) and (3.44), we have

$$-12(q_{1113} + q_{1223}) + 4(3)^{1/2}(q_{2223} + q_{1213}) = 0 \tag{3.46}$$

Adding Eqs. (3.42) and (3.45), we have

$$4(3)^{1/2}(q_{1113} + q_{1223}) + 12(q_{2223} + q_{1213}) = 0 \tag{3.47}$$

Equations (3.46) and (3.47) yield

$$q_{1113} = -q_{1223}, \qquad q_{2223} = -q_{1213} \tag{3.48}$$

Substituting the q_{ijkl} equalities from Eq. (3.48) in Eq. (3.47) and Eq. (3.45), respectively, we find

$$-(3)^{1/2}(q_{1123} + q_{2223}) + 3(q_{1113} + q_{2213}) = 0 \tag{3.49}$$

$$3(q_{1123} + q_{2223}) + (3)^{1/2}(q_{1113} + q_{2213}) = 0 \tag{3.50}$$

From Eqs. (3.49) and (3.50) we get

$$q_{1113} = -q_{2213}, \qquad q_{1123} = -q_{2223} \tag{3.51}$$

Putting Eqs. (3.48) and (3.51) together, we obtain

$$q_{1113} = -q_{1223} = -q_{2213}, \qquad q_{2223} = -q_{1213} = -q_{1123} \tag{3.52}$$

Next, starting with q'_{1133}, we obtain the following three equations:

$$-\frac{3}{4}q_{1133} + \frac{3}{4}q_{2233} - \frac{2(3)^{1/2}}{4}q_{1233} = 0 \tag{3.53}$$

$$\frac{3}{4}q_{1133} - \frac{3}{4}q_{2233} + \frac{2(3)^{1/2}}{4}q_{1233} = 0 \tag{3.54}$$

$$\frac{(3)^{1/2}}{4}q_{1133} - \frac{(3)^{1/2}}{4}q_{2233} - \frac{6}{4}q_{1233} = 0 \tag{3.55}$$

Equations (3.53) and (3.54) are equivalent. However, combining either of them with Eq. (3.55), we get

$$q_{1133} = q_{2233} \qquad \text{and} \qquad q_{1233} = 0 \tag{3.56}$$

$$q'_{3333} = \alpha_{3m}\alpha_{3n}\alpha_{3o}\alpha_{3p}q_{mnop} = \alpha_{33}^4 q_{3333} \tag{3.57}$$

Therefore $q_{3333} = q_{3333}$, and it survives.

Starting with q'_{3211}, we get the following six equations:

$$-(3)^{1/2}q_{3111}+6q_{3112}-3(3)^{1/2}q_{3122}-9q_{3211}+2(3)^{1/2}q_{3212}-3q_{3222}=0 \quad (3.58)$$

$$-9q_{3111}+2(3)^{1/2}q_{3112}-3q_{3122}+(3)^{1/2}q_{3211}-6q_{3212}+3(3)^{1/2}q_{3222}=0 \quad (3.59)$$

$$-3(3)^{1/2}q_{3111}-6q_{3112}-(3)^{1/2}q_{3122}-3q_{3211}-2(3)^{1/2}q_{3212}-9q_{3222}=0 \quad (3.60)$$

$$-3q_{3111}-2(3)^{1/2}q_{3112}-9q_{3122}+3(3)^{1/2}q_{3211}+6q_{3212}+(3)^{1/2}q_{3222}=0 \quad (3.61)$$

$$-3q_{3111}+2(3)^{1/2}q_{3112}+3q_{3122}-(3)^{1/2}q_{3211}-6q_{3212}+(3)^{1/2}q_{3222}=0 \quad (3.62)$$

$$-(3)^{1/2}q_{3111}-6q_{3121}+(3)^{1/2}q_{3122}+3q_{3211}-2(3)^{1/2}q_{3212}-3q_{3222}=0 \quad (3.63)$$

Adding Eqs. (3.59) and (3.62), we get

$$-12(q_{3111}+q_{3221})+4(3)^{1/2}(q_{3121}+q_{3222})=0 \quad (3.64)$$

Adding Eqs. (3.60) and (3.63), we obtain

$$4(3)^{1/2}(q_{3111}+q_{3221})+12(q_{3121}+q_{3222})=0 \quad (3.65)$$

Equations (3.64) and (3.65) together yield

$$q_{3111}=-q_{3221} \quad \text{and} \quad q_{3121}=-q_{3222} \quad (3.66)$$

Substituting the equalities from Eq. (3.66) in Eqs. (3.62) and (3.63), we obtain

$$3(q_{3111}+q_{3122})-(3)^{1/2}(q_{3211}-q_{3121})=0 \quad (3.67)$$

$$(3)^{1/2}(q_{3111}+q_{3122})+3(q_{3211}-q_{3121})=0 \quad (3.68)$$

Equations (3.67) and (3.68) together yield

$$q_{3111}=-q_{3122} \quad \text{and} \quad q_{3211}=q_{3121} \quad (3.69)$$

Summing up the results of Eqs. (3.66) and (3.69), we have

$$q_{3111}=-q_{3221}=-q_{3122}, \quad q_{3121}=-q_{3222}=q_{3211} \quad (3.70)$$

Now, starting with q'_{3311}, we obtain the following three equations:

$$-3(q_{3311}-q_{3322})-2(3)^{1/2}q_{3312}=0 \quad (3.71)$$

$$3(q_{3311}-q_{3322})+2(3)^{1/2}q_{3312}=0 \quad (3.72)$$

$$(3)^{1/2}(q_{3311}-q_{3322})-6q_{3312}=0 \quad (3.73)$$

Equations (3.71) and (3.72) are not independent. However, by combining either of them with Eq. (3.73), we get

$$q_{3311} = q_{3322} \quad \text{and} \quad q_{3321} = 0 \tag{3.74}$$

Next, starting with q'_{3332}, we have

$$(3)^{1/2}q_{3331} + 3q_{3332} = 0 \tag{3.75}$$

$$3q_{3331} - (3)^{1/2}q_{3332} = 0 \tag{3.76}$$

From Eqs. (3.75) and (3.76), we get

$$q_{3331} = 0 \quad \text{and} \quad q_{3332} = 0 \tag{3.77}$$

Starting with q'_{2333} we find

$$-3^{1/2}q_{1333} - 3q_{2333} = 0 \tag{3.78}$$

$$-3(3)^{1/2}q_{1333} + 3q_{2333} = 0 \tag{3.79}$$

Equations (3.78) and (3.79) together yield

$$q_{1333} = 0 \quad \text{and} \quad q_{2333} = 0 \tag{3.80}$$

Finally let us consider q_{2323}. Here we obtain only two independent equations, and they are

$$3(q_{1313} - q_{2323}) + (3)^{1/2}(q_{1323} + q_{2313}) = 0 \tag{3.81}$$

$$(3)^{1/2}(q_{1313} - q_{2323}) - 3(q_{1323} + q_{2313}) = 0 \tag{3.82}$$

By combining Eqs. (3.81) and (3.82), we get

$$q_{1313} = q_{2323} \quad \text{and} \quad q_{1323} = -q_{2313} \tag{3.83}$$

In conclusion, we have thus established that six q_{ijkl} vanish and some of the remaining 30 q_{ijkl} exist either independently or are related to some others among themselves. A comparison of the vanishing and nonvanishing q_{ijkl} with their equivalent q_{mn} in the two-suffix notation listed on page 80 gives the q_{mn} matrix for class 3 (C_3) given on page 60.

Starting with the orthorhombic class 222 (D_2) and the trigonal class 3 (C_3), it is easy to derive the q_{ijkl} matrices for all the classes of higher symmetry. We shall, however, consider below only three classes, namely, class 32 (D_3), class 6 (C_6), and class 23 (T); all the three classes are characterized by each having as the generating elements a three-fold axis and a two-fold axis (but not with the same orientation).

8. Class 32 (D_3); Symmetry Elements $E\ 2C_3\ 3C_2'$; Generating Elements $C_3\ C_2'$. We arrive at the nonvanishing q_{ijkl} for this class by considering the influence of a diad axis parallel to OX on the nonvanishing q_{ijkl} for class 3 (C_3) because C_3 and C_2' form the generating elements of this class. The transformation matrix for a diad parallel to OX is given by

$$\begin{bmatrix} 1 & 0 & 0 \\ 0 & -1 & 0 \\ 0 & 0 & -1 \end{bmatrix}$$

Following the argument given in the case of a diad axis parallel to OZ (see page 81), all those q_{ijkl} vanish for which α_{11} appears once or thrice in the transformation equation $q'_{ijkl} = \alpha_{im}\alpha_{jn}\alpha_{ko}\alpha_{lp}q_{mnop}$. Accordingly, from among the nonvanishing q_{ijkl} for class 3 (C_3), the following should vanish for class 32 (D_3):

$$\begin{matrix} q_{1131} & q_{1112} & q_{2231} & q_{2212} & q_{2331} & q_{2312} \\ q_{3111} & q_{3122} & q_{3123} & q_{1211} & q_{1222} & q_{1223} \end{matrix}$$

and their equivalents. Then the resulting q_{mn} matrix represents that for class 32 (D_3) (given on page 60).

Hexagonal System

9. Class 6 (C_6); Symmetry Elements $E\,2C_6\,2C_3\,C_2$; Generating Elements $C_3\ C_2$. The nonvanishing q_{ijkl} for this class can be obtained by considering the effect of a diad axis parallel to OZ on the nonvanishing q_{ijkl} for class 3 (C_3), because these two axes C_3 and C_2 form the generating elements of this class. The transformation matrix for a diad parallel to OZ or OX_3 is given by

$$\begin{bmatrix} -1 & 0 & 0 \\ 0 & -1 & 0 \\ 0 & 0 & 1 \end{bmatrix}$$

Following the procedure given earlier, we observe that all those q_{ijkl} vanish for which α_{33} appears once or thrice in the transformation equation:

$$q'_{ijkl} = \alpha_{im}\alpha_{jn}\alpha_{ko}\alpha_{lp}q_{mnop}$$

Therefore, from among the nonvanishing q_{ijkl} for class 3 (C_3), the following should vanish for this class:

$$q_{1123} \quad q_{1131} \quad q_{2223} \quad q_{2231} \quad q_{2311} \quad q_{2322}$$
$$q_{2312} \quad q_{3111} \quad q_{3122} \quad q_{3112} \quad q_{1223} \quad q_{1231}$$

and their equivalents. Then the resulting q_{mn} matrix represents that for class 6 (C_6) given on page 62).

Cubic System

10. Class 23 (T); Symmetry Elements $E\,3C_2\,8C_3'$; Generating Elements $C_2\,C_3'$. Here we can obtain the nonvanishing q_{ijkl} from those for class 222 (D_2) by applying an additional symmetry element of a triad axis equally inclined to the crystallographic axes of the orthorhombic system, class 222 (D_2). The transformation matrix for this triad axis which makes $X \rightarrow Y$, $Y \rightarrow Z$, and $Z \rightarrow X$ is given by

$$\begin{bmatrix} 0 & 1 & 0 \\ 0 & 0 & 1 \\ 1 & 0 & 0 \end{bmatrix}$$

Now we shall examine some of the 12 nonvanishing independent q_{ijkl} for class 222 (D_2):

$$q'_{1111} = \alpha_{12}\alpha_{12}\alpha_{12}\alpha_{12}q_{2222} = q_{2222} = q_{1111} \qquad \text{(by Neumann's principle)}$$
$$q'_{2222} = \alpha_{23}\alpha_{23}\alpha_{23}\alpha_{23}q_{3333} = q_{3333} = q_{1111} = q_{2222}$$
$$q'_{1122} = \alpha_{12}\alpha_{12}\alpha_{23}\alpha_{23}q_{2233} = q_{2233} = q_{1122}$$
$$q'_{2233} = \alpha_{23}\alpha_{23}\alpha_{31}\alpha_{31}q_{3311} = q_{3311} = q_{2233} = q_{1122}$$
$$q'_{1133} = \alpha_{12}\alpha_{12}\alpha_{31}\alpha_{31}q_{2211} = q_{2211} = q_{1133}$$
$$q'_{2211} = \alpha_{23}\alpha_{23}\alpha_{12}\alpha_{12}q_{3322} = q_{3322} = q_{2211} = q_{1133}$$
$$q'_{2323} = \alpha_{23}\alpha_{31}\alpha_{23}\alpha_{31}q_{3131} = q_{3131} = q_{2323}$$
$$q'_{1212} = \alpha_{12}\alpha_{23}\alpha_{12}\alpha_{23}q_{2323} = q_{2323} = q_{1212} = q_{3131}$$

The above equations exhaust all the 12 q_{ijkl}. The remaining q'_{ijkl}, namely, q'_{3333}, q'_{3311}, q'_{3322}, and q'_{3131}, do not give any new information.

When the q_{ijkl} equalities derived above are substituted in the q_{mn} matrix for class 222 (D_2), we obtain the q_{mn} matrix for class 23 (T) (given on page 62).

Isotropic Solid

11. We shall study in detail the case of an isotropic solid. This can be considered as a special case of a cubic crystal of class $m3m$ (O_h), with the additional condition that here the tensor properties are invariant under a transformation of a rotation about any axis through any angle θ.

Assume a rotation through angle θ around the Z axis; then the transformation matrix is

	X_1	X_2	X_3
X_1'	$\cos\theta$	$\sin\theta$	0
X_2'	$-\sin\theta$	$\cos\theta$	0
X_3'	0	0	1

Let us express q'_{1111} of the rotated system in terms of q_{ijkl} of the original system:

$$\begin{aligned}
q'_{1111} &= \alpha_{1i}\alpha_{1j}\alpha_{1k}\alpha_{1l}q_{ijkl} \\
&= \alpha_{1j}\alpha_{1k}\alpha_{1l}[\alpha_{11}q_{1jkl} + \alpha_{12}q_{2jkl}] \\
&= \alpha_{1k}\alpha_{1l}[\alpha_{11}\{\alpha_{11}q_{11kl} + \alpha_{12}q_{12kl}\} + \alpha_{12}\{\alpha_{11}q_{21kl} + \alpha_{12}q_{22kl}\}] \\
&= \alpha_{1l}[\alpha_{11}^2\{\alpha_{11}q_{111l} + \alpha_{12}q_{112l}\} + \alpha_{11}\alpha_{12}\{\alpha_{11}q_{121l} + \alpha_{12}q_{122l}\}] \\
&\quad + \alpha_{1l}[\alpha_{12}\alpha_{11}\{\alpha_{11}q_{211l} + \alpha_{12}q_{212l}\} + \alpha_{12}^2\{\alpha_{11}q_{221l} + \alpha_{12}q_{222l}\}] \\
&= \alpha_{11}^3[\alpha_{11}q_{1111} + \alpha_{12}\cancel{q_{1112}}] + \alpha_{11}^2\alpha_{12}[\alpha_{11}\cancel{q_{1121}} + \alpha_{12}q_{1122}] \\
&\quad + \alpha_{11}^2\alpha_{12}[\alpha_{11}\cancel{q_{1211}} + \alpha_{12}q_{1212}] + \alpha_{11}\alpha_{12}^2[\alpha_{11}q_{1221} + \alpha_{12}\cancel{q_{1222}}] \\
&\quad + \alpha_{12}\alpha_{11}^2[\alpha_{11}\cancel{q_{2111}} + \alpha_{12}q_{2112}] + \alpha_{12}^2\alpha_{11}[\alpha_{11}q_{2121} + \alpha_{12}\cancel{q_{2122}}] \\
&\quad + \alpha_{12}^2\alpha_{11}[\alpha_{11}q_{2211} + \alpha_{12}\cancel{q_{2212}}] + \alpha_{12}^3[\alpha_{11}\cancel{q_{2221}} + \alpha_{12}q_{2222}] \\
&= C^4q_{1111} + C^2S^2q_{1122} + C^2S^2q_{1212} + C^2S^2q_{1221} \\
&\quad + C^2S^2q_{2112} + S^2C^2q_{2121} + S^2C^2q_{2211} + S^4q_{2222} \\
&= C^4q_{1111} + S^4q_{2222} + C^2S^2q_{1122} + S^2C^2q_{2211} + 4C^2S^2q_{1212}
\end{aligned}$$

But by Neumann's principle $q'_{1111} = q_{1111}$ for an isotropic body for a rotation of the axes through θ about any line; here we have taken the Z axis as the axis of rotation. In the above equations $C = \cos\theta$ and $S = \sin\theta$. Then

$$q_{1111} = C^4q_{1111} + S^4q_{2222} + C^2S^2(q_{1122} + q_{2211}) + 4C^2S^2q_{1212} \tag{3.84}$$

For a cubic crystal of $m3m$ class, $q_{1111} = q_{2222}$ and $q_{1122} = q_{2211}$. Obviously for an isotropic solid these equalities should also be satisfied. Hence from Eq. (3.84) we obtain

$$4C^2S^2q_{1212} = (1 - C^4 - S^4)q_{1111} - 2C^2S^2q_{1122} \tag{3.85}$$

which on simplification yields

$$2q_{1212} = q_{1111} - q_{1122}$$

$$\therefore q_{66} = (q_{11} - q_{12}) \qquad \text{or} \qquad q_{44} = (q_{11} - q_{12})$$

Instead of starting with q_{1111} as in the above case, we can also deduce the same relation starting with q_{1122}.

3.3.3. Group Theoretical Method

3.3.3.1. Introduction

In crystalline solids the particles (atoms, ions, or molecules) comprising a crystal are arranged in space in a three-dimensional periodic lattice. This periodic lattice consists of a set of identical adjacent fundamental units called unit cells. The symmetry elements of a crystal are defined by a set of transformation operations which leave the unit cell in a state indistinguishable from its original state. These operations consist of rotations about certain axes, reflections through planes or points, and combinations of these operations, which may be collectively called symmetry operations. They can all be represented by rotations and rotation–reflections or rotation–inversions, which are also called proper and improper rotations respectively. Pure reflections and center of symmetry are improper rotations.

A symmetry operation applied to a crystal results in a movement of the frame of the unit cell to a new position which, however, coincides with the original one and is physically indistinguishable from it.

Crystallographers have shown by geometrical considerations that only 32 point groups or groups of combination of the fundamental elements of symmetry can be derived, no less and no more (see, for example, pages 33–36). Group theory also leads to the same result [134; 208; 832].

A symmetry operation can have no effect on any physical property of a crystal, and all the applications of the theory of symmetry to crystalline solids are based on this fact and the consequences that follow from it.

3.3.3.2. Symmetry Elements and Their Matrices

We now relate each one of the geometric operations (performed by applying the elements of symmetry) to the appropriate matrix in order to obtain the symmetry element and the corresponding matrix.

Examples

a. Let us use two-dimensional geometry: The coordinates of a point P change from the old system to the new system as shown in Fig. 3.11. If the direction cosine-scheme is given by

	X	Y
X'	α_{11}	α_{12}
Y'	α_{21}	α_{22}

then

$$\begin{bmatrix} x' \\ y' \end{bmatrix} = \begin{bmatrix} \alpha_{11} & \alpha_{12} \\ \alpha_{21} & \alpha_{22} \end{bmatrix} \begin{bmatrix} x \\ y \end{bmatrix}$$

Here $[\alpha_{ij}]$ is the required matrix and its elements α_{ij} are given by the direction cosines thus:

$$\alpha_{11} = \cos \phi_{11}, \qquad \alpha_{12} = \cos \phi_{12}, \qquad \alpha_{21} = \cos \phi_{21}, \qquad \alpha_{22} = \cos \phi_{22}$$

b. Taking a concrete example, say, a mirror reflection normal to OX, we obtain the transformation matrix

$$A = \begin{bmatrix} -1 & 0 \\ 0 & 1 \end{bmatrix}$$

Therefore $A^2 = I_2$ [see Eq. (2.21b)].

c. The transformation matrix for a four-fold rotation about an axis normal to the XOY plane is given thus:

$$B = \begin{bmatrix} 0 & 1 \\ -1 & 0 \end{bmatrix}$$

Therefore $B^4 = I_2$.

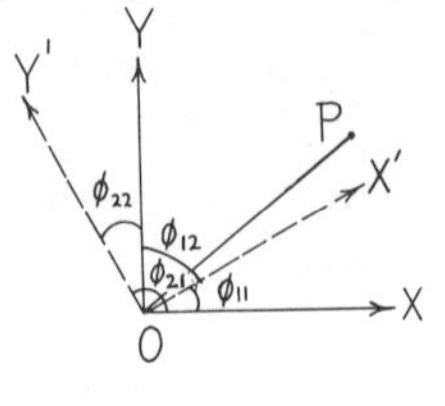

Fig. 3.11. Transformation of the coordinates of a point P from the XY system to the $X'Y'$ system in the same plane.

The matrices corresponding to some symmetry operations in three-dimensional geometry are given on page 38.

3.3.3.3. *Different Types of Symmetry Operations and Their Interrelations*

a. The Identity Operation E. This operation changes nothing and leaves all the particles in the unit cell in their original positions.

b. Rotation about an Axis of Symmetry, C_n. If the crystal is rotated about an axis through an angle θ ($= 2\pi/n$) so as to bring it into a position indistinguishable from the original position, then the axis is said to be of n-fold symmetry. The result of applying the symmetry operation C_n exactly n times is to bring the crystal back into its original position after a rotation through 2π radians about the symmetry axis. This is equivalent to applying the symmetry operation E, and hence $(C_n)^n = C_n{}^n = E$.

c. Reflection in a Plane of Symmetry. The effect of reflecting a crystal twice in the same plane is to restore it to its original position. Thus $\sigma^2 = E$.

d. Rotation–Reflection (*or Rotary Reflection*); *Improper Rotation.* A rotation about the Z axis is one symmetry operation; a reflection in the XY plane constitutes another symmetry operation. The two operations combined result in a composite symmetry operation called rotary reflection or improper rotation, denoted by S_n. If σ_h is a reflection in the plane perpendicular to the rotary-reflection axis, then (in Schönflies notation) $S_n = \sigma_h C_n = C_n \sigma_h$.

If $n = 1$, then $S_1 = \sigma_h$. Thus a simple reflection is an improper rotation. By applying S_n twice we do two rotations C_n and two reflections σ_h; since the double reflection is equivalent to the identity operation E, the result is a simple rotation through an angle $2(2\pi/n)$. Hence

$$S_n{}^2 = C_n{}^2$$

In general k applications of S_n will give $S_n{}^k = \sigma_h C_n{}^k$ if k is odd, and $S_n{}^k = C_n{}^k$ if n is even. If n is odd, $S_n{}^n = \sigma_h C_n{}^n = \sigma_h$. This means that a simple reflection in the plane must be a symmetry operation. Also when n is odd,

$$S_n{}^{n+1} = S_n{}^n S_n = \sigma_h \sigma_h C_n = \sigma_h{}^2 C_n = C_n$$

and a simple rotation about the C_n axis is a single symmetry operation.

e. Inversion about a Point, i.e., a Center of Symmetry, i. Inversion, i, is another case of improper rotation in which the angle of rotation is 180°. So

$$i = S_2 = \sigma_h C_2$$

The effect of inverting twice around the same center of symmetry is to return the crystal back to its original position, so that

$$i^2 = E$$

In a similar way we can describe still other symmetry operations.

3.3.3.4. Product of Two Symmetry Operations

A succession of symmetry operations is called the product of the symmetry operations. The product of two symmetry operations P and Q may result in a new symmetry operation R which could alternatively be applied directly to give the same result. We cite below some cases discussed earlier.

Examples

a. $C_2 \cdot \sigma_h = i$;

b. $i^2 = E$;

c. $\sigma_h{}^2 = E$;

d. $\sigma_h = \sigma_h{}^{-1}$ ($\because \sigma_h \cdot \sigma_h = E = \sigma_h \sigma_h{}^{-1} \therefore \sigma_h = \sigma_h{}^{-1}$);

e. $i = i^{-1}$ ($\because i \cdot i = E = i \cdot i^{-1} \therefore i = i^{-1}$).

Thus σ_h and i are their own inverses.

f. Rotation: To describe the rotation completely it is necessary to state the direction of the rotation axis, for instance, in the sense of a right-handed screw. If $C_n{}^k$ is a rotation through a positive angle $2\pi k/n$, there can be a rotation $-2\pi k/n$ (a negative angle) around the same axis.

$$C_n{}^k C_n{}^{-k} = C_n{}^{k-k} = C_n{}^0 = E$$
$$C_n{}^{n-k} = C_n{}^n C_n{}^{-k} = EC_n{}^{-k} = C_n{}^{-k}$$

which is consistent, since a positive rotation through the angle $2\pi(n-k)/n = 2\pi - 2\pi k/n$ is the same thing as a negative rotation through an angle $(2\pi k/n)$.

g. Improper rotation: $S_n{}^k = \sigma_h C_n{}^k$; $S_n{}^{-k} = \sigma_h C_n{}^{-k}$.

3.3.3.5. Symmetry Operations of Regular Bodies and Multiplication Tables of Their Symmetry Elements

We examine now the symmetry operations of two regular bodies as examples.

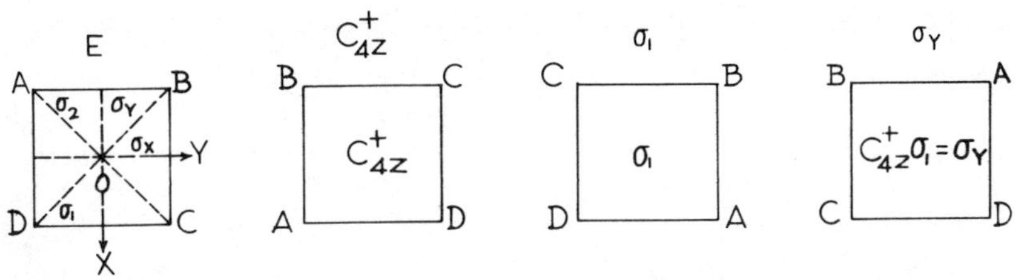

Fig. 3.12. The product of the two symmetry operations C_{4z}^{+} and σ_1 is equal to the single symmetry operation σ_y: geometric representation.

a. Consider the case of a square *ABCD*, with $AD \parallel OX$ axis and $AB \parallel OY$ axis. It has the following symmetry operations.

E	identity
C_{4z}^{+}	anticlockwise rotation by 90° about OZ, and $\perp$ plane XOY
C_{2z}	anticlockwise rotation by 180° about OZ, and $\perp$ plane XOY
C_{4z}^{-}	anticlockwise rotation by −90° about OZ, and $\perp$ plane XOY
σ_x	reflection in a plane through O, $\perp OX$ and $\parallel OY$
σ_y	reflection in a plane through O, $\perp OY$ and $\parallel OX$
σ_1	reflection in a plane through O, $\perp AC$ and $\parallel BD$
σ_2	reflection in a plane through O, $\perp BD$ and $\parallel AC$

The symmetry operations E, C_{4z}^{+}, σ_1, and $C_{4z}^{+}\sigma_1 = \sigma_y$ are shown in Fig. 3.12. The compound operation $C_{4z}^{+}\sigma_1$ means that the first operation to be performed is on the right. This is a universally adopted convention. This compound operation is termed the product of C_{4z}^{+} and σ_1. Table 3.5 gives

Table 3.5. *Multiplication Table of the Symmetry Elements of a Square*

Second operation	First operation							
	E	C_{4z}^{+}	C_{2z}	C_{4z}^{-}	σ_x	σ_y	σ_1	σ_2
E	E	C_{4z}^{+}	C_{2z}	C_{4z}^{-}	σ_x	σ_y	σ_1	σ_2
C_{4z}^{+}	C_{4z}^{+}	C_{2z}	C_{4z}^{-}	E	σ_1	σ_2	σ_y	σ_x
C_{2z}	C_{2z}	C_{4z}^{-}	E	C_{4z}^{+}	σ_y	σ_x	σ_2	σ_1
C_{4z}^{-}	C_{4z}^{-}	E	C_{4z}^{+}	C_{2z}	σ_2	σ_1	σ_x	σ_y
σ_x	σ_x	σ_2	σ_y	σ_1	E	C_{2z}	C_{4z}^{-}	C_{4z}^{+}
σ_y	σ_y	σ_1	σ_x	σ_2	C_{2z}	E	C_{4z}^{+}	C_{4z}^{-}
σ_1	σ_1	σ_x	σ_2	σ_y	C_{4z}^{+}	C_{4z}^{-}	E	C_{2z}
σ_2	σ_2	σ_y	σ_1	σ_x	C_{4z}^{-}	C_{4z}^{+}	C_{2z}	E

Table 3.6. Multiplication Table of the Symmetry Elements of an Isosceles Triangle

Second operation	First operation			
	E	C_2	σ_v	$\sigma_{v'}$
E	E	C_2	σ_v	$\sigma_{v'}$
C_2	C_2	E	$\sigma_{v'}$	σ_v
σ_v	σ_v	$\sigma_{v'}$	E	C_2
$\sigma_{v'}$	$\sigma_{v'}$	σ_v	C_2	E

the results of the compound operations, taking two of the eight symmetry elements at a time. Such a table is called a multiplication table. Note also that some of the products are not commutative; for example, $C_{4z}^{+}\sigma_x = \sigma_1$ but $\sigma_x C_{4z}^{+} = \sigma_2$. (In forming the product AB, we take A from the extreme left column and B from the topmost row.)

b. An isosceles triangle has the elements of symmetry E, C_2, σ_v, and $\sigma_{v'}$, and its multiplication table is given in Table 3.6, where C_2 is the diad in the plane of the triangle and bisecting the vertex, σ_v is reflection in the plane of the triangle, and $\sigma_{v'}$ is reflection in a plane normal to the plane of the triangle and containing the C_2 axis.

3.3.3.6. Symmetry Operations, Matrices, and Groups

Group theory greatly facilitates the study of the symmetry properties of a system. The symmetry elements (axis, plane, and center) are called the operators in the language of group theory. The symmetry operations are (a) rotation, (b) reflection, (c) inversion, and (d) their combinations. They can all be represented or described by rotation, rotation–inversion, or rotation–reflection. All the 32 point groups are neatly represented by the Schönflies notation. Schönflies treated crystallographic symmetries as mathematical groups or operators. Every symmetry operation, say, C_1, C_2, or σ_h, can be replaced or described by a corresponding matrix of transformation. The set of symmetry operations and hence the corresponding set of matrices representing the symmetry operations of a point group forms a group of finite order.* This set of matrices representing the symmetry operations of a point group is called a representation of the group.

* The details follow in the next section.

The set of basic vectors x_1, x_2, x_3 of the coordinate system with reference to which the matrices are defined forms the basis of the representation.

The operation of any of the symmetry elements (i.e., the geometric operations in crystallography) results in a change of the coordinates of the point on which the operation is affected. A change of coordinates can be brought about by a transformation of coordinates. A transformation of coordinates can be considered as a multiplication of matrices. Matrices are applicable in two-, three-, or even higher-dimensional space; in crystallography we are concerned only with three-dimensional space. These matrices form the elements of a group of finite order.

3.3.3.7. Group Postulates and Theorems

We now study the group postulates and some relevant theorems. The abstract theory of groups deals with certain sets of discrete or distinct elements $A, B, C, \ldots$ called the group elements, forming a group G, under a single law of composition. Conventionally the notation of multiplication is used to express the composition of the group elements; for example, if any two elements A, B of a group G possess a unique product C, we write $AB = C$. (The arrangement of the sequence of the elements $A, B, C, \ldots$ of a group G is immaterial, except when we compare two groups which are isomorphic.*)

a. The Postulates. A set G of a finite or infinite number of elements forms a group under a particular law of composition if the following conditions are satisfied:

Closure. For every ordered pair A, B of elements of G, there belongs a unique element C of G, written $C = AB$, which is called the product of A and B.

Associative Law. If A, B, C are any three elements of G which need not be distinct, then $(AB)C = A(BC)$. The pairs AB and BA are regarded as distinct if $A \neq B$.

Unit Element or Identity. G contains an element E called the unit element or identity, such that for every element A of G, $AE = EA = A$.

Reciprocal Element or Inverse. Corresponding to every element A of G there exists in G an element A^{-1} such that $AA^{-1} = A^{-1}A = E$.

b. Commutative or Abelian Group. This is one in which for every pair of elements A, B the products AB and BA are equal, i.e., $AB = BA$. The

* Defined later.

two elements A, B are said to commute if $AB = BA$. Note that the group cannot be called an Abelian group if the law of commutation is satisfied for only some of the elements of the group.

The associative law is not confined just to some three elements A, B, C of this group G, but is applicable in a general form. Thus, $(A_1 A_2 \cdots A_r) \times (A_{r+1} A_{r+2} \cdots A_n) = (A_1 A_2 \cdots A_s)(A_{s+1} A_{s+2} \cdots A_n)$ even when $r \neq s$. Obviously here r and s are less than n. In particular, when all the elements are identical, we shall, as in ordinary algebra, write $AA = A^2$, $(AA)A = A(AA) = A^3$, and also $A^m A^n = A^n A^m = A^{m+n}$, $(A^m)^n = A^{mn} = (A^n)^m$. If A and B commute, obviously $A^m B^n = B^n A^m$.

The inverse of every element of a group is also an element of the group. Thus if A is an element of the group G, then A^{-1} is also an element. If A and B are any two elements of G, there exist elements X and Y such that

$$AX = B, \qquad YA = B$$

Then

$$X = A^{-1}B, \qquad Y = BA^{-1}$$

Thus post-multiplication and pre-multiplication by the inverse of an element is also permissible. Note that A commutes with A^{-1}; i.e., every element commutes with its inverse.

3.3.3.8. Examples of Infinite Groups

Any group containing an infinite number of group elements is called an infinite group or group of infinite order.

a. All positive rational numbers (p/q) excluding zero form a group with respect to ordinary multiplication. (Note that the product of two rational numbers is a rational number.) The identity element is the rational number 1. The inverse operation is $1/A$ for an element A. This group is Abelian, of infinite order. It should be noted, however, that a set of positive integers does not form a group with respect to multiplication.

b. The set of all integers forms an infinite Abelian group, the law of composition being addition. The unit element or identity is zero because $A + 0 = 0 + A = A$, and the inverse of A is $-A$.

3.3.3.9. Examples of Finite Groups

a. A set of all linear orthogonal transformations forms a group. The composition rule is the application of transformations successively.

b. The finite set of p-fold rotation transformations, which are rotations through angles $2\pi/p$ (p being an integer), forms a cyclic group (defined on page 108).

c. The sets of matrices that correspond to the groups of transformations (given under Example b), corresponding to the p-fold rotation transformations, form groups under the composition rule of matrix multiplication. Here the identity element and the inverse element are those matrices which correspond to the identity transformation and the inverse transformation.

We note that the following sets of matrices form groups with respect to matrix multiplication, provided, however, they obey the group postulates:

1. all nonsingular matrices of fixed order n
2. all orthogonal matrices of fixed order n
3. all orthogonal matrices of fixed order and determinant $+1$

3.3.3.10. Multiplication Tables of Groups of Finite Order

a. We consider a group of elements comprising the numbers $1, i, -1, -i$, which form an Abelian group of finite order 4 with respect to ordinary multiplication. [Here $i = (-1)^{1/2}$.] See Table 3.7.

b. Consider a group of order 6, with the elements represented by E, A, B, C, D, F respectively. Let their multiplication table be as given in Table 3.8.

It is easy to see that the group postulates are fully satisfied in Table 3.8. Table 3.8 is not symmetrical with respect to the leading or main diagonal, which illustrates the fact that in general $XY \neq YX$; for example, $BC = D$ but $CB = F$; also $DF \neq FD$. Hence the group is non-Abelian. But it will

***Table** 3.7. Multiplication Table of a Group of Four Elements* $1, i, -1, -i$

Second operation	First operation			
	1	i	-1	$-i$
1	1	i	-1	$-i$
i	i	-1	$-i$	1
-1	-1	$-i$	1	i
$-i$	$-i$	1	i	-1

Table 3.8. Multiplication Table of a Group of Six Elements E, A, B, C, D, F (with Certain Interrelations among Themselves)

Second operation	First operation					
	E	*A*	*B*	*C*	*D*	*F*
E	*E*	*A*	*B*	*C*	*D*	*F*
A	*A*	*B*	*E*	*F*	*C*	*D*
B	*B*	*E*	*A*	*D*	*F*	*C*
C	*C*	*D*	*F*	*E*	*A*	*B*
D	*D*	*F*	*C*	*B*	*E*	*A*
F	*F*	*C*	*D*	*A*	*B*	*E*

be observed that the element E always lies either on the leading diagonal or else occupies positions which are symmetrically placed in pairs about the diagonal. It is seen that $AB = BA = E$, showing that A and B are inverse to each other.

The following remarks are important with regard to all multiplication tables of finite groups.

1. A given element appears in a given row (or column) once and once only.
2. Every row (or column) is different from every other row (or every other column).
3. For an Abelian group, the table is symmetric across the principal diagonal.
4. Every element has a finite order n (i.e., $A^n = E$).
5. Multiplication of each element A_i in G by one element B in G (i.e., $BA_i = A_i'$) reproduces the group G but generally in a different order (as can be seen in the multiplication table). This is called the *rearrangement theorem*.
6. Identity and inversion commute with all the elements of rotation groups.

3.3.3.11. Isomorphic Groups

Two groups $G\,(= A, B, C, \ldots)$ and $G'\,(= A', B', C', \ldots)$ are said to be isomorphic if a one-to-one correspondence can be established between their elements, i.e., if A corresponds to A' and B to B' and if $AB = C$, then $A'B'$ is equal to C'. Here G and G' must be of the same order but need not be of the same composition rule. Furthermore, if G' is a group of nonsingular matrices satisfying the group multiplication table of G, then it is usually represented by Γ and is called a representation of G. Thus G and Γ are isomorphic and Γ contains only nonsingular matrices.

Examples

The following groups of order four are isomorphic, the law of composition for each case being given in parentheses:

a. The numbers $(1, i, -1, -i)$ (under ordinary multiplication).

b. The matrices

$$\begin{bmatrix} 1 & 0 \\ 0 & 1 \end{bmatrix}, \quad \begin{bmatrix} 0 & 1 \\ -1 & 0 \end{bmatrix}, \quad \begin{bmatrix} -1 & 0 \\ 0 & -1 \end{bmatrix}, \quad \begin{bmatrix} 0 & -1 \\ 1 & 0 \end{bmatrix}$$

(under matrix multiplication). If the elements of each set are renamed E, A, B, C, the multiplication table is as shown in Table 3.9.

c. Consider six functions $f_1(z) = z, f_2(z) = 1/(1 - z), f_3(z) = (z - 1)/z$, $f_4(z) = 1/z, f_5(z) = (1 - z), f_6(z) = z/(z - 1)$. Let the law of composition be defined as substitution of one function in another, e.g., in the multiplication below:

$$f_4 f_3 = f_4 f_3(z) = f_4(z - 1)/z = 1/(z - 1)/z = z/(z - 1) = f_6(z) = f_6$$

It is found that the functions form a closed set and $f_1(z)$ corresponds to the identity or unit element. The multiplication table is given in Table 3.10.

Table 3.9. *Multiplication Table of Two Isomorphic Groups with Their Elements E, A, B, C*

Second operation	First operation			
	E	*A*	*B*	*C*
E	*E*	*A*	*B*	*C*
A	*A*	*B*	*C*	*E*
B	*B*	*C*	*E*	*A*
C	*C*	*E*	*A*	*D*

Table 3.10. Multiplication Table of a Group of Six Elements $f_1, f_2, f_3, f_4, f_5, f_6$

Second operation	First operation					
	f_1	f_2	f_3	f_4	f_5	f_6
f_1	f_1	f_2	f_3	f_4	f_5	f_6
f_2	f_2	f_3	f_1	f_6	f_4	f_5
f_3	f_3	f_1	f_2	f_5	f_6	f_4
f_4	f_4	f_5	f_6	f_1	f_2	f_3
f_5	f_5	f_6	f_4	f_3	f_1	f_2
f_6	f_6	f_4	f_5	f_2	f_3	f_1

We note that this group is isomorphic with the earlier group of order 6, having the elements E, A, B, C, D, F on page 106.

3.3.3.12. *The Order (or Period) of an Element*

Let A be an element of a group G of order n. Consider the set of powers of A as $E, A, A^2, A^3, \ldots$, all of which are, of course, elements of G. Since G is a finite group, these elements cannot all be distinct. The least power of A which is equal to the unit element or identity is called the order of A. Here G itself need not be a cyclic group.

3.3.3.13. *Cyclic Group*

A group whose elements can all be expressed as powers of a single element is called a cyclic group. The general form of a cyclic group of order n is $C_n = EAA^2 \cdots A^{n-1}$. All cyclic groups are Abelian, but the converse may not be true. Each element of a group gives rise to a cyclic subgroup, as defined in the next section. The order of the element is the order of this cyclic subgroup.

In crystallography, the groups of interest are the cyclic groups. A cyclic group is a group of operations that, when repeated indefinitely, reproduces the original result at intervals of m operations, where m is the order of the group.

If the original effect (or the initial position) is reproduced every other time, the order m is two. If it is reproduced once every third operation, the order m is three, e.g., for a mirror reflection the order is two, for a diad axis the order is two, and for a triad axis it is three.

Every one of the symmetry operations, caused by the application of axis, plane, or center of symmetry or their combinations, when repeated the requisite and finite number of times brings the crystal back to its original position. Thus each of the symmetry elements representing a symmetry operation forms a member of a finite cyclic group.

Repetition of operations of the symmetry element is equivalent to multiplication of the corresponding transformation matrix successively. Since each of the symmetry operations is represented by a discrete matrix, naturally each of such matrices forms an element of a cyclic group of finite order.

Furthermore, if we know the characteristics of transformation or the character of each of the groups, we also know the behavior of the particular crystal class or point group under consideration.

3.3.3.14. Subgroups

The elements E, a, a^2, b, ab, a^2b form a group defined by the relations $a^3 = E$, $b^2 = E$, and $a^2b = ba^2$. A subset of elements in a group forming a group by itself is defined as a subgroup of the original group, the law of composition being the same for both the main group and the subgroup. For example, the group defined by the above relations has the following subgroups:

1. E, b form a subgroup of order two.
2. E, a, a^2 form a subgroup of order three.
3. Identity forms a subgroup of order one in every group.

The order of a subgroup is a factor of the order of the group. In the extreme case the whole group is one subgroup; similarly, E, the identity element, alone is a subgroup. These subgroups are called the trivial subgroups. The multiplication table of this group of order six is given in Table 3.11.

3.3.3.15. Cosets

Let g, having the elements $E, A, B, C, \ldots$, be a subgroup of the group G, containing the elements $E, A, B, C, D, F, \ldots$. Then taking D, the left coset is given by $DE, DA, DB, DC, \ldots$ and the right coset is given by $ED, AD, BD, CD, \ldots$. If the right coset is equal to the left coset, the subgroup is called an invariant subgroup. Left and right cosets can be formed

Table 3.11. Multiplication Table of a Group of Order Six Whose Elements a and b Are the Two Matrices

$$a = \begin{bmatrix} 0 & 1 & 0 \\ 0 & 0 & 1 \\ 1 & 0 & 0 \end{bmatrix} \qquad b = \begin{bmatrix} 1 & 0 & 0 \\ 0 & 0 & 1 \\ 0 & 1 & 0 \end{bmatrix}$$

Second operation	First operation					
	E	a	a^2	b	ab	a^2b
E	E	a	a^2	b	ab	a^2b
a	a	a^2	E	ab	a^2b	b
a^2	a^2	E	a	a^2b	b	ab
b	b	a^2b	ab	E	a^2	a
ab	ab	b	a^2b	a	E	a^2
a^2b	a^2b	ab	b	a^2	a	E

only by taking an element, say, D or F of the main group, which does not form a member of the subgroup g. It may be noted that cosets themselves are not subgroups.

3.3.3.16. Conjugate Elements and Conjugate Classes

Let s and t be any two elements of a group. Then from the fundamental group postulates, $t^{-1}st$, which may be denoted by u, will also be an element of the group. Here u is called the transform of s by the element t. Consequently $tut^{-1} = s$ [because $tut^{-1} = tt^{-1}stt^{-1} = (tt^{-1})s(tt^{-1}) = s$]. Then s is the transform of u by the element t^{-1}. Two such elements s and u which may be transformed into one another by one element or another of the group are called conjugate elements. If $u, v, w, \ldots$ are the transforms of s by the elements t, x, y (where t, x, y and u, v, w are elements of the group G), then s, u, v, w form the elements of a conjugate class. Thus the complete set of distinct elements which are transforms of s with all the elements of the group G forms a conjugate class of elements. The order of the conjugate class is defined by the number of conjugate elements in the class; the order of a conjugate element is the order of the cyclic group generated by the element. The elements of a conjugate class are conjugate to one another; i.e., $s, u, v, w, \ldots$ are conjugate to one another. All the elements of a conjugate class have the same order; this order is a divisor of the order of the main group. In every group the identity element forms a class by itself.

A main group may have any number of conjugate classes. However, a conjugate class does not form a subgroup. In a group of orthogonal matrices, the elements of a conjugate class are equivalent matrices; their traces* are equal. The trace of a matrix is unchanged by conjugation of the matrix. Hence the trace is the same for all the matrices of the same conjugate class.

Let M and N be two matrices. Let K be another matrix equal to $N^{-1}MN$. Then K is a matrix equivalent to M. Both K and M have the same trace or spur.

Self-Conjugate Subgroups. If H is a subgroup of G and if $t^{-1}at$ is an element of H for all elements a of H and t of G, then H is called the self-conjugate subgroup of G. We note that all the elements of symmetry of the crystallographic point group D_{2h} are self-conjugate elements, and each symmetry element forms a conjugate class by itself.

In an Abelian group, every element is a transform of itself, i.e., self-conjugate, and so constitutes a class by itself. In an Abelian group, each element commutes with every other element of the group, even if the Abelian group is not cyclic. Thus every cyclic group is an Abelian group, but every Abelian group need not be cyclic.

3.3.3.17. Direct-Product Groups

If G_1 and G_2 are two groups with the same composition rule and if the elements of G_1 commute with the elements of G_2, there being no common element in G_1 and G_2 except the identity element, the totality of the elements $K = a_1a_2$ forms a group G which is called the direct product of G_1 and G_2. The direct products are commutative, i.e., $G_1 \times G_2 = G = G_2 \times G_1$.

3.3.3.18. Permutation Groups

Let there be n identical objects arranged in n boxes, and let some or all of the objects be rearranged into different boxes so that at the end of the operation there is only one object in each. Then this operation is said to be a symmetry operation of the permutation group of n identical objects. This is called a symmetric group of degree n and is represented by S_n. The number of all possible permutations or arrangements of the n identical objects in the boxes is $n!$. This set of all permutations forms a group of the permutations of n identical objects. The permutation group contains $n!$ elements. As an example of the permutation group, let $n = 3$. Then there

* Spur or trace is the sum of the elements on the leading diagonal of a matrix.

are $3! = 6$ elements in the group of permutation. Let P, Q, R be three balls:

E	P	Q	R	All are in their own places
C	P	R	Q	P is fixed; Q, R are interchanged
D	R	Q	P	Q is fixed; R, P are interchanged
F	Q	P	R	R is fixed; P, Q are interchanged
A	R	P	Q	All are altered (say, clockwise)
B	Q	R	P	All are altered (anticlockwise)

There are in all six permutations or operations which change (P, Q, R) into (P, Q, R), (P, R, Q), (R, Q, P), (Q, P, R), (R, P, Q), (Q, R, P). The set of these six permutations forms a finite group of order 6. The six elements are

$$E = \begin{pmatrix} P & Q & R \\ P & Q & R \end{pmatrix}, \quad C = \begin{pmatrix} P & Q & R \\ P & R & Q \end{pmatrix}, \quad D = \begin{pmatrix} P & Q & R \\ R & Q & P \end{pmatrix}$$

$$F = \begin{pmatrix} P & Q & R \\ Q & P & R \end{pmatrix}, \quad A = \begin{pmatrix} P & Q & R \\ R & P & Q \end{pmatrix}, \quad B = \begin{pmatrix} P & Q & R \\ Q & R & P \end{pmatrix}$$

Here the operation BA is defined thus: First perform the permutation A on the original (or identity) arrangement; then perform the permutation B on this result. The procedure is explained in Fig. 3.13. Here E, A, B are three boxes, each divided into three compartments 1, 2, 3. Box E is not shown separately in the figure. Ball P in compartment 1 of box E, after passing through A, exits from its compartment 2. So B receives P into its compartment 2 and ejects it out through compartment 1. Hence the combined effect of BA is to allow P back to its original position.

The arrows indicate the paths followed by the three balls P, Q, and R due to the operation A followed by B. Thus it is seen that finally the three balls regain the original arrangement E. Hence $BA = E$.

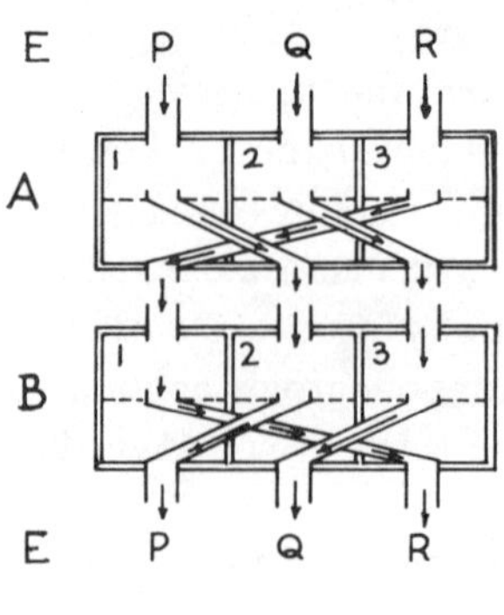

Fig. 3.13. Product of two elements of a permutation group: schematic representation. The arrows indicate the path taken by each ball under each permutation operation.

Table 3.12. *Multiplication Table of a Permutation Group of Order Six*

Second operation	First operation					
	E	*A*	*B*	*C*	*D*	*F*
E	*E*	*A*	*B*	*C*	*D*	*F*
A	*A*	*B*	*E*	*F*	*C*	*D*
B	*B*	*E*	*A*	*D*	*F*	*C*
C	*C*	*D*	*F*	*E*	*A*	*B*
D	*D*	*F*	*C*	*B*	*E*	*A*
F	*F*	*C*	*D*	*A*	*B*	*E*

Thus we may derive the entire group multiplication table shown in Table 3.12. Here A and B are third-order elements, while C, D, F are second-order elements. This group is isomorphic with the group of symmetry operations of an equilateral triangle. It is non-Abelian, e.g., $DB \neq BD$. Mathematically, it is the group multiplication table which characterizes any group. All groups with the same group multiplication table are all isomorphic with one another. Note that this group multiplication table is the same as that given on page 106.

Cayley's Theorem. Every finite group is isomorphic with a suitable group of permutations. Every crystallographic point group is a finite group, hence it has a corresponding isomorphic permutation group. Conversely, taking suitable permutation groups, all the crystallographic point groups can be derived by group theoretical methods. Hence the importance of permutation groups in the study of crystallographic point groups.

3.3.3.19. *Group Theoretical Method of Determining the Number of Non-vanishing Independent Constants*

The logic of applying the group theoretical method to enumerate the number of constants is as follows:

1. Every crystal class has a finite number of symmetry elements, comprised of rotations and rotation-inversions. Thus each of the symmetry operations is a geometric operation.

2. Each of the geometric operations, when applied once, brings the body into congruence with itself, and, what is more important, when the

geometric operation is applied successively a definite number of times, it brings the body back to its initial position. That is, the operations become cyclic after a fixed number of steps; this number of steps depends upon the nature of the symmetry operations, e.g., a tetrad symmetry becomes cyclic after four operations. A reflection becomes cyclic after two applications, as does an inversion or what is commonly called a center.

3. The several symmetry operations which govern a point group form the elements of a finite group called the symmetry group, whose order is equal to the number of elements of the symmetry operations. The finite group thus formed may contain some subgroups (which may be cyclic groups or permutation groups).

4. Since the symmetry operations form a finite group, all the properties and characteristics of finite groups are applicable to these symmetry groups also.

5. Every one of the symmetry operations can be represented by a matrix which results in the formation of new coordinate axes. The transformation of coordinates is accomplished by matrix multiplication of the original or old axes by a suitable matrix, uniquely determined by the symmetry operation, to yield the transformed or new axes. The rank of the matrix depends upon the rank of the tensor property. Thus a tensor property of rank one has a column matrix of 3×1 elements. A tensor property of rank two has a corresponding 3×3 matrix, and so on. The matrices of the symmetry operations of a crystal form an orthogonal three-dimensional matrix representation of the appropriate point group.

6. Let G be a finite group describing the symmetry operations of a point group. Corresponding to this symmetry group G, we can find an equivalent group G' whose elements are given by the transformation matrices representing the various symmetry elements of the group G.

7. Each one of the components of the physical-property tensor is transformed into linear combinations of all the components, under the symmetry transformations of the crystal. This is due to the law of transformation of tensors. For example, for a second-rank tensor $T'_{ij} = \alpha_{ik}\alpha_{jl}T_{kl}$, thereby implying that T'_{ij} is a linear function of all the T_{kl}.

8. The matrices of transformation of the tensor components of any property relevant to the symmetry elements of the point group also form a multidimensional matrix representation of the same group.

9. The multidimensional representation is reducible. For example, the linear substitutions constitute a reducible representation of the group G of symmetry operations of any point group. By a proper choice of basis,

we may get a completely reducible representation of the group G. The character appropriate to any symmetry element R in this representation will be the same as that obtained before, since the two are equivalent representations.

10. By the application of Neumann's principle, we conclude that in the case of any specified physical property (see Table 3.14), only those linear combinations of tensor components which have the entire symmetry of the crystal can survive. Thus, in order to know the number of combinations that remain invariant for all operations R of the crystallographic group G, we need only find the total number n_i of the irreducible representations for which the character $\chi_i(R) = 1$ and which are all contained in the reducible representations; such irreducible representations for which $\chi_i(R) = 1$ are said to be total symmetric irreducible representations.* Customarily $\chi_\varrho(R)$ stands for the character of the element R belonging to the ϱth conjugate class in the completely reducible representation.

Hence the problem reduces to finding n_i, the number of times the total symmetric irreducible representation Γ_i occurs in the completely reducible representation Γ. Thus the number of nonvanishing independent components of a tensor, subject to the point group symmetry of a crystal, is given by n_i, the number of times the total symmetric irreducible representation Γ_i occurs in the completely reducible representation Γ formed by the transformation matrices of the tensor components corresponding to the symmetry transformations of the point group. Obviously n_i can be calculated from the equation

$$n_i = \frac{1}{N} \sum_{\varrho} h_\varrho \chi_\varrho(R) \chi_i(R)$$

in which h_ϱ is the order of the ϱth conjugate class, and N is the order of the group G. This equation is derived below.

3.3.3.20. Derivation of the Relation $n_i = (1/N) \sum h_\varrho \chi_\varrho(R) \chi_i(R)$

We assume that the character for the transformation representing the symmetry operation R is $\chi(R)$ for the original reducible representation. Since the process of reduction consists of a mere change of the coordinates, the completely reducible representation, taken as a whole, is equivalent to the original representation and will therefore have the same characters. The character $\chi(R)$ for the completely reducible representation will, how-

* Defined in a later section.

ever, be the sum of the characters of the various irreducible representations of which it is composed. If the representation Γ_i occurs n_i times in the completely reduced form of the original representation, then

$$\begin{aligned} \chi(R) &= n_1\chi_1(R) + n_2\chi_2(R) + \cdots \\ &= \sum_i n_i\chi_i(R) \end{aligned} \tag{3.86}$$

for each operation R, $\chi_i(R)$ being the character of the irreducible representation Γ_i for the operation R. Since the quantities $\chi(R)$ can easily be obtained by inspection of the original irreducible representation, while the $\chi_i(R)$ are given in the standard tables of characters of irreducible representations, it is possible to solve the set of equations for the unknowns n_1, n_2, etc. We can also calculate n_i and express it in the desired form by combining Eq. (3.86) above with the orthogonality relation connecting the characters of irreducible representations of any finite group G.

Statement of the Orthogonality Relations. If Γ_i and Γ_j are any two irreducible representations of the same finite group, then it can be shown that

$$\begin{aligned} \sum_R \chi_i(R)\bar{\chi}_j(R) &= 0 \qquad \text{if} \quad i \neq j \\ \sum_R \chi_i(R)\bar{\chi}_j(R) &= N \qquad \text{if} \quad i = j \end{aligned} \tag{3.87}$$

where $\chi_i(R)$ is the character of the irreducible representation Γ_i of G, and $\chi_j(R)$ that of Γ_j of G. In other words, the characters form an orthogonal system. That is,

$$\sum_R \chi_i(R)\bar{\chi}_j(R) = N\delta_{ij}$$

the summation being only over R; here $\bar{\chi}_j(R)$ is the complex conjugate of $\chi_j(R)$ and N is the order of the group; δ_{ij} is the Kronecker delta.

Multiplying both sides of Eq. (3.86) by $\chi_j(R^{-1})$, we have

$$\chi(R)\chi_j(R^{-1}) = \sum_i n_i\chi_i(R)\chi_j(R^{-1}) \tag{3.88}$$

Now summing each side over all symmetry operations R, we obtain

$$\sum_R \chi(R)\chi_j(R^{-1}) = \sum_R \Big(\sum_i n_i\chi_i(R)\Big)\chi_j(R^{-1}) \tag{3.89}$$

Since R is an element of the group G, its inverse R^{-1} must also be a member of the same group. Furthermore, the character of the inverse operation is equal to $\bar{\chi}_i(R)$, the complex conjugate of the character of operation, i.e., $\bar{\chi}_i(R) = \chi_i(R^{-1})$.

After rearranging the right-hand side of Eq. (3.89), we get

$$\sum_R \chi(R)\chi_j(R^{-1}) = \sum_i n_i\left\{\sum_R \chi_i(R)\chi_j(R^{-1})\right\}$$

$$= \sum_i n_i\left\{\sum_R \chi_i(R)\bar{\chi}_j(R)\right\} \tag{3.90}$$

Now for any particular value of n_i, the need for summation over i does not arise. Hence, combining Eq. (3.90) with the orthogonality relation, we have simply

$$\sum_R \chi(R)\chi_j(R^{-1}) = \sum_i n_i N\delta_{ij} = n_i N \tag{3.91}$$

because when $j = i$, $\delta_{ij} = 1$ and $\chi_j(R^{-1}) = \chi_i(R^{-1})$. Thus

$$\sum_R \chi(R)\chi_i(R^{-1}) = n_i N \tag{3.92}$$

Now for every crystallographic point group each element of symmetry generates a cyclic group. Thus, if $R^m = E$, then $R^{m-1} \cdot R = E$. Hence R^{m-1} is the inverse of R. But R^{m-1} is conjugate with R since both belong to the same cyclic group. Hence R^{-1} and R form a conjugate class. Whenever R^{-1} and R belong to the same conjugate class, Eq. (3.92) changes to

$$n_i N = \sum_R \chi(R)\chi_i(R)$$

or

$$n_i = (1/N)\sum_\varrho h_\varrho \chi_\varrho(R)\chi_i(R) \tag{3.93}$$

The summation is now over all the conjugate classes, and h_ϱ is the order of the ϱth conjugate class. The suffix ϱ in $\chi_\varrho(R)$ indicates that R belongs to the ϱth conjugate class.

3.3.3.21. Explanation of Some of the Terms Such as Representations and Characters

a. Representations. The symmetry operations of a point group can be described by matrices. These matrices are formed by associating a group of linear operators with the symmetry operations of the point group and by calculating the effects of these operations on the base vectors of an n-dimensional vector space. The set of matrices describing the set of symmetry operations of a point group is called a representation. By changing the base vectors (the reference system of coordinates) we get different representations of the same group. The set of matrices a_{ij}, b_{ij}, etc. must have the same dimensions, i.e., if a_{ij} is a 2×2 matrix, then b_{ij}, c_{ij}, etc. are also of the same type, i.e., 2×2 matrices. This is warranted because the product cannot be formed between two matrices of different dimensions, such as one 3×3 matrix and another 2×2. A faithful representation is one in which each element of the group is associated with a different matrix. In a general representation, more than one element may be represented by the same matrix.

b. Equivalent Representations. In order to find the matrix representation of a group of linear operators (symmetry elements) acting on a set of coordinate axes of a reference system, first we have to choose a definite basis (set of coordinate axes) for the space and then calculate the effect of each of the operators on each base vector. Obviously, if we choose a different basis, we get a numerically different set of matrices or representation. Since any set of n linearly independent vectors will provide a basis for an n-dimensional space, the choice of the base vectors is arbitrary; in fact, the choice can be made in an infinite number of ways. Thus, there will be many sets of matrices, and hence representations, that can represent a group of operators (describing the group of symmetry operations) in a given vector space. All these possible sets are, however, equivalent to one another, because all give the same information, namely the effect of the operators on a general vector of the space. Take, for example, two sets of matrices $[D_f(R)]$ and $[D_g(R)]$. These two representations are equivalent since there exists a matrix $[A]$ with its inverse $[A]^{-1}$, for which the following relation holds for every operation R of the group:

$$[D_g(R)] = [A][D_f(R)][A]^{-1}$$

Here $[A]$ is the transformation matrix between $[D_g(R)]$ and $[D_f(R)]$.

c. Characters of Representations and Importance of the Trace of a Matrix. We have seen in the previous section that $D_g(R)$ and $D_f(R)$ are two equivalent representations of the same point group, but the basis of the reference (or the reference coordinate system) is different for the two. So we know how the matrices of representation change when the basis of representation is changed, and how the matrices of two equivalent representations are related to each other.

We now describe a representation in general without referring it to any specific basis. For this purpose we look for quantities associated with a representation that are the same for all equivalent representations, and different for nonequivalent representations. Such quantities naturally must be associated with the matrices of the representation and must remain invariant even if the basis of the representation is changed. Suppose that such an invariant quantity associated with the matrix $[D(R)]$ of a representation is denoted by $\chi[D(R)]$. Since the change of basis transforms $[D(R)]$ to $[A][D(R)][A]^{-1}$, where $[A]$ is the transformation matrix, the invariant quantity χ must be defined in such a way that

$$\chi[D(R)] = \chi[[A][D(R)][A]^{-1}]$$

$[A]$ is between matrices f and g. The trace or spur of the matrix $\mathbf{D}(R)$, denoted by $\mathrm{Tr}[(D)R]$ or spur $[(D)R]$, is such an appropriate invariant quantity. We define

$$\mathrm{Tr}[(D)R] = \sum_{i=1}^{n} d_{ii}$$

where d_{ii} is a diagonal element. Hence if two representations of a group with matrices $D_f(R)$ and $D_g(R)$ are equivalent, it follows that for every element R of the group

$$\mathrm{Tr}[D_f(R)] = \mathrm{Tr}[D_g(R)]$$

This is the test for the equivalence of two representations. Thus the trace of a matrix is unaltered by any transformation from one set of base vectors to another (i.e., by conjugation of the matrix). Hence the trace is the same for all the conjugate classes in a given representation.

To sum up, we see that the chief characteristic property of a representation in general terms, without having to refer it to a specific basis, is that for each element R of the group the value of the trace of the matrix $[D(R)]$ characterizes the representation without any specific basis. This gives rise to a set of numbers which is characteristic of the representation in the sense that all equivalent representations give rise to the same set of numbers, while nonequivalent representations have different sets. For this reason, $\mathrm{Tr}[D(R)]$ is called the character of the group element R in the representation with matrix $[D(R)]$. The character is usually denoted by the symbol $\chi(R) = \mathrm{Tr}[D(R)]$. The complete set of numbers (characters), one for each element of the group, is called the character of representation, and we say that two representations are equivalent if and only if they have the same character. The elements belonging to the same conjugate class have the same character. Since the trace of a matrix is unaltered by a transformation of the matrix by any element of the full linear group, two facts are evident:

(1) The characters are unaffected by any change of the basis. This leads to the concept of equivalent representations. (2) The characters $\chi(R)$ are unaffected by a transformation of R by any element of G. This leads to the concept of conjugate classes.

3.3.3.22. Determination of $\chi_\varrho(R)$

We shall now determine the value of $\chi_\varrho(R)$, the character of the transformation matrix for tensor properties of different ranks, for various symmetry elements, where ϱ defines the conjugate classes.

We now consider in detail a few typical cases.

a. For a first-rank tensor property, transformation being a rotation of θ about the OZ axis (Fig. 3.14), we first form the transformation matrix for the given symmetry element, namely, rotation of θ about the Z axis. Then we express the tensor property in the new system in terms of the tensor property in the old system. Then we determine the character, which is the trace (or spur) for the transformation matrix of the given property, which is equal to the sum of the matrix elements (of the transformation) along the leading diagonal (top left to bottom right).

From the law of transformation of a first-rank tensor property, we have $p_i' = \alpha_{ij}p_j$; here p_j is the tensor property:

$$\begin{aligned} p_1' &= \alpha_{11}p_1 + \alpha_{12}p_2 + \alpha_{13}p_3 \\ p_2' &= \alpha_{21}p_1 + \alpha_{22}p_2 + \alpha_{23}p_3 \\ p_3' &= \alpha_{31}p_1 + \alpha_{32}p_2 + \alpha_{33}p_3 \end{aligned} \tag{3.94}$$

Therefore we have

$$\underset{\text{Tensor property in the new system}}{\begin{bmatrix} p_1' \\ p_2' \\ p_3' \end{bmatrix}} = \underset{\text{Transformation matrix}}{\begin{bmatrix} \alpha_{11} & \alpha_{12} & \alpha_{13} \\ \alpha_{21} & \alpha_{22} & \alpha_{23} \\ \alpha_{31} & \alpha_{32} & \alpha_{33} \end{bmatrix}} \times \underset{\text{Tensor property in the old system}}{\begin{bmatrix} p_1 \\ p_2 \\ p_3 \end{bmatrix}} \tag{3.95}$$

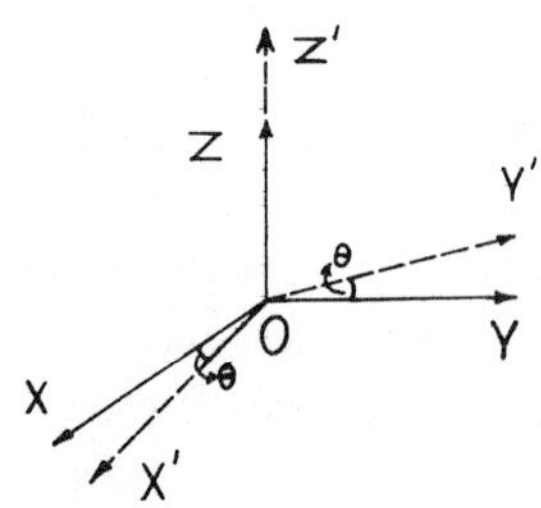

Fig. 3.14. Transformation of axes: a rotation of the X system through an angle θ about the Z axis; here $OZ' = OZ$.

Here the character of the transformation matrix is $\alpha_{11} + \alpha_{22} + \alpha_{33} = \cos\theta + \cos\theta + 1 = 2\cos\theta + 1$.

We can sum up the above procedure thus: In the transformation matrix, the character is obtained as the sum of the elements along the leading diagonal. Any typical element α_{ij} relates p_i' to p_j, but i must be equal to j in order to determine the character. Therefore the character

$$\begin{aligned}\chi_\varrho(R) &= \alpha_{ii} \text{ (summed up from 1 to 3)}\\ &= \alpha_{11} + \alpha_{22} + \alpha_{33} = \cos\theta + \cos\theta + 1 = 2\cos\theta + 1 \end{aligned} \tag{3.96}$$

b. For a second-rank tensor property (general tensor and not symmetric) we show that $\chi_\varrho(R) = (2\cos\theta + 1)^2 = 4\cos^2\theta + 4\cos\theta + 1$. For a rotation θ about the OZ axis, the direction-cosine scheme and the procedure are same as in (a). To determine the character $\chi_\varrho(R)$ or the spur of the transformation matrix of 9×9 elements relating p'_{ij} with p_{kl}, we have first to frame the matrix of 9×9 elements and then add up the diagonal terms, or, instead, it is sufficient to determine the terms along the leading diagonal alone and add them up.

We can sum up the conclusions as was done in case (a) thus: The character $\chi_\varrho(R)$ of this transformation matrix is obtained from the relation

$$p'_{ij} = \alpha_{ik}\alpha_{jl}p_{kl}$$

The terms on the leading diagonal will be the coefficients of p_{kl} when $k = i$ and $l = j$. Hence

$$\chi_\varrho(R) = \sum \alpha_{ii}\alpha_{jj} \qquad (i = 1, 2, 3 \text{ and } j = 1, 2, 3)$$

$$\begin{aligned}\chi_\varrho(R) &= \alpha_{11}(\alpha_{11} + \alpha_{22} + \alpha_{33}) + \alpha_{22}(\alpha_{11} + \alpha_{22} + \alpha_{33}) + \alpha_{33}(\alpha_{11} + \alpha_{22} + \alpha_{33})\\ &= (\alpha_{11} + \alpha_{22} + \alpha_{33})^2 = (2\cos\theta + 1)^2 = 4\cos^2\theta + 4\cos\theta + 1\end{aligned} \tag{3.97}$$

c. On the basis of the above two results, we conclude that $\chi_\varrho(R)$, the trace of the matrix of transformation for a general tensor of rank n, is equal to $(\alpha_{11} + \alpha_{22} + \alpha_{33})^n$.

d. The character $\chi_\varrho(R)$ is left unchanged regardless of the sequence of arrangement of p'_{ij}, provided, however, that p_{ij} are also arranged in the same sequence in the matrix equation relating p'_{ij} and p_{ij}.

e. For the second-rank symmetric tensor, with the same element of symmetry as above, i.e., rotation θ about the OZ axis, $\chi_\varrho(R) = (4\cos^2\theta + 2\cos\theta)$, as shown below.

Table 3.13. The Character $\chi_\varrho(R)$ for a Second-Rank Tensor for Some Typical Symmetry Elements

Symmetry element	Nature of tensor	$\chi_\varrho(R)$
E (identity)	General	9
i (inversion)	Symmetric	6
C_2 (diad: OX_3 axis)	General	1
	Symmetric	2
σ_h (plane $\perp$ to OX_3 axis)	General	1
	Symmetric	2
C_4 (tetrad: OX_3 axis)	General	1
	Symmetric	0

For a symmetric tensor of second rank, we have $p'_{ij} = p'_{ji}$. Hence we find only six tensor components, namely, p_{11}, p_{12}, p_{31}, p_{22}, p_{23}, p_{33} because $p_{12} = p_{21}$; $p_{23} = p_{32}$; $p_{31} = p_{13}$. Thus $\chi_\varrho(R)$ can be deduced from that for the general tensor of rank 2 by simply dropping out the diagonal terms corresponding to p_{21}, p_{32}, and p_{13}. Here

$$\chi_\varrho(R) = 4\cos^2\theta + 2\cos\theta. \tag{3.98}$$

f. Table 3.13 shows $\chi_\varrho(R)$ for a second-rank tensor for some typical symmetry elements.

g. In order to determine $\chi_\varrho(R)$, one prepares the direction-cosine scheme corresponding to each one of the elements of symmetry. This is irrespective of the crystal system and crystal class. The direction-cosine schemes are given on page 38. The method of determining $\chi_\varrho(R)$ for a proper rotation through θ is given in detail on page 120.

h. We show that $\chi_\varrho(R)$ of a fourth-rank symmetric tensor, taking elasticity as an example, is given by

$$\chi_\varrho(R) = 16C^4 \pm 8C^3 - 4C^2 + 1 \tag{3.99}$$

The law of transformation for a fourth-rank tensor is given by

$$T'_{ijkl} = \alpha_{im}\alpha_{jn}\alpha_{ko}\alpha_{lp}T_{mnop}$$

giving thereby a total number of 81 tensor components for the elastic tensor which relates the nine stress components P_{ij} and nine strain components ε_{kl} (see pages 137, 140). However, since in the classical theory of elasticity both P_{ij} and ε_{kl} are symmetric second-rank tensors, there exist only six P_{ij} and six ε_{kl}, yielding 36 T_{ijkl}. Furthermore, in the classical theory of elasticity, we have $T_{ijkl} = T_{klij}$ due to the existence of the strain-energy function (see page 144, where it has been proved that in the two-suffix notation $c_{ij} = c_{ji}$). Hence the 36 T_{ijkl} reduce to 21 T_{ijkl} independent components even for the most anisotropic crystal.

Thus the transformation matrix is a 21×21 matrix. The character, therefore, is the sum of 21 terms along the leading diagonal.

We now consider the simple transformation matrix for a rotation through θ about the OX_3 axis (Fig. 3.15), with C and S used for $\cos\theta$ and $\sin\theta$ respectively:

	X_1	X_2	X_3
X_1'	C	S	0
X_2'	$-S$	C	0
X_3'	0	0	1

Now, from the law of transformation of a fourth-rank tensor component, we know that each one of the tensor components T'_{ijkl} in the rotated coordinate system is a linear function of all the T_{mnop} of the elastic tensor, but it is only one of the 21 independent components, namely the T_{ijkl} satisfying the condition $m = i$, $n = j$, $o = k$, and $p = l$, that contributes toward $\chi_\varrho(R)$. We give below the individual contributions of each one of the 21 independent T_{ijkl}, indicating its contribution by an arrow. Thus we find

$$T_{1111} \to C^4; \quad T_{1122} \to (C^4 + S^4); \quad T_{1133} \to C^2; \quad T_{1123} \to C^3; \quad T_{1131} \to C^3;$$
$$T_{1112} \to (C^4 - 3C^2S^2); \quad T_{2222} \to C^4; \quad T_{2233} \to C^2; \quad T_{2223} \to C^3; \quad T_{2231} \to C^3;$$
$$T_{2212} \to (C^4 - 3C^2S^2); \quad T_{3333} \to 1; \quad T_{3323} \to C; \quad T_{3331} \to C; \quad T_{3312} \to (C^2 - S^2);$$
$$T_{2323} \to C^2; \quad T_{3131} \to C^2; \quad T_{1212} \to (C^4 - 2C^2S^2 + S^4); \quad T_{2331} \to (C^2 - S^2);$$
$$T_{2312} \to (C^3 - CS^2); \quad T_{3112} \to (C^3 - CS^2)$$

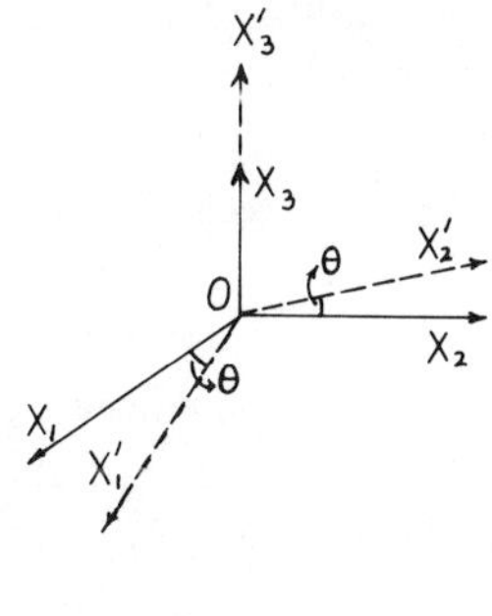

Fig. 3.15. Transformation of axes: a rotation of the X_i system through an angle θ about thc X_3 axis; here $OX_3' = OX_3$.

The character $X_{\varrho}(R)$ of the transformation matrix for a fourth-rank symmetric tensor (relating a second-rank symmetric tensor with another second-rank symmetric tensor, with the additional condition that $T_{ijkl} = T_{klij}$) is given by the sum of the contributions by the $21 T_{mnop}$. Hence

$$\begin{aligned}
\chi_{\varrho}(R) &= C^4 + (C^4 + S^4) + C^2 + C^3 + C^3 + (C^4 - 3C^2S^2) + C^4 \\
&\quad + C^2 + C^3 + C^3 + (C^4 - 3C^2S^2) + 1 + C + C \\
&\quad + (C^2 - S^2) + C^2 + C^2 + (C^4 - 2C^2S^2 + S^4) + (C^2 - S^2) \\
&\quad + (C^3 - CS^2) + (C^3 - CS^2) \\
&= 6C^4 + 2S^4 + 6C^2 + 6C^3 + 1 - 8C^2S^2 + 2C - 2S^2 - 2CS^2 \\
&= 16C^4 + 8C^3 - 4C^2 + 1 \qquad \text{(3.100a)}
\end{aligned}$$

Thus $\chi_{\varrho}(R) = 16C^4 + 8C^3 - 4C^2 + 1$ for a proper rotation θ about the OX_3 axis.

Similarly it can be shown that for an improper rotation (such as a rotation–inversion and a reflection), we have

$$\chi_{\varrho}(R) = 16C^4 - 8C^3 - 4C^2 + 1 \qquad \text{(3.100b)}$$

Combining Eqs. (3.102a) and (3.102b), we have

$$\chi_{\varrho}(R) = 16 \cos^4 \theta \pm 8 \cos^3 \theta - 4 \cos^2 \theta + 1 \qquad \text{(3.101)}$$

for a tensor property such as elasticity, which is a fourth-rank symmetric tensor (relating a second-rank symmetric tensor with another second-rank symmetric tensor), where $T_{ijkl} = T_{klij}$.

3.3.3.23. *Description of the General Method of Determining $\chi_{\varrho}(R)$*

We shall consider the transformation properties under any symmetry operation of a scalar, a vector, a second-rank symmetric tensor, and a second-rank unsymmetric or general tensor, and the determination of $\chi_{\varrho}(R)$ for several cases.

A scalar remains unchanged, the components of a vector transform as the Cartesian coordinates, the components of a symmetric tensor transform as the products of Cartesian coordinates with some additional conditions, and the components of an unsymmetric tensor transform as simple products of Cartesian coordinates. Thus, under an operation R_{ϕ} consisting of a rotation about the Z axis through ϕ or a rotation–inversion through ϕ, the components p_x, p_y, and p_z of a vector, $a_{xx}, \ldots, a_{xy}, \ldots$ of a symmetric tensor, and $b_{xx}, \ldots, b_{yx}, \ldots$ of an unsymmetric tensor transform according to the equations

$$\begin{aligned} p_1' &= p_1 \cos\phi + p_2 \sin\phi \\ p_2' &= -p_1 \sin\phi + p_2 \cos\phi \\ p_3' &= \pm p_3 \end{aligned} \tag{3.102}$$

$$\begin{aligned} a_{11}' &= a_{11}\cos^2\phi + a_{22}\sin^2\phi + 2a_{12}\sin\phi\cos\phi \\ a_{22}' &= a_{11}\sin^2\phi + a_{22}\cos^2\phi - 2a_{12}\sin\phi\cos\phi \\ a_{33}' &= a_{33} \\ a_{23}' &= \pm a_{23}\cos\phi \mp a_{31}\sin\phi \\ a_{31}' &= \pm a_{23}\sin\phi \pm a_{31}\cos\phi \\ a_{12}' &= -a_{11}\sin\phi\cos\phi + a_{22}\sin\phi\cos\phi + a_{12}(\cos^2\phi - \sin^2\phi) \end{aligned} \tag{3.103}$$

$$\begin{aligned} b_{11}' &= b_{11}\cos^2\phi + b_{22}\sin^2\phi + (b_{12} + b_{21})\sin\phi\cos\phi \\ b_{22}' &= b_{11}\sin^2\phi + b_{22}\cos^2\phi - (b_{12} + b_{21})\sin\phi\cos\phi \\ b_{33}' &= b_{33} \\ b_{23}' &= \mp b_{13}\sin\phi \pm b_{23}\cos\phi \\ b_{32}' &= \mp b_{31}\sin\phi \pm b_{32}\cos\phi \\ b_{31}' &= \pm b_{31}\cos\phi \pm b_{32}\sin\phi \\ b_{13}' &= \pm b_{13}\cos\phi \pm b_{23}\sin\phi \\ b_{12}' &= -b_{11}\sin\phi\cos\phi + b_{22}\sin\phi\cos\phi + b_{12}\cos^2\phi - b_{21}\sin^2\phi \\ b_{21}' &= -b_{11}\sin\phi\cos\phi + b_{22}\sin\phi\cos\phi - b_{12}\sin^2\phi + b_{21}\cos^2\phi \end{aligned} \tag{3.104}$$

The upper and lower signs, where an alternative occurs, refer respectively to a pure rotation and a rotation–inversion. These transformations may be regarded as linear substitutions. The characters of the three transformation matrices for an operation R_ϕ are easily seen to be respectively

$$2\cos\phi \pm 1$$

$$4\cos^2\phi \pm 2\cos\phi$$

$$4\cos^2\phi \pm 4\cos\phi + 1$$

Similarly, the transformation matrix and the respective character for any tensor of a higher order representing any given physical property can be deduced. However, there is a relation among the characters for various tensors. Thus, the character for an unsymmetric tensor having nine components is the square of the character for the vector of three components. Similarly, the character for a third-rank tensor (for example, piezoelectric constants and linear electro-optic constants in the Pockels effect, giving the relation between a vector and a symmetric tensor of the second rank) is the product of the characters appropriate to a vector and a symmetric tensor. The character for the case of photoelasticity and the quadratic electro-optic constants in the Kerr electro-optic effect (involving the relation between two symmetric tensors of the second rank, namely the stress or the strain tensor, on the one hand, and the optical coefficients, on the other, for photoelasticity; and the relation between the square of a vector, namely electric field, and the optical coefficients for the Kerr quadratic electro-optic effect) is the square of the character for the second-rank symmetric tensor, and so on. Obviously, when special relations exist between the various components, the character has to be deduced by writing the full transformation matrix or those terms in the matrix which contribute to the character. For example, the elastic coefficients show how a stress tensor gives rise to the strain tensor (here both are second-rank symmetric), but the maximum number of such coefficients is only 21 and not 36 because of the special relations $c_{ik} = c_{ki}$ (in the two-suffix notation) on account of the existence of the strain-energy function, and the character in such a case has to be calculated independently from the coefficients of the respective tensor components in the product transformation, remembering that $c_{ik} = c_{ki}$. The full details of calculating $\chi_\varrho(R)$ for the elasticity tensor are given earlier on page 122.

The characters for the various properties are given in column 4 of Table 3.14. In the third column of Table 3.14 are given the names of the physical properties, if any, corresponding to the tensor rank noted in column 2. The physical properties listed in any row are only representative and not complete. Pyroelectricity is the phenomenon in which a change in the temperature (a scalar) produces an electric moment (a vector). The converse effect is known as the electrocaloric effect. Production of opposite charges on opposite faces (i.e., an electric moment) on the application of hydrostatic pressure (a scalar) is a similar phenomenon. Thermal expansion is an elastic deformation (second-order symmetric tensor) caused by a change in temperature. Optical refraction arises out of an electric moment vector induced in a general direction by an incident electric field vector

Serial number	Tensor property as relation between	Physical property as example	Character $\chi_\varrho(R)$	Value of $\chi_i(R)$	Max. number of constants
1	Scalar and scalar	Density	1	1	1
2	Scalar and vector	Pyroelectricity (production of charges by hydrostatic pressure)	$2\cos\phi \pm 1$	1	3
3	a. Scalar and symmetric tensor of second rank	Thermal expansion	$4\cos^2\phi \pm 2\cos\phi$	1	6
	b. Vector and vector ($a_{ik} = a_{ki}$)	Optical, dielectric and magnetic polarizations, thermal and electrical conductivities	$4\cos^2\phi \pm 2\cos\phi$	1	6
4	Scalar and unsymmetric tensor of second rank	—	$4\cos^2\phi \pm 4\cos\phi + 1 = (2\cos\phi \pm 1)^2$	1	9
5	Vector and vector	—	$(2\cos\phi \pm 1)^2$	1	9
6	Vector and symmetric tensor of second rank	Piezoelectricity and linear electro-optic effect (Pockels effect)	$8\cos^3\phi \pm 8\cos^2\phi + 2\cos\phi = (2\cos\phi \pm 1)(4\cos^2\phi \pm 2\cos\phi)$	1	18
7	Vector and unsymmetric tensor of second rank	—	$8\cos^3\phi \pm 12\cos^2\phi + 6\cos\phi \pm 1 = (2\cos\phi \pm 1)(2\cos\phi \pm 1)^2$	1	27
8	Second-rank symmetric tensor and second-rank symmetric tensor ($c_{ik} = c_{ki}$)	Elasticity	$16\cos^4\phi \pm 8\cos^3\phi - 4\cos^2\phi + 1$	1	21
9	Second-rank symmetric tensor and second-rank symmetric tensor	Photoelasticity; effect of pressure on electrical conductivity	$16\cos^4\phi \pm 16\cos^3\phi + 4\cos^2\phi = (4\cos^2\phi \pm 2\cos\phi)^2$	1	36
10	Square of vector and symmetric tensor of second rank	Quadratic electro-optic effect (Kerr effect); electrostriction (quadratic)	$(4\cos^2\phi \pm 2\cos\phi)^2$	1	36

but with the additional relation $a_{ik} = a_{ki}$, etc. The piezoelectric stress and strain constants connect the components of the electric moment vector and the components of the second-rank stress or strain tensors. The converse piezoelectric effects also have the same number of independent constants. The linear electro-optic constants in the Pockels effect connect the components of the electric field vector and the six optical constants. The constants, as pointed out earlier, in these latter cases transform as products of a vector and the components of a second-rank symmetric tensor. The phenomena of elasticity, photoelasticity, and the quadratic electro-optic effect (Kerr effect) have been dealt with earlier.

We list in Table 3.14 properties which may be termed linear effects, in that the components of the resultant tensor are linear functions of the components of the applied parameter (tensors of rank one, two, etc.). The higher-order phenomena, such as the third-order elastic coefficients, have been deliberately left out.

3.3.3.24. *Computation of n_i for Some Typical Tensor Properties of a Crystal by Group Theoretical Methods Using the Relation $n_i = (1/N)\sum_\varrho h_\varrho \chi_\varrho(R)\chi_i(R)$*

a. First we consider the piezoelectric and Pockels effects as applied to three different crystal classes.

In the case of a third-rank tensor property such as piezoelectricity and Pockels' linear electro-optic effect relating a vector with a second-rank symmetric tensor, for any symmetry operation R we have

$$\chi_\varrho(R) = 8\cos^3\phi \pm 8\cos^2\phi + 2\cos\phi$$

(see Table 3.14, serial number 6).

Case I. Cubic System, Class 432 (O); Symmetry Elements E $3C_2$ $8C_3'$ $6C_2''$ $6C_4$

This gives $N = 1 + 8 + 3 + 6 + 6 = 24$. From the values of the characters $\chi_\varrho(R)$ for the symmetry elements of this class (see Table 3.15), we have

$$n_i = \tfrac{1}{24}(18 + 8\times 0 + 3\times(-2) + 6\times(-2) + 6\times 0) = \tfrac{1}{24}\times 0 = 0$$

Hence crystals belonging to class 432 (O) do not exhibit the piezoelectric and Pockels effects, in spite of the fact that they do not possess a center of inversion.

Table 3.15. *Table of R, ϕ, and $\chi_\varrho(R)$ for a Third-Rank Tensor Property of the Class 432 (O) of the Cubic System*

R	ϕ	$\chi_\varrho(R)$
E	0°	$2+8+8=\ 18$
C_3'	120°	$-1+2-1=\ 0$
C_2	180°	$-2+8-8=-2$
C_2''	180°	$-2+8-8=-2$
C_4	90°	$0+0+0=\ 0$

Case II. Trigonal System, Class 3m (C_{3v}); Symmetry Elements E $2C_3$ 3 σ_v

This gives $N = 1 + 2 + 3 = 6$. The character $\chi_\varrho(R)$ for E and C_3 have already been calculated in Case I. Now to calculate $\chi_\varrho(R)$ for σ_v, we note that σ_v comes under the category of improper rotations with $\phi = 0°$. Hence take the lower sign in the expressions for $\chi_\varrho(R)$ in terms of $\cos\phi$. Thus for σ_v we get

$$\chi_\varrho(R) = 2\cos\phi - 8\cos^2\phi + 8\cos^3\phi = 2 - 8 + 8 = 2$$

Therefore

$$n_i = \tfrac{1}{6}(18 + 2\times 0 + 3\times 2) = 4$$

a result which agrees with the matrices on pages 343 and 367.

Case III. Triclinic System, Class $\bar{1}$ (C_i); Symmetry Elements E i

This gives $N = 2$. Here i (being equal to $C_2\sigma_h$) is an improper rotation with $\phi = 180°$. (Note that σ_h and σ_v are improper rotations with $\phi = 0°$.) Thus for i we get

$$\chi_\varrho(R) = 2(-1) - 8(-1)^2 + 8(-1)^3 = -18$$

Therefore

$$n_i = \tfrac{1}{2}\{18 + (-18)\} = 0$$

Hence class C_i ($\bar{1}$), which possesses a center of symmetry, cannot exhibit piezoelectric and Pockels effects. In fact, similar calculations will show that all crystal classes possessing a center of inversion should necessarily fail to exhibit the piezoelectric and Pockels effects.

Table 3.16. Table of R, ϕ, and $\chi_\varrho(R)$ for a Second-Rank Tensor Property of the Class mm2 (C_{2v}) of the Orthorhombic System

R	ϕ	$\chi_\varrho(R)$
E	0°	$4 + 2 = 6$
C_2	180°	$4 - 2 = 2$
σ_v'	0°	$4 - 2 = 2$
σ_v''	0°	$4 - 2 = 2$

b. We shall next study four different types of tensor properties for the same crystal class. As an example, we take class $mm2$ (C_{2v}) and study four of its tensor properties: (1) optical polarization, (2) piezoelectric and Pockels effects, (3) elasticity, and (4) photoelasticity.

Orthorhombic System, Class mm2 (C_{2v}); Symmetry Elements E, C_2, σ_v', σ_v''

This gives $N = 4$. We consider four cases here:

Case I. Optical polarization:

$$\chi_\varrho(R) = 4\cos^2\phi \pm 2\cos\phi \qquad \text{(see Table 3.14, No. 3)}$$

From the values of the characters $\chi_\varrho(R)$ for the symmetry elements of this class (see Table 3.16) we have $n_i = \frac{1}{4}(6 + 2 + 2 + 2) = 3$.

Table 3.17. Table of R, ϕ, and $\chi_\varrho(R)$ for a Third-Rank Tensor Property of the Class mm2 (C_{2v}) of the Orthorhombic System

R	ϕ	$\chi_\varrho(R)$
E	0°	$8 + 8 + 2 = 18$
C_2	180°	$-8 + 8 - 2 = -2$
σ_v'	0°	$8 - 8 + 2 = 2$
σ_v''	0°	$8 - 8 + 2 = 2$

Table 3.18. Table of R, ϕ, and $\chi_Q(R)$ for a Fourth-Rank Tensor Property of the Class mm2 (C_{2v}) of the Orthorhombic System[a]

R	ϕ	$\chi_Q(R)$
E	0°	$16 + 8 - 4 + 1 = 21$
C_2	180°	$16 - 8 - 4 + 1 = 5$
σ_v'	0°	$16 - 8 - 4 + 1 = 5$
σ_v''	0°	$16 - 8 - 4 + 1 = 5$

[a] Here $c_{ij} = c_{ji}$.

Case II. Piezoelectric and Pockels effects:

$$\chi_Q(R) = 8\cos^3\phi \pm 8\cos^2\phi + 2\cos\phi \qquad \text{(see Table 3.14, No. 6)}$$

Using Table 3.17 we find $n_i = \frac{1}{4}(18 - 2 + 2 + 2) = 5$.

Case III. Elasticity:

$$\chi_Q(R) = 16\cos^4\phi \pm 8\cos^3\phi - 4\cos^2\phi + 1 \qquad \text{(see Table 3.14, No. 8)}$$

Using Table 3.18 we have $n_i = \frac{1}{4}(21 + 5 + 5 + 5) = 9$.

Case IV. Photoelasticity:

$$\chi_Q(R) = 16\cos^4\phi \pm 16\cos^3\phi + 4\cos^2\phi \qquad \text{(see Table 3.14, No. 9)}$$

Using Table 3.19 we have $n_i = \frac{1}{4}(36 + 4 + 4 + 4) = 12$.

Table 3.19. Table of R, ϕ, and $\chi_Q(R)$ for a Fourth-Rank Tensor Property of the Class mm2 (C_{2v}) of the Orthorhombic System[a]

R	ϕ	$\chi_Q(R)$
E	0°	$16 + 16 + 4 = 36$
C_2	180°	$16 - 16 + 4 = 4$
σ_v'	0°	$16 - 16 + 4 = 4$
σ_v''	0°	$16 - 16 + 4 = 4$

[a] Here $q_{ij} \neq q_{ji}$.

Table 3.20. Table of R, ϕ, and $\chi_Q(R)$ for a Fourth-Rank Tensor Property of the Class m3 (T_h) of the Cubic System[a]

R	ϕ	$\chi_Q(R)$
E	0°	$16 + 8 - 4 + 1 = 21$
C_3'	120°	$1 - 1 - 1 + 1 = 0$
C_2	180°	$16 - 8 - 4 + 1 = 5$
i	180°	$16 + 8 - 4 + 1 = 21$
$\tilde{S}_3$	60°	$1 - 1 - 1 + 1 = 0$
σ_h	0°	$16 - 8 - 4 + 1 = 5$

[a] Here $c_{ij} = c_{ji}$.

c. Next we study the cubic crystal class $m3$ (T_h) for which Pockels assigns three photoelastic constants q_{ij} while Bhagavantam assigns four q_{ij}; but both of them give the same number of c_{ij}, namely three, for elasticity.

Class m3 (T_h); Symmetry Elements $E\ 3C_2\ 8C_3'\ i\ 8\tilde{S}_3\ 3\sigma_h$

We consider two cases here:

Case I. Elasticity:

$$\chi_Q(R) = 16 \cos^4 \phi \pm 8 \cos^3 \phi - 4 \cos^2 \phi + 1$$

Table 3.21. Table of R, ϕ, and $\chi_Q(R)$ for a Fourth-Rank Tensor Property of the Class m3 (T_h) of the Cubic System[a]

R	ϕ	$\chi_Q(R)$
E	0°	$16 + 16 + 4 = 36$
C_3'	120°	$1 - 2 + 1 = 0$
C_2	180°	$16 - 16 + 4 = 4$
i	180°	$16 + 16 + 4 = 36$
$\tilde{S}_3$	60°	$1 - 2 + 1 = 0$
σ_h	0°	$16 - 16 + 4 = 4$

[a] Here $q_{ij} \neq q_{ji}$.

Here $\tilde{S}_3$ (being equal to $C_3 \cdot i$) has to be split up into $C_3 \cdot (C_2 \cdot \sigma_h)$ in order to find the value of ϕ. Thus $\tilde{S}_3$ involves a C_6 (which can include both C_3 and C_2) and a reflection plane. Hence the smallest value of ϕ is 60°. Using Table 3.20 we have $n_i = \frac{1}{24}(21 + 0 + 15 + 21 + 0 + 15) = 3$.

Case II. Photoelasticity:

$$\chi_\varrho(R) = 16 \cos^4 \phi \pm 16 \cos^3 \phi + 4 \cos^2 \phi \qquad \text{(see Table 3.14, No. 9)}$$

Using Table 3.21 we have $n_i = \frac{1}{24}(36 + 0 + 12 + 36 + 0 + 12) = 4$; this confirms Bhagavantam's scheme for q_{ij} and p_{ij}.

4

Elasticity of Crystals

4.1. Introduction

The importance of elasticity in the study of photoelasticity was pointed out in the preceding chapter. For that reason we devote the present chapter to studying the elasticity of crystals before proceeding to describe the experimental methods of photoelasticity in the next chapter. Furthermore, it may be pointed out that when the crystal under investigation is transparent, some of the optical techniques employed for the measurement of the mechanical properties can also be utilized to study the photoelastic behavior of the crystal, with only the addition of some polarization equipment.

A solid is deformed under the influence of an external stress, and when the applied stress is removed, the strain disappears, provided, however, that the strain is not unduly large. This ability on the part of the solid to regain its normal size and shape after the removal of stress is called elasticity. Different solids respond differently to external stress, i.e., the deformation is not the same for all bodies under the same stress; also for the same crystalline solid, the relation between stress and strain may be different in different directions.

So long as the strain produced is small, the ratio of stress to strain is a constant, called the modulus of elasticity. Thus, for a wire or a rod, in particular, the ratio of linear stress to the resulting linear strain is called Young's modulus Y, which is expressed in units of dynes per square centimeter per unit strain. Similar definitions for n (rigidity modulus) and K (bulk modulus) hold. These three moduli of elasticity, Y, n, and K, together with σ (Poisson's ratio) are sufficient to describe completely the elastic behavior of isotropic solids such as glass, vitreous silica, or plastics. Furthermore, Y, n, K, and σ are interrelated among themselves and, in fact, each can be expressed in terms of any two out of the remaining three elastic moduli; for example, for glasses, K and σ can be expressed in terms of Y and n thus:

$$K = \frac{nY}{3(3n - Y)}, \qquad \sigma = \frac{Y - 2n}{2n} \tag{4.1}$$

For crystals, including those possessing highest symmetry, namely cubic crystals, the elastic behavior cannot be described completely in terms of only two constants. For cubic crystals three constants are required. The elastic behavior of even cubic crystals is an anisotropic property.

The ratio stress/strain defines c_{ijkl}, called the stiffness coefficient, and the ratio strain/stress defines s_{ijkl}, called the compliance coefficient. Note that the moduli of elasticity Y, n, and K are not to be confused with s_{ijkl}; Y, n, and K are of the dimensions of c_{ijkl} (reciprocals of s_{ijkl}).

4.2. *Stress and Strain as Tensors*

4.2.1. *Stress as a Second-Rank Tensor*

Stress is defined as force per unit area. Force is a vector, and hence a force F has three components, say, F_x, F_y, and F_z, acting along the three axes X, Y, and Z of a Cartesian system of axes (Fig. 4.1a). The area A on which the force F acts can also be considered as a vector having three components A_x, A_y, and A_z which are, respectively, normal to the X, Y, and Z axes. The vector nature of A can be appreciated if we consider its components as projections on the YZ, ZX, and XY planes (Fig. 4.1a).

Now each of the force components acts on each of the areal components, thereby giving nine stress components. Thus the force F_x acting on the area A_x gives rise to a stress $P_{xx} = F_x/A_x$. The stress P_{xx} may now be defined as that caused by a force acting along the X axis on a plane normal to the X axis. The quantity P_{xx} is therefore a linear stress (Fig. 4.1b). Similarly P_{yz} may be defined as that caused by a force acting along the Y axis on a plane normal to the Z axis. Such a stress P_{yz} $(= F_y/A_z)$ is a shearing or tangential stress. We thus have nine components of stress

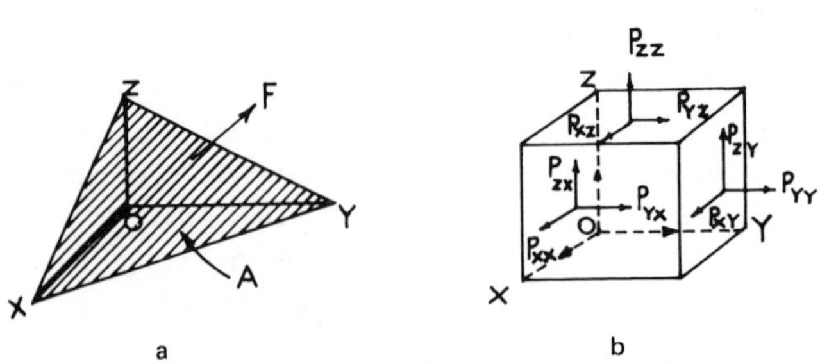

Fig. 4.1. Stress as a second-rank tensor: geometric representation. (a) The force F acts on a plane A. (b) The nine stress components P_{ij} acting on the solid.

forming the components of a second-rank tensor given by the following scheme:

	A_x	A_y	A_z
F_x	P_{xx}	P_{xy}	P_{xz}
F_y	P_{yx}	P_{yy}	P_{yz}
F_z	P_{zx}	P_{zy}	P_{zz}

In each of these nine stress components P_{ij} the first letter in the suffixes denotes the direction of the stress component (or of force F_i) and the second letter defines the normal to the plane A_j on which F_i is acting. The quantity P_{xx}, called a tensile or linear component of stress, tends to produce an elongation in the X direction, and contractions in the Y and Z directions. The quantities P_{xy} and P_{yx}, called shear components, tend to displace opposite faces of the cube in opposite parallel directions. The component P_{xy} is a stress due to a force F_x acting on a plane perpendicular to the Y axis, and hence it is a shearing force. The significance of the other components of stress is similar.

These nine components of stress, however, are not all independent of one another, as can be seen below. Consider a rectangular parallelepiped *ABCDEFGH* of sides x, y, z on which stress components P_{ij} are acting (Fig. 4.2). The force F_y acting on the face $ABCD = P_{yz} \cdot y \cdot x$. The force F_z acting on the face $FBCG = P_{zy} \cdot x \cdot z$. We assume now in problems in elasticity that there are no external forces capable of exerting a couple which produces body rotation since such a situation does not deform the body. Hence equating the moments of F_y and F_z about EH in order to avoid body rotation, we note that

$$(P_{yz} \cdot y \cdot x)z = (P_{zy} \cdot x \cdot z)y$$

Therefore $P_{yz} = P_{zy}$. Similarly $P_{xy} = P_{yx}$ and $P_{zx} = P_{xz}$.

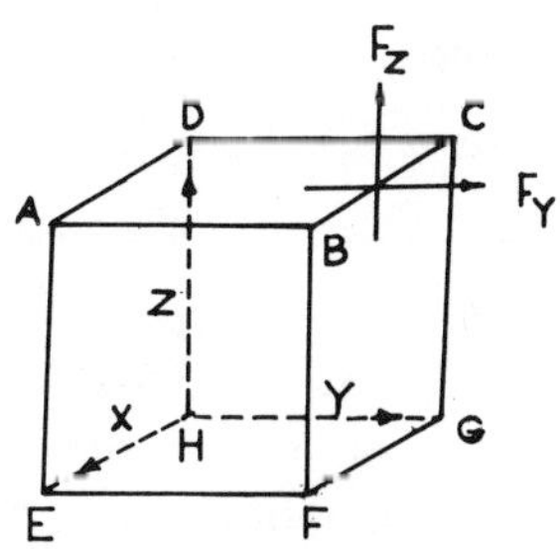

Fig. 4.2. A solid under shearing forces F_y and F_z (to prove the equivalence of P_{yz} and P_{zy}): geometric representation.

The nine components of a stress tensor P_{ij} can be now represented by the matrix

	A_x	A_y	A_z
F_x	P_{xx}	P_{xy}	P_{xz}
F_y	P_{xy}	P_{yy}	P_{yz}
F_z	P_{xz}	P_{yz}	P_{zz}

(a)

which is symmetric with respect to the diagonal; this describes a second-rank symmetric tensor.

It is usual to represent the six independent components P_{ij} of the second-rank symmetric tensor of stress in terms of P_i, which is in the single-suffix notation. Hence

$$P_{xx} = P_1, \quad P_{yy} = P_2, \quad P_{zz} = P_3, \quad P_{yz} = P_4, \quad P_{zx} = P_5, \quad P_{xy} = P_6$$

The matrix

	A_x	A_y	A_z
F_x	P_1	P_6	P_5
F_y	P_6	P_2	P_4
F_z	P_5	P_4	P_3

(b)

represents the P_i matrix.

Symmetric tensors of second rank can be studied with reference to a triaxial ellipsoid described by a second-degree equation (called a quadric). Particular significance attaches to the principal directions which are identified with the axes of the triaxial ellipsoid. When described with reference to these principal directions, there are only three components of stress, say, $P'_{x'x'}$, $P'_{y'y'}$, and $P'_{z'z'}$ in terms of the three principal axes X', Y', Z' of the triaxial ellipsoid. The stress components are then represented by the matrix

$$\begin{bmatrix} P'_{x'x'} & 0 & 0 \\ 0 & P'_{y'y'} & 0 \\ 0 & 0 & P'_{z'z'} \end{bmatrix} \qquad \text{(c)}$$

which can be obtained from the earlier symmetric matrix (a) by diagonalization.

The stress tensor was the first example of a tensor to be recognized in physics, and the name *tensor* itself recalls this origin, since it was a question of a system of tensions or elastic stresses. Gradually a whole series of quantities of the same nature was discovered. It was W. Voigt, the great German crystallographer, who discovered tensors, identified them, and gave them this name. Since then the idea of a tensor has become classic, and has acquired considerable importance in both physics and mathematics.

Certain tensor properties of some physical systems can be represented by relations which appreciably simplify the appearance of the tensor components. We have defined earlier the stress tensors P_{ij} acting on a continuous medium. For an ideal liquid medium, i.e., one with no viscosity, the tangential stresses vanish; whatever be the orientation of a surface element traced in the midst of a fluid at rest, the force across this surface element dS is always normal to dS. If p is the pressure, the tensile force per unit area will be $-p$ (treating compressional stresses as positive). Then writing N for tensile stress and T for shearing stress, the stress tensor simplifies to

$$\begin{bmatrix} P_{11} & P_{12} & P_{13} \\ P_{21} & P_{22} & P_{23} \\ P_{31} & P_{32} & P_{33} \end{bmatrix} = \begin{bmatrix} N_1 & T_3 & T_2 \\ T_3 & N_2 & T_1 \\ T_2 & T_1 & N_3 \end{bmatrix} = \begin{bmatrix} p & 0 & 0 \\ 0 & p & 0 \\ 0 & 0 & p \end{bmatrix}$$

Only one independent component is left, the pressure p which may be called the hydrostatic pressure. Thus the tensor is characterized by only a single number which can be treated ordinarily as a scalar.

We note that it is not the number of components alone that characterizes a tensor; the essential point is the nature of the transformations which the components undergo when there is a change of reference system. (The laws of transformation of tensors of different ranks have been dealt with in Chapter 2.)

4.2.2. *Strain as a Second-Rank Tensor*

In order to define the strain at a point in a body under deformation, we must consider how its position relative to adjacent points has changed. For this purpose consider two points P, Q close to each other in the body (Fig. 4.3). When the body is deformed, the line PQ in the body before deformation becomes $P'Q'$ under deformation. Let P be a point (x, y, z), and Q a point $(x + dx, y + dy, z + dz)$. Due to the strain, let P be displaced to P' so that its coordinates become $(x + u, y + v, z + w)$. Similarly let Q be displaced to Q', whose coordinates can be put in the form $(x + u + dx + (\partial u/\partial x)\, dx + (\partial u/\partial y)\, dy + (\partial u/\partial z)\, dz,\ \ y + v + dy + (\partial v/\partial x)\, dx + (\partial v/\partial y)\, dy + (\partial v/\partial z)\, dz,\ \ z + w + dz + (\partial w/\partial x)\, dx + (\partial w/\partial y)\, dy + (\partial w/\partial z)\, dz)$.

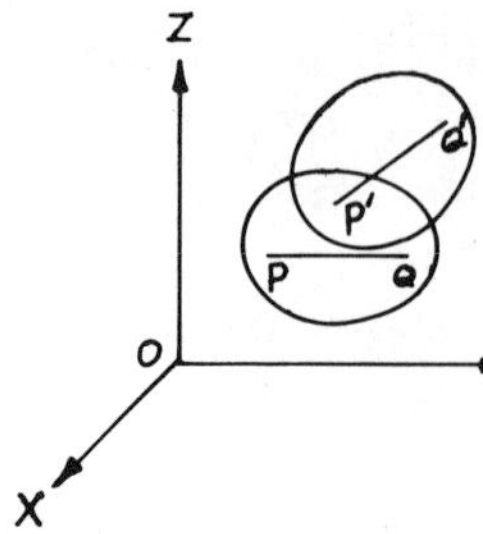

Fig. 4.3. Plane lamina under strain: geometric representation.

Let the displacement of Q with respect to P be given by a vector with components du, dv, dw. Then

$$du = \Delta(Q - P)_x = (Q' - P')_x - (Q - P)_x$$
$$= (Q' - Q)_x - (P' - P)_x$$
$$= \left(u + \frac{\partial u}{\partial x}dx + \frac{\partial u}{\partial y}dy + \frac{\partial u}{\partial z}dz\right) - u = \frac{\partial u}{\partial x}dx + \frac{\partial u}{\partial y}dy + \frac{\partial u}{\partial z}dz$$

and similarly for dv and dw. Thus

$$\begin{aligned} du &= \frac{\partial u}{\partial x}dx + \frac{\partial u}{\partial y}dy + \frac{\partial u}{\partial z}dz \\ dv &= \frac{\partial v}{\partial x}dx + \frac{\partial v}{\partial y}dy + \frac{\partial v}{\partial z}dz \\ dw &= \frac{\partial w}{\partial x}dx + \frac{\partial w}{\partial y}dy + \frac{\partial w}{\partial z}dz \end{aligned} \tag{4.2}$$

Now du, dv, dw are components of a vector defining the displacement of Q with respect to that of P, and dx, dy, dz are also components of a vector defining the position of Q with respect to P. Hence the nine terms or coefficients $\partial u/\partial x, \partial u/\partial y, \ldots, \partial w/\partial y, \partial w/\partial z$ are second-rank tensor components. Furthermore, we see here that both ∂u and ∂x measure lengths; hence $\partial u/\partial x$ is a measure of strain. Thus all the nine coefficients of $\partial u/\partial x$ are strain components. Let $\partial u/\partial x = \varepsilon_{xx}$, $\partial u/\partial y = \varepsilon_{xy}$, etc. Then the matrix of the nine strain components is represented by

$$\begin{bmatrix} \varepsilon_{xx} & \varepsilon_{xy} & \varepsilon_{xz} \\ \varepsilon_{yx} & \varepsilon_{yy} & \varepsilon_{yz} \\ \varepsilon_{zx} & \varepsilon_{zy} & \varepsilon_{zz} \end{bmatrix}$$

Here $\partial u/\partial x$ represents a linear strain, as do $\partial v/\partial y$ and $\partial w/\partial z$, whereas $\partial u/\partial y$, $\partial v/\partial z$, etc. are shearing strains.

The nine strain components are not all independent. We shall now show that $\varepsilon_{23} = \varepsilon_{32}$, $\varepsilon_{31} = \varepsilon_{13}$, and $\varepsilon_{21} = \varepsilon_{12}$. We shall argue out the case by taking a two-dimensional body, say a rectangular lamina $ABCD$ of sides dy, dz (Fig. 4.4). Let the rectangle $ABCD$ be sheared into $A'B'C'D'$. Here A is displaced to A', and AB to $A'B_1$ and then rotated through the angle θ_1, and AD is displaced to $A'D_1$ and then rotated through the angle θ_2. Let the coordinates of A' be (v, w). Then the coordinates of B' will be $(dy + v + (\partial v/\partial y)\,dy;\ w + (\partial w/\partial y)\,dy)$. Similarly, the coordinates of D' will be $(v + (\partial v/\partial z)\,dz;\ dz + w + (\partial w/\partial z)\,dz)$.

We are not so much interested in the coordinates of B' and D' as in the angles θ_1 and θ_2, which are the shears. We see from the figure that $\tan\theta_1 = \partial w/\partial y$ and $\tan\theta_2 = \partial v/\partial z$. When θ_1 and θ_2 are small, we can put $\theta_1 \cong \tan\theta_1$ and $\theta_2 \cong \tan\theta_2$. The total shear is obviously $(\theta_1 + \theta_2)$. Next, due to θ_1 the diagonal AC is rotated clockwise through $\theta_1/2$, and due to θ_2 AC is rotated anticlockwise through $\theta_2/2$.

If $\theta_1 \neq \theta_2$, the diagonal is rotated and hence the body is rotated as a whole through $\theta_1 - \theta_2$. But a rotation is not a deformation. Here in problems in elasticity the body rotation is ruled out. Hence we put $\theta_1 - \theta_2 = 0$ or $\theta_1 = \theta_2$. Therefore $\varepsilon_{zy} = \varepsilon_{yz}$.

The total shearing strain in the YOZ plane is

$$\theta_1 + \theta_2 = 2\theta_1 = 2\,\frac{\partial w}{\partial y} = 2\varepsilon_{zy} = 2\varepsilon_{yz}$$

(since $\partial w/\partial y = \theta_1 = \theta_2 = \partial v/\partial z$).

Hence the strain-tensor components can be arranged thus:

$$\begin{bmatrix} \varepsilon_{xx} & \varepsilon_{xy} & \varepsilon_{xz} \\ \varepsilon_{xy} & \varepsilon_{yy} & \varepsilon_{yz} \\ \varepsilon_{xz} & \varepsilon_{yz} & \varepsilon_{zz} \end{bmatrix} \quad \text{or} \quad \begin{bmatrix} \varepsilon_1 & \frac{1}{2}\varepsilon_6 & \frac{1}{2}\varepsilon_5 \\ \frac{1}{2}\varepsilon_6 & \varepsilon_2 & \frac{1}{2}\varepsilon_4 \\ \frac{1}{2}\varepsilon_5 & \frac{1}{2}\varepsilon_4 & \varepsilon_3 \end{bmatrix} \tag{4.3}$$

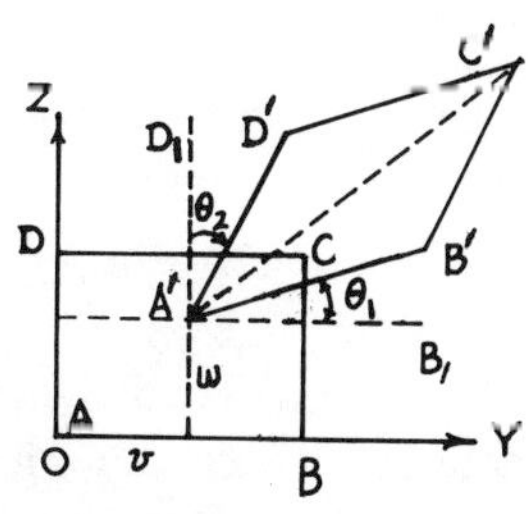

Fig. 4.4. Equivalence of ε_{yz} and ε_{zy} of a strained lamina: geometric representation.

where

$$\varepsilon_1 = \varepsilon_{xx}, \quad \varepsilon_4 = 2\varepsilon_{yz}$$
$$\varepsilon_2 = \varepsilon_{yy}, \quad \varepsilon_5 = 2\varepsilon_{zx}$$
$$\varepsilon_3 = \varepsilon_{zz}, \quad \varepsilon_6 = 2\varepsilon_{xy}$$

The first three quantities ε_{xx}, ε_{yy}, ε_{zz} correspond to the fractional expansions and contractions of infinitesimal line elements passing through P parallel to the x, y, z axes, respectively. The second three, ε_{yz}, ε_{zx}, ε_{xy}, correspond to the components of shearing strain in the planes denoted by their suffixes.

4.3. *Hooke's Law*

Whenever a body is subjected to a stress, it will deform in a manner determined by the components of the stress tensor and the elastic properties of the body. If a linear relationship exists between the stress and the strain components, the material is said to be elastic, and the constants involved in the linear equations are the elastic constants of the body.

It is found experimentally that for most solids the observed strains are proportional to the applied load, provided the load does not exceed a certain value which is known as the elastic limit.

4.3.1. *Generalized Form of Hooke's Law with Elastic Constants c_{ij} and s_{ij} and the Matrices of c_{ij} and s_{ij} for the 32 Point Groups*

This experimental law is stated mathematically thus: Each of the components of stress is at any point a linear function of all the components of strain. Denoting stress and strain in single-suffix notation, this generalized form of Hooke's law may be written thus:

$$\begin{aligned}
P_1 &= c_{11}\varepsilon_1 + c_{12}\varepsilon_2 + c_{13}\varepsilon_3 + c_{14}\varepsilon_4 + c_{15}\varepsilon_5 + c_{16}\varepsilon_6 \\
P_2 &= c_{21}\varepsilon_1 + c_{22}\varepsilon_2 + c_{23}\varepsilon_3 + c_{24}\varepsilon_4 + c_{25}\varepsilon_5 + c_{26}\varepsilon_6 \\
P_3 &= c_{31}\varepsilon_1 + c_{32}\varepsilon_2 + c_{33}\varepsilon_3 + c_{34}\varepsilon_4 + c_{35}\varepsilon_5 + c_{36}\varepsilon_6 \\
P_4 &= c_{41}\varepsilon_1 + c_{42}\varepsilon_2 + c_{43}\varepsilon_3 + c_{44}\varepsilon_4 + c_{45}\varepsilon_5 + c_{46}\varepsilon_6 \\
P_5 &= c_{51}\varepsilon_1 + c_{52}\varepsilon_2 + c_{53}\varepsilon_3 + c_{54}\varepsilon_4 + c_{55}\varepsilon_5 + c_{56}\varepsilon_6 \\
P_6 &= c_{61}\varepsilon_1 + c_{62}\varepsilon_2 + c_{63}\varepsilon_3 + c_{64}\varepsilon_4 + c_{65}\varepsilon_5 + c_{66}\varepsilon_6
\end{aligned} \tag{4.4a}$$

where the 36 coefficients c_{ij} are called the elastic stiffness constants of the material. These equations are more concisely expressed in the dummy-suffix notation thus:

$$P_i = c_{ij}\varepsilon_j \tag{4.4b}$$

These 36 constants are not all independent. We show below that the condition for the elastic energy to be a single-valued function of the strain is that any coefficient c_{ij} be equal to the coefficient c_{ji}.

We show that $c_{ij} = c_{ji}$: Let $\varepsilon_1, \varepsilon_2, \varepsilon_3, \varepsilon_4, \varepsilon_5$, and ε_6 be the strain components developed in a body which has undergone a deformation. Let the body be further strained by an infinitesimal amount, and let the strain components now be $\varepsilon_1 + d\varepsilon_1$, $\varepsilon_2 + d\varepsilon_2$, $\varepsilon_3 + d\varepsilon_3$, $\varepsilon_4 + d\varepsilon_4$, $\varepsilon_5 + d\varepsilon_5$, and $\varepsilon_6 + d\varepsilon_6$. The potential energy ϕ stored in the body is a function of the strains. No two strains are interrelated. Hence we can express $d\phi$, the increase in the strain energy, in terms of partial differentials. Thus

$$d\phi = \frac{\partial\phi}{\partial\varepsilon_1} d\varepsilon_1 + \frac{\partial\phi}{\partial\varepsilon_2} d\varepsilon_2 + \frac{\partial\phi}{\partial\varepsilon_3} d\varepsilon_3 + \frac{\partial\phi}{\partial\varepsilon_4} d\varepsilon_4 + \frac{\partial\phi}{\partial\varepsilon_5} d\varepsilon_5 + \frac{\partial\phi}{\partial\varepsilon_6} d\varepsilon_6 \tag{4.5}$$

Now from Hooke's law the stress components P_1 and P_2 are given by

$$\begin{aligned} P_1 &= c_{11}\varepsilon_1 + c_{12}\varepsilon_2 + c_{13}\varepsilon_3 + c_{14}\varepsilon_4 + c_{15}\varepsilon_5 + c_{16}\varepsilon_6 \\ P_2 &= c_{21}\varepsilon_1 + c_{22}\varepsilon_2 + c_{23}\varepsilon_3 + c_{24}\varepsilon_4 + c_{25}\varepsilon_5 + c_{26}\varepsilon_6 \end{aligned} \tag{4.6}$$

Since the strains are independent of one another, differentiating P with respect to ε, we have

$$\partial P_1/\partial\varepsilon_2 = c_{12}, \qquad \partial P_2/\partial\varepsilon_1 = c_{21} \tag{4.7}$$

Now, the work done per unit volume is equal to each stress component times the corresponding strain component (since, by the definition of work, stress $\times$ perpendicular strain $= 0$). Therefore

$$d\phi = P_1\, d\varepsilon_1 + P_2\, d\varepsilon_2 + P_3\, d\varepsilon_3 + \cdots + P_6\, d\varepsilon_6 \tag{4.8}$$

Comparing (4.5) and (4.8), we see that

$$P_1 = \partial\phi/\partial\varepsilon_1, \qquad P_2 = \partial\phi/\partial\varepsilon_2$$

Therefore

$$c_{12} = (\partial/\partial\varepsilon_2)P_1 = (\partial/\partial\varepsilon_2)(\partial\phi/\partial\varepsilon_1) = (\partial^2\phi/\partial\varepsilon_2\partial\varepsilon_1) \tag{4.9a}$$

Similarly,

$$c_{21} = \partial^2\phi/\partial\varepsilon_1\partial\varepsilon_2 \tag{4.9b}$$

Now ϕ, the potential energy, depends upon the final strains alone but not on their sequence. Therefore

$$\partial^2\phi/\partial\varepsilon_2\partial\varepsilon_1 = \partial^2\phi/\partial\varepsilon_1\partial\varepsilon_2{}^* \quad (4.10)$$

and

$$\therefore c_{12} = c_{21} \quad (4.11)$$

This relation reduces the number of independent coefficients from 36 to 21, which are listed in Table 4.2.

In a completely aeolotropic material where no plane or axis of symmetry exists (as in a triclinic crystal) the values of 21 different quantities must be known in order to define the elastic properties of the medium. But when the material has axes or planes of symmetry, relations may be established between these coefficients, and the number of independent elastic constants is consequently reduced. The method of reduction of the number of fourth-rank tensor components (under which come the elastic constants) is discussed in Chapter 3. Thus for a cubic crystal there are only three independent constants of elasticity. In an isotropic solid the values of the coefficients must be independent of the set of rectangular axes chosen, and if this condition is applied to the equations only two independent constants remain. We thus have, for a cubic crystal,

$$c_{11} = c_{22} = c_{33}$$
$$c_{12} = c_{13} = c_{21} = c_{23} = c_{31} = c_{32}$$
$$c_{44} = c_{55} = c_{66}$$

and all the other 24 out of the 36 stiffness constants become zero.

* The elastic energy of a strained crystal can be described in terms of an initial state and a strained state. In straining the crystal, it is possible to apply the strain and remove it as slowly as desired, and in so doing the energy goes from one equilibrium state to another. These changes are thermodynamically reversible. We can put the same information in a slightly different way thus: Let a crystal receive two strains, one after another, say ε_1 and ε_5. The energy stored per unit volume is governed by the strains. A certain amount of energy is stored due to the first strain, say ε_1, and then, in addition to this, a certain other amount of energy is stored due to a second strain, say ε_5. Each strain is independent of any other strain, so each amount of energy stored due to a particular strain is independent of its sequence in the series of operations, namely whether ε_1 first and then ε_5, or ε_5 first and then ε_1. That is, the total amount of strain energy is independent of the sequence of its occurrence. Hence the strain energy function is a perfect differential with respect to ε_1 and ε_5 or ε_i and ε_j. Thus $\partial^2\phi/\partial\varepsilon_i\partial\varepsilon_j = \partial^2\phi/\partial\varepsilon_j\partial\varepsilon_i$.

For isotropic solids, c_{12} and c_{44} are replaced by λ and μ, called Lamé's constants. Then

$$c_{11} = \lambda + 2\mu \qquad \text{of} \qquad c_{44} = \frac{c_{11} - c_{12}}{2} \tag{4.12}$$

Alternately, the generalized form of Hooke's law may be expressed thus:

$$\begin{aligned}
\varepsilon_1 &= s_{11}P_1 + s_{12}P_2 + s_{13}P_3 + s_{14}P_4 + s_{15}P_5 + s_{16}P_6 \\
\varepsilon_2 &= s_{21}P_1 + s_{22}P_2 + s_{23}P_3 + s_{24}P_4 + s_{25}P_5 + s_{26}P_6 \\
\varepsilon_3 &= s_{31}P_1 + s_{32}P_2 + s_{33}P_3 + s_{34}P_4 + s_{35}P_5 + s_{36}P_6 \\
\varepsilon_4 &= s_{41}P_1 + s_{42}P_2 + s_{43}P_3 + s_{44}P_4 + s_{45}P_5 + s_{46}P_6 \\
\varepsilon_5 &= s_{51}P_1 + s_{52}P_2 + s_{53}P_3 + s_{54}P_4 + s_{55}P_5 + s_{56}P_6 \\
\varepsilon_6 &= s_{61}P_1 + s_{62}P_2 + s_{63}P_3 + s_{64}P_4 + s_{65}P_5 + s_{66}P_6
\end{aligned} \tag{4.13a}$$

where the coefficients s_{ij} are called the elastic compliance constants.

As in the case of the elastic stiffness constants, here also we can express Eq. (4.13a) more concisely in the dummy suffix notation; thus

$$\varepsilon_i = s_{ij}P_j \tag{4.13b}$$

Here again, as proved in the case of c_{ij}, it can be shown that $s_{ij} = s_{ji}$, thus reducing the total number of constants from 36 to 21. The number of nonvanishing independent s_{ij} is further reduced depending upon the nature and number of elements of symmetry of the crystal specimen under study. Thus, as in the case of c_{ij}, for a cubic crystal there are only three non-vanishing independent compliance constants s_{ij}, namely, s_{11}, s_{12}, and s_{44}.

From the expressions for ε_i and P_j related by c_{ij} and s_{ij} given in the above two sets of equations (4.4a) and (4.13a), we note that c_{ij} and s_{ij} are interrelated among themselves [see Eq. (3.11)]. Thus c_{ij} can be expressed in terms of s_{ij} and vice versa.

The matrices of c_{ij} and s_{ij} in the two-suffix notation for all the 32 classes of crystals are given in Table 4.1.

4.3.2. Generalized Form of Hooke's Law with Elastic Constants c_{ijkl} and s_{ijkl}

In tensor notation, Hooke's law can be stated as follows: If a general homogeneous stress P_{ij} is applied to a crystal, the resulting homogeneous strain ε_{kl} is such that each component is linearly related to all the stress

Table 4.1. *Elastic Constants (Matrices of c_{ij} and s_{ij} of the 32 Crystallographic Point Groups)*

Triclinic system: classes C_1, C_i $(1, \bar{1})$ (21 constants)

c_{11}	c_{12}	c_{13}	c_{14}	c_{15}	c_{16}	s_{11}	s_{12}	s_{13}	s_{14}	s_{15}	s_{16}
	c_{22}	c_{23}	c_{24}	c_{25}	c_{26}		s_{22}	s_{23}	s_{24}	s_{25}	s_{26}
		c_{33}	c_{34}	c_{35}	c_{36}			s_{33}	s_{34}	s_{35}	s_{36}
			c_{44}	c_{45}	c_{46}				s_{44}	s_{45}	s_{46}
				c_{55}	c_{56}					s_{55}	s_{56}
					c_{66}						s_{66}

Monoclinic system: classes C_s, C_2, C_{2h} $(m, 2, 2/m)$ (13 constants); diad $\parallel X_3$

c_{11}	c_{12}	c_{13}	0	0	c_{16}	s_{11}	s_{12}	s_{13}	0	0	s_{16}
	c_{22}	c_{23}	0	0	c_{26}		s_{22}	s_{23}	0	0	s_{26}
		c_{33}	0	0	c_{36}			s_{33}	0	0	s_{36}
			c_{44}	c_{45}	0				s_{44}	s_{45}	0
				c_{55}	0					s_{55}	0
					c_{66}						s_{66}

Monoclinic system: Classes C_s, C_2, C_{2h} $(m, 2, 2/m)$ (13 constants); diad $\parallel X_2$ according to IEEE Standards on Piezoelectric Crystals (1949)

c_{11}	c_{12}	c_{13}	0	c_{15}	0	s_{11}	s_{12}	s_{13}	0	s_{15}	0
	c_{22}	c_{23}	0	c_{25}	0		s_{22}	s_{23}	0	s_{25}	0
		c_{33}	0	c_{35}	0			s_{33}	0	s_{35}	0
			c_{44}	0	c_{46}				s_{44}	0	s_{46}
				c_{55}	0					s_{55}	0
					c_{66}						s_{66}

Orthorhombic system: classes D_2, C_{2v}, D_{2h} (222, *mm*2, *mmm*) (9 constants)

c_{11}	c_{12}	c_{13}	0	0	0	s_{11}	s_{12}	s_{13}	0	0	0
	c_{22}	c_{23}	0	0	0		s_{22}	s_{23}	0	0	0
		c_{33}	0	0	0			s_{33}	0	0	0
			c_{44}	0	0				s_{44}	0	0
				c_{55}	0					s_{55}	0
					c_{66}						s_{66}

Tetragonal system: classes C_4, S_4, C_{4h} (4, $\bar{4}$, 4/*m*) (7 constants)

c_{11}	c_{12}	c_{13}	0	0	c_{16}	s_{11}	s_{12}	s_{13}	0	0	s_{16}
	c_{11}	c_{13}	0	0	$-c_{16}$		s_{11}	s_{13}	0	0	$-s_{16}$
		c_{33}	0	0	0			s_{33}	0	0	0
			c_{44}	0	0				s_{44}	0	0
				c_{44}	0					s_{44}	0
					c_{66}						s_{66}

Tetragonal system: classes D_{2d}, C_{4v}, D_4, D_{4h} ($\bar{4}$2*m*, 4*mm*, 422, 4/*mmm*) (6 constants)

c_{11}	c_{12}	c_{13}	0	0	0	s_{11}	s_{12}	s_{13}	0	0	0
	c_{11}	c_{13}	0	0	0		s_{11}	s_{13}	0	0	0
		c_{33}	0	0	0			s_{33}	0	0	0
			c_{44}	0	0				s_{44}	0	0
				c_{44}	0					s_{44}	0
					c_{66}						s_{66}

continued overleaf

Table 4.1 *(continued)*

Trigonal system: classes C_3, C_{3i} $(3, \bar{3})$ (7 constants)

c_{11}	c_{12}	c_{13}	c_{14}	$-c_{25}$	0	s_{11}	s_{12}	s_{13}	s_{14}	$-s_{25}$	0
	c_{11}	c_{13}	$-c_{14}$	c_{25}	0		s_{11}	s_{13}	$-s_{14}$	s_{25}	0
		c_{33}	0	0	0			s_{33}	0	0	0
			c_{44}	0	c_{25}				s_{44}	0	$2s_{25}$
				c_{44}	c_{14}					s_{44}	$2s_{14}$
					$\frac{1}{2}(c_{11}-c_{12})$						$2(s_{11}-s_{12})$

Trigonal system: classes C_{3v}, D_3, D_{3d} $(3m, 32, \bar{3}m)$ (6 constants)

c_{11}	c_{12}	c_{13}	c_{14}	0	0	s_{11}	s_{12}	s_{13}	s_{14}	0	0
	c_{11}	c_{13}	$-c_{14}$	0	0		s_{11}	s_{13}	$-s_{14}$	0	0
		c_{33}	0	0	0			s_{33}	0	0	0
			c_{44}	0	0				s_{44}	0	0
				c_{44}	c_{14}					s_{44}	$2s_{14}$
					$\frac{1}{2}(c_{11}-c_{12})$						$2(s_{11}-s_{12})$

Hexagonal system: classes C_{3h}, C_6, C_{6h}, D_{3h}, C_{6v}, D_6, D_{6h} $(\bar{6}, 6, 6/m, \bar{6}2m, 6mm, 622, 6/mmm)$ (5 constants)

c_{11}	c_{12}	c_{13}	0	0	0	s_{11}	s_{12}	s_{13}	0	0	0
	c_{11}	c_{13}	0	0	0		s_{11}	s_{13}	0	0	0

		c_{33}	0	0	0			s_{33}	0	0	0
			c_{44}	0	0				s_{44}	0	0
				c_{44}	0					s_{44}	0
					$\frac{1}{2}(c_{11}-c_{12})$						$2(s_{11}-s_{12})$

Cubic system: classes T, T_h, T_d, O, O_h (23, $m3$, $\bar{4}3m$, 432, $m3m$) (3 constants)

c_{11}	c_{12}	c_{12}	0	0	0	s_{11}	s_{12}	s_{12}	0	0	0
	c_{11}	c_{12}	0	0	0		s_{11}	s_{12}	0	0	0
		c_{11}	0	0	0			s_{11}	0	0	0
			c_{44}	0	0				s_{44}	0	0
				c_{44}	0					s_{44}	0
					c_{44}						s_{44}

Isotropic solids (2 constants)

c_{11}	c_{12}	c_{12}	0	0	0	s_{11}	s_{12}	s_{12}	0	0	0
	c_{11}	c_{12}	0	0	0		s_{11}	s_{12}	0	0	0
		c_{11}	0	0	0			s_{11}	0	0	0
			$\frac{1}{2}(c_{11}-c_{12})$	0	0				$2(s_{11}-s_{12})$	0	0
				$\frac{1}{2}(c_{11}-c_{12})$	0					$2(s_{11}-s_{12})$	0
					$\frac{1}{2}(c_{11}-c_{12})$						$2(s_{11}-s_{12})$

components. Thus, for example,

$$\varepsilon_{11} = s_{1111}P_{11} + s_{1112}P_{12} + s_{1113}P_{13} + s_{1121}P_{21} + s_{1122}P_{22} + s_{1123}P_{23} + s_{1131}P_{31} + s_{1132}P_{32} + s_{1133}P_{33} \tag{4.14}$$

Similar relations hold for $\varepsilon_{12}, \varepsilon_{13}, \varepsilon_{22}$, etc.

Another form of Hooke's law is the following: If a general homogeneous stress P_{ij} is applied to a crystal, the resulting homogeneous strain is such that each component of stress P_{ij} is linearly related to all the strain components ε_{kl}. Thus, for example,

$$P_{11} = c_{1111}\varepsilon_{11} + c_{1112}\varepsilon_{12} + c_{1113}\varepsilon_{13} + c_{1121}\varepsilon_{21} + c_{1122}\varepsilon_{22} + c_{1123}\varepsilon_{23} + c_{1131}\varepsilon_{31} + c_{1132}\varepsilon_{32} + c_{1133}\varepsilon_{33} \tag{4.15}$$

Similar relations hold for P_{12}, P_{13}, P_{21}, etc.

From the expressions for ε_{ij} and P_{kl} given in the above two sets of equations, we find that c_{ijkl} and s_{ijkl} are interrelated among themselves. Thus c_{ijkl} can be expressed in terms of s_{ijkl}, and also s_{ijkl} can be expressed in terms of c_{ijkl}.*

We note here that the basic elastic properties of a solid are the elastic rigidity or the stiffness $\partial P/\partial\varepsilon$ and the elastic compliance $\partial\varepsilon/\partial P$; both are fourth-rank tensor properties. But they are not reciprocals in the sense that scalars are, even though they bear an inverse relationship to each other. The relationship between the components of $\partial P/\partial\varepsilon$ and $\partial\varepsilon/\partial P$ is given through two sets of elastic constants which are defined above. Denoting the rigidity tensor or stiffness by c_{ijkl} and the compliance tensor by s_{ijkl}, we can relate the stresses and strains by either of the following sets of equations which are shortened forms of Eqs. (4.14) and (4.15):

$$\varepsilon_{ij} = s_{ijkl}P_{kl} \tag{4.16}$$

$$P_{ij} = c_{ijkl}\varepsilon_{kl} \tag{4.17}$$

Both of these involve nine equations, each of which contains nine terms. However, both P_{kl} and ε_{ij} are second-rank symmetric tensors. From the symmetry of P_{kl} we have

$$\varepsilon_{ij} = s_{ijkl}P_{kl} = s_{ijlk}P_{lk}$$

and therefore

$$s_{ijkl} = s_{ijlk}$$

* When the elastic constants are expressed in the 2-suffix notation, their interrelation becomes much simpler and is given by Eq. (3.11).

Similarly, the symmetry of ε_{ij} requires that

$$s_{ijkl} = s_{jikl}$$

and also, considering the symmetry of ε_{ij} and of P_{kl} simultaneously we find

$$s_{ijkl} = s_{jilk}$$

Therefore, only 36 of the components of the compliance tensor need be distinct.

This is also true for the components of the rigidity tensor c_{ijkl}.

We can reduce the number of distinct components still further by considering the fact that $c_{ijkl} = c_{klij}$ (which was proved on page 144 for stiffness constants in the two-suffix notation). The net effect of the conditions deduced above, namely,

$$s_{ijkl} = s_{ijlk}, \qquad s_{ijkl} = s_{jikl}, \qquad s_{ijkl} = s_{jilk}, \qquad s_{ijkl} = s_{klij}$$

and similarly,

$$c_{ijkl} = c_{ijlk}, \qquad c_{ijkl} = c_{jikl}, \qquad c_{ijkl} = c_{jilk}, \qquad c_{ijkl} = c_{klij}$$

will enable us to reduce the number of distinct components of s_{ijkl} and c_{ijkl} from 81 to 21. These are listed in Tables 4.2 and 4.3. Only the most asymmetric type of crystal will have this number of 21 distinct components. Many of the components become equal to each other or vanish for crystals of symmetry higher than triclinic.

The effect of crystal symmetry on the compliance and stiffness components can be studied along the lines adopted in Chapter 3 to study the effect of crystal symmetry on the stress-optical and strain-optical constants.

4.3.3. *Interrelation between c_{ijkl} and c_{mn} and between s_{ijkl} and s_{mn}*

The relationship between the elastic constants c_{ij} and the components of the rigidity tensor (stiffness constants) can be shown by expanding Eq. (4.17) (i.e., writing down all terms explicitly),

$$P_{ij} = c_{ijkl}\varepsilon_{kl}$$

and (bearing in mind the notational change from P_{ij} to P_i and ε_{kl} to ε_k) comparing the results with the corresponding member of the set of equations (4.4a). Thus, for example,

$$P_{12} = c_{1211}\varepsilon_{11} + c_{1222}\varepsilon_{22} + c_{1233}\varepsilon_{33} + 2c_{1223}\varepsilon_{23} + 2c_{1231}\varepsilon_{31} + 2c_{1212}\varepsilon_{12} \quad (4.18)$$

Table 4.2. Relation[a] *between s_{ij} and s_{ijkl}*

$$s_{11} = s_{1111}$$
$$s_{12} = s_{1122} = s_{2211}$$
$$s_{13} = s_{1133} = s_{3311}$$
$$\tfrac{1}{2}s_{14} = s_{1123} = s_{1132} = s_{2311} = s_{3211}$$
$$\tfrac{1}{2}s_{15} = s_{1131} = s_{1113} = s_{3111} = s_{1311}$$
$$\tfrac{1}{2}s_{16} = s_{1112} = s_{1121} = s_{1211} = s_{2111}$$
$$s_{22} = s_{2222}$$
$$s_{23} = s_{2233} = s_{3322}$$
$$\tfrac{1}{2}s_{24} = s_{2223} = s_{2232} = s_{2322} = s_{3222}$$
$$\tfrac{1}{2}s_{25} = s_{2231} = s_{2213} = s_{3122} = s_{1322}$$
$$\tfrac{1}{2}s_{26} = s_{2212} = s_{2221} = s_{1222} = s_{2122}$$
$$s_{33} = s_{3333}$$
$$\tfrac{1}{2}s_{34} = s_{3323} = s_{3332} = s_{2333} = s_{3233}$$
$$\tfrac{1}{2}s_{35} = s_{3331} = s_{3313} = s_{3133} = s_{1333}$$
$$\tfrac{1}{2}s_{36} = s_{3312} = s_{3321} = s_{1233} = s_{2133}$$
$$\tfrac{1}{4}s_{44} = s_{2323} = s_{2332} = s_{3223} = s_{3232}$$
$$\tfrac{1}{4}s_{45} = s_{2331} = s_{2313} = s_{3231} = s_{3213} = s_{3123} = s_{3132} = s_{1323} = s_{1332}$$
$$\tfrac{1}{4}s_{46} = s_{2312} = s_{2321} = s_{3212} = s_{3221} = s_{1223} = s_{1232} = s_{2123} = s_{2132}$$
$$\tfrac{1}{4}s_{55} = s_{3131} = s_{3113} = s_{1331} = s_{1313}$$
$$\tfrac{1}{4}s_{56} = s_{3112} = s_{3121} = s_{1312} = s_{1321} = s_{1231} = s_{1213} = s_{2131} = s_{2113}$$
$$\tfrac{1}{4}s_{66} = s_{1212} = s_{1221} = s_{2112} = s_{2121}$$

[a] As already pointed out, it is important to note that while $c_{ijkl} = c_{mn}$ for all values of i, j, k, l (1, 2, 3) and m, n (1 to 6), the relationship between s_{ijkl} and s_{mn} is not as simple. Utmost care is necessary when substituting the numerical values of s_{ijkl} or s_{mn} in order to calculate the individual values of q_{ijkl} or q_{mn} and p_{ijkl} or p_{mn} from the experimentally observed data.

Comparison of Eq. (4.18) with the last equation of (4.4a) gives

$$c_{1211} = c_{61}, \qquad c_{1222} = c_{62}, \qquad c_{1233} = c_{63}$$
$$c_{1223} = c_{64}, \qquad c_{1231} = c_{65}, \qquad c_{1212} = c_{66}$$

In this manner, we can construct the relationships between c_{ijkl} and c_{mn}, which can be put in the compact form

$$c_{ijkl} = c_{mn} \tag{4.19}$$

where

$$i, j, k, l = 1, 2, 3 \qquad \text{and} \qquad m, n = 1, 2, 3, 4, 5, 6$$

Table 4.3. *Relation between c_{ij} and c_{ijkl}*

$c_{11} = c_{1111}$

$c_{12} = c_{1122} = c_{2211}$

$c_{13} = c_{1133} = c_{3311}$

$c_{14} = c_{1123} = c_{1132} = c_{2311} = c_{3211}$

$c_{15} = c_{1131} = c_{1113} = c_{3111} = c_{1311}$

$c_{16} = c_{1112} = c_{1121} = c_{1211} = c_{2111}$

$c_{22} = c_{2222}$

$c_{23} = c_{2233} = c_{3322}$

$c_{24} = c_{2223} = c_{2232} = c_{2322} = c_{3222}$

$c_{25} = c_{2231} = c_{2213} = c_{3122} = c_{1322}$

$c_{26} = c_{2212} = c_{2221} = c_{1222} = c_{2122}$

$c_{33} = c_{3333}$

$c_{34} = c_{3323} = c_{3332} = c_{2333} = c_{3233}$

$c_{35} = c_{3331} = c_{3313} = c_{3133} = c_{1333}$

$c_{36} = c_{3312} = c_{3321} = c_{1233} = c_{2133}$

$c_{44} = c_{2323} = c_{2332} = c_{3223} = c_{3232}$

$c_{45} = c_{2331} = c_{2313} = c_{3231} = c_{3213} = c_{3123} = c_{3132} = c_{1323} = c_{1332}$

$c_{46} = c_{2312} = c_{2321} = c_{3212} = c_{3221} = c_{1223} = c_{1232} = c_{2123} = c_{2132}$

$c_{55} = c_{3131} = c_{3113} = c_{1331} = c_{1313}$

$c_{56} = c_{3112} = c_{3121} = c_{1312} = c_{1321} = c_{1231} = c_{1213} = c_{2131} = c_{2113}$

$c_{66} = c_{1212} = c_{1221} = c_{2112} = c_{2121}$

In the matrix or two-suffix notation, for the stiffness constant c_{ij} we have

$$P_i = c_{ij}\varepsilon_j \qquad (i, j = 1, 2, \ldots, 6) \tag{4.20}$$

Similarly the compliance tensor components s_{ijkl} can be shortened to the matrix or two-suffix notation. Here again we have

$$\varepsilon_i = s_{ij}P_j \qquad (i, j = 1, 2, \ldots, 6) \tag{4.21}$$

We can illustrate the relationship between s_{ijkl} and s_{mn} as in the case of c_{ijkl} and c_{mn}. Upon expanding

$$\varepsilon_{ij} = s_{ijkl}P_{kl} \tag{4.22}$$

and comparing it with the appropriate member of the set of equations

(4.13a), which is an expansion of Eq. (4.21), we note that

$$\varepsilon_5 = s_{51}P_1 + s_{52}P_2 + s_{53}P_3 + s_{54}P_4 + s_{55}P_5 + s_{56}P_6 \tag{4.23a}$$

$$\varepsilon_{31} = s_{3111}P_{11} + s_{3122}P_{22} + s_{3133}P_{33} + 2s_{3123}P_{23} + 2s_{3131}P_{31} + 2s_{3112}P_{12} \tag{4.23b}$$

Consequently, since $\varepsilon_5 = 2\varepsilon_{31}$, we have

$$s_{51} = 2s_{3111}, \quad s_{52} = 2s_{3122}, \quad s_{53} = 2s_{3133}$$

$$s_{54} = 4s_{3123}, \quad s_{55} = 4s_{3131}, \quad s_{56} = 4s_{3112}$$

It is important to note that while $c_{ijkl} = c_{mn}$ for all values of $ijkl$ (1, 2, 3) and mn (1 to 6), the relationship between s_{ijkl} and s_{mn} is not as simple. From the set of relationships for all the 21 s_{mn} in terms of s_{ijkl}, we find

$$\begin{aligned} s_{ijkl} &= s_{mn} && \text{when } m \text{ and } n \text{ are 1, 2, or 3} \\ 2s_{ijkl} &= s_{mn} && \text{when either } m \text{ or } n \text{ is 4, 5, or 6} \\ 4s_{ijkl} &= s_{mn} && \text{when both } m \text{ and } n \text{ are 4, 5, or 6} \end{aligned} \tag{4.24}$$

Bearing in mind Eqs. (4.3), which relate ε_{ij} and ε_m (where $i, j = 1, 2, 3$ and $m = 1$–6), we have defined c_{ij} by Eqs. (4.4a) and s_{ij} by Eqs. (4.13a). Hence we have similar matrices for c_{ij} and s_{ij}, as given in Table 4.1.

We should not overlook the fact that, even though the c_{ij} and s_{ij} have only two suffixes, they are not the components of a second-rank tensor, and hence they do not transform like the components of a second-rank tensor. To transform them to other axes, it is necessary to go back to the tensor notation, and here it is a tensor of fourth rank.

It is important to note that the moduli of elasticity Y, n, and k are not to be confused for s_{ijkl}. On the other hand, Y, n, and k are of the dimensions of c_{ijkl} but not of s_{ijkl}.

The relationship between the stiffness constants c_{ij} in the two-suffix or matrix notation and c_{ijkl} in the four-suffix or tensor notation is given in Table 4.2, and that between the compliance constants s_{ij} in the two-suffix or matrix notation and s_{ijkl} in the four-suffix notation is given in Table 4.3.

4.4. Experimental Methods of Determining c_{ij} and s_{ij}; Christoffel's Equation and Its Use in Determining c_{ij} of Crystals

Since the bulk of the literature gives the elastic constants c and s in the two-suffix notation, we present below the elastic constants in that notation.

The elastic constants of solids, isotropic or anisotropic, can be determined by either static or dynamic methods. Earlier methods are given on pages 256–263 of Wooster's book [1398]. Reviews of the most important methods have been given by Hearmon [500–504], and also by Bhagavantam [130], Krishnan [685], and McSkimin [800; 801].

Until recently the method of determining c_{ij} and s_{ij} was only by static experiments as used by Voigt [1334], Bridgman [198], and others for a number of crystals.

These methods fall into the following types:

a. Young's modulus or extension type

b. Rigidity modulus or twisting type

c. Flexural vibrations or vibrations of cantilever type

The elastic constants obtained by the above methods give the isothermal values, whereas those values obtained by dynamic or ultrasonic methods give the adiabatic values. The two sets, however, differ very little for solids.

The more recent investigations for determining c_{ij} and s_{ij} use dynamic methods. The principle is as follows: Newton derived the relation $V = (E/\varrho)^{1/2}$, where V is the velocity of propagation of a mechanical disturbance through a medium of density ϱ and the appropriate elastic constant E. A mechanical disturbance can be propagated in two forms: longitudinal and transverse. The value of E in terms of the elastic constants depends upon the mode of vibration, longitudinal or transverse. For an arbitrary direction of propagation of elastic waves in solids there are in general three types of waves, simultaneously propagated but traveling with different velocities given by

$$E_1 = \varrho V_1^2, \qquad E_2 = \varrho V_2^2, \qquad E_3 = \varrho V_3^2 \tag{4.25}$$

Hence E can have three values, one of them leading to velocity of longitudinal waves and the other two to velocity of transverse waves. A general theory of the propagation of elastic waves in crystalline solids was developed by Christoffel in 1877, a theory we now use to evaluate the elastic constants from experimental data by dynamic methods.

The derivation and application of Christoffel's equation* is based on the fact that when an elastic wave is propagated in an anisotropic infinite medium, there are associated three independent, mutually orthogonal displacement vectors **D** which are governed by the direction of wave

* For the derivation of Christoffel's equation, see [225; 663].

propagation. In general, these three **D** indicate that three different plane waves are propagated in the direction of the wave vector **k**, with three different velocities. Since the three **D** are mutually orthogonal, it follows that if one of the **D** is parallel to the direction of **k** one pure longitudinal wave results which is associated with two pure transverse waves. Otherwise the waves are quasilongitudinal and quasitransverse.

Let the wave vector **k** have the direction cosines l, m, and n with reference to the orthogonal coordinate system defined with respect to the crystal axes. The directions of the three possible **D** are, in general, different from the direction of **k**. In Green's theory the equations dealing with these directions contain a set of relations between the phase velocity of the elastic wave and the direction cosines l, m, and n. These general relations, adapted to the particular point group of the crystal under study, are used in this connection. In a crystalline medium of density ϱ the three possible phase velocities V associated with the propagation vector **k** and the direction cosines l, m, n are given by the roots of the cubic equation

$$\begin{vmatrix} \lambda_{11} - \varrho V^2 & \lambda_{12} & \lambda_{13} \\ \lambda_{12} & \lambda_{22} - \varrho V^2 & \lambda_{23} \\ \lambda_{13} & \lambda_{23} & \lambda_{33} - \varrho V^2 \end{vmatrix} = 0 \qquad (4.26)$$

where $\lambda_{ij} = c_{ikjl}\alpha_k\alpha_l$. Here $i, j, k, l = 1, 2$, or 3, the c_{ikjl} are elastic constants, and the α's are direction cosines of the wave normal. Where the suffixes are repeated on one side of an equation, they are to go through all values 1, 2, and 3 in turn, and the sum over all such terms is to be taken.

Equation (4.26) is a secular equation of ϱV^2 in terms of λ_{ij}, which are functions of c_{ij}; but $c_{ij} = c_{ji}$. Hence $\lambda_{ij} = \lambda_{ji}$. The values of the six possible λ_{ij} $(i, j = 1, 2, 3)$ are given below. Note that the stiffness constants are in the two-suffix notation for the sake of compactness. Replacing $\alpha_1, \alpha_2, \alpha_3$ by l, m, n, respectively, we have

$$\begin{aligned}
\lambda_{11} &= l^2c_{11} + m^2c_{66} + n^2c_{55} + 2mnc_{56} + 2nlc_{51} + 2lmc_{16} \\
\lambda_{22} &= l^2c_{66} + m^2c_{22} + n^2c_{44} + 2mnc_{24} + 2nlc_{46} + 2lmc_{26} \\
\lambda_{33} &= l^2c_{55} + m^2c_{44} + n^2c_{33} + 2mnc_{34} + 2nlc_{35} + 2lmc_{45} \\
\lambda_{23} &= l^2c_{56} + m^2c_{24} + n^2c_{34} + mn(c_{23}+c_{44}) + nl(c_{45}+c_{36}) + lm(c_{46}+c_{25}) \\
\lambda_{31} &= l^2c_{15} + m^2c_{46} + n^2c_{35} + mn(c_{45}+c_{36}) + nl(c_{31}+c_{55}) + lm(c_{56}+c_{14}) \\
\lambda_{12} &= l^2c_{16} + m^2c_{26} + n^2c_{45} + mn(c_{46}+c_{25}) + nl(c_{56}+c_{14}) + lm(c_{12}+c_{66})
\end{aligned} \qquad (4.27)$$

As examples, we examine now the application of Christoffel's equation to determine the c_{ij} of crystals in two special cases.

1. The point group to which the experimental crystal belongs determines the number of constants c_{ij} different from zero, as well as the arrangement of the nonzero terms in the appropriate matrix. Let us take the case of calcite, which belongs to the crystal class D_{3d}; the c_{ij} matrix becomes

$$\begin{bmatrix} c_{11} & c_{12} & c_{13} & c_{14} & 0 & 0 \\ c_{12} & c_{11} & c_{13} & -c_{14} & 0 & 0 \\ c_{13} & c_{13} & c_{33} & 0 & 0 & 0 \\ c_{14} & -c_{14} & 0 & c_{44} & 0 & 0 \\ 0 & 0 & 0 & 0 & c_{44} & c_{14} \\ 0 & 0 & 0 & 0 & c_{14} & c_{66} \end{bmatrix}$$

where $c_{66} = \frac{1}{2}(c_{11} - c_{12})$.

Only six constants are present in the above matrix, which, when substituted in Eq. (4.27) together with the appropriate direction cosines, yield λ_{ij}. Equation (4.26) can now be solved for ϱV^2 in terms of c_{ij}. One will note that Eq. (4.26) contains 21 c_{ijkl}, which for calcite are reduced to six owing to the symmetry properties of this class of crystals. For the D_{3d} class of the trigonal system the following λ_{ij} are obtained:

$$\begin{aligned} \lambda_{11} &= c_{11}l^2 + c_{66}m^2 + c_{44}n^2 + 2c_{14}mn \\ \lambda_{12} &= 2c_{14}nl + (c_{66} + c_{12})ml \\ \lambda_{13} &= (c_{44} + c_{13})nl + 2c_{14}ml \\ \lambda_{22} &= c_{66}l^2 + c_{11}m^2 + c_{44}n^2 - 2c_{14}mn \\ \lambda_{23} &= c_{14}(l^2 - m^2) + (c_{13} + c_{44})mn \\ \lambda_{33} &= c_{44}(l^2 + m^2) + c_{33}n^2 \end{aligned} \tag{4.28}$$

Let ultrasonic waves be propagated with **k** parallel to the Z axis of the crystal. We substitute the resulting direction cosines $l = 0$, $m = 0$, $n = 1$ into Eqs. (4.28). All λ_{ij} vanish except

$$\lambda_{11} = \lambda_{22} = c_{44}, \qquad \lambda_{33} = c_{33}$$

This gives the solution of Eq. (4.26) for the three values of V:

$$\varrho V_1^2 = c_{44}, \qquad \varrho V_2^2 = c_{44}, \qquad \varrho V_3^2 = c_{33}$$

Similarly, one can select suitable orientations to obtain the remaining c_{ij}.

2. We next examine the case of NaCl, as an example of cubic crystals (all the classes of which have the same c_{ij} matrix). The Christoffel moduli

λ_{ij} for cubic crystals are as follows:

$$\begin{aligned}
\lambda_{11} &= l^2c_{11} + (m^2 + n^2)c_{44} \\
\lambda_{12} &= lm(c_{12} + c_{44}) \\
\lambda_{13} &= ln(c_{12} + c_{44}) \\
\lambda_{22} &= m^2c_{11} + (l^2 + n^2)c_{44} \\
\lambda_{23} &= mn(c_{12} + c_{44}) \\
\lambda_{33} &= n^2c_{11} + (l^2 + m^2)c_{44}
\end{aligned} \tag{4.29}$$

This can be treated as a special case of the trigonal crystal calcite, with the restriction that here, for the cubic crystals,

$$c_{14} = 0, \qquad c_{13} = c_{12}, \qquad c_{33} = c_{11}, \qquad c_{66} = c_{44}$$

The rest of the procedure remains the same as for calcite.

For the sake of completeness, we give in Table 4.4 the expressions for ultrasonic velocities in cubic crystals in terms of c_{ij}. We note that the

Table 4.4. *Velocity and Direction of Displacement of Acoustic Waves in Cubic Crystals*

Nature of wave	Direction of propagation	Direction of displacement	Velocity $= (E/\varrho)^{1/2}$
L	[001]	[001]	$(c_{11}/\varrho)^{1/2}$
T_1	[001]	[010]	$(c_{44}/\varrho)^{1/2}$
T_2	[001]	[100]	$(c_{44}/\varrho)^{1/2}$
L	[110]	[110]	$[(c_{11} + c_{12} + 2c_{44})/2\varrho]^{1/2}$
T_1	[110]	[001]	$(c_{44}/\varrho)^{1/2}$
T_2	[110]	$[1\bar{1}0]$	$[(c_{11} - c_{12})/2\varrho]^{1/2}$
L	[111]	[111]	$[(c_{11} + 2c_{12} + 4c_{44})/3\varrho]^{1/2}$
T_1	[111]	Any direction normal to [111]	$[(c_{11} - c_{12} + c_{44})/3\varrho]^{1/2}$
T_2	[111]	Any direction normal to [111]	$[(c_{11} - c_{12} + c_{44})/3\varrho]^{1/2}$

velocity of propagation of acoustic waves in a crystal depends on the nature of the waves and also on their direction of propagation in the crystal. Essentially we use Newton's formula $V = (E/\varrho)^{1/2}$, where E is the relevant stiffness constant. Note the fact that the sum of the squares of the velocities of sound waves moving in any direction in a cubic crystal is constant and given by

$$\sum_{\mu=1}^{3} \varrho V_\mu^2 = c_{11} + 2c_{44} \tag{4.30}$$

This relation can be deduced from Christoffel's equation when applied to cubic crystals [241].

The values of compliance constants s_{ij} for any crystal can be obtained from its c_{ij} values [see Eq. (3.11)]. Thus, for example, for a cubic crystal of any class,

$$s_{11} = \frac{c_{11} + c_{12}}{(c_{11} + 2c_{12})(c_{11} - c_{12})}, \quad s_{12} = \frac{-c_{12}}{(c_{11} - c_{12})(c_{11} + 2c_{12})}, \quad s_{44} = \frac{1}{c_{44}}$$

Similar relationships exist for s_{ij} in terms of c_{ij} of all crystal classes [225; 782; 885; 1397].

When the crystal under investigation is transparent, there are a number of optical techniques which provide convenient methods of measuring its mechanical properties. Some of these depend on the fact that when a transparent specimen is vibrating at a very high frequency the periodic fluctuations in density will result in the crystal behaving like an optical-diffraction grating, and if a beam of monochromatic light is passed through it, a series of diffracted beams appear. From the diffraction patterns observed, the grating element and hence the wavelength of the mechanical waves may be calculated using the Raman–Nath theory,* and if the frequency of the oscillations is known, the velocity of propagation may be determined. This method of obtaining the diffraction pattern was first developed in liquids by Debye and Sears (1932) [286], and by Lucas and Biquard (1932) [747], and it has been applied by Hiedemann and Hoesch (1935) [514; 515], Hiedemann (1935) [512], and Schaefer and Bergmann (1934) [1086] to evaluate the elastic constants of transparent solids. The same experimental setup combined with some polarization equipment enables one to study the photoelastic behavior of the given transparent solid.

* See Section 4.5.4 in this chapter.

4.5. Ultrasonics

4.5.1. Introduction

The reversible character of the piezoelectric effect in crystals, such as quartz or Rochelle salt, forms the basis for the use of piezoelectric crystals to produce ultrasonic waves.

Figure 4.5 shows a quartz plate between electrodes. Here OZ is the optic axis, and the direction OX is normal to the flat surfaces of the quartz plate. When an alternating potential difference is applied to the electrodes in contact with the flat surfaces of the quartz plate, two types of periodic deformation will occur. One of them will be a change in thickness along the X direction and the other a change in length along the Y direction. By choosing the proper frequency, longitudinal stationary waves can be set up in the crystal along either the X or Y direction. If we consider the fundamental modes only, thickness vibrations will obviously generate the higher frequencies; length vibrations are more suitable when lower frequencies are required. In either case, one face of the crystal becomes an acoustic radiator of longitudinal waves. The orientation and magnitude of its thickness govern the resonant frequency of the crystal. For other orientations of quartz plates see Cady [225].

A simplified circuit for the piezoelectric oscillator is shown in Fig. 4.6. Here, as in the case of every type of oscillator, the piezoelectric resonance of the quartz crystal essentially determines the frequency of oscillation. By changing the LC combination to one of the higher modes of vibration of the crystal, the quartz plate may be made to generate waves whose frequencies are odd multiples of its fundamental frequency.

It may be noted here that in thickness vibrations the crystal does not move as a unit like a piston. On the other hand, there are variations in phase and amplitude due to the simultaneous existence of longitudinal waves in the crystal parallel to the flat surfaces. These waves are the result

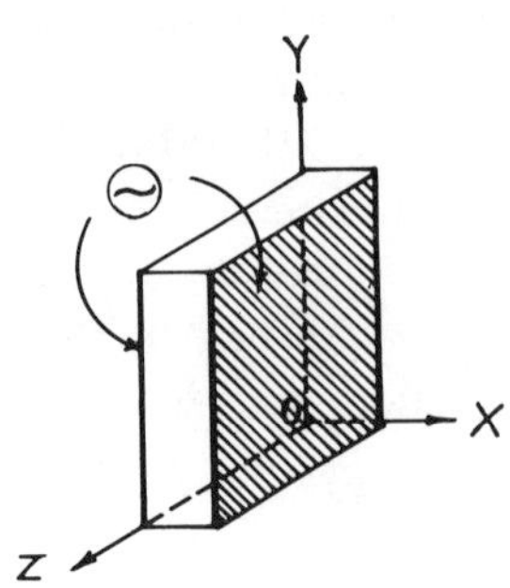

Fig. 4.5. A piezoelectric plate of quartz between electrodes.

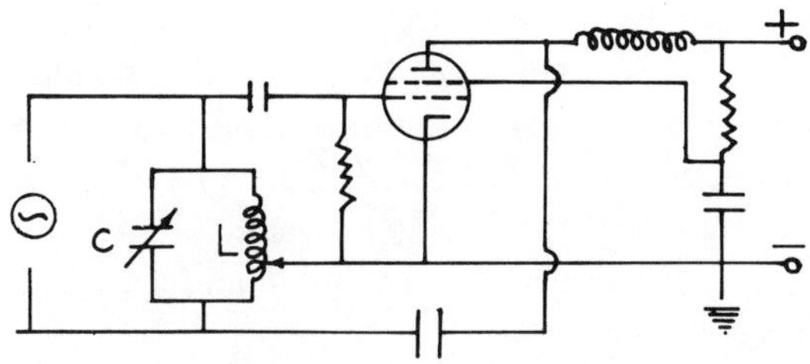

Fig. 4.6. Hartley oscillator.

of shrinkages and expansions in the length. Great care must be exercised to prevent such vibrations from building up in order to prevent the crystal from shattering.

4.5.2. Diffraction of Light by Liquids Excited Ultrasonically

In 1932, Debye and Sears [286] in the U.S.A. and Lucas and Biquard [746; 747] in France reported the diffraction of light by liquids under ultrasonic excitation. The experimental arrangement, in its simplest form, is shown in Fig. 4.7. A parallel beam of light from the lens L_1 is allowed to pass through a rectangular glass cell C containing a liquid. A lens L_2 on the other side of the cell focuses an image of the slit on a photographic plate or on the screen Sc. At the top or bottom of the container C is a quartz ultrasonic generator Q suitably insulated and excited by an oscillator circuit of the required frequency range. The transducer propagates longitudinal waves into the liquid in a direction normal to the light beam. In the presence of the ultrasonic waves in the liquid, a diffraction pattern appears on the screen. The pattern is characterized by a central maximum with symmetrically situated diffraction orders on either side, similar to

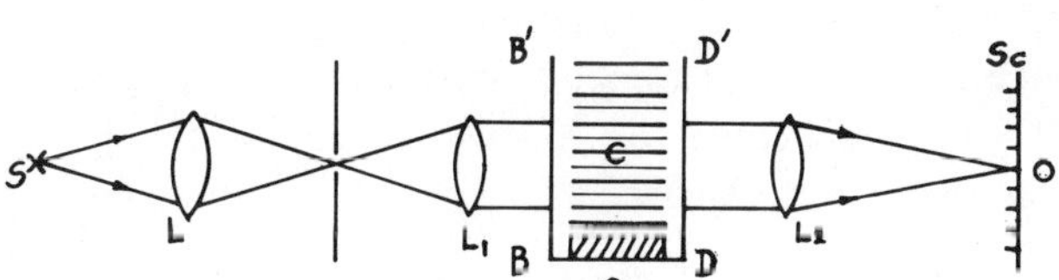

Fig. 4.7. Experimental setup to demonstrate the diffraction of light by an ultrasonically excited liquid (see Fig. 4.8 for the formation of diffraction orders). S, source of monochromatic light; Q, quartz transducer; C, ultrasonic cell—a rectangular glass cell $BB'D'D$ containing a transparent liquid; compressions and rarefactions are shown by thick and thin lines, respectively; Sc, screen, showing the grating spectra, on either side of the central image O.

the usual pattern caused by an ordinary plane transmission type diffraction grating. An increase of frequency of the transducer results in increasing the diffraction angles from the central undiffracted image of the slit, thereby resulting in an increased spacing of the diffraction orders. The diffraction images are sufficiently sharp to provide a reasonably accurate estimate of the acoustic velocity if the angle of diffraction θ and the acoustic frequency ν_a of the transducer are known.

The formation of the diffraction pattern was predicted in 1922 by Brillouin [201], according to whom the diffraction pattern is due to the presence (in the liquid) of layers of variable density, periodically situated along the direction of propagation of the ultrasonic (acoustic) wave. The variations in density are accompanied by variations in the index of refraction. Brillouin argues that the periodic variation in the optical properties of the liquid in this case are analogous to the periodic structure of a crystal lattice. The fact that the lattice in this case is changing rapidly is not of serious concern, since this state of affairs will not change the essential angular relationship that governs the formation of the diffraction maxima in view of the very high value of the velocity of light in the liquid compared with that of sound.

The simple picture presented above by Brillouin, namely that the diffraction phenomenon is caused by the scattering of light from the periodically spaced lattices (layers), the grating structure being sinusoidal, would, for monochromatic light, lead one to expect the formation of only one order, flanking on either side of the central maximum. But actually several orders (sometimes eight or nine) could be seen, their diffraction angles obeying the well-known formula

$$e \sin \theta = \pm n\lambda \tag{4.31a}$$

for a plane transmission grating, where e is the grating element, n an integer which defines the order of the diffraction maximum, and λ the wavelength of light in vacuum. Note that for both stationary and progressive acoustic waves in the liquid, e is equal to λ_a, the wavelength of the acoustic wave disturbance in the liquid. Hence

$$\lambda_a \sin \theta = \pm n\lambda \tag{4.31b}$$

Experimentally, one obtains the value of λ_a from a measurement of θ_n, the angle of diffraction,* and f, the focal length of the converging lens

* For the sake of clarity we use θ_n instead of θ to represent the angle of diffraction of the nth-order image.

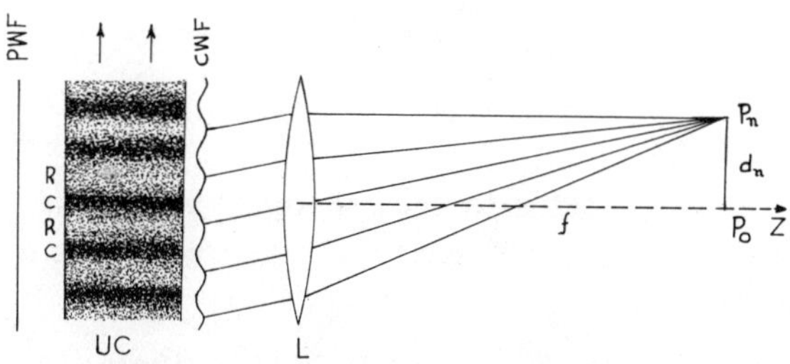

Fig. 4.8. An ultrasonically excited medium, acting as a phase grating, converting a plane optical wave front (PWF) into a corrugated wave front (CWF), and its grating action. *UC*, ultrasonic cell; *R*, rarefaction; *C*, compression; P_0, central image or zero-order spectral line; P_n, diffracted image of *n*th order; *f*, focal length of the converging lens; $d_n = P_0P_n$.

employed to focus the diffraction pattern (Figs. 4.8 and 4.11). If θ_n is small, as is usually the case, then $\theta_n = d_n/f$. Hence

$$\lambda_a = \pm n\lambda/\theta_n = \pm n\lambda f/d_n$$

Now d_n can be determined by placing first a ground-glass plate to catch the pattern and then focusing a cathetometer on the pattern. The glass plate is then removed and measurements of d_n are taken. Alternatively, a photographic plate can be substituted for the ground-glass plate and a permanent record of the pattern obtained.

A complete theory of the diffraction phenomena in a medium excited by progressive and stationary acoustic waves has been developed by Raman and Nath [1007–1011]. Their theory is closely analogous to Rayleigh's theory of the diffraction of a plane wave (optical or acoustic) incident on a periodically corrugated surface, and considers not only normal incidence but also oblique incidence. It explains satisfactorily the observations of Debye and Sears [286], Lucas and Biquard [746; 747], Hiedemann and Siefen [511], Bär [76; 77], and Parthasarathy [905; 906]. It explains in an interesting manner many subtleties of the ultrasonic diffraction patterns observed by Bär, such as (a) the wandering of intensities of the different orders including the central or zero order as the experimental conditions are varied, and (b) the asymmetric pattern due to obliquity in incidence.

From the Raman–Nath theory we can also calculate the values of the light frequencies due to the Doppler effect for both progressive and stationary waves. The conclusions are listed below.

1. In the case of progressive acoustic waves traveling in a medium perpendicular to the direction of the incident light propagation, the emergent

light beam will suffer diffraction at angles θ_n, where

$$\lambda_a \sin \theta_n = \pm n\lambda$$

and the light beam diffracted to the nth order will possess a frequency $(\nu \pm n\nu_a)$, where ν and ν_a are, respectively, the frequencies of the incident optical and acoustic radiations.

2. When the acoustic waves are stationary, the diffraction of light occurs at the same angles as those given by θ_n, where $\lambda_a \sin \theta_n = \pm n\lambda$. But in this case an even order would have light frequencies ν, $\nu \pm 2\nu_a$, $\nu \pm 4\nu_a, \ldots, \nu \pm 2r\nu_a$, and an odd order would have frequencies $\nu \pm \nu_a$, $\nu \pm 3\nu_a, \ldots, \nu \pm (2r + 1)\nu_a$.

3. The light amplitudes in the diffracted orders can be represented by a differential-difference equation, whose approximate solution is satisfied by certain Bessel functions. One may note that, since the acoustic velocity is very small compared with that of light in the medium, the resultant changes in the frequencies due to the Doppler effect are indeed very small. Nevertheless, the existence of the Doppler frequencies was first experimentally demonstrated by Debye, Sack, and Coulon [287].

The experimental technique first devised by Debye and Sears and by Lucas and Biquard was suitably modified by Bachem, Hiedemann, and Asbach [55; 57] in order to render visible the stationary acoustic wave pattern formed in the medium. The optical arrangement is essentially the same; however, here it is not the diffracted images of the slit but a plane cross section of the stationary acoustic waves that is seen on the screen or the photographic plate. By this method Hiedemann and Siefen [511] measured the wavelength spacings of the acoustic grating and thus obtained the values of the overtones of a quartz transducer. The technique was applied for both stationary and progressive waves by Hiedemann and his collaborators. To study the acousto-optic phenomena they employed a stroboscope whose frequency was the same as that of the quartz transducer; this was facilitated by passing polarized light through a Kerr cell operated by the same ac supply as that applied to the transducer.

Schaefer and Bergmann [1086; 1087] have employed a cross-grating technique (or multiple beam method) for studying the interaction of acoustic waves in the medium with the incident light waves (Fig. 4.9a). In this technique the source of light is a point, and the solid specimen can be taken in the form of a parallelepiped, the exact geometrical shape being of no importance. The sample is placed in good acoustic contact with the transducer, using a film of oil between. The frequency of the applied rf voltage

Fig. 4.9. Optical arrangements to study optical effects of ultrasonically excited solids by (a) the Schaefer–Bergmann method, (b) the Hiedemann line diffraction method, and (c) the visibility method due to Bachem and Hiedemann. *M*, mercury arc; L_1, L_2, L_3, lenses; *F*, green filter; *SS*, slit source; *PS*, point source; *CA*, circular aperture; *S*, sample solid; *Q*, quartz transducer; P_1, P_2, P_3, periods of interference; *SC*, scale; *C*, camera; *A* indicates the distance between the components as shown in the figure. From [1534].

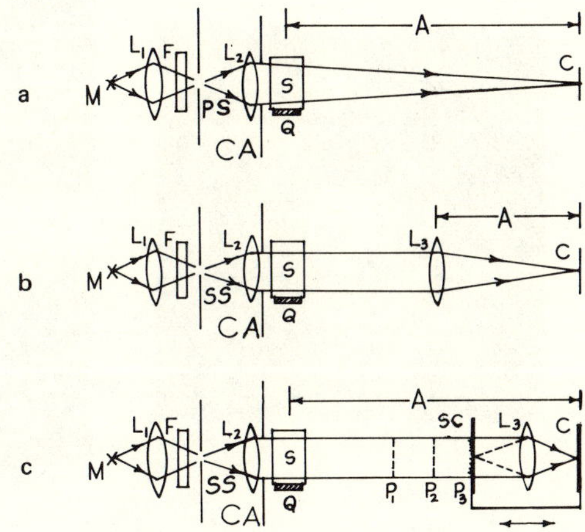

is then carefully manipulated until the sample is thrown into strong resonant vibrations. As a result of the grating action of the ultrasonically excited sample, the optical diffraction pattern takes a form characteristic of the elastic properties of the experimental sample. For amorphous or isotropic solids like glass, one obtains two concentric rings of sharp interference points (since the source of light is a point). The inner and outer rings are caused by the space gratings resulting from the longitudinal and transverse (or shear) waves respectively. From the measurements of the radii of the two rings, one can compute the elastic constants of the isotropic solid sample. Schaefer and Bergmann have applied their technique to study the elastic properties of a number of transparent solids. The ultrasonic diffraction patterns obtained in this way are reproduced in Fig. 4.10A for glass, fluorspar, rock salt, sylvine, lithium fluoride, and α-quartz.

4.5.3. *Optical Methods of Determining the Ultrasonic Velocities and Elastic Constants of Transparent Solids Employing the Schaefer–Bergmann Pattern, the Hiedemann Pattern, and the Lucas–Biquard Effect*

We now describe in detail some of the optical methods since these methods have a bearing on the studies of the photoelastic behavior of solids by ultrasonic methods.

1. The simultaneous but independent discovery by Debye and Sears and by Lucas and Biquard that ultrasonic waves can act as optical-diffraction gratings opened an entirely new technique for determining the elastic constants of solids, transparent or opaque.

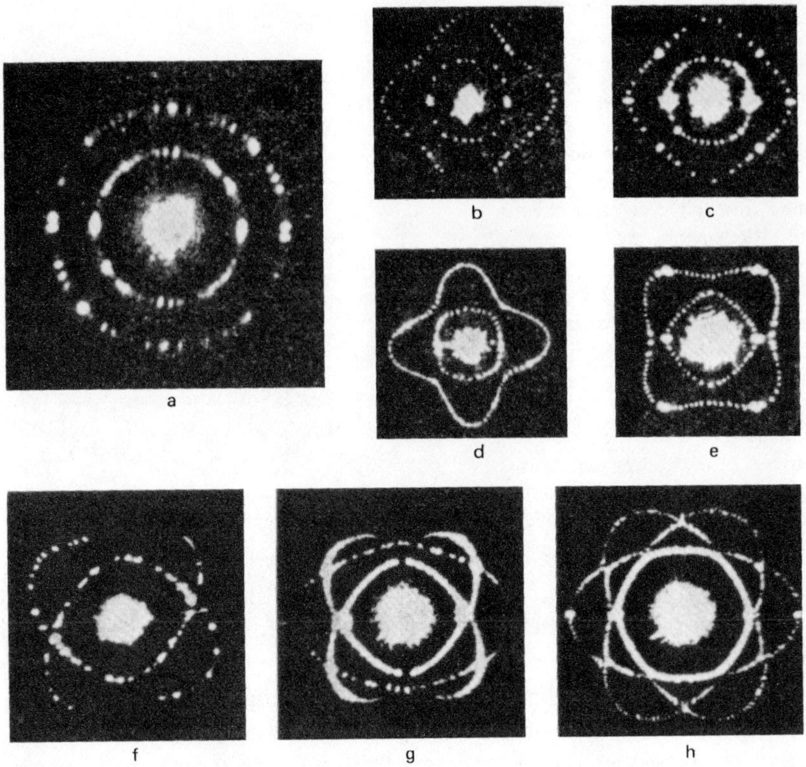

Fig. 4.10A. Schaefer–Bergmann patterns of monochromatic light from a point source diffracted by ultrasonic waves propagated in solids: (a) glass; (b) fluorspar; (c) rock salt; (d) sylvine; (e) lithium fluoride; (f)–(h) quartz, for acoustic waves propagated in the X, Y, and Z directions. From [110].

The use of high ultrasonic frequencies allows one to obtain the condition for an infinite medium with rather small samples. It is only necessary that the dimensions of the sample be large compared with the acoustic wavelengths. Usually the vibrations in the ultrasonically excited crystal or glass are in the form of standing waves of the same frequency as that of the radiofrequency voltage applied to the quartz transducer. Then the spacing of the grating corresponds to the length of the sound wave in the medium.

2. Two different methods (based on the ultrasonically excited optical-diffraction grating) can be applied for systematic studies of ultrasonic velocities in transparent solids for the purpose of determining elastic constants.

Schaefer, Bergmann, and Goehlich [1089] used the method of Schaefer–Bergmann diffraction patterns. Hiedemann and Hoesch [519] used a method in which one could make a direct measurement of the wavelength of both longitudinal and shear waves.

The Schaefer–Bergmann patterns are obtained if one replaces the slit in the usual Debye–Sears arrangement with a small circular aperture (Fig. 4.9a). In isotropic media (like optical glasses), these patterns consist of numerous images of the aperture lying on two concentric circles. The diameter of one circle corresponds to the wavelength of longitudinal waves, and the other diameter corresponds to the shear waves. Hiedemann and Hoesch [513–519] have developed a technique for obtaining sharp line spectra with a slit source and separating the effects of longitudinal and shear waves by means of a photoelastic method (Fig. 4.9b). They claim better accuracy by this technique.

The formation of the Schaefer–Bergmann pattern and the Hiedemann pattern is explained as follows: The type of source aperture used determines the type of the diffraction pattern obtained. A point source consisting of a small circular aperture will produce a diffraction pattern consisting of a set of points of light. A slit source produces a pattern which is a set of parallel lines.

The diffracted points of a point source lie along a straight line passing through the central order. The direction of this line is the direction of propagation of the particular wave which is causing the diffraction. If more than one standing wave of sufficient intensity to produce diffraction occurs within the ultrasonically excited glass solid, say in the form of a cube, there will occur several lines of points of equal spacing but with different directions depending on the direction of the diffracting ultrasonic wave. With enough of these waves in different directions, one obtains the characteristic Schaefer–Bergmann circular diffraction patterns.

A slit source may be considered as a large number of point sources set in a row. Each diffracted point, as already described for the point source, becomes a diffracted line. In order that the distance between the diffracted lines be a true measure of the diffraction spacing, the wave producing the diffraction must be propagated in a direction perpendicular to the slit.

3. Barnes and Hiedemann [79] have used two light sources simultaneously in taking photographs, one a point source and the other a slit source. The point-source pattern showed the direction of the waves while the slit-source pattern was used for measurements because of its better definition.

Either of the patterns can be permanently recorded on a photographic plate. A calibration grating greatly facilitates calculation of the ultrasonic wavelength. For this purpose, a calibration grating, prepared by photographic reduction, is set on a cross-slide on the optical bench in such a way that the experimental specimen (say glass sample) could be moved out of the way and the comparison grating moved into its place. For each photographic plate, measurements are made of the spacing resulting from the ultrasonic diffraction and of the spacing resulting from the comparison-grating diffraction. This eliminates the necessity of calculating the constants of the optical system. The two sets of measurements on each plate can be directly compared, and, using the known (or measured) value d of the grating element of the comparison grating, the wavelengths of the ultrasonic waves are calculated using the relation

$$\lambda_a = \frac{d}{r} x$$

where d is the comparison-grating element and r and x are the separations of the diffraction orders due to comparison grating and ultrasonic grating, respectively. This relation holds since the separations of the diffraction orders in the present case are linearly proportional to the grating elements causing them. Hiedemann has greatly improved the definition and hence the selectivity of the type of pattern to be recorded by the addition of a polarizer and an analyzer in the optical system, although it is possible to take diffraction pictures without them. The nature of the strains due to the longitudinal or shear wave in the block is such that plane-polarized light passing through the experimental specimen emerges with a change in polarization depending on whether the diffraction orders are due to the longitudinal or shear wave. A polarizer angle, with the analyzer crossed with respect to it, is selected such that the intensity of the orders of the two types of diffraction is about equal.

For the determination of elastic and photoelastic properties of transparent solids by ultrasonic-diffraction methods, care is necessary to avoid multiple diffraction of light. An elastic wave may be the result of a purely longitudinal or a purely transverse vibration; in anisotropic solids these waves are frequently coupled or mixed. Transverse waves have a lower velocity than longitudinal waves, which means that at about the same frequency their respective wavelengths are quite different and the diffraction patterns produced will differ accordingly. Ramavataram [1024] points out that the intensity of the obtainable diffraction patterns is closely connected with the density and index of refraction of the glass samples he investigated. In addition to this observation, one finds that whenever the propagation characteristics of a particular elastic wave are such that the sample is in resonance with that mode of vibration, the diffraction pattern produced by that wave will be most intense. In many cases, it will be the only one visible. Multiple diffraction patterns (comprising both longitudinal and transverse modes) originate if more than one mode of vibration is excited. If a longitudinal and a shear wave are present in the sample with sufficient intensities (which occur when some overtone of one type of vibration is nearly the same as some overtone of the other type of vibration), then each wave produces a periodic change of the index of refraction, and hence the resulting diffraction pattern is a superposition of the individual patterns. Transverse mode patterns are less intense than those of the longitudinal modes.

4. Light passing through a medium with varying refractive index will be refracted toward the direction of increase in the refractive index. When the discovery was made that ultrasonic waves can act like optical gratings, Lucas and Biquard [746; 747] developed a theory for this phenomenon based on the curving of the light rays in the sound field produced by the gradients of the refractive index in the field. To demonstrate the reality of the deflection of light beams in a sound field, they limited the width of the light beam passing through the sound field to a value smaller than the ultrasonic wavelength. In this case a broadening of the slit image was noted instead of the usual diffraction pattern observed with wider light beams. Thus we see that the optical methods for studying the influence of ultrasonic waves on a medium are based either on the diffraction or the refraction of light by ultrasonic waves. The diffraction method consists in measuring the relative intensity in the different orders of diffraction, whereas the refraction method is based on the periodic variation of the gradient of the refractive index in a sound field. Lucas and Biquard discovered that the image of a slit was broadened if the light beam forming the image passed through an ultrasonic field in which λ_a is larger than the width of the light beam. In a later theoretical paper Lucas [748] discussed in detail the relation between the broadening of the image and the sound pressure and pointed out that the refraction of light in a sound field might be used for measuring sound absorption and reflection coefficients. The first use of the possibilities proposed by Lucas [748] appears to have been made much later (1949) by Hueter and Pohlman [543] who applied this method for measurement of sound pressure due to progressive ultrasonic waves in liquids, and by Porreca [966] who measured the distribution of light intensity across the slit image under various conditions. The Lucas–Biquard effect was applied by Iyengar [567] to study the photoelastic behavior of glasses.

Based on the Lucas–Biquard effect, measurement of sound velocity in liquids was made by Kolb and Loeber [662] by visually observing and also photographically recording the width of the slit image at various points of a stationary ultrasonic wave. Subsequently, Loeber and Hiedemann [743] developed a complementary method in which an additional slit was placed in the path of the light beam after its emergence from the liquid and coinciding with the position of the slit image. The width of the second slit is exactly equal to the width of the image of the first slit. In the presence of sound the intensity of the light passing through the additional slit decreases due to the broadening of the beam. The intensity of the light passing through the second slit can be measured qualitatively by means of a photomultiplier microphotometer. A continuous recording of the effects of light refraction of the various parts of the stationary wave field as a function of the distance from the sound source has been used by Loeber and Hiedemann for the determination of sound velocity and intensity and for the study of the wave form. For this purpose the cell was moved across the light beam by means of a micrometer screw. The intensity of the diffraction as a function of the distance y measured from the center of slit S_2 of the photomultiplier tube is given by

$$I_y = I_0(\sin \pi ay/r\lambda)^2/(\pi ay/r\lambda)^2$$

which is the usual formula for diffraction through a single slit, where a is the width of the slit S_1 placed in front of the ultrasonic cell, and r is the distance between S_1 and S_2. The variation of I_y with y enables one to count the number of half-waves in a given distance taken perpendicular to the light beam. Hence, knowing the frequency of the oscillator and λ_a, the wavelength of sound, the velocity can be computed. The rf current into the transducer was measured by a thermocouple-type milliammeter. For further details of the experiment, see the original paper [743]. This method gives directly the value of $\lambda_a/2$ for the stationary-wave pattern.

4.5.4. Mayer and Hiedemann's Experiments

Mayer and Hiedemann [789; 790] have described three different optical techniques to determine the elastic constants of transparent crystals and have evaluated the attainable accuracies of these methods under identical experimental conditions. The methods are (a) line diffraction, (b) thickness resonance, and (c) visibility methods. The necessary optical arrangements are given in Fig. 4.9.

Depending on the geometry of the light source and on the type of elastic vibrations in the solid sample (Fig. 4.9a,b), one can observe either the interference pattern found by Schaefer and Bergmann (Fig. 4.10A) or a line diffraction spectrum due to Hiedemann (Fig. 4.10B). In order to obtain a Schaefer–Bergmann pattern with all the details, it is desirable to excite as many elastic modes as possible for a given frequency. The Schaefer–Bergmann method yields symmetrical figures which represent the elastic anisotropy of the crystal in a plane normal to the diffracted light beam. This method is extensively described in a number of publications, for example References [1086–1090].

The main advantage of the method is the possibility of using a single sample of a crystal for the determination of its elastic constants, while the main limitation is the rather unsatisfactory accuracy. Schaefer and Bergmann do not expect an accuracy of better than 1% even under optimum conditions.

4.5.4.1. The Line-Diffraction Method

This method was developed by Hiedemann and Hoesch [513–519] for the study of isotropic transparent solids. In order to obtain line spectra

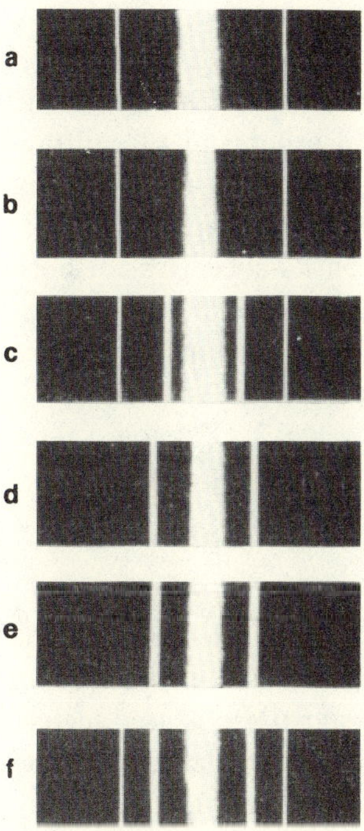

B

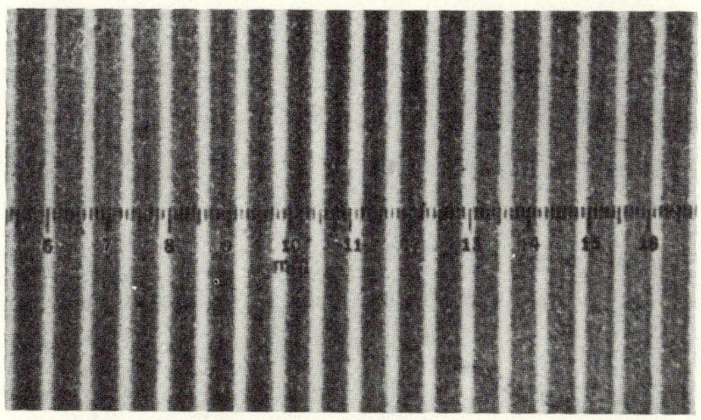

C

Fig. 4.10B,C. (B) Hiedemann patterns of monochromatic light from a slit source diffracted by ultrasonic waves in glass under different experimental conditions. Patterns (a) and (b) are due to shear waves only, (d) and (e) are due to longitudinal waves only, and (c) and (f) are due to both shear and longitudinal waves. (Note that the longitudinal pattern is always more intense than the shear pattern.) (C) Optical interference pattern showing standing longitudinal acoustic waves in the Z direction in sapphire by the visibility method due to Mayer and Hiedemann. Part (B) is from [190] and part (C) from [790].

which can be used for precise measurements, this method requires the excitation of a stationary plane wave in one direction. This can be achieved by using sharp resonances for compressional and shear waves in isotropic solids. With this method Barnes and Hiedemann [79] have been able to obtain an accuracy better than that obtainable with the Schaefer–Bergmann method and comparable with that by Hoesch [531], who used the visibility method. In order to obtain plane waves in a crystal, the size of the sample should be large enough so that the crystal can be considered an infinite medium.

The technique used for obtaining plane stationary longitudinal and shear waves in isotropic solids is also successful in crystallographic directions where the plane waves are not of the purely compressional or purely shear type.

The optical arrangement used is the standard setup of Debye–Sears illustrated in Fig. 4.7. A strong line pattern is obtained whenever a standing wave is set up in the sample. Figure 4.10B shows a photograph of a line-diffraction pattern in glass [190] in which both the longitudinal and the shear diffractions are visible and from which the sound velocity is calculated using the approximate formula

$$V = n\lambda FA/D_n \tag{4.32}$$

where n is the diffraction order, λ is the wavelength of light, F is the acoustic frequency, D_n is the distance between the central order and the diffraction order n, and A is the distance shown in Fig. 4.9b. The accuracy of a diffraction method depends on the size of the effective grating constant. In an ultrasonic grating this is equal to the ultrasonic wavelength for both progressive and stationary waves. The ultrasonic velocity in crystals like diamond, MgO, and sapphire is exceedingly high, and even at frequencies of around 10 MHz the distance between the diffraction orders is so small that the finite width of the spectral lines is not negligible. For low values of sound velocity this is fairly accurate.

4.5.4.2. Thickness Resonance Method

The optical arrangement used in this method is the same as that for the diffraction method. However, here one is not interested in the spacing between the diffraction orders because the diffraction pattern is used merely to indicate that a standing wave is set up in the sample. To start with, one finds a frequency at which maximum transmission through the sample occurs, and then the frequency is varied until another maximum is observed. If a longitudinal elastic wave causes the occurrence of both maxima in such a manner that the number of loops (each loop being $\frac{1}{2}\lambda_a$) set up between the parallel faces (distant D apart) of the crystal by the standing wave produced at the two frequencies has changed by exactly one due to the change in frequency, one can find the total number of loops present in the specimen for each maximum. This is obvious because the velocity is constant in a given direction. Therefore the crystal will vibrate in resonance if the distance D (between the faces) parallel to the direction of sound propagation is equal to an integral number of $\frac{1}{2}\lambda_a$. Thus $n\lambda_a = 2D$. It is easily seen that this condition can be met only at certain frequencies which depend on the velocity V in the sample and on the distance D between the parallel faces according to

$$\frac{V}{D} = \frac{2F_0}{n} = \frac{2F_1}{n+1} = \frac{2F_2}{n+2} \tag{4.33}$$

where $n, n + 1, \ldots$ are integers and $F_0 < F_1 < F_2$.

Resonant frequencies can be determined by optical effects. The transparent sample is placed in the path of collimated light in such a position that the acoustic wave vector $\mathbf{k}$ is perpendicular to the direction of light propagation. A standing wave, which occurs at a resonant frequency, will act as an ultrasonic grating for the collimated light, and hence at each F_i a strong diffraction pattern will be visible which will disappear very rapidly with a slight change in frequency owing to the small radiation damping in air which surrounds the crystal. It is therefore possible to determine the resonant frequency with an accuracy of better than 1 part in 1000. It can be seen that at least two neighboring F_i have to be found if one is to determine V. The accuracy of the method is increased if a series of measurements is taken to reduce the influence of possible small errors in locating the F_i. It is advisable to find more than two resonance frequencies for the calculation of V, particularly when $n > \approx 10$. Because of the small radiation damping in air, the resonance is very sharp; the diffraction pattern disappears for very small deviations from the resonance frequencies, which can therefore be determined with an accuracy of better than 0.1%. Both longitudinal and shear waves produce diffraction patterns. It is relatively easy to differentiate between them since the shear-wave pattern is much weaker than the pattern produced by a standing compressional wave. Furthermore, at about the same frequency, shear waves having a lower velocity produce lines at a larger spacing. The longitudinal velocity in a particular direction is found by noting a series of resonant frequencies and substituting the values of F_i into Eq. (4.33) to find first the appropriate values of $n, n + 1,$

. . . , disregarding the quotient V/D for the moment. After the number of half-waves has been found, one obtains the velocity. A similar procedure yields the shear velocity.

One can partly immerse the crystal in a liquid in which a diffraction pattern will be observed if maximum transmission of acoustic energy takes place through the sample. The wedge method, developed by Bhagavantam and Bhimasenachar [118], is based on this principle. However, in this case the sharpness of the resonant frequency is decreased owing to the considerable radiation damping caused by the liquid. The diffraction pattern does not vanish as rapidly with a slight change in frequency; hence the determination of the resonant frequency suffers a loss in accuracy. For the high velocities found in sapphire by Mayer and Hiedemann, the frequencies required in diffraction measurements of comparable accuracy become undesirably high. Nevertheless, for opaque substances the wedge method is equally applicable. Much higher accuracies can be obtained by vibrating the specimen in air and determining the resonance frequencies by electronic instead of optical methods. Even slight variations in the thickness of the plates causes the sharpness of the thickness resonance to decrease appreciably.

4.5.4.3. Visibility Method

Bachem and Hiedemann [56] have developed a method by which ultrasonic waves can be made visible in transparent media. For large values of ultrasonic velocity, the wavelength is correspondingly large in such cases. The quantity λ_a can be measured by the visibility (secondary interference) method directly with much greater accuracy than by means of line-diffraction (or primary interference) pattern.

The experimental arrangement is shown in Fig. 4.9c. The visibility is based on the principle of secondary interference and visibility of ultrasonic waves. The phenomenon of secondary interference is familiar from optical gratings, with the important difference that the grating is caused here by the change of the index of refraction in the medium produced by an ultrasonic wave. The simplest form of the method uses stationary ultrasonic waves. If parallel light traverses the ultrasonic field, the secondary interference produces a system of interference fringes, whose spacings are equal to the distribution between the nodes or antinodes of the standing wave in this sample. The lines in the interference pattern are exactly one-half wavelength apart.

If, for example, the wavelength of the elastic wave in the specimen is

1 mm, the lines produced at $P_1, P_2, P_3, \ldots$ are 0.5 mm apart. One can observe and measure these lines either with a traveling microscope focused on one of the image planes, or one can photograph directly a section of the planes containing the interference lines. Depending on the dimensions of the illuminated sections of the crystal and on the wavelength of the standing wave, one can observe from 10 to more than 50 lines. If we place a transparent scale in the plane to be photographed, we can eliminate a number of possible errors in relating distances measured on the enlarged photograph to actual distances on the film negative. If we move the assembly consisting of the camera L_3 and a scale along the optic axis, we obtain more lines without increasing the frequency. At various distances between the planes, $P_1, P_2, P_3, \ldots,$ the interference of diffraction orders is such that the number of visible lines doubles, as was pointed out by Hiedemann and Schreuer [521]. The fact that the shear waves have a lower velocity can be used to distinguish them from the longitudinal waves because at the same frequency the former produce lines more closely spaced than those due to longitudinal waves. Sharp secondary interference patterns are formed in successive planes $P_1, P_2, P_3, \ldots$ behind the grating, analogous to the phenomena obtained by ruled gratings. These planes P_1, P_2, P_3, etc., are equally spaced along the axis of the optical system. The periodicity of visibility of the ultrasonic grating is given by a relation derived by Nath [860] and by Nomoto [878] and very similar to that mentioned by Lord Rayleigh for the periodicity of ruled optical gratings. Bachem and Hiedemann [56] developed a measurement procedure designed to compensate for any maladjustment of the optical system. The optical axis remained fixed, and the measuring cell, mounted on a compensator, was moved normally to the optical axis. For a known motion of the comparator, the number of interference lines which moved through a reference mark in the ocular of the microscope used for this purpose was counted. The λ_a in liquids can be measured with high accuracy. Schreuer [1104] reports an accuracy of better than 1 part in 10,000. The same procedure was used by Hoesch [531] for his measurements of the elastic properties of optical glasses. In this visibility method, use is made of the fact that an ultrasonic wave in a transparent substance acts very much like a ruled grating; the ultrasonic waves are made visible, and their grating constants can be measured. In an ultrasonic grating, the grating constant is equal to λ_a for both progressive waves and stationary waves.

Mayer and Hiedemann [789; 790] modified the above method by setting up a transparent scale exactly in the image plane for a sharp secondary interference and photographing the scale and the "grating" together. In this manner a permanent record is obtained on the film. Comparison of the distance between a large number of lines of the fringe pattern with an appropriate distance on the scale can be made with great reliability and accuracy.

The visibility method in its modified form was applied by Mayer and Hiedemann [789; 790] to determine the elastic constants of sapphire, a trigonal crystal of point group D_{3d} ($\bar{3}m$). An accuracy of almost 0.01% can be obtained by this method. The use of polarized light increases the definition of the fringes. A typical interference pattern obtained by Mayer and Hiedemann [790] showing a standing longitudinal wave in the Z direction of the sapphire crystal at a frequency of 9.006 MHz is reproduced in Fig. 4.10C, from which one obtains a value of 11215 m/sec for the longitudinal velocity.

At high sound velocities the distance between the fringes is large; the measurement of only a few fringes gives, therefore, the required accuracy even in samples of small size. At low sound velocities the spacing of the fringes is small, but this is compensated by the greater number of fringes per unit length available for measurement.

Note that every c_{ij} (in the two-suffix notation) in which $i = j$ corresponds to some directly measurable velocity while the c_{ij} in which $i \neq j$ must be obtained indirectly by substitutions into equations containing sums and differences of measurable constants. Small errors in the determination of measurable constants may affect the calculated values quite significantly. (This difficulty is present to a greater degree in the case of photoelasticity.)

In the visibility method the distance between the interference fringes is one-half the wavelength for a standing wave. However, one has to be careful that he is looking at exactly the right position behind the sample. If one goes to other places outside the repeat periodicity, one does introduce additional fringes which are usually fainter and, in general, are recognizable. Hiedemann and Breazeale [524] have published pictures describing the conditions under which these extra lines become visible. But it is safe to say that the largest distance one gets between the lines corresponds to the simplest situation, and in that case one-half the wavelength is measurable directly.

4.5.5. Raman–Nath Theory of Diffraction of Light by Ultrasonic Waves

An important new field of research was opened up by Langevin [716] in 1918 when he showed that ultrasonic waves can be generated in liquids using a plate of quartz as a piezoelectric oscillator. The diffraction of light by a liquid excited by high-frequency sound waves was predicted by Brillouin [201] in 1922. Debye and Sears [286] in the U.S.A. and Lucas and Biquard [746; 747] in France have (in 1932) demonstrated experimentally the phenomenon of diffraction of light by a liquid excited by ultrasonic waves; Bär [76; 77] has carried out a thorough investigation of this phenomenon. The optical diffraction by ultrasonic waves in a liquid can be observed by the experimental setup shown in Fig. 4.7. A parallel beam of monochromatic light is incident normally on one face of a rectangular glass cell containing a transparent liquid, say carbon tetrachloride or benzene. Since the liquid is homogeneous and optically isotropic, the emergent light beam will be undeviated from the incident beam. If, however, the liquid is excited by high-frequency ultrasonic waves generated by introducing a piezoelectric oscillator Q, say of quartz, at the top or bottom of the cell, the liquid medium becomes stratified into parallel layers of varying indices of refraction. Now let us consider the case in which the incident light beam is normal to the sound field (and hence parallel to the plane wave front of the sound waves). The light emerging from the ultrasonic cell will now be found to be split up into various beams traveling in different directions, in the same way as a parallel beam of light incident normally on a plane transmission grating is found to be split up and diffracted by the plane transmission grating. In fact, if λ is the wavelength of incident light, and if θ_n is the angle of diffraction of the emergent diffracted beams, it has been found experimentally that they are related as in Eq. (4.31b)*:

$$\lambda_a \sin \theta_n = \pm n\lambda$$

Here n is an integer equal to or greater than zero, and λ_a is the wavelength of the incident sound waves in the liquid. The positive and negative orders are symmetric in both intensity and angle of diffraction. In contrast with the plane transmission grating, Bär found that when the ultrasonic power is raised, several orders of the diffraction spectra are produced; furthermore, a wandering of the intensity distribution among the several orders of the spectra has also been noticed as the experimental conditions are varied. Various theories of the phenomenon have been put forward, among which may be mentioned those by Debye and Sears, by Lucas and Biquard, and by Brillouin, none of which could satisfactorily explain all the features of the phenomenon. Raman and Nath [1007–1011] have satisfactorily ex-

* Here θ_n represents the angle of diffraction of the nth-order image.

plained all the essential features in their theory. Their theory enables one to predict correctly the intensities of the various orders of diffraction for experimental investigations covering moderate ultrasonic frequencies and intensities. We shall now discuss the Raman and Nath theory, especially because this theory forms the basis for Mueller's theory [829] for the experimental methods of studying the photoelastic behavior of amorphous and crystalline transparent solids; it also forms the basis for some experimental methods of determining the elastic constants of liquids and solids. The Raman–Nath theory is based on the well-known fact that the transmission of light in a liquid medium is accompanied by phase changes. Hence a plane wave front of light incident on one side, after transmission through the medium excited by the ultrasonic waves, no longer remains a plane wave but assumes a corrugated form, owing to the periodic fluctuations in the density and, consequently, also in the refractive index of the medium.

The Raman–Nath theory bears a very close analogy to Lord Rayleigh's theory of the diffraction of a plane wave (optical or acoustic) incident normally on a periodically corrugated surface; in the latter case Lord Rayleigh showed that the resulting diffraction phenomenon would consist of several components given by Eq. (4.31b) with regard to their diffraction angles. Even with regard to the relative intensities of the various diffraction spectra the Raman–Nath theory is similar to Lord Rayleigh's theory.

Let a collimated beam of monochromatic light of wavelength λ be incident normally on a plane face BB' of a liquid medium with a rectangular cross section $BB'D'D$, and let the light emerge from the opposite face DD' of the liquid medium parallel to the face BB' on which the light is incident (Fig. 4.7). In the undisturbed condition the liquid has the same index of refraction at all points, and hence the emerging beam will be parallel to the incident beam. Now let the piezoelectric oscillator of quartz Q excite plane ultrasonic waves in the liquid, thus creating in the liquid fluctuations in density which produce layers of varying index of refraction. If the physical length of the path of light within the sound field is small, the incident light can be regarded as experiencing variations in phase corresponding to the varying refractive index at the different parts of the medium.* We can calculate the phase change of the emerging light at different parts from a knowledge of the length of the optical path during the passage of the light within the sound field. Owing to the ultrasonic excitation, the liquid medium becomes stratified into parallel layers with

* In other words, the plane wave front of light is phase modulated while passing through the acoustic field, thus becoming a corrugated wave front.

periodically varying refractive index. Let us now assume a set of Cartesian coordinates with its origin at the center of the emerging plane wave front in the unexcited state of the liquid. As in Fig. 4.11, the X axis is parallel to the direction of propagation of the sound waves and the Z axis is along the direction of the incident light beam.

Let the incident light wave have a displacement ξ_i at the point of incidence O_i of the sound field in the medium, where

$$\xi_i = A \exp(2\pi i \nu t) \tag{4.34}$$

A being the amplitude of the incident electric vector and ν the frequency of the light beam. When the light beam reaches the other end of the sound field at distance L from the point of incident O_i, the displacement of the optical wave will be

$$\xi = A \exp[2\pi i \nu \{t - L\mu_x/c\}] \tag{4.35}$$

where μ_x is the index of refraction of the ultrasonically disturbed medium at the height x from the origin O, and c is the velocity of light in vacuum. Here it is tacitly assumed that the radii of curvature of the corrugated wave front are large compared with L, the width of the sound field. If μ_0 is the index of refraction of the liquid medium in its undisturbed state, we can define the quantity μ_x thus:

$$\mu_x = \mu_0 - \mu \sin(2\pi x/\lambda_a) \tag{4.36}$$

Here μ defines the maximum variation (i.e., the amplitude of variation) of the refractive index from the normal value μ_0 of the undisturbed liquid (μ depends on the maximum acoustic pressure, but not on x); the variation of μ_x with time is ignored. Since the liquid is periodically disturbed by a sinusoidally varying ultrasonic wave, we are justified in including the sine term as the multiplying factor of μ, the amplitude of variation of the index of refraction. Because of such a periodic variation of the change in index of refraction of the medium with height x from the origin, the incident plane wave front PWF becomes deformed and takes the shape of a corrugated wavefront CWF by the time the light wave front emerges out of the liquid medium (Fig. 4.8). The difference between the maximum and the minimum optical paths indicates the extent of corrugation of the emerging wave front, which in its turn causes the formation of diffraction spectra, and they can be viewed through a low-power telescope or focused on a screen as in Fig. 4.7.

Now we examine the formation of the diffraction spectra, i.e., the distribution of light intensity on the screen caused by the corrugated wave front *CWF* (on emergence from the liquid medium). For this purpose, let us choose a point P_n on a screen parallel to the plane face DD' from which light is emerging. Let the line OP_n be inclined to the OZ axis at an angle θ so that $\sin\theta = l$ (i.e., the direction cosine of OP_n with the X axis is l). The intensity distribution of light at P_n is proportional to the square of the resulting amplitude of light vibrations reaching P_n from the entire corrugated wave front *CWF*. Consider an element dx (measured along the X axis) of the corrugated wave front at the point Q at a distance x from the origin O. Let QR be drawn normal to OP_n. Then, if f is the distance of O from the screen SP_n, we have

$$\begin{aligned} QP_n \simeq RP_n = OP_n - OR &= \frac{f}{\cos\theta} - x\cos\left(\frac{\pi}{2} - \theta\right) \\ &\simeq f - x\sin\theta \simeq f - xl \end{aligned} \tag{4.37}$$

We have already seen that the light vibration at the point Q is obtained from Eq. (4.35). Thus

$$\xi_Q = A\exp[2\pi i\nu\{t - L\mu_x/c\}]$$

Therefore the light vibration at P_n due to an element dx at Q is given by

$$\begin{aligned} \xi_{P_n} &= A\exp(2\pi i\nu[\{t - (L\mu_x/c)\} - QP_n/c])\,dx \\ &= A\exp(2\pi i\nu[\{t - (L\mu_x/c)\} - (f - xl)/c])\,dx \\ &= A\exp(2\pi i\nu t)\exp(-2\pi i\nu f/c)\exp(-2\pi i\nu[L\mu_x - xl]/c)\,dx \\ &= A'\exp(-2\pi i[L\{\mu_0 - \mu\sin(2\pi x/\lambda_a)\} - xl]/\lambda)\,dx\exp(2\pi i\nu t) \\ &= A'\exp(-2\pi iL\mu_0/\lambda)\exp(2\pi i\{\mu L\sin(2\pi x/\lambda_a) + xl\}/\lambda)\,dx\exp(2\pi i\nu t) \\ &= A''\exp(2\pi i\{\mu L\sin(2\pi x/\lambda_a) + xl\}/\lambda)\,dx\exp(2\pi i\nu t) \end{aligned} \tag{4.38}$$

where $A' = A\exp(-2\pi i\nu f/c)$ and $A'' = A'\exp(-2\pi iL\mu_0/\lambda)$ are constants.

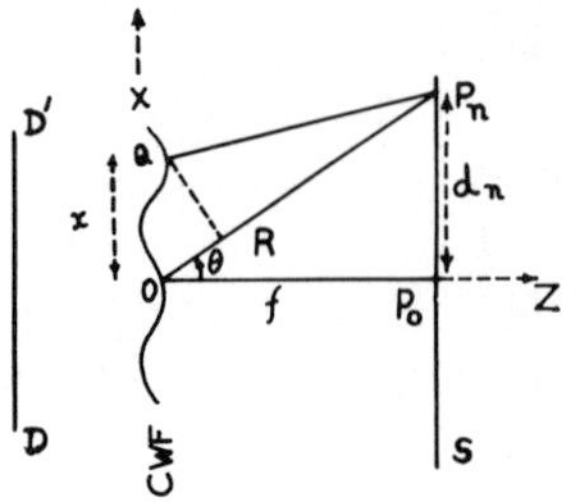

Fig. 4.11. Grating action of a corrugated optical wave front in the Raman–Nath theory (see the text above).

Hence the total amplitude of light vibrations reaching P_n due to the entire corrugated wave front is proportional to

$$\int_{-p/2}^{+p/2} \exp(2\pi i\{\mu L \sin(2\pi x/\lambda_a) + xl\}/\lambda)\, dx \tag{4.39}$$

where p is the height of the optical beam along the X axis. The real and imaginary parts of the definite integral above are respectively

$$\int_{-p/2}^{+p/2} \cos[2\pi\{\mu L \sin(2\pi x/\lambda_a) + xl\}/\lambda]\, dx \tag{4.40}$$

and

$$\int_{-p/2}^{+p/2} \sin[2\pi\{\mu L \sin(2\pi x/\lambda_a) + xl\}/\lambda]\, dx \tag{4.41}$$

If we set

$$\frac{2\pi}{\lambda} = u, \qquad \frac{2\pi}{\lambda_a} = b, \qquad \frac{2\pi\mu L}{\lambda} = v = u\mu L \tag{4.42}$$

the real part of the definite integral is given by

$$\int_{-p/2}^{+p/2} \{\cos ulx \cos(v \sin bx) - \sin ulx \sin(v \sin bx)\}\, dx \tag{4.43}$$

and the imaginary part of the definite integral is given by

$$\int_{-p/2}^{+p/2} \{\sin ulx \cos(v \sin bx) + \cos ulx \sin(v \sin bx)\}\, dx \tag{4.44}$$

To evaluate the definite integrals (4.43) and (4.44) we express $\cos(v \sin bx)$ and $\sin(v \sin bx)$ in terms of the series expansions

$$\cos(v \sin bx) = 2 \sum_0^{\infty}{}' J_{2r}(v) \cos 2rbx$$

(here all even terms exist, and all odd terms vanish) and

$$\sin(v \sin bx) = 2 \sum_0^{\infty} J_{2r+1}(v) \sin\{(2r+1)bx\}$$

(here all odd terms exist, and all even terms vanish).

In the above series expansions $J_{2r}(v) = J_n(v)$ is the Bessel function of the nth order and the prime on the summation sign indicates that the coefficient of J_0 is half that of the others. Then the definite integral (4.43) describing the real part of the amplitude of light vibrations at P_n is given by

$$2\sum_0^{\infty}{}' J_{2r}(v) \int_{-p/2}^{+p/2} \cos(ulx)\cos(2rbx)\,dx - 2\sum_0^{\infty} J_{2r+1}(v) \int_{-p/2}^{+p/2} \sin(ulx) \times \sin\{(2r+1)bx\}\,dx \tag{4.45}$$

$$= \sum_0^{\infty}{}' J_{2r}(v) \int_{-p/2}^{+p/2} [\cos(ul+2rb)x + \cos(ul-2rb)x]\,dx + \sum_0^{\infty} J_{2r+1}(v) \int_{-p/2}^{+p/2} [\cos\{ul+(2r+1)b\}x - \cos\{ul-(2r+1)b\}x]\,dx \tag{4.46}$$

Integrating the above, we obtain the value of the definite integral:

$$p\sum_0^{\infty}{}' J_{2r}(v)\left\{\frac{\sin[(ul+2rb)p/2]}{[(ul+2rb)p/2]} + \frac{\sin[(ul-2rb)p/2]}{[(ul-2rb)p/2]}\right\} + p\sum_0^{\infty} J_{2r+1}(v)\left\{\frac{\sin[\{ul+(2r+1)b\}p/2]}{[\{ul+(2r+1)b\}p/2]} - \frac{\sin[\{ul-(2r+1)b\}p/2]}{[\{ul-(2r+1)b\}p/2]}\right\} \tag{4.47}$$

The value of the definite integral corresponding to the imaginary part of the diffraction integral becomes zero.

Discussion. We shall now examine closely expression (4.47) above.

1. The magnitude of each individual term among the four terms inside the brackets is of the form $(\sin\alpha)/\alpha$ and hence has a maximum value of unity as $\alpha \to 0$.

2. When any one term attains its maximum, the values of all the remaining terms become negligibly small compared to unity. This happens because for each of the remaining three terms, the numerator cannot exceed unity, whereas the denominator is some integral nonvanishing multiple of b, which is a sufficiently large number.

3. Consequently expression (4.47), which is proportional to the amplitude of the optical-wave disturbance reaching P, becomes a maximum whenever *any one* of its four terms of the type $\sin\alpha/\alpha$ reaches maximum, which is unity. Hence the amplitude becomes maximum when

$$ul \pm nb = 0 \tag{4.48}$$

where n is an integer equal to or greater than zero, i.e., n is any even or odd positive integer. Here n takes the place of $2r$ or $(2r+1)$ in expressions

(4.45)–(4.47). Substituting back the values of $b = 2\pi/\lambda_a$ and $u = 2\pi/\lambda$, we get

$$\frac{2\pi \sin \theta}{\lambda} = \mp n \frac{2\pi}{\lambda_a}, \qquad \lambda_a \sin \theta = \mp n\lambda \tag{4.49}$$

which is the familiar formula describing optical diffraction through a plane transmission grating for normal incidence.

4. Now, to calculate the relative intensities of the different orders of spectra given by Eq. (4.49), we shall take some typical order, say m. For this order, as seen in paragraph 2 above, the amplitude of optical disturbance

$$J_m(v)\left[\frac{\sin(ul + mb)p/2}{(ul + mb)p/2}\right] \propto J_m(v) \tag{4.50}$$

since the term in brackets is unity for the particular order under consideration. Hence the intensity of the mth component relatively to the nth component is given by

$$J_m^2(v)/J_n^2(v) \qquad \text{where} \qquad v = 2\pi\mu L/\lambda \tag{4.51}$$

In the absence of ultrasonic excitation, the liquid medium is left undisturbed, and hence there is no variation of the index of refraction, i.e., $\mu = 0$. In this case all the diffraction orders vanish except the zero order, for $J_0(0) = 1$, and $J_m(0) = 0$ for all $m \neq 0$. Under ultrasonic excitation, the relative intensities depend on the parameter v or $(2\pi\mu L/\lambda)$ and hence also on the following: on μ, the amplitude of variation of the refractive index (thus also on the ultrasonic power); on L, the thickness of the ultrasonically excited liquid medium traversed by the light waves; and on λ, the wavelength of light.

It is now a matter of simple calculation to show that the intensity of the first order in the diffraction spectrum increases from zero to a maximum as the parameter v increases from zero to about 2.5, while the second, third, and fourth orders gradually gain in intensity, each one being fainter than its immediate lower neighbor. Further increase in v results in a wandering of intensities of the different orders. This information is of special importance to us in discussing the ultrasonic methods of studying the photoelastic behavior of solids. The Raman–Nath theory has satisfactorily explained the results of experimental investigations by Bär [76; 77].

Thus we have shown that a simply periodic corrugated wave is equivalent to a number of plane waves traveling in different directions inclined at angles θ_n with the incident light beam, where according to Eq. (4.49)

$$\sin\theta_n = \pm n\lambda/\lambda_a \qquad (n \text{ being an integer} \geq 0)$$

In fact, if the emerging corrugated wave front is subjected to Fourier analysis, then it automatically reveals the diffraction effects that are observed through a low-power telescope when viewing the emergent waves.

The action of a corrugated wave front as a grating has been experimentally demonstrated by Subbaramaiya [1183a] by producing ripples on the surface of water.

Sanders [1080] experimentally confirmed the light intensities predicted by the Raman–Nath theory for both progressive and stationary waves. The necessary conditions are:

1. The acoustic waves are plane.
2. The height of the light beam is small compared to the distance of the screen.
3. The radii of curvature of the corrugated wave front are very large compared with the width of the sound field in the ultrasonic cell.
4. The sound intensities are low so that the change in refractive index $\Delta\mu$ is proportional to the pressure of the acoustic wave.
5. The angle θ is very small for far-field observations.
6. The width L should be small, especially in the case of oblique incidence, otherwise the bending of light rays in the medium due to refraction becomes prominent.

4.5.6. Doppler Effect and Coherence Phenomenon

Recalling Eq. (4.36), if we do not ignore the variation of μ_x with time, we find the following interesting results:

1. The light diffracted by progressive ultrasonic waves exhibits frequency shifts due to the Doppler effect on the light waves caused by the superposition of ultrasonic waves. The nth order has the frequency $(\nu - n\nu_a)$, where ν is the frequency of light, ν_a that of sound, and n is an integer, positive or negative. Consequently the light waves from the different orders are incoherent with each other.

2. The light diffracted by standing ultrasonic waves exhibits Doppler shifts differently from those by progressive waves. In the latter case the diffraction orders can be classified under two groups, namely all even orders under one group and all odd orders under the other group. Any

even order, say $2n$, contains light waves with frequencies $\nu \pm 2r\nu_a$, where r is an integer including zero; the relative intensity* of a typical subcomponent is

$$I_{n,r} = J_{s-r}^2(\pi\mu L/\lambda)J_{s+r}^2(\pi\mu L/\lambda), \qquad s = n/2 \tag{4.52a}$$

Any odd order, say $(2n + 1)$, contains light waves with frequencies $\nu \pm (2r + 1)\nu_a$; here the relative intensity of a typical subcomponent is

$$I_{n,r} = J_{s-r}^2(\pi\mu L/\lambda)J_{s+r+1}^2(\pi\mu L/\lambda), \qquad s = (n - 1)/2 \tag{4.52b}$$

Consequently, any pair of even orders or odd orders can partly cohere, and an even order and an odd order are incoherent.

We note from the above discussion that the zero-order spectrum due to progressive ultrasonic waves is strictly the same as the incident monochromatic radiation, whereas even the zero-order spectrum due to stationary ultrasonic waves has a structure with frequencies $\nu \pm 2r\nu_a$.

3. Whether it is progressive sound waves or stationary sound waves in the medium, the incident plane wave of light becomes corrugated as it emerges from the ultrasonically excited medium, and the light will be diffracted at the angles given by $\sin^{-1}(\pm n\lambda/\lambda_a)$. These angles of diffraction are the same for both types of waves, hence the diffraction phenomenon cannot distinguish between the two types of waves, progressive and stationary.

For a general periodic supersonic disturbance in the medium, when a plane wave front of monochromatic light is incident normal to the propagation of sound waves, the intensity (or the square of the amplitude) of the light will be periodic in x and t on the optical wave front. In the case of a standing sound wave, the average intensity I on the light wave front with respect to time is periodic in $\lambda_a/2$; that is, the stationary sound waves render themselves observable, their separation being $\lambda_a/2$; this was first noted by Hiedemann and subsequently applied by Mayer to determine the elastic constants of sapphire [789; 790].

A review of the various methods for determining the ultrasonic velocities in liquids and solids has been given by McSkimin [800; 801]. Two methods involving the interaction between electromagnetic radiation and acoustic waves of thermal origin in a crystal, first conceived by Debye [285], have been used in recent times for determining the elastic constants

* Henceforth, the constant of proportionality between the light intensity of a diffraction order at a point P_n and the terms involving the Bessel functions describing the amplitude of vibration at P_n will be assumed to be unity.

of crystals. The first method involves measurement of the intensity of diffuse x rays scattered by the acoustic waves in a solid [974; 999; 1000]. The second is the light-scattering method which involves an analysis of the fine structure of the spectra of the scattered light, called the Brillouin spectra [201]. The frequency separation between the central line due to Rayleigh scattering and the shifted Brillouin components is used to calculate the acoustic velocity [238; 373; 685] in crystals.

We shall discuss now, in some detail, the light-scattering method, particularly because it affords a method of determining both elastic and photoelastic constants of crystals.

4.6. Brillouin Effect and Crystal Elasticity

4.6.1. Introduction

Based on the classical electromagnetic theory of light, Lord Rayleigh [1036–1038] deduced the famous result that the intensity of the scattered light varies inversely as the fourth power of the wavelength of the incident radiation. Combining the Bragg reflection condition with Debye's concept [285] of acoustic waves of thermal origin, Brillouin [201] predicted that the light is scattered also by these acoustic waves present in the scattering medium. For scattering in a given direction, owing to the effect of interference, only the acoustic waves of a certain wave-number vector should be operative. Hence the scattered light radiation should consist of altered frequencies besides that of the unmodified or incident radiation which is, of course, due to the classical Rayleigh scattering. Furthermore, these frequency shifts, which are in the nature of a Doppler effect, should vary with the direction of observation of the scattered light, and also with the frequency of the incident light. The frequencies of the components, henceforth to be referred to as Brillouin components, are identical with the frequencies of the acoustic waves, which are effective in scattering along the direction of observation with respect to the orientation of the crystalline medium.

In the case of the simplest scattering medium, the Brillouin scattering should contain only a doublet with their frequencies $\nu \pm \Delta\nu$ where

$$\frac{\Delta\nu}{\nu} = \pm 2 \frac{v_a}{c} n \sin\frac{\theta}{2} \tag{4.53}$$

Here v_a and c are the acoustic and optical velocities in the liquid or solid medium of refractive index n, the angle of scattering being θ and the incident light radiation frequency being ν.

Equation (4.53) indicates the possibility of determining the acoustic velocity in a solid medium, the frequencies of the acoustic waves in solids being of the order of 10–50 GHz, and their wavelengths, of the same size as those of optical waves.

Although Brillouin's prediction was made as early as 1922, an experimental confirmation of the same was not forthcoming till very much later. This was mainly due to the difficulties in the experimental detection and measurements of such low-frequency shifts of the Brillouin components in the light scattered by crystals. In most of the crystals the frequency shifts of the Brillouin components are of the order of 0.5–3 cm^{-1}, which are 10^3–10^4 times smaller than those occurring in the Raman effect, with the added disadvantage that the Brillouin components are very much weaker in intensity compared with the Raman lines. Consequently, it is absolutely necessary to utilize an instrument of a sufficiently high resolving power in order to measure the small Brillouin shifts. Furthermore, the crystal sample should show the greatest clarity and freedom from inclusions, because the inclusions in the crystal result in the existence of a large number of scattering centers, which in turn cause the unmodified (Rayleigh) line to spread on either side or develop tails having more intensity than the Brillouin components, and also overlap and mask the latter which are situated close by, thus making it extremely difficult to make an accurate assessment of the very small Brillouin shifts between the weak Brillouin components and the broad and strong Rayleigh line. Nevertheless, the fine structure of the scattered radiation predicted by Brillouin's theory was first observed by Gross [453; 454] in quartz, later by Raman and Venkateswaran [1012] in gypsum, and by Sibaiya [1131] in Rochelle salt. Using the resonance radiation at 2537 Å of a quartz mercury arc in conjunction with a three-meter quartz spectrograph, Krishnan [683] and Krishnan and Chandrasekharan [684] have recorded unambiguously the Brillouin components in diamond, quartz, alumina, and quite a few other crystals. They have eliminated the Rayleigh line from the scattered radiation by the use of a mercury vapor filter following the Rasetti technique [1032]. Furthermore, as the intensity of the scattered radiation is proportional to λ^{-4} and n^8, the enormously increased scattering power in the ultraviolet region is of added advantage and hence shortens the time of exposure necessary for recording the Brillouin spectra. Over and above these advantages, the separation of the Brillouin components increases with the frequency of the incident radiation. The use of the mercury resonance radiation, therefore, combined with a Hilger three-meter quartz spectrograph which has a dispersion of about 14 cm^{-1} in the 2537 Å region, was adequate for Krishnan [685] to make accurate measurements of the frequency shifts and to evaluate the elastic constants of several crystals.

Studies on the Brillouin effect in crystals, in order to investigate their elastic and photoelastic properties, have come into the limelight only recently, largely owing to the development of lasers whose high monochromaticity, intensity, and directivity make them ideal sources of light in experimental investigations on the Brillouin effect in crystals. Although the unavailability of large-size single crystals free from inclusions was a major hurdle for use with the more familiar types of light sources, quite a few crystals like diamond, quartz, calcite, sodium chloride, and lithium fluoride were studied quite satisfactorily during the period 1940–1960. But, with the advent of lasers, there has been a phenomenal growth in these investigations since 1965. See the Bibliography for more references.

The theory of light scattering by acoustic waves of thermal origin in crystals has been discussed by several investigators including Mandelstamm, Landsberg, and Leontowitsch [767], Tamm [1200], Mueller [830], Gross [453; 454], Bhatia and Krishnan [136], and more recently by Benedek and Fritsch [98], Hope [538], Vacher and Boyer [1246; 1247], and Humphreys and Maradudin [547]. Leontowitsch and Mandelstamm [733] have given a general theory of the thermal scattering of light by crystals from the classical standpoint. They considered the medium macroscopically as an elastic medium and obtained an intensity formula depending only on the elastic and elasto-optical constants of the scattering crystal. The order of magnitude of the effect as given by their theory has been confirmed by the experiments carried out by Krishnan [683] and Ramachandran [993; 996; 997] on diamond crystals. From the standpoint of the quantum theory, Tamm [1200] has developed a theory where both the crystal and the radiation are treated quantum mechanically. In his treatment Tamm considered only the optical dielectric effects due to density fluctuations. Theimer [1209; 1210] has given a macroscopic theory on the basis of the lattice model. However, an error crept into his calculations, which was pointed out by Born and Huang [173]. These authors [173] have tackled the problem, treating the radiation classically and the lattice quantum mechanically, and derived the intensity formula which can readily give numerical estimates for concrete examples. Since, except at the very lowest temperatures, the quantum effects are negligible, their treatment is essentially the same as the theory of Leontowitsch and Mandelstamm [733].

According to the simple theory due to Brillouin, one concludes that in a crystal only three pairs of the Brillouin components should occur in the scattered light due, respectively, to the three types of acoustic waves which are propagated with different velocities in any given direction in the crystal. By taking into consideration the symmetry of the crystal and the influence of birefringence on the Brillouin components, one should expect, in general, not three but twelve components to appear in the case of a birefringent crystal. The most general expression for a birefringent crystal is given by

$$\frac{\Delta\nu}{\nu} = \pm \frac{v_a}{c} (n_i^2 + n_s^2 - 2n_i n_s \cos\theta)^{1/2} \tag{4.54}$$

where n_i and n_s are the two refractive indices of the crystal for the directions of incidence and scattering. In the case of singly refracting crystals like NaCl and KCl, Eq. (4.54) reduces to Eq. (4.53) because n_i and n_s become equal to n of the singly refracting crystal.

4.6.2. Theory of Light Scattering in Birefringent Crystals

We give below the theory of light scattering in birefringent crystals, along the lines of Chandrasekharan [240].

4.6.2.1. Classical Method

Here we treat the thermal scattering as being of macroscopic origin, which justifies the classical method as adopted by Brillouin [201].

The propagation of acoustic waves of thermal origin, first envisaged by Debye [285], causes periodic stratifications in the crystal, represented by $N'N$, $N_1'N_1$, etc. (Fig. 4.12a). Consequently there are changes in density resulting in changes in refractive index, both of which vary periodically with a frequency equal to that of the acoustic waves. Let a monochromatic light radiation of wavelength λ be incident along $I'CI$, and let CS be the direction of the scattered light waves. Then the plane of $I'CS$ determines the plane of scattering which obviously should be perpendicular to the plane of stratification for reflections to occur. Then the volume elements lying on a plane of stratification, say NN', will contribute to the scattering in the direction CS only if there is coherence of phase between the wavelets scattered by the stratifications $N'N$, $N_1'N_1$, etc. The net effect will be appreciable only if the individual contributions are all in phase.

Let λ_a be the distance between two consecutive stratifications $N'N$ and $N_1'N_1$ of a like nature (say compressions). Then for constructive

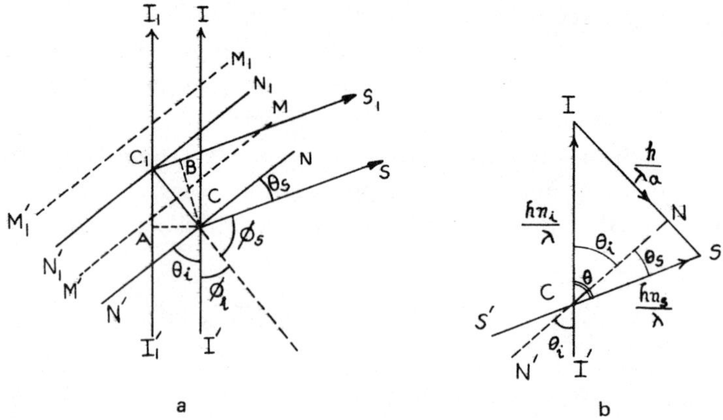

Fig. 4.12. Brillouin scattering of light in crystals: geometric representation (see the text on pages 189 and 191). (a) Wave picture of scattering; (b) quantum picture of scattering. From [240].

interference to occur the path difference should be an integral multiple of the optical wavelength λ of the incident light wave. Let CA and CB be perpendicular to C_1I_1 and C_1S_1, respectively. Then, if p is an integer,

$$p\lambda = n_i(AC_1) + n_s(C_1B) = n_i\lambda_a \sin\theta_i + n_s\lambda_a \sin\theta_s$$
$$= (n_i \sin\theta_i + n_s \sin\theta_s)\lambda_a \tag{4.55}$$

[Note that $n_i(AC_1)$ is the optical path for a ray AC_1 in a medium of refractive index n_i for the incident ray; similarly, for C_1B we have $n_s(C_1B)$.]

Now there is a sinusoidal variation of density of the medium as we go from $N'N$ to $N_1'N_1$, with the change in refractive index also varying sinusoidally due to the propagation of the elastic wave. The propagation of the acoustic waves thus produces a three-dimensional grating in the medium, and it is well known [201] that in such a case there can only be one maximum produced, namely the first. That is, $p = 1$ and hence

$$\lambda_a(n_i \sin\theta_i + n_s \sin\theta_s) = \lambda$$

or

$$n_i \sin\theta_i + n_s \sin\theta_s = \lambda/\lambda_a \tag{4.56}$$

When a light wave goes from a medium of refractive index n_1 to another of refractive index n_2 and if i_1 and i_2 are the angles of incidence and refraction at the surface of separation, then it is well known that

$$n_1 \sin i_1 = n_2 \sin i_2 \tag{4.57}$$

Thus if ϕ_i and ϕ_s are the angles of incidence and scattering at the point C on the plane $N'N$ then

$$n_i \sin \phi_i = n_s \sin \phi_s$$

or, from the geometry of the figure, we note that

$$n_i \cos \theta_i = n_s \cos \theta_s \tag{4.58}$$

Furthermore the angle of scattering is given by

$$\theta = \theta_i + \theta_s \tag{4.59}$$

Squaring and adding Eqs. (4.58) and (4.56), we obtain

$$n_i^2 + n_s^2 - 2n_i n_s(\cos \theta_i \cos \theta_s - \sin \theta_i \sin \theta_s) = \lambda^2/\lambda_a^2$$

Therefore

$$\frac{\lambda}{\lambda_a} = (n_i^2 + n_s^2 - 2n_i n_s \cos \theta)^{1/2} \tag{4.60}$$

taking Eq. (4.59) into consideration.

Now $\Delta\nu$, the frequency shift of each of the Brillouin components, is equal to $(\nu_i - \nu_s)$, that is, the difference between the frequencies of the incident and scattered light radiations. This $\Delta\nu$, being in the nature of a Doppler shift, is equal to ν_a, the frequency of the effective acoustic wave of thermal origin in the solid. Thus

$$\Delta\nu = (\nu_i - \nu_s) = \nu_a$$

Now $\nu_a = v_a/\lambda_a$ for acoustic waves and $\nu = c/\lambda$ for optical waves. Therefore

$$\frac{\Delta\nu}{\nu} = \frac{\nu_a}{\nu} = \frac{v_a/\lambda_a}{c/\lambda} = \frac{v_a}{c}\,\frac{\lambda}{\lambda_a}$$

$$= \frac{v_a}{c}(n_i^2 + n_s^2 - 2n_i n_s \cos \theta)^{1/2} \tag{4.54}$$

which is the desired equation.

4.6.2.2. Method Based on Quantum Theory

Brillouin scattering of light was treated by Tamm [1200] on the basis of quantum theory and applied to singly refracting media. Following Tamm, the theory has been developed for birefringent crystals by Chandrasekharan [240], which is briefly given below.

We consider here a collision process between a quantum of light energy (photon) and a quantum of the acoustic or elastic energy (phonon).

As in Fig. 4.12b, $I'CI$ is the direction of the incident light wave, and $S'CS$ that of the scattered wave, the angle of scattering $\angle ICS$ being θ. The momentum of the photon is h/λ and is in the direction of the wave normal inside the birefringent medium, where h is Planck's constant and λ is the wavelength of the incident light in vacuum.

For any particular direction of propagation of light in the birefringent medium, there are two possible states of polarization (and of the electric vector); hence there are two possible wavelengths equal to λ/n_i (n_i being one of the two possible refractive indices for the incident radiation inside the birefringent medium), corresponding to the two states of polarization. The wavelength of the scattered radiation can also have two values depending on the direction of scattering inside the crystal; let this wavelength be given by λ/n_s, where n_s is one of the two possible indices of refraction of the scattered radiation inside the crystal.

Then $h(n_i/\lambda)$ and $h(n_s/\lambda)$ are the momenta of the incident and scattered photons respectively. Similarly the momentum of the phonon is h/λ_a, where λ_a is the wavelength of the elastic wave.

Let these momenta be represented vectorially by $\overrightarrow{CI}$, $\overrightarrow{CS}$, and $\overrightarrow{IS}$ in Fig. 4.12b. Then the law of conservation of momentum demands

$$\overrightarrow{CS} = \overrightarrow{CI} + \overrightarrow{IS} \tag{4.61}$$

Draw CN perpendicular to $\overrightarrow{IS}$, and let $\angle ICN = \theta_i$ and $\angle NCS = \theta_s$. Then applying the condition of the law of conservation of momentum to the resolved components of $\overrightarrow{CI}$ and $\overrightarrow{CS}$ parallel and perpendicular to $\overrightarrow{IS}$, we have

$$\frac{h}{\lambda_a} = \frac{hn_s \sin \theta_s}{\lambda} + \frac{hn_i \sin \theta_i}{\lambda} \tag{4.62}$$

and

$$\frac{hn_i \cos \theta_i}{\lambda} + \frac{h}{\lambda_a} \cos 90 = \frac{h}{\lambda} n_s \cos \theta \tag{4.63}$$

From Eqs. (4.62) and (4.63) we obtain

$$n_i \sin \theta_i + n_s \sin \theta_s = \lambda/\lambda_a \tag{4.64}$$

and

$$n_i \cos \theta_i - n_s \cos \theta_s = 0 \tag{4.65}$$

Squaring and adding Eqs. (4.64) and (4.65), we have

$$n_i^2 + n_s^2 - 2n_i n_s \cos(\theta_i + \theta_s) = \lambda^2/\lambda_a^2 \tag{4.66}$$

and since the scattering angle $\theta = \theta_i + \theta_s$ from the geometry of Fig. 4.12b, Eq. (4.66) simplifies to

$$\lambda/\lambda_a = (n_i^2 + n_s^2 - 2n_i n_s \cos\theta)^{1/2} \tag{4.67}$$

which is the same as Eq. (4.60), obtained by the classical method.

From the law of conservation of energy,

energy of scattered photon = energy of incident photon ± energy of phonon

Thus

$$h\nu_s = h\nu_i \pm h\nu_a \qquad \text{from which} \qquad \nu_s = (\nu_i \pm \nu_a) \tag{4.68}$$

Now, the optical frequency shift $\Delta\nu$ is equal to the incident optical frequency minus the scattered optical frequency, i.e.,

$$\Delta\nu = (\nu_i - \nu_s) = \pm\nu_a \tag{4.69}$$

The implication of Eq. (4.69) above, in terms of classical wave theory, is that the scattered radiation is a consequence of the Doppler effect of the acoustic wave on the incident optical wave, giving rise to a modified scattered optical wave which is called the Brillouin component.

As in the previous case,

$$\frac{\Delta\nu}{\nu} = \frac{\text{optical frequency shift}}{\text{incident optical frequency}} = \pm\frac{\nu_a}{\nu}$$

$$= \pm\frac{v_a/\lambda_a}{c/\lambda} = \pm\frac{v_a}{c}\,\frac{\lambda}{\lambda_a}$$

$$= \pm\frac{v_a}{c}(n_i^2 + n_s^2 - 2n_i n_s \cos\theta)^{1/2} \tag{4.70}$$

Along any specific direction in a crystal there are three types of acoustic waves traveling with different velocities, and the shifts of the Brillouin components depend on these acoustic velocities. Furthermore, since the waves effective in scattering along a particular direction are either approaching or receding, there can be three doublets, or pairs, of the Brillouin components in the light scattered in a nonbirefringent crystal. However, the situation is different for a birefringent crystal.

For a given direction of incidence of light on a birefringent crystal and for a given angle of scattering, the direction of scattering is also fixed; then, in general, n_i and n_s can each take two values. We can then have four pairs of values for (n_i, n_s); hence four sets of values for θ_i, θ_s, and λ_a are possible. Thus the wavelengths as well as the direction of the acoustic wave propagation effective in light scattering are fixed for any particular pair of values of n_i and n_s. But along this direction (of propagation of acoustic waves) there are three types of acoustic waves with the same wavelength λ_a but with different velocities v_a and frequencies ν_a. From Eqs. (4.54) and (4.70), it is obvious that there must in general be $2\times2\times3 = 12$ values for the frequency shifts $\Delta\nu$. Therefore, the light scattered by a birefringent crystal, like barite or sulfur, must consist of 12 pairs of Brillouin components; however, the number of components that appear in any specific situation depends on the symmetry of the experimental crystal and the directions of incidence and of scattering of light.

In the case of singly refracting crystals like NaCl and KCl, Eq. (4.70) reduces to the familiar expression

$$\frac{\Delta\nu}{\nu} = \pm \frac{v_a}{c} (2n^2 - 2n^2 \cos\theta)^{1/2} = \pm \frac{1}{c} v_a 2n \sin\frac{\theta}{2} \tag{4.71}$$

because n_i and n_s can each take only one value, equal to n of the singly refracting crystal. Hence there can be only three pairs of Brillouin components. In this situation each individual Brillouin component can be regarded as having a degeneracy of four.

The effective acoustic wave front giving rise to the Brillouin components always bisects the internal angle between the directions of incidence and of scattering of light, and the scattering phenomenon can thus be regarded as "specular reflection" from the acoustic waves.

From Eq. (4.70) it is obvious that the fractional change of frequency is very small, being of the order of the ratio of the elastic wave velocities to the light velocity.

4.6.3. Concluding Remarks

More rigorous theory shows that, even for cubic crystals, there is a slight asymmetry in the frequency shift of the Brillouin pair around the unmodified (Rayleigh) line of frequency ν (Benedek and Fritsch [98]). The high-frequency line is shifted by an amount $\delta\nu'$ farther away from ν while the low-frequency line is shifted by $\delta\nu'$ toward ν. But the effect is too small to be detected with the existing experimental techniques.

In order to obtain the directional variation of the acoustic velocities in a crystal, one studies the spectral character of the transversely scattered light for different orientations of the crystal with respect to the direction of the incident beam and of the observation. Both the shift and the intensity of the Brillouin components vary with the orientation of the crystal specimen. Then, using Eq. (4.54), we can calculate the acoustic velocities for different directions, and using Christoffel's equations, we can readily evaluate the elastic constants. Thus the thermal scattering of light in crystals affords an experimental determination of the elastic constants of crystals.

The elastic constants of solids determined by the Brillouin light scattering method in the microwave region of sound waves (their frequencies ranging from 8 to 15 GHz) are in very good agreement with those obtained by ultrasonic methods using externally generated acoustic waves of frequencies of the order of 10 MHz. This experimental fact confirms the absence of dispersion in the acoustic velocity over three orders of change of magnitude in the acoustic frequency (between 10 MHz and 10 GHz).

It is interesting to note that in the Brillouin scattering method, acoustic waves are generated by heat and detected by light!

5

Experimental Methods of Determining the Photoelastic Constants

5.1. Optical Behavior of a Solid under a Mechanical Stress, and Neumann's Constants

The earliest attempts to study quantitatively the change of the optical properties of solids under the influence of a mechanical stress were made by Neumann [871]. He described the optical changes in amorphous solids by introducing two constants p and q. These constants (which may be called the strain-optical constants due to Neumann) are defined in terms of a linear strain, say ε_{zz}, thus:

$$\Delta n_z = n_z - n = -n^2 q \varepsilon_{zz}, \qquad \Delta n_x = n_x - n = -n^2 p \varepsilon_{zz}$$

Here p and q relate the strain ε_{zz} to the change produced by it in the index of refraction for light vibrations perpendicular and parallel to the direction of the linear strain; n is the refractive index of the solid in its undeformed state. It is the difference $(p - q)$ that can be determined using a Babinet compensator. Their individual values can be determined by interferometric methods, though not to a high degree of accuracy. The ratio p/q can be determined by ultrasonic methods (Bergmann and Fues [107]; Hiedemann and Hoesch [514–516]), and combining the values of $(p - q)$ and p/q, one can get the individual values of p and q.

It was not until the time of Pockels (1889) that it was realized that crystals behave differently from glasses with regard to their photoelastic properties; it was again Pockels who discovered that even the simplest type of crystals, namely cubic crystals, showed photoelastic anisotropy. He gave the phenomenological theory of the photoelastic behavior of crystals in 1889 and published a series of papers on the behavior of a variety of crystals, thereby proving the validity of his phenomenological theory.

In what follows we shall describe, by the classical method, originally due to Pockels himself, the derivation of expressions for stress birefringence in terms of q_{ij} for a cubic crystal of the NaCl type when the crystal is subjected to a uniaxial stress for three simple, yet important, orientations. Later, we shall develop a method of obtaining expressions for the stress birefringence for a few typical orientations of an orthorhombic crystal. This is important in two ways. First, this will indicate the method of approach to tackle an initially birefringent crystal. Secondly, we can deduce the expressions for tetragonal (C_{4v}, S_4, D_4, and D_{4h}) and cubic crystals as special cases of the orthorhombic type by introducing certain simplifications arising out of additional symmetry considerations. We describe also a simple method due to Mandarino [765] of relating Δn with q_{ij} and apply his method to some trigonal and orthorhombic crystal classes. Finally, we shall indicate the tensor method of obtaining the stress birefringence for some typical orientations of an orthorhombic crystal.

5.2. Derivation of Expressions for the Stress Birefringence in Terms of q_{ij} for Cubic and Noncubic Crystals

5.2.1. Stress Birefringence in Cubic Crystals

The traditional method of measuring the photoelastic constants in crystals is to take rectangular parallelepipeds of the crystal with their edges along known directions, compress a suitable prism in one direction, and measure the amount of birefringence produced using a Babinet compensator. The experimental arrangement will be described later on. The expressions for the observed artificial double refraction for cubic crystals are derived below. The set of equations (3.9d) of Chapter 3 when applied to a cubic crystal of the T and T_h classes, which have four stress-optical constants q_{ij} according to Bhagavantam's scheme, reduce to Eq. (5.1) below, in which the B_{ij} and P_{ij} are put in the tensor notation to facilitate their transformation from one coordinate system to another, while the q_{mn} are retained in their two-suffix notation because the numerical values of p_{mn} and q_{mn} are given in the two-suffix notation in the literature (for example, see [83; 84; 1526]):

$$\begin{aligned} B_{11} - B_{11}^0 &= -(q_{11}P_{xx} + q_{12}P_{yy} + q_{13}P_{zz}), \qquad & B_{23} &= -q_{44}P_{yz} \\ B_{22} - B_{22}^0 &= -(q_{13}P_{xx} + q_{11}P_{yy} + q_{12}P_{zz}), \qquad & B_{31} &= -q_{44}P_{zx} \\ B_{33} - B_{33}^0 &= -(q_{12}P_{xx} + q_{13}P_{yy} + q_{11}P_{zz}), \qquad & B_{12} &= -q_{44}P_{xy} \end{aligned} \tag{5.1}$$

Here, for the cubic crystal, $B_{11}^0 = B_{22}^0 = B_{33}^0 = 1/n_o^2$, where n_o is the refractive index of the unstressed (and hence undeformed) crystal, and $B_{23}^0 = B_{31}^0 = B_{12}^0 = 0$.

Case 1. Let a stress P be applied in the Z direction. Then $P_{zz} = P$, and all other components vanish. Therefore

$$B_{11} - B_{11}^0 = -q_{13}P, \qquad B_{22} - B_{22}^0 = -q_{12}P,$$

$$B_{33} - B_{33}^0 = -q_{11}P, \qquad B_{23} = B_{31} = B_{12} = 0$$

Hence the equation of the optical index ellipsoid in the deformed state of the crystal is

$$B_{11}x^2 + B_{22}y^2 + B_{33}z^2 = 1 \tag{5.2}$$

If the observation is made along the X axis, the section perpendicular to the X axis is given by

$$B_{22}y^2 + B_{33}z^2 = 1 \tag{5.3}$$

The axes of the ellipse thus formed are parallel to the axes of the reference system. Hence there is no tilt of the axes.

Then an expression for the stress birefringence of the crystal (i.e., the difference between the indices of refraction for the two polarized waves, one with its vibration direction along and the other perpendicular to the direction of stress) can be obtained from the equation

$$(B_{33} - B_{22}) = -(q_{11} - q_{12})P \tag{5.4}$$

The second-rank tensor nature of a stress P is elaborated in Chapter 4. But $(B_{33} - B_{22}) = 1/n_{33}^2 - 1/n_{22}^2$ since the optical parameters B_{33} and B_{22} are, respectively, equal to $1/n_{33}^2$ and $1/n_{22}^2$, where n_{33} and n_{22} are the indices of refraction of the stressed crystal for light vibrations along and perpendicular to the stress direction. Therefore

$$\begin{aligned}(B_{33} - B_{22}) &= (n_{22}^2 - n_{33}^2)/(n_{33}^2 \cdot n_{22}^2) \\ &\cong -(n_{33} - n_{22})(n_{33} + n_{22})/n_o^4 \\ &\cong -\Delta n \cdot 2n_o/n_o^4 = -2\Delta n/n_o^3\end{aligned} \tag{5.5}$$

where n_o is the refractive index of the crystal in the undeformed state. Comparing Eqs. (5.4) and (5.5), we have

$$-2\Delta n/n_o^3 = -(q_{11} - q_{12})P \qquad \text{or} \qquad \Delta n/P = +(n_o^3/2)(q_{11} - q_{12})$$

Thus the stress birefringence or the birefringence produced per unit stress is $\Delta n/P = (n_o^3/2)(q_{11} - q_{12})$.

Since the optical-path retardation is equal to the birefringence times the thickness of a crystal, the optical-path retardation δ_x per unit stress and per unit thickness for observation along the X axis is given by

$$\delta_x = (n_o{}^3/2)(q_{11} - q_{12}) \tag{5.6}$$

Similarly, if the observation direction is along the Y axis,

$$\delta_y = (n_o{}^3/2)(q_{11} - q_{13}) \tag{5.7}$$

In an actual experiment, a small correction term has to be applied to the thickness of the crystal, since the thickness changes under a linear stress along the length. But this correction term is of a second order and hence is usually ignored for cubic crystals.

Case 2. Let the orientation of the prism be such that its length is parallel to [001], say Z', its breadth parallel to [110], say X', and the thickness parallel to $[1\bar{1}0]$, say Y'. Let the stresses P_{xx}, P_{yy}, etc. be transformed to the new axes, defined by the directions of the edges of the prism. The direction-cosine scheme

	X	Y	Z
X'	$\dfrac{1}{2^{1/2}}$	$\dfrac{1}{2^{1/2}}$	0
Y'	$\dfrac{-1}{2^{1/2}}$	$\dfrac{1}{2^{1/2}}$	0
Z'	0	0	1

gives the relation between X', Y', Z', and the X, Y, Z axes. Since P is a stress, its components are second-rank tensor components. Hence the law of transformation of second-rank tensor components holds. Now, bearing in mind that the suffixes x, y, and z can be interchanged with 1, 2, and 3 wherever necessary, for the purpose of transformation of $P_{xx}, \ldots,$ $P_{yz}, \ldots,$ we proceed thus:

$$P_{ij} = \alpha_{ki}\alpha_{lj}P'_{kl} \qquad \text{Note here that } i = x;\ j = x$$

$$\begin{aligned} P_{xx} &= \alpha_{lj}[\alpha_{11}P'_{1l} + \alpha_{21}P'_{2l} + \alpha_{31}P'_{3l}] \\ &= \alpha_{11}(\alpha_{11}P'_{11} + \alpha_{21}P'_{12} + \alpha_{31}P'_{13}) + \alpha_{21}(\alpha_{11}P'_{21} + \alpha_{21}P'_{22} + \alpha_{31}P'_{23}) \\ &= \alpha_{11}^2P'_{11} + 2\alpha_{11}\alpha_{21}P'_{21} + \alpha_{21}^2P'_{22} \\ &= \tfrac{1}{2}P'_{x'x'} + \tfrac{1}{2}P'_{y'y'} - P'_{x'y'} \end{aligned} \tag{5.8a}$$

Similarly,

$$P_{yy} = \alpha_{k2}\alpha_{l2}P'_{kl} = \alpha_{l2}(\alpha_{12}P'_{1l} + \alpha_{22}P'_{2l} + \alpha_{32}P'_{3l}) = \tfrac{1}{2}P'_{x'x'} + \tfrac{1}{2}P'_{y'y'} + P'_{x'y'}$$

$$P_{zz} = \alpha_{k3}\alpha_{l3}P'_{kl} = \alpha_{l3}(\alpha_{13}P'_{1l} + \alpha_{23}P'_{2l} + \alpha_{33}P'_{3l}) = P'_{z'z'}$$

$$P_{yz} = \alpha_{k2}\alpha_{l3}P'_{kl} = \alpha_{l3}(\alpha_{12}P'_{1l} + \alpha_{22}P'_{2l} + \alpha_{32}P'_{3l}) = \frac{1}{2^{1/2}}P'_{z'x'} + \frac{1}{2^{1/2}}P'_{y'z'}$$

$$P_{zx} = \alpha_{k3}\alpha_{l1}P'_{kl} = \alpha_{l1}(\alpha_{33}P'_{3l}) = \alpha_{33}(\alpha_{11}P'_{31} + \alpha_{21}P'_{32}) = \frac{1}{2^{1/2}}P'_{z'x'} - \frac{1}{2^{1/2}}P'_{y'z'}$$

$$P_{xy} = \alpha_{k1}\alpha_{l2}P'_{kl} = \alpha_{l2}(\alpha_{11}P'_{11} + \alpha_{21}P'_{21}) = \tfrac{1}{2}P'_{x'x'} - \tfrac{1}{2}P'_{y'y'} \tag{5.8b}$$

The set of equations (5.1) in this chapter takes the form

$$\begin{aligned}
B_{11} - B_{11}^0 &= -[q_{11}(\tfrac{1}{2}P'_{x'x'} + \tfrac{1}{2}P'_{y'y'} - P'_{x'y'}) + q_{12}(\tfrac{1}{2}P'_{x'x'} + \tfrac{1}{2}P'_{y'y'} + P'_{x'y'}) \\
&\quad + q_{13}P'_{z'z'}] \\
B_{22} - B_{22}^0 &= -[q_{13}(\tfrac{1}{2}P'_{x'x'} + \tfrac{1}{2}P'_{y'y'} - P'_{x'y'}) + q_{11}(\tfrac{1}{2}P'_{x'x'} + \tfrac{1}{2}P'_{y'y'} + P'_{x'y'}) \\
&\quad + q_{12}P'_{z'z'}] \\
B_{33} - B_{33}^0 &= -[q_{12}(\tfrac{1}{2}P'_{x'x'} + \tfrac{1}{2}P'_{y'y'} - P'_{x'y'}) + q_{13}(\tfrac{1}{2}P'_{x'x'} + \tfrac{1}{2}P'_{y'y'} + P'_{x'y'}) \\
&\quad + q_{11}P'_{z'z'}] \\
B_{23} &= -q_{44}\left(\frac{1}{2^{1/2}}P'_{y'z'} + \frac{1}{2^{1/2}}P'_{z'x'}\right) \\
B_{31} &= -q_{44}\left(\frac{1}{2^{1/2}}P'_{z'x'} - \frac{1}{2^{1/2}}P'_{y'z'}\right) \\
B_{12} &= -q_{44}(\tfrac{1}{2}P'_{x'x'} - \tfrac{1}{2}P'_{y'y'})
\end{aligned} \tag{5.9}$$

Let the equation for the transformed index ellipsoid be represented by

$$A_{11}x'^2 + A_{22}y'^2 + A_{33}z'^2 + 2A_{23}y'z' + 2A_{31}z'x' + 2A_{12}x'y' = 1$$

The quantity A_{ij} can now be expressed in terms of B_{kl} by the law of transformation of a second rank symmetric tensor. Thus $T'_{ij} = \alpha_{ik}\alpha_{jl}T_{kl}$. Hence

$$\begin{aligned}
A_{11} &= \tfrac{1}{2}B_{11} + \tfrac{1}{2}B_{22} + B_{12}, & A_{22} &= \tfrac{1}{2}B_{11} + \tfrac{1}{2}B_{22} - B_{12}, \\
A_{33} &= B_{33}, & A_{23} &= \frac{1}{2^{1/2}}B_{23} - \frac{1}{2^{1/2}}B_{31}, \\
A_{31} &= \frac{1}{2^{1/2}}B_{31} + \frac{1}{2^{1/2}}B_{23}, & A_{12} &= -\tfrac{1}{2}B_{11} + \tfrac{1}{2}B_{22}
\end{aligned} \tag{5.10}$$

1. Let the stress be applied along the new Z' axis, which is the same as the old Z axis. Then for observations along the new X' axis, the section of the optical index ellipsoid normal to the X' axis is

$$A_{22}y'^2 + A_{33}z'^2 + 2A_{23}y'z' = 1$$

But $A_{23} = (1/2^{1/2})B_{23} - (1/2^{1/2})B_{31} = 0$ from Eqs. (5.9) and (5.10), because $P'_{z'z'}$ is the only stress not equal to zero. Therefore there is no tilt of the axes of the ellipse with respect to the Y', Z' axes of the reference coordinate system.

Now the stress birefringence can be obtained from the term $(A_{33}-A_{22})$ thus

$$\begin{aligned}(A_{33} - A_{22}) &= B_{33} - \{\tfrac{1}{2}(B_{11} + B_{22}) - B_{12}\} \\ &= \tfrac{1}{2}[2B_{33} - B_{11} - B_{22}] \quad \text{since} \quad B_{12} = 0 \quad \text{from} \quad \text{Eq. (5.9)} \\ &= -\tfrac{1}{2}(2q_{11} - q_{12} - q_{13})P'_{z'z'}\end{aligned}$$

Also, $A_{33}-A_{22} = 1/n_{33}^2 - 1/n_{22}^2 = -2(n_{33}-n_{22})/n_o^3 = -2\Delta n/n_o^3$, Δn being the difference between the indices of refraction for light vibrations, respectively, parallel and perpendicular to the stress direction.

The stress birefringence per unit stress is

$$\frac{\Delta n}{P} = +\left(\frac{n_o^3}{4}\right)(2q_{11} - q_{12} - q_{13})$$

Therefore the path retardation per unit stress and per unit thickness is

$$\delta_{x'} = +\tfrac{1}{4}n_o^3(2q_{11} - q_{12} - q_{13}) \tag{5.11a}$$

Similarly, for observations along the new Y' axis

$$\delta_y' = +\tfrac{1}{4}n_o^3(2q_{11} - q_{12} - q_{13}) \tag{5.11b}$$

2. Let the stress be along the X' axis, and the observations along the Y' axis. Then it can be shown that there is no tilt of the axes of the ellipse. In this case, the stress birefringence is

$$\Delta n = +\frac{n_o^3}{2}\,\frac{1}{4}(2q_{11} - q_{12} - q_{13} + 2q_{44})P'_{x'x'}$$

and the path retardation per unit stress and per unit thickness is

$$\delta_{x'} = \frac{1}{4}\,\frac{n_o^3}{2}(2q_{11} - q_{12} - q_{13} + 2q_{44}) \tag{5.11c}$$

3. Let the stress be along the X' axis, and the observations along the Z' axis. Then the equation of the optical index ellipsoid becomes

$$A_{11}x'^2 + A_{22}y'^2 + A_{33}z'^2 + 2A_{12}x'y' = 1 \tag{5.12}$$

since $A_{23} = A_{31} = 0$ from Eqs. (5.9) and (5.10).

Therefore the section of the optical index ellipsoid normal to the Z' axis is

$$A_{11}x'^2 + A_{22}y'^2 + 2A_{12}x'y' = 1 \tag{5.13}$$

thereby indicating that the principal axes of the ellipse are inclined to the X' axis by angles θ and $(\theta + 90°)$, where $\tan 2\theta = 2A_{12}/(A_{11} - A_{22})$, a relation between θ, A_{11}, A_{12}, and A_{22} which is proved below.

To derive an expression for the tilt of axes in terms of the q_{ij}, i.e., the stress-optical constants, we proceed as follows: Let a stress act along the OX direction and the observation be parallel to OZ.

Then for crystals of all systems except the monoclinic and triclinic, the optical index ellipsoid for the undeformed crystal is

$$B_{11}^0x^2 + B_{22}^0y^2 + B_{33}^0z^2 = 1 \tag{5.14}$$

since $B_{12}^0 = B_{23}^0 = B_{31}^0 = 0$. The optical index ellipsoid for the stressed crystal (i.e., in the deformed state) is

$$B_{11}x^2 + B_{22}y^2 + B_{33}z^2 + 2B_{23}yz + 2B_{31}zx + 2B_{12}xy = 1 \tag{5.15}$$

Its cross section normal to the direction of observation OZ is given by

$$B_{11}x^2 + B_{22}y^2 + 2B_{12}xy = 1 \tag{5.16}$$

Now we can determine the tilt of the axes thus: Let OX rotate in the positive direction, i.e., toward positive OY in the XOY plane (Fig. 5.1). Then if (x, y) and (x', y') are the coordinates of the same point P, we have

$$x = x' \cos\theta - y' \sin\theta$$

$$y = x' \sin\theta + y' \cos\theta$$

Substituting these values of x and y in Eq. (5.16), we have

$$B_{11}(x'\cos\theta - y'\sin\theta)^2 + B_{22}(x'\sin\theta + y'\cos\theta)^2$$
$$+ 2B_{12}(x'\cos\theta - y'\sin\theta)(x'\sin\theta + y'\cos\theta) = 1$$

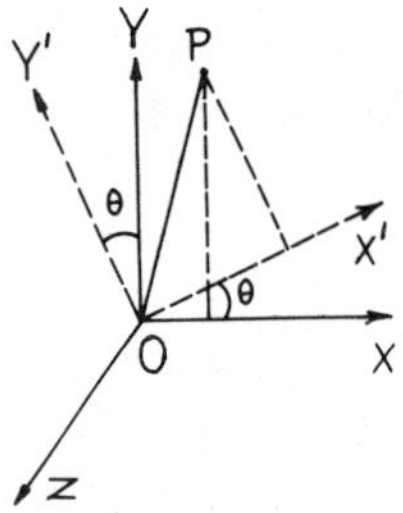

Fig. 5.1. The orientation or the tilt θ of the principal axes of an ellipse to relate θ with q_{ij} (see page 203).

Therefore

$$\begin{aligned}&B_{11}(x'^2\cos^2\theta + y'^2\sin^2\theta - 2x'y'\cos\theta\sin\theta)\\&\quad + B_{22}(x'^2\sin^2\theta + y'^2\cos^2\theta + 2x'y'\cos\theta\sin\theta)\\&\quad + 2B_{12}\{x'^2\cos\theta\sin\theta - y'^2\cos\theta\sin\theta + x'y'(\cos^2\theta - \sin^2\theta)\} = 1\end{aligned}$$

Taking OX' and OY' along the principal axes of the tilted ellipse, the coefficient of the cross term $x'y'$ should vanish. Therefore

$$-2B_{11}\cos\theta\sin\theta + 2B_{22}\cos\theta\sin\theta + 2B_{12}(\cos^2\theta - \sin^2\theta) = 0$$

$$(B_{22} - B_{11})\sin 2\theta = -2B_{12}\cos 2\theta$$

$$\tan 2\theta = 2B_{12}/(B_{11} - B_{22})$$

The above derivation and relations are equally applicable to Eq. (5.13).

This is the procedure adopted to derive the several expressions for the tilt of the principal axes of the elliptically polarized light in terms of the q_{ij} for cubic crystals as well as for crystals of other systems.

If the lengths of the principal semiaxes are represented by a_{11} and a_{22}, it can be shown that

$$a_{11} - a_{22} = \pm[(A_{11} - A_{22})^2 + 4A_{12}^2]^{1/2}$$

Expressing the A_{ij} in terms of the B_{ij} [see Eqs. (5.10)], we find from Eq. (5.13) that

$$\begin{aligned}\tan 2\theta &= -(B_{11} - B_{22})/2B_{12}\\(a_{11} - a_{22}) &= \pm[(B_{11} - B_{22})^2 + 4B_{12}^2]^{1/2}\end{aligned} \tag{5.17}$$

Hence

$$\begin{aligned}\tan 2\theta &= -(q_{12} - q_{13})/2q_{44}\\(a_{11} - a_{22}) &= \pm\tfrac{1}{2}P'_{x'x'}[(q_{12} - q_{13})^2 + 4q_{44}^2]^{1/2}\end{aligned} \tag{5.18}$$

Therefore

$$\Delta n = (n_{11} - n_{22}) = \pm\tfrac{1}{4}n_o{}^3 P'_{x'x'}[(q_{12} - q_{13})^2 + 4q_{44}^2]^{1/2} \qquad (5.19)$$

where n_{11} and n_{22} are the refractive indices of the crystal for light polarized in the direction of the principal axes.

It is evident from Eq. (5.18) that the principal axes of the index ellipsoid will not coincide with the X and Y axes so long as $q_{12} \neq q_{13}$. For the T_d, O, and O_h classes, the principal axes will be along the X and Y axes, and the amount of stress birefringence for this case, namely stress parallel to X' and observation parallel to Z', is given by

$$\Delta n / P'_{x'x'} = (n_{x'} - n_{y'})/P'_{x'x'} = \tfrac{1}{2}n_o{}^3 q_{44} \qquad (5.20)$$

Stress along Y' and observation along Z' again yield the same results. Expressions thus derived for prisms of different orientations of the T and T_h classes are given in Table 5.1.

Some Important Consequences. It is easily seen from Table 5.1 that as a result of the nonequivalence of q_{12} and q_{13} in the T and T_h classes, a prism belonging to either of these classes becomes optically orthorhombic, i.e., biaxial, by a simple compression along a cube axis, instead of becoming tetragonal or uniaxial as in the case of the T_d, O, and O_h classes, for which $q_{12} = q_{13}$. This result was theoretically predicted by Bhagavantam and experimentally confirmed by Bhagavantam and collaborators [117; 119; 121–129].

A simple linear stress along certain directions produces a tilt of the axes of the optical index ellipsoid for observations along certain other directions. For example, for a stress along a face diagonal and observation along a cube axis, the index ellipsoid will have its axes tilted with reference to the direction of stress (see cases 9 and 11 in Table 5.1). This result for the T and T_h classes was predicted by Bhagavantam [125] and confirmed experimentally by Bhagavantam and Krishna Rao [126]. Expressions for T_d, O, and O_h of cubic crystals are obtained from Table 5.1 by putting $q_{12} = q_{13}$.

5.2.2. *Stress Birefringence in Noncubic Crystals*

5.2.2.1. *Derivation of Expressions for δ: Traditional Method Due to Pockels*

Uniaxial and biaxial crystals are naturally birefringent, hence the simple method applied to derive expressions for cubic crystals cannot be applied here. Furthermore, the dimensions of the crystal undergo a change

Table 5.1. *Expressions for Stress Birefringence in Cubic Crystals of Classes 23 (T) and m3 (T_h)*

Direction of stress	Direction of observation	Birefringence[a] proportional to $\Delta n = \Delta n_\parallel - \Delta n_\perp$	Tilt of axes $\tan 2\theta$	Condition for $\tan 2\theta = 0$
[001]	[100]	$q_{11} - q_{12}$	0	—
[001]	[010]	$q_{11} - q_{13}$	0	—
[100]	[001]	$q_{11} - q_{13}$	0	—
[100]	[010]	$q_{11} - q_{12}$	0	—
[010]	[100]	$q_{11} - q_{13}$	0	—
[010]	[001]	$q_{11} - q_{12}$	0	—
[001]	[110]	A	0	—
[001]	$[1\bar{1}0]$	A	0	—
[110]	[001]	$\frac{1}{2}[4q_{44}^2 + (q_{12} - q_{13})^2]^{1/2}$	$\frac{q_{12} - q_{13}}{2q_{44}}$	$q_{12} = q_{13}$
[110]	$[\bar{1}10]$	$\frac{1}{2}A + \frac{1}{2}q_{44}$	0	—
$[\bar{1}10]$	[001]	$\frac{1}{2}[4q_{44}^2 + (q_{12} - q_{13})^2]^{1/2}$	$\frac{q_{12} - q_{13}}{2q_{44}}$	$q_{12} = q_{13}$
$[\bar{1}10]$	[110]	$\frac{1}{2}A + \frac{1}{2}q_{44}$	0	—
[111]	$[\bar{2}11]$	q_{44}	0	—
[111]	$[01\bar{1}]$	q_{44}	0	—
$[\bar{2}11]$	[111]	$\frac{1}{3}[(A + 2q_{44})^2 + \frac{3}{4}(q_{12} - q_{13})^2]^{1/2}$	$\frac{(3^{1/2}/2)(q_{12} - q_{13})}{A + 2q_{44}}$	$q_{12} = q_{13}$
$[\bar{2}11]$	$[01\bar{1}]$	$\frac{1}{3}[(A + 5q_{44})^2 + 2(A - q_{44})^2]^{1/2}$	$\frac{2(2)^{1/2}(A - q_{44})}{A + 5q_{44}}$	$A = q_{44}$
$[01\bar{1}]$	$[\bar{2}11]$	$\frac{1}{6}[(A + 5q_{44})^2 + 6(q_{12} - q_{13})^2]^{1/2}$	$\frac{6^{1/2}(q_{12} - q_{13})}{A + 5q_{44}}$	$q_{12} = q_{13}$
$[01\bar{1}]$	[111]	$\frac{1}{3}[(A + 2q_{44})^2 + \frac{3}{4}(q_{12} - q_{13})^2]^{1/2}$	$\frac{(3/2)^{1/2}(q_{12} - q_{13})}{A + 2q_{44}}$	$q_{12} = q_{13}$

[a] [illegible] for any orientation is obtained by multiplying the expression in the fourth column by the factor $\frac{1}{2}n_o^3$.

in their magnitudes due to the *elastic response* under the applied stress, causing thereby an additional path retardation. In the case of cubic crystals, this is generally negligible. In the actual experiment one observes the sum total of the path retardations, due to (a) the stress birefringence, and (b) the change in dimensions. Hence the method of deriving expressions for stress birefringence and the observed path retardation is of particular interest.

To illustrate the method, we take the case of an orthorhombic crystal. It may be noted here that all the three classes of the orthorhombic system come under one group with regard to their photoelastic behavior. Topaz, barite, and Rochelle salt are examples of orthorhombic crystals which have been fully studied for all their 12 p_{ij} and 12 q_{ij} constants.

Let a rectangular prism, say of barite, be cut so that its length is parallel to the X axis, its breadth parallel to the Y axis, and its thickness parallel to the Z axis. Let a linear compressional stress P_{xx} exist alone, all other components being zero. Fresnel's optical index ellipsoid is given by

$$B_{11}^0 x^2 + B_{22}^0 y^2 + B_{33}^0 z^2 + 2B_{23}^0 yz + 2B_{31}^0 zx + 2B_{12}^0 xy = 1 \tag{5.21}$$

in the undeformed state, and by

$$B_{11} x^2 + B_{22} y^2 + B_{33} z^2 + 2B_{23} yz + 2B_{31} zx + 2B_{12} xy = 1 \tag{5.22}$$

in the deformed state. For orthorhombic crystals, when B_{11}^0, B_{22}^0, and B_{33}^0 are referred to the crystallographic axes of the crystal, then $B_{23}^0 = 0$, $B_{31}^0 = 0$, and $B_{12}^0 = 0$. According to Pockels' assumption that the differences between the corresponding optical parameters (also called the polarization constants) are linearly related to all the stress components, we have, from a knowledge of Pockels' scheme of q_{ij} (see p. 53), the following equations:

$$\begin{aligned} B_{11} - B_{11}^0 &= -q_{11} P_{xx}, \qquad & B_{23} - B_{23}^0 &= 0 \\ B_{22} - B_{22}^0 &= -q_{21} P_{xx}, \qquad & B_{31} - B_{31}^0 &= 0 \\ B_{33} - B_{33}^0 &= -q_{31} P_{xx}, \qquad & B_{12} - B_{12}^0 &= 0 \end{aligned} \tag{5.23}$$

Let the observation be along the Z axis. Then the cross section of the index ellipsoid perpendicular to the Z axis is

$$B_{11}^0 x^2 + B_{22}^0 y^2 = 1 \qquad \text{(under no stress)}$$

$$B_{11} x^2 + B_{22} y^2 = 1 \qquad \text{(under stress)}$$

Hence there is no tilt of the principal axes of the ellipse. Here the procedure adopted in the case of cubic crystals cannot be followed because $n_x \neq n_y \neq n_z$. That is, there is an initial natural birefringence, and it is the additional birefringence caused by the application of mechanical stress that has to be estimated. To this end, we proceed as indicated below.

The change in the refractive index, say Δn_x, for electric vibrations along the X axis is calculated; similarly, Δn_y for electric vibrations along the Y axis is obtained. Then Δn, the stress birefringence produced, is equal to $(\Delta n_x - \Delta n_y)$, and the path retardation δ per centimeter thickness due to the stress birefringence alone is given by

$$\delta = (\Delta n_x - \Delta n_z)$$

Now

$$B_{11} - B_{11}^0 = -q_{11} P_{xx}$$

from (5.23). But

$$B_{11} - B_{11}^0 = 1/n_{11}^2 - 1/(n_{11}^0)^2 = -2\Delta n_{11}/(n_{11}^0)^3$$

Comparing the above two equations,

$$\Delta n_x = \Delta n_{11} = \tfrac{1}{2}(n_{11}^0)^3 q_{11}$$

per unit stress. Similarly

$$\Delta n_y = \tfrac{1}{2}(n_{22}^0)^3 q_{21}$$

per unit stress. Therefore

$$\Delta n = (\Delta n_{11} - \Delta n_{22}) = \tfrac{1}{2}\{(n_{11}^0)^3 q_{11} - (n_{22}^0)^3 q_{21}\}$$

per unit stress, and

$$\left\{\begin{matrix}\delta\text{, the path retardation per centimeter}\\ \text{thickness per unit stress}\end{matrix}\right\} = \tfrac{1}{2}(n_x{}^3 q_{11} - n_y{}^3 q_{21}) \qquad (5.24)$$

In an actual experiment, the value of δ_o observed requires a correction in order to give δ in the above equation, because of the change in thickness under stress. There is an inherent birefringence of the value $(n_x - n_y)$ for observations along the Z axis; now owing to the compressional stress P_{xx}, the dimension along the Z axis changes by the amount $(-s_{13}P_{xx})$. This introduces additional birefringence of magnitude $-(n_x - n_y)s_{13}P_{xx}$ or of magnitude $-(n_x - n_y)s_{13}$ per unit stress.

$$\delta_{(\text{observed})} = \delta_{(\text{due to stress birefringence})} - (n_x - n_y)s_{13}$$

Therefore

$$\delta_o = \tfrac{1}{2}(n_x^3 q_{11} - n_y^3 q_{21}) - (n_x - n_y)s_{13} \tag{5.25}$$

The absolute path retardation alone for the electric vector parallel to, say the X axis, can be obtained thus:

$$\left\{\begin{matrix}\delta \text{ due to the change in refractive index} \\ \text{when the crystal is under stress}\end{matrix}\right\} = \Delta n_{11} \text{ per unit thickness per unit stress} = \tfrac{1}{2} n_x^3 q_{11}$$

$$\left\{\begin{matrix}\text{Path retardation due to change of thickness} \\ \text{caused by the compressional stress}\end{matrix}\right\} = -s_{13} \cdot n_x$$

Therefore

$$\left\{\begin{matrix}\delta_0 \text{ (absolute) for light propagation along the} \\ Z \text{ axis and electric vector along the } X \text{ axis}\end{matrix}\right\} = \tfrac{1}{2} n_x^3 q_{11} - s_{13} n_x \tag{5.26}$$

This value of the absolute path retardation, $\frac{1}{2}n_x^3 q_{11} - n_x s_{13}$, enters into the calculation when the light radiation is allowed to pass through the crystal only once; this is the case when one uses a Jamin interferometer or a Rayleigh refractometer or the like, as Pockels [956] did for almost all his studies, or as Eppendahl [321] did for topaz. When the light travels through the stressed crystal twice, as in the case of the localized-fringe method adopted in recent times by several investigators, the absolute path retardation will be twice the value for single transit; thus $\delta_o = n_x^3 q_{11} - 2n_x s_{13}$ for the particular case under consideration.

Expressions similar to Eqs. (5.25) and (5.26) for different orientations are presented in Table 5.2 for orthorhombic crystals and in Table 5.3 for trigonal crystals of classes C_{3v}, D_3, and D_{3d} forming the fifth group. Expressions for path retardation for tetragonal crystals of the seventh group (C_{4v}, D_{2d}, D_4, and D_{4h}) having seven constants each, and also expressions for cubic crystals of the 10th group (T and T_h) and of the 11th group (T_d, O, and O_h), can be simply deduced from Table 5.2 as special cases by introducing certain simplifications arising out of additional symmetry considerations. Similarly, the expressions for the hexagonal crystals of the ninth group (D_{3h}, C_{6v}, and D_{6h}) having six constants follow from Table 5.3.

Table 5.2. *Expressions for Stress Birefringence in Orthorhombic Crystals (Traditional Method Due to Pockels)*[a,b]

Direction of: stress	Direction of: observation	Nature of observation	Expression for path retardation δ_o (observed) in terms of q_{ij} (stress birefringence $\Delta n = \Delta n_\parallel - \Delta n_\perp$)
[100]	[001]	Absolute	$(n_x{}^3 q_{11} - 2n_x s_{13})$ for electric vector parallel to stress
[100]	[001]	Absolute	$(n_y{}^3 q_{21} - 2n_y s_{13})$ for electric vector perpendicular to stress
[100]	[001]	Relative	$\frac{1}{2}(n_x{}^3 q_{11} - n_y{}^3 q_{21}) + s_{13}(n_y - n_x)$
[100]	[010]	Relative	$\frac{1}{2}(n_x{}^3 q_{11} - n_z{}^3 q_{31}) + s_{12}(n_z - n_x)$
[010]	[100]	Absolute	$(n_y{}^3 q_{22} - 2n_y s_{12})$ for electric vector parallel to stress
[010]	[100]	Absolute	$(n_z{}^3 q_{32} - 2n_z s_{12})$ for electric vector perpendicular to stress
[010]	[100]	Relative	$\frac{1}{2}(n_y{}^3 q_{22} - n_z{}^3 q_{32}) + s_{12}(n_z - n_y)$
[010]	[001]	Relative	$\frac{1}{2}(n_y{}^3 q_{22} - n_x{}^3 q_{12}) + s_{23}(n_x - n_y)$
[001]	[100]	Absolute	$(n_z{}^3 q_{33} - 2n_z s_{13})$ for electric vector parallel to stress
[001]	[100]	Absolute	$(n_y{}^3 q_{23} - 2n_y s_{13})$ for electric vector perpendicular to stress
[001]	[100]	Relative	$\frac{1}{2}(n_z{}^3 q_{33} - n_y{}^3 q_{23}) + s_{13}(n_y - n_z)$
[001]	[010]	Relative	$\frac{1}{2}(n_z{}^3 q_{33} - n_x{}^3 q_{13}) + s_{23}(n_x - n_z)$
[100]	M or M'	Relative	$\frac{1}{2}n_x{}^3 q_{11} - \frac{1}{4}n_{yz}^3(q_{21} + q_{31}) - \frac{1}{2}(n_x - n_{yz})(s_{12} + s_{13})$
[010]	L or L'	Relative	$\frac{1}{2}n_y{}^3 q_{22} - \frac{1}{4}n_{xz}^3(q_{32} + q_{12}) - \frac{1}{2}(n_y - n_{xz})(s_{12} + s_{23})$
[001]	N or N'	Relative	$\frac{1}{2}n_z{}^3 q_{33} - \frac{1}{4}n_{xy}^3(q_{13} + q_{23}) - \frac{1}{2}(n_z - n_{xy})(s_{13} + s_{23})$
N	N'	Relative	$\frac{1}{8}n_{xy}^3(q_{11} + q_{12} + q_{21} + q_{22} + 2q_{66}) - \frac{1}{4}n_z{}^3(q_{31} + q_{32}) - \frac{1}{4}(n_{xy} - n_z)(s_{11} + 2s_{12} + s_{22} - s_{66})$
L	L'	Relative	$\frac{1}{8}n_{zx}^3(q_{11} + q_{13} + q_{31} + q_{33} + 2q_{55}) - \frac{1}{4}n_y{}^3(q_{21} + q_{23}) - \frac{1}{4}(n_{xz} - n_y)(s_{11} + 2s_{13} + s_{33} - s_{55})$
M	M'	Relative	$\frac{1}{8}n_{yz}^3(q_{22} + q_{23} + q_{32} + q_{33} + 2q_{44}) - \frac{1}{4}n_x{}^3(q_{12} + q_{13}) - \frac{1}{4}(n_{yz} - n_x)(s_{22} + 2s_{23} + s_{33} - s_{44})$

[a] *M* indicates a direction equally inclined to *Y* and *Z* in the *YZ* plane. *L* indicates a direction equally inclined to *Z* and *X* in the *ZX* plane. *N* indicates a direction equally inclined to *X* and *Y* in the *XY* plane. *M'*, *L'*, *N'* are, respectively, perpendicular to *M*, *L*, and *N*, lying in the *YZ*, *ZX*, and *XY* planes, respectively.

[b] The expressions for absolute path retardation hold for light passing *twice* through the crystal.

Table 5.3. *Expressions for Stress Birefringence in Trigonal Crystals of Classes* $3m\,(C_{3v})$, $32\,(D_3)$, *and* $\bar{3}m\,(D_{3d})$ *(Traditional Method Due to Pockels)*[a,b]

Direction of stress	Direction of observation	Nature of observation	Expression for path retardation δ_o (observed) in terms of q_{ij} (stress birefringence $\Delta n = \Delta n_{\parallel} - \Delta n_{\perp}$)
[100]	[001]	Relative	$\frac{1}{2}n_o{}^3(q_{11} - q_{12}) - s_{13}(n_x - n_y)$
[100]	[010]	Relative	$\frac{1}{2}(n_x{}^3 q_{11} - n_z{}^3 q_{31}) - s_{12}(n_x - n_z)$
[001]	[100]	Relative	$\frac{1}{2}(n_z{}^3 q_{33} - n_y{}^3 q_{13}) - s_{13}(n_z - n_y)$
[001]	[010]	Relative	$\frac{1}{2}(n_z{}^3 q_{33} - n_x{}^3 q_{13}) - s_{13}(n_z - n_x)$
[100]	$[M]$	Relative	$\frac{1}{2}n_x{}^3 q_{11} - \frac{1}{4}n_{yz}^3(q_{12} + q_{31} - 2q_{41}) - \frac{1}{2}(n_x - n_{yz})(s_{12} + s_{13} + s_{14})$
[100]	$[M']$	Relative	$\frac{1}{2}n_x{}^3 q_{11} - \frac{1}{4}n_{yz}^3(q_{12} + q_{31} + 2q_{41}) - \frac{1}{2}(n_x - n_{yz})(s_{12} + s_{13} - s_{14})$
$[M]$	$[M']$	Relative	$\frac{1}{8}n_{yz}^3(q_{11} + q_{13} - q_{14} + q_{31} + q_{33} + 2q_{44} - 2q_{41}) - \frac{1}{4}n_x{}^3(q_{12} + q_{13} + q_{14})$ $- \frac{1}{4}(n_{yz} - n_x)(s_{11} + 2s_{13} + s_{33} - s_{44})$
$[M']$	$[M]$	Relative	$\frac{1}{8}n_{yz}^3(q_{11} + q_{13} + q_{14} + q_{31} + q_{33} + 2q_{44} + 2q_{41}) - \frac{1}{4}n_x{}^3(q_{12} + q_{13} - q_{14})$ $- \frac{1}{4}(n_{yz} - n_x)(s_{11} + 2s_{13} + s_{33} - s_{44})$
$[M']$	$[M]$	Absolute	$\frac{1}{4}n_x{}^3(q_{12} + q_{13} - q_{14}) - \frac{1}{2}n_x(s_{11} + 2s_{13} + s_{33} - s_{44})$ for electric vector normal to stress
$[M']$	$[M]$	Absolute	$\frac{1}{4}n_{yz}^3(q_{11} + q_{13} + q_{14} + q_{31} + q_{33} + 2q_{44} + 2q_{41}) - \frac{1}{2}n_{yz}(s_{11} + 2s_{13} + s_{33} - s_{44})$ for electric vector parallel to stress

[a] M indicates a direction equally inclined to $+Y$ and $+Z$ in the YZ plane. M' indicates a direction equally inclined to $-Y$ and $+Z$ in the YZ plane.
[b] The expressions for absolute path retardation hold for light passing twice through the crystal.

5.2.2.2. *Mandarino's Method of Expressing the Path Retardation* δ *in Terms of* q_{ij}

We have indicated in the foregoing pages the traditional method, due to Pockels, of deriving a relation between the stress-optical constants q_{ij} and the observed stress birefringence $\Delta n = \Delta n_{\parallel} - \Delta n_{\perp}$. We describe below a more direct method, due to Mandarino [765], of relating Δn with q_{ij}. Mandarino followed this method for sapphire, a trigonal crystal, for only one orientation, namely the stress along the Z axis, and for observation along the Y axis. The interesting feature in this method of derivation is that the final expressions thus deduced considerably reduce the computational difficulties encountered in arriving at the individual values of q_{ij} from the observed data on Δn, the observed stress birefringence.

We now consider the case of sapphire, which belongs to the crystal class $\bar{3}m$ (D_{3d}) of the trigonal system.

Let a stress $P_3 = P_{zz}$ exist alone on a rectangular parallelepiped having its edges along the OX, OY, OZ axes of the crystal. Then from Pockels' scheme of the q_{ij} matrix,

$$B_{11} - B_{11}^0 = -q_{13}P_3, \qquad B_{22} - B_{22}^0 = -q_{13}P_3, \qquad B_{33} - B_{33}^0 = -q_{33}P_3$$

$$B_{23} - B_{23}^0 = 0, \qquad B_{31} - B_{31}^0 = 0, \qquad B_{12} - B_{12}^0 = 0$$

Let the observation be along the OY axis. Since $B_{31} - B_{31}^0 = 0$, there is no tilt of the axes of the elliptic section when the crystal experiences a mechanical stress P_3. We then have

$$\frac{1}{n_{11}^2} - \frac{1}{(n_{11}^0)^2} = -q_{13}P_3, \qquad \frac{1}{n_{33}^2} - \frac{1}{(n_{33}^0)^2} = -q_{33}P_3$$

Therefore

$$(q_{33} - q_{13})P_3 = \left(\frac{1}{n_{11}^2} - \frac{1}{n_{33}^2}\right) - \left[\frac{1}{(n_{11}^0)^2} - \frac{1}{(n_{33}^0)^2}\right]$$

$$= \frac{(n_{33} + n_{11})(n_{33} - n_{11})}{n_{11}^2 \cdot n_{33}^2} - \frac{(n_{33}^0 + n_{11}^0)(n_{33}^0 - n_{11}^0)}{(n_{11}^0)^2 \cdot (n_{33}^0)^2}$$

$$= \frac{n_{33}^0 + n_{11}^0}{(n_{11}^0)^2 \cdot (n_{33}^0)^2} [(n_{33} - n_{11}) - (n_{33}^0 - n_{11}^0)]$$

since $(n_{33}^0 + n_{11}^0) \approx (n_{33} + n_{11})$ and $\{(n_{11}^0)^2 \cdot (n_{33}^0)^2\} \approx (n_{11}^2 \cdot n_{33}^2)$; therefore

$$(q_{33} - q_{13})P_3 = \frac{n_{33}^0 + n_{11}^0}{(n_{11}^0)^2 \cdot (n_{33}^0)^2} [(n_{33} - n_{33}^0) - (n_{11} - n_{11}^0)]$$

$$= \frac{n_{33}^0 + n_{11}^0}{(n_{11}^0)^2 \cdot (n_{33}^0)^2} (\Delta n_{33} - \Delta n_{11}) = \frac{n_{33}^0 + n_{11}^0}{(n_{11}^0)^2 \cdot (n_{33}^0)^2} \Delta n$$

$$\left\{\begin{matrix}\text{Photoelastic birefringence } \Delta n \\ \text{per unit stress}\end{matrix}\right\} = \left(\frac{n_x^2 n_z^2}{n_x + n_z}\right)(q_{33} - q_{13}) \qquad (5.27)$$

from which $(q_{33} - q_{13})$ can be directly calculated, whereas in the traditional method due to Pockels, we get, as shown on page 211,

$$\Delta n = \Delta n_{33} - \Delta n_{11} = \tfrac{1}{2}[(n_{33}^0)^3 q_{33} - (n_{11}^0)^3 q_{13}]$$

Thus we see that Mandarino's method is preferable to the traditional Pockels' method, which, however, has been followed without any modification by all investigators ever since.

Proceeding along the lines indicated above, we derive the expressions shown in Tables 5.4 and 5.5 corresponding, respectively, to Tables 5.3 and 5.2.

A correction for the change in thickness has to be applied here too, as in the traditional method.

5.2.2.3. *Optical Behavior of Crystals under Linear and Tangential Stresses: Some Typical New Results Regarding a Few Other Classes*

Let a stress P act in an arbitrary direction (having direction cosines α_1, α_2, and α_3 with respect to the principal axes—the crystallographic axes) of the crystal. Then the stress components are given by

$$\begin{aligned} P_{xx} &= \alpha_1^2 P, \quad & P_{yy} &= \alpha_2^2 P, \quad & P_{zz} &= \alpha_3^2 P \\ P_{yz} &= \alpha_2\alpha_3 P, \quad & P_{zx} &= \alpha_3\alpha_1 P, \quad & P_{xy} &= \alpha_1\alpha_2 P \end{aligned} \qquad (5.28)$$

If these values of the stress components are substituted in the fundamental photoelastic equations (3.9d), one can calculate the orientation of the principal axes of the optical index ellipsoid of the deformed crystal along the lines indicated in Sections 5.2 and 5.3. The general case of an arbitrarily directed stress is very complicated, and we shall mention below some special cases of interest.

Table 5.4. Expressions for Stress Birefringence in Trigonal Crystal Classes $3m$ (C_{3v}), 32 (D_3), *and* $\bar{3}m$ (D_{3d}) *by Mandarino's Method*[a]

Direction of: stress	Direction of: observation	Expression for the relative path retardation per unit stress per unit thickness (stress birefringence $\Delta n = \Delta n_{\parallel} - \Delta n_{\perp}$)
[100]	[001]	$[(n_x{}^2 n_y{}^2)/(n_x + n_y)](q_{11} - q_{12}) - (n_x - n_y)s_{13}$
[100]	[010]	$[(n_x{}^2 n_z{}^2)/(n_x + n_z)](q_{11} - q_{31}) - (n_x - n_z)s_{12}$
[001]	[010] [100]	$[(n_x{}^2 n_z{}^2)/(n_x + n_z)](q_{33} - q_{13}) - (n_z - n_x)s_{13}$
[100]	M	$[(n_x{}^2 n_{yz}^2)/(n_x + n_{yz})](q_{11} - \frac{1}{2}q_{12} - \frac{1}{2}q_{31} + q_{41}) - \frac{1}{2}(n_x - n_{yz})(s_{12} + s_{13} + s_{14})$
[100]	M'	$[(n_x{}^2 n_{yz}^2)/(n_x + n_{yz})](q_{11} - \frac{1}{2}q_{12} - \frac{1}{2}q_{31} - q_{41}) - \frac{1}{2}(n_x - n_{yz})(s_{12} + s_{13} - s_{14})$
M	M'	$[(n_{yz}^2 n_x{}^2)/(n_{yz} + n_x)][(q_{11} + q_{13} - q_{14} + q_{31} + q_{33} + 2q_{44} - 2q_{41}) - \frac{1}{2}(q_{12} + q_{13} + q_{14})] - \frac{1}{4}(n_{yz} - n_x)(s_{11} + 2s_{13} + s_{33} - s_{44})$
M'	M	$[(n_{yz}^2 n_x{}^2)/(n_{yz} + n_x)][\frac{1}{4}(q_{11} + q_{13} + q_{14} + q_{31} + q_{33} + 2q_{44} + 2q_{41}) - \frac{1}{2}(q_{12} + q_{13} - q_{14})] - \frac{1}{4}(n_{yz} - n_x)(s_{11} + 2s_{13} + s_{33} - s_{44})$

[a] M and M' have the same significance as in Table 5.2.

Table 5.5. *Expressions for Stress Birefringence in Orthorhombic Crystals by Mandarino's Method*[a]

Direction of stress	Direction of observation	Expression for the relative path retardation per unit stress per unit thickness (stress birefringence $\Delta n = \Delta n_{\parallel} - \Delta n_{\perp}$)
[100]	[001]	$[(n_x^2 n_y^2)/(n_x + n_y)](q_{11} - q_{21}) - (n_x - n_y)s_{13}$
[100]	[010]	$[(n_x^2 n_z^2)/(n_x + n_z)](q_{11} - q_{31}) - (n_x - n_z)s_{12}$
[010]	[100]	$[(n_y^2 n_z^2)/(n_y + n_z)](q_{22} - q_{32}) - (n_y - n_z)s_{12}$
[010]	[001]	$[(n_y^2 n_x^2)/(n_y + n_x)](q_{22} - q_{12}) - (n_y - n_x)s_{23}$
[001]	[100]	$[(n_z^2 n_y^2)/(n_z + n_y)](q_{33} - q_{23}) - (n_z - n_y)s_{13}$
[001]	[010]	$[(n_z^2 n_x^2)/(n_z + n_x)](q_{33} - q_{13}) - (n_z - n_x)s_{23}$
[100]	M or M'	$[(n_x^2 n_{yz}^2)/(n_x + n_{yz})](q_{11} - \frac{1}{2}q_{21} - \frac{1}{2}q_{31}) - \frac{1}{2}(n_x - n_{yz})(s_{12} + s_{13})$
[010]	L or L'	$[(n_y^2 n_{xz}^2)/(n_y + n_{xz})](q_{22} - \frac{1}{2}q_{32} - \frac{1}{2}q_{12}) - \frac{1}{2}(n_y - n_{xz})(s_{12} + s_{23})$
[001]	N or N'	$[(n_z^2 n_{xy}^2)/(n_z + n_{xy})](q_{33} - \frac{1}{2}q_{13} - q_{23}) - \frac{1}{2}(n_z - n_{xy})(s_{13} + s_{23})$
N	N	$[(n_{xy}^2 n_z^2)/(n_{xy} + n_z)][\frac{1}{4}(q_{11} + q_{12} + q_{21} + q_{22} + 2q_{66}) - \frac{1}{2}(q_{31} + q_{32})] - \frac{1}{4}(n_{xy} - n_z)(s_{11} + 2s_{12} + s_{22} - s_{66})$
L	L'	$[(n_{zx}^2 n_y^2)/(n_{zx} + n_y)][\frac{1}{4}(q_{11} + q_{13} + q_{31} + q_{33} + 2q_{55}) - \frac{1}{2}(q_{21} + q_{23})] - \frac{1}{4}(n_{zx} - n_y)(s_{11} + 2s_{13} + s_{33} - s_{55})$
M	M'	$[(n_{yz}^2 n_x^2)/(n_{yz} + n_x)][\frac{1}{4}(q_{22} + q_{23} + q_{32} + q_{33} + 2q_{44}) - \frac{1}{2}(q_{12} + q_{13})] - \frac{1}{4}(n_{yz} - n_x)(s_{22} + 2s_{23} + s_{33} - s_{44})$

[a] M, L, N and M', L', N' have the same significance as in Table 5.2.

a. Cubic Crystals. It is possible to distinguish between the two photoelastic groups 10 and 11 by observing the tilt of the axes of the optical index ellipsoid. Thus, if a linear stress is equally inclined to the OX and OY axes and observation is made along the OZ axis, then a tilt will be observed only in the tenth group consisting of classes T and T_h [125].

b. Noncubic Crystals. Let a stress P_{xx} act on a crystal belonging to any photoelastic group other than groups 1 and 2 (i.e., other than triclinic and monoclinic systems). Let us calculate the resulting tilt of the principal axes of the elliptic section of the optical index ellipsoid for observations along the OY and OZ axes. From the fundamental photoelastic equations, we have

$$\begin{aligned} B_{11} - B_{11}^0 &= -q_{11}P_{xx}, \quad B_{23} - B_{23}^0 = B_{23} = -q_{41}P_{xx} \quad (\because B_{23}^0 = 0) \\ B_{22} - B_{22}^0 &= -q_{21}P_{xx}, \quad B_{31} - B_{31}^0 = B_{31} = -q_{51}P_{xx} \quad (\because B_{31}^0 = 0) \\ B_{33} - B_{33}^0 &= -q_{31}P_{xx}, \quad B_{12} - B_{12}^0 = B_{12} = -q_{61}P_{xx} \quad (\because B_{12}^0 = 0) \end{aligned} \tag{5.29}$$

The cross sections of the deformed optical index ellipsoid perpendicular to OZ and OY are, respectively,

$$\begin{aligned} B_{11}x^2 + B_{22}y^2 + 2B_{12}xy &= 1 \\ B_{11}x^2 + B_{33}z^2 + 2B_{13}xz &= 1 \end{aligned} \tag{5.30}$$

The tilts of the principal axes in the two cases are given, respectively, by

$$\begin{aligned} \tan 2\theta_1 &= 2B_{12}/(B_{11} - B_{22}) = 2q_{61}/(q_{11} - q_{12}) \\ \tan 2\theta_2 &= 2B_{31}/(B_{11} - B_{33}) = 2q_{51}/(q_{11} - q_{13}) \end{aligned} \tag{5.31}$$

These tilts can occur, for both directions of observation, only with crystals belonging to the photoelastic group 6 (comprising the C_3 and C_{3i} classes), for which alone, barring groups 1 and 2, q_{51} and q_{61} do not vanish in the revised scheme due to Bhagavantam. Hence this test gives a simple method of distinguishing C_3 and C_{3i} ($= S_6$) classes from the remaining photoelastic groups.

Further interesting results have been pointed out by Bhagavantam [124] as arising from the existence of new constants in the revised scheme of q_{ij}. For example, if to a prism belonging to any of the crystal classes (C_3, C_{3i}) and (C_{3h}, C_6, C_{6h}) for observations along the Z axis, birefringence is caused by a shearing stress P_{xy}. This result arises from the nonvanishing constant q_{61} in the revised scheme due to Bhagavantam, whereas if $q_{61} = 0$ no birefringence should appear even in the sheared crystal.

If a prism of a crystal belonging to classes C_4, S_4, and C_{4h} is subjected to a shearing stress P_{zx}, then according to Bhagavantam's scheme there is a tilt of the principal axes in the YZ plane due to the existence of the constant q_{45}.

5.2.3. Tensor Method of Deriving q'_{ijkl} in Terms of q_{mnop}

In the earlier sections where we derived expressions for the stress birefringence of a crystal, Pockels' stress-optical constants appeared in the form of q_{ij}, in the contracted or two-suffix notation. No mention has hitherto been made regarding the fourth-rank tensor nature of the q_{ij}. We shall make use of the fourth-rank tensor nature of the q_{ij} and derive in this section expressions for the resultant stress-optical constants for an orthorhombic crystal for stress in a specific direction and observation in another assumed direction. In deriving the expressions for the path retardations, due consideration should of course be given to the change in dimensions due to elastic response. We can take advantage, in the following derivations, of the fact that both the elastic and the photoelastic constants are coefficients of fourth-rank tensors, and hence the law of transformation is the same for both properties, remembering that in elasticity $s_{ijkl} = s_{jikl} = s_{klij}$, etc.

To derive an expression for path retardation in terms of q_{ij} for an orthorhombic crystal for a specific orientation, we proceed as follows: Let $X'Y'Z'$ be the new coordinate system, rotated through 45° about the Z axis of the old coordinate system XYZ (Fig. 5.2). Their direction-cosine scheme is

	X	Y	Z
X'	$1/2^{1/2}$	$1/2^{1/2}$	0
Y'	$-1/2^{1/2}$	$1/2^{1/2}$	0
Z'	0	0	1

Let the direction of stress be along OX', and observation along OY'. Then $P'_{x'x'} = P'_{11}$ alone exists. Now

$$
\begin{aligned}
(A_{11} - A^0_{11}) &= -[q'_{1111}P'_{11} + q'_{1122}P'_{22} + q'_{1133}P'_{33} + q'_{1123}P'_{23} + \cdots] \\
&\;\vdots \\
(A_{33} - A^0_{33}) &= -[q'_{3311}P'_{11} + q'_{3322}P'_{22} + q'_{3333}P'_{33} + q'_{3323}P'_{23} + \cdots] \\
&\;\vdots
\end{aligned}
\tag{5.32}
$$

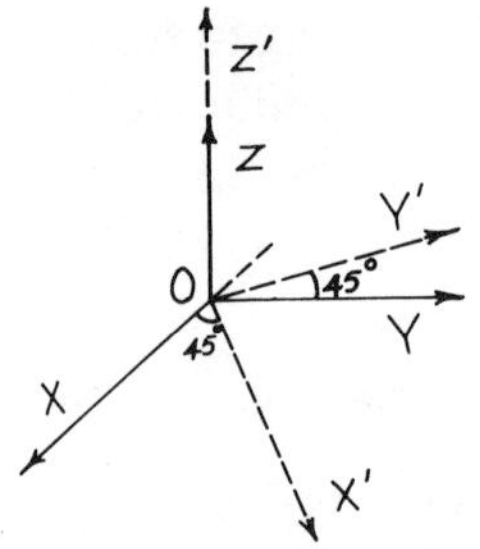

Fig. 5.2. Transformation of axes: a rotation of the X system through an angle of 45° about the Z axis; here $OZ' = OZ$.

Since only $P'_{x'x'}$ exists, we deal with only the term containing $P'_{x'x'}$ $(= P'_{11})$.

Expressions for the absolute and relative path retardations are derived as follows:

$$(A_{11} - A^0_{11}) - (A_{33} - A^0_{33}) = -(q'_{1111} - q'_{3311})P'_{11}$$

Now

$$A^0_{11} = 1/n^2_{x'} = \tfrac{1}{2}(1/n_x{}^2 + 1/n_y{}^2), \qquad A^0_{33} = 1/n^2_{z'} = 1/n_z{}^2$$

Let us represent the OX, OY, OZ axes as X_1, X_2, X_3 and, likewise, OX', OY', OZ' as X_1', X_2', X_3' and then apply the law of transformation of fourth-rank tensor coefficients.

1. We determine q'_{1111} (i.e., for electric vibrations parallel to the X' axis):

$$\begin{aligned}
q'_{1111} &= \alpha_{1i}\alpha_{1j}\alpha_{1k}\alpha_{1l}q_{ijkl} \\
&= \alpha_{1j}\alpha_{1k}\alpha_{1l}(\alpha_{11}q_{1jkl} + \alpha_{12}q_{2jkl} + \alpha_{13}q_{3jkl}) \\
&= \alpha_{1k}\alpha_{11}[\alpha_{11}(\alpha_{11}q_{11kl} + \alpha_{12}q_{12kl} + \alpha_{13}q_{13kl}) \\
&\quad +\alpha_{12}(\alpha_{11}q_{21kl} + \alpha_{12}q_{22kl} + \alpha_{13}q_{23kl})] + \cdots \\
&= \alpha_{11}\alpha_{11}\alpha_{11}(\alpha_{11}q_{111l} + \alpha_{12}q_{112l} + \alpha_{13}q_{113l}) + \cdots \\
&= \alpha^3_{11}[(\alpha_{11}q_{1111} + \alpha_{12}q_{1112} + \alpha_{13}q_{1113}) + \cdots] \\
&= \tfrac{1}{4}q_{1111} + \tfrac{1}{4}q_{1112} + \tfrac{1}{4}q_{1121} + \tfrac{1}{4}q_{1122} + \tfrac{1}{4}q_{1211} + \tfrac{1}{4}q_{1212} \\
&\quad + \tfrac{1}{4}q_{1221} + \tfrac{1}{4}q_{1222} + \tfrac{1}{4}q_{2111} + \tfrac{1}{4}q_{2112} + \tfrac{1}{4}q_{2121} + \tfrac{1}{4}q_{2122} \\
&\quad + \tfrac{1}{4}q_{2211} + \tfrac{1}{4}q_{2212} + \tfrac{1}{4}q_{2221} + \tfrac{1}{4}q_{2222} \qquad (5.33a) \\
&= \tfrac{1}{4}(q_{11} + q_{12} + \tfrac{1}{2}q_{66} + \tfrac{1}{2}q_{66} + \tfrac{1}{2}q_{66} + \tfrac{1}{2}q_{66} + q_{21} + q_{22}) \\
&= \tfrac{1}{4}(q_{11} + q_{12} + q_{21} + q_{22} + 2q_{66}) \qquad (5.33b)
\end{aligned}$$

The other q_{ij} coefficients are each equal to zero, and

$$\text{absolute path retardation} = (n_{x'}^3/2) \cdot \tfrac{1}{4}(q_{11} + q_{12} + q_{21} + q_{22} + 2q_{66}) \tag{5.33c}$$

2. We determine q'_{3311} (i.e., for electric vibrations parallel to the Z' axis):

$$q'_{3311} = \alpha_{3i}\alpha_{3j}\alpha_{1k}\alpha_{1l}q_{ijkl}$$

Therefore

$$\begin{aligned}
q'_{3311} &= \alpha_{3j}\alpha_{1k}\alpha_{1l}(\alpha_{33}q_{3jkl}) \qquad \text{(other terms are zero)}\\
&= \alpha_{1k}\alpha_{1l}\alpha_{33}(\alpha_{33}q_{33kl}) \qquad \text{(other terms are zero)}\\
&= (\alpha_{33})^2\alpha_{1l}(\alpha_{11}q_{331l} + \alpha_{12}q_{332l} + \alpha_{13}q_{333l})\\
&= 1^2[\alpha_{11}(\alpha_{11}q_{3311} + \alpha_{12}q_{3312}) + \alpha_{12}(\alpha_{11}q_{3321} + \alpha_{12}q_{3322})] \qquad (5.34a)\\
&= \tfrac{1}{2}q_{31} + \tfrac{1}{2}q_{32} = \tfrac{1}{2}(q_{31} + q_{32}) \qquad (\text{since } q_{3312} = q_{3321} = 0) \qquad (5.34b)
\end{aligned}$$

$$\text{Absolute path retardation} = (n_{z'}^3/2) \cdot \tfrac{1}{2}(q_{31} + q_{32}) = \tfrac{1}{4}n_{z'}^3(q_{31} + q_{32}) \tag{5.34c}$$

Therefore

$$\{\delta, \text{relative path retardation}\} = \tfrac{1}{8}n_{x'}^3(q_{11} + q_{12} + q_{21} + q_{22} + 2q_{66}) - \tfrac{1}{4}n_{z'}^3(q_{31} + q_{32}) \tag{5.35}$$

5.2.4. *Expression for the Change of Thickness in Terms of s_{ij} for an Orthorhombic Crystal for a Specific Orientation*

We take the same orientation as in the earlier case of q_{ijkl}. Let OX' be the direction of stress and OY' the direction of observation. Let a stress P'_{xx} act on a rectangular parallelepiped measuring 1 cm along OY'. Let e'_{yy} be the extension or change of thickness along OY'. Then, by definition, for a stress P'_{xx} along OX',

$$e'_{yy} = s'_{yyxx}P'_{xx}$$

where s'_{yyxx} is the compliance constant along OY'. Hence $s'_{yyxx} = e'_{yy}$ numerically if $P'_{xx} = 1$ dyne/cm². Now we can obtain s'_{yyxx} in terms of s_{ijkl} in the original coordinate system by the transformation law for a fourth-rank tensor. Thus, putting $s'_{yyxx} = s'_{2211}$, we obtain

$$
\begin{aligned}
s'_{2211} &= \alpha_{2i}\alpha_{2j}\alpha_{1k}\alpha_{1l}s_{ijkl}, \qquad i, j, k, l = 1, 2, \text{ or } 3 \\
&= \alpha_{2i}\alpha_{2j}\alpha_{1k}(\alpha_{11}s_{ijk1} + \alpha_{12}s_{ijk2} + \alpha_{13}s_{ijk3}) \qquad (\text{since } \alpha_{13} = 0) \\
&= \alpha_{2i}\alpha_{2j}[\alpha_{11}(\alpha_{11}s_{ij11} + \alpha_{12}s_{ij12}) + \alpha_{12}(\alpha_{11}s_{ij21} + \alpha_{12}s_{ij22})] \\
&= \alpha_{2i}\alpha_{21}[\alpha_{11}(\alpha_{11}s_{i111} + \alpha_{12}s_{i112}) + \alpha_{12}(\alpha_{11}s_{i121} + \alpha_{12}s_{i122})] + \cdots \\
&= \alpha_{21}\{\alpha_{21}[\alpha_{11}(\alpha_{11}s_{1111} + \alpha_{12}s_{1112}) + \alpha_{12}(\alpha_{11}s_{1121} + \alpha_{12}s_{1122})]\} + \cdots
\end{aligned} \tag{5.36a}
$$

Putting $s_{1111} = s_{11}$, etc. (from page 152), we obtain

$$
\begin{aligned}
s'_{2211} &= \alpha_{21}^2[\alpha_{11}^2 s_{11} + 2\alpha_{11}\alpha_{12} \cdot \tfrac{1}{2}s_{16} + \alpha_{12}^2 s_{12}] \\
&\quad + \alpha_{21}\alpha_{22}[\alpha_{11}^2 \cdot \tfrac{1}{2}s_{61} + 2\alpha_{11}\alpha_{12} \cdot \tfrac{1}{4}s_{66} + \alpha_{12}^2 \cdot \tfrac{1}{2}s_{62}] \\
&\quad + \alpha_{22}\alpha_{21}[\alpha_{11}^2 \cdot \tfrac{1}{2}s_{61} + 2\alpha_{11}\alpha_{12} \cdot \tfrac{1}{4}s_{66} + \alpha_{12}^2 \cdot \tfrac{1}{2}s_{62}] \\
&\quad + \alpha_{22}\alpha_{22}[\alpha_{11}^2 s_{21} + 2\alpha_{11}\alpha_{12} \cdot \tfrac{1}{2}s_{26} + \alpha_{12}^2 s_{22}] \\
&= \tfrac{1}{2}(\tfrac{1}{2}s_{11} + \tfrac{1}{2}s_{12}) + (-\tfrac{1}{2}) \cdot 2 \cdot \tfrac{1}{2}s_{66} \cdot \tfrac{1}{4} + (-\tfrac{1}{2}) \cdot 2 \cdot \tfrac{1}{2}s_{66} \cdot \tfrac{1}{4} \\
&\quad + \tfrac{1}{2}(\tfrac{1}{2}s_{21} + \tfrac{1}{2}s_{22}) \qquad (\text{since } s_{16} = s_{61} = 0 \text{ and } s_{26} = s_{62} = 0) \\
&= \tfrac{1}{4}(s_{11} + s_{12} + s_{21} + s_{22}) - 2 \cdot \tfrac{1}{2}s_{66} \cdot \tfrac{1}{4} \\
&= \tfrac{1}{4}(s_{11} + 2s_{12} + s_{22} - s_{66})
\end{aligned} \tag{5.36b}
$$

Conclusion. For the particular directions of a compressional stress and of observation given above, the observed path retardation δ_0 for an orthorhombic crystal is given by

$$
\begin{aligned}
\delta_0 &= \tfrac{1}{8}n_{x'}^3(q_{11} + q_{12} + q_{21} + q_{22} + 2q_{66}) - \tfrac{1}{4}n_{z'}^3(q_{31} + q_{32}) \\
&\quad - (n_{x'} - n_{z'}) \cdot \tfrac{1}{4}(s_{11} + 2s_{12} + s_{22} - s_{66})
\end{aligned} \tag{5.37}
$$

(as given on page 210).

5.3. Experimental Determination of q_{ij} and p_{ij} by Optical Methods

We now describe the well-established experimental methods and then give a review, followed by a discussion, of other methods. Let a rectangular parallelepiped (i.e., a rectangular prism) of an isotropic transparent solid, say of glass, be subjected to a linear (uniaxial or unidirectional) stress parallel to the length of the prism (Fig. 5.3). Then the solid behaves as a uniaxial crystal, with its optic axis parallel to the stress direction. In the

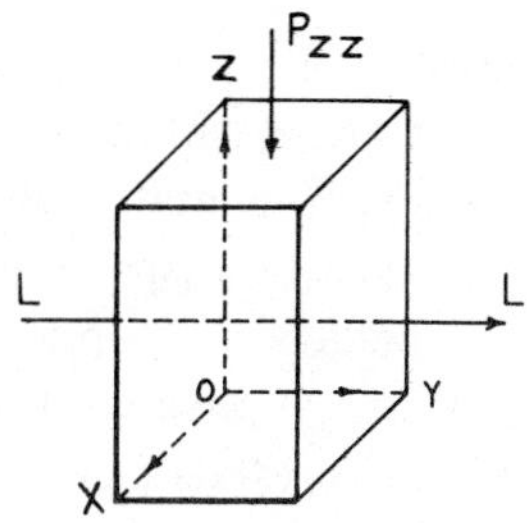

Fig. 5.3. Rectangular parallelepiped for an experimental study of stress birefringence in solids. (P_{zz} is the applied stress and LL is the direction of light propagation and observation.)

undeformed state, let n be the index of refraction. In the deformed state, under stress P_{zz} the refractive index changes from n to n_1 ($= n_{\parallel}$ or n_p) for the electric vector parallel to OZ, and from n to n_2 ($= n_{\perp}$ or n_s) for the electric vector perpendicular to OZ (stress direction). So for the electric vector in the XOY plane the refractive index is the same ($= n_s$) irrespective of the direction of the propagation of light.

In the present experimental studies we are interested in determining $\Delta n_1 = (n_p - n)$ and $\Delta n_2 = (n_s - n)$. Neumann introduced two constants p and q to relate Δn_1 and Δn_2 with the elastic strains produced in the solid.

$$\Delta n_1 = \Delta n_z = -n^2 q \varepsilon_{zz}, \qquad \Delta n_2 = \Delta n_x = -n^2 p \varepsilon_{zz} \tag{5.38}$$

(Note that it is ε_{zz} that enters both equations; ε_{zz} is a linear strain.)

The two constants p and q are sufficient to describe the photoelastic behavior of all isotropic solids like glasses, fused silica, and plastics. Having realized the inadequacy of Neumann's constants for describing the photoelastic behavior of crystals, Pockels developed his phenomenological theory of photoelasticity of crystals, according to which even the simplest type of crystals (namely the T_d, O, and O_h classes of the cubic system) requires three constants, q_{11}, q_{12}, and q_{44}. We have defined these q_{ij} in Chapter 3 and have used them in the earlier sections in this chapter; now Δn_p and Δn_s are related to the q_{ij} (see, for example, Tables 5.1–5.3 for the case of cubic, orthorhombic, and trigonal crystals). Neumann's p and q are particular cases of Pockels' p_{ij} (strain-optical constants). In order to obtain the several q_{ij} for any crystal, therefore, one has to determine experimentally Δn_p and Δn_s for at least as many orientations of the specimen as there are q_{ij}. For an isotropic solid like glass, two experiments will be sufficient to determine Neumann's p and q or Pockels' q_{ij} or p_{ij}. For a cubic crystal of the NaCl type, we have to make at least three measurements; for ADP, seven; and for barite, 12. Whatever be the type of crystal and consequently the number of measurements, the essential point to remember is that one is concerned only with the measurement of Δn_p and Δn_s. Hence for a large number of crystals the basic experimental requirements and the methods of observations are essentially the same. It is only when special problems arise that the experimental techniques have to be changed. For example, even the photoelastic dispersion of glasses demands a special technique. Studies of the behavior of optically active crystals, even cubic, are complicated. Experimental investigation of the photoelastic dispersion of uniaxial and biaxial crystals is unique to itself.

It is important to note here that Pockels worked out the entire theory of stress birefringence in terms of strain, but it is easy to develop the theory in terms of stress and double refraction. This has the advantage of avoiding the otherwise necessary introduction of all the elastic constants of the crystal.

We shall now proceed to describe Pockels' method of experimentally determining the q_{ij} of a cubic crystal like NaCl. Pockels' method of making the stress-birefringence measurements is important not only because of historical interest but also because it remains until today the standard method for most photoelastic studies. The experiment consists of two parts: (a) to determine the stress birefringence produced for different orientations of the crystals, and (b) to determine the absolute path retardation, and hence to calculate Δn_p and Δn_s when the crystal is under stress.

5.3.1. Measurement of Stress Birefringence, and Relative Path Retardation

The traditional method, from Pockels' time, is to use a Babinet compensator† to determine the relative path retardation. The optical arrangement for measuring the relative path retardation is shown in Fig. 5.4, and the compressing arrangement to produce static stress is shown in Figs. 5.5 and 5.6.

A polished prism (in the form of a rectangular parallelepiped) of known orientation is placed on the compressing arrangement and the amount of artificial double refraction produced for a known load is measured using a Babinet or a Soleil–Babinet compensator. The compression is produced by a lever, shown schematically in Fig. 5.5. Four fine V grooves G_1, G_2, G_3, and G_4 perpendicular to its length are marked on a steel bar AB of dimensions $50 \times 1.5 \times 1.5$ cm, two on the upper and two on the lower

† Its modification in the form of a Soleil–Babinet compensator can be used with equal facility. In fact, where the ordinary Babinet compensator cannot be used at all, the Soleil–Babinet compensator has found special application by Pettersen [922] and by Veerabhadra Rao and Narasimhamurty [1303].

Fig. 5.4. Optical arrangement to measure the relative path retardation in a solid under a static load. *S*, source of monochromatic light; *P*, polarizer; *A*, analyzer; *C*, solid under a static linear stress; *B*, Babinet compensator.

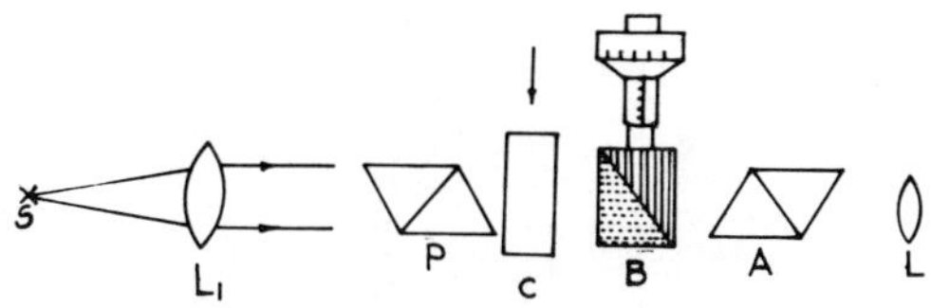

side. The groove G_1 engages a knife edge K_1 fixed to a metal frame (Fig. 5.5) rigidly fixed to a massive table *T*. The groove G_4 nearly 40 cm from the first groove is used for the knife edge carrying the weights. The two grooves on the lower side of the bar, to take the knife edge K_2 on the crystal prism *C*, give mechanical advantage of two different values. The lower end of the prism rests on a plane metal block *M* or another knife edge. The double knife-edge system holding the crystal ensures better uniformity of stress distribution. Lead and rubber washers are used to help distribute the load uniformly. With this experimental arrangement, the direction of stress is

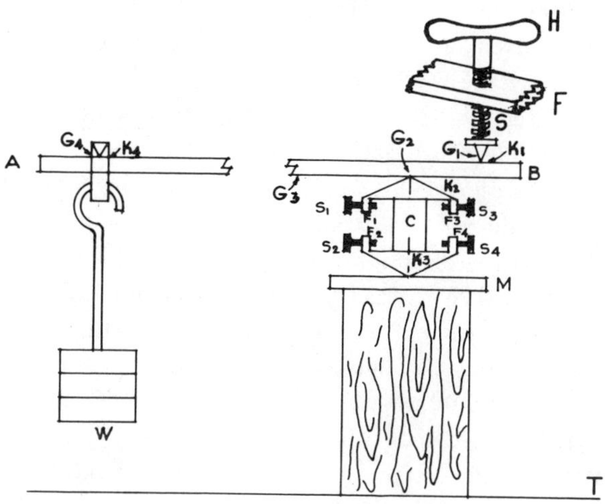

Fig. 5.5. Compressing arrangement (lever type) to apply a static stress on a solid for stress-optical measurements. *C*, solid specimen between two knife edges K_2, K_3; *AB*, lever, pressing against the knife edge K_1 at one end and carrying a load *W* at the other end using the knife edge K_4; *H*, handle on the top of the screw *S*, pressing K_1 in order to maintain the lever *AB* horizontal; *F*, horizontal top of a rectangular frame rigidly fixed vertically over the table *T*; *S*, screw working through the frame *F*; G_1, G_2, G_3, G_4, grooves on *AB* to take the knife edges in order to give different mechanical advantages; S_1, S_2, S_3, S_4, screws working through the flanges F_1, F_2, F_3, F_4 to facilitate the adjustment of the crystal *C* in position between the double knife edges K_2 and K_3.

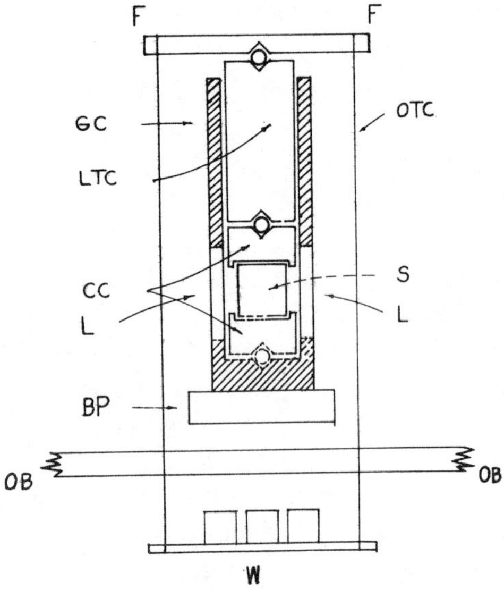

Fig. 5.6. Compressing arrangement (direct loading type) to apply a static stress on a solid for stress optical measurements—Pettersen's method. *S*, solid sample; *L*, *L*, light ports; *C*, *C*, contact cylinders; *W*, load suspended by three cables from a frame *FF*; *OTC*, one of three cables spaced at 120°, which support the load; *GC*, guide cylinder; *LTC*, load transmission cylinder; *BP*, brass plate; *OB*, optical bench. From [922].

vertical. By rotating the handle *H* the screw *S* moves up and down the frame *F* which is rigidly fixed on the table *T*. The knife edge K_1 is kept pressed down on the bar by the screw *S*. This arrangement helps in maintaining the lever *AB* perfectly horizontal even with crystals of different lengths.

Light from a monochromatic source (e.g., a sodium vapor lamp) is condensed by a lens on the slit *S* of a collimator (Fig. 5.4). The emergent parallel beam of light passes through a Nicol *P*, mounted on a circular scale, with the vibration direction set at $+45°$ or $-45°$ to the vertical. The parallel beam of light then passes through the crystal prism *C* into a Babinet compensator *B*, the principal axes of which are vertical and horizontal. The light, after passing through the Babinet compensator, is observed through an eyepiece containing a Nicol *A*, crossed with the polarizer; the usual Babinet fringes are then obtained. When the crystal prism is compressed, the fringes shift to one side or the other, depending on the sign of stress birefringence. The magnitude of the shift is a measure of the path difference produced between the horizontally polarized (vertically vibrating) and vertically polarized (horizontally vibrating) beams of light passing through the compressed crystals.

After the lever is placed in position, it is loaded. Following Pockels, the shift or displacement of the fringes is measured at the middle of the length of the prism at three places along its breadth, namely near the left, the middle, and near the right edge. Before taking each set of observations, a preliminary measurement of the fringe displacement at the three places on the crystal is made, and only when the distribution of the stress is fairly uniform are the final measurements recorded. In each of the three positions the polarizing and the analyzing Nicols are rotated through 90° so that in the two positions the incident light has its electric vector at +45° and −45° to the vertical. This procedure, as Pockels has shown, eliminates any small errors due to the direction of the stress in the crystal being not strictly parallel to the principal axis of the compensator. For each of the measurements the initial and final positions of the Babinet fringes are taken thrice before placing the load, and thrice after removing the load. The mean of the several observations for shifts is adopted for the calculations. The crystal is subjected to loads of the order of 1 kg/mm^2 depending upon the crystal, and it is ascertained that the fringes return to their original position, thus indicating that the crystals return to their original state of no stress birefringence. Initially a dead load is applied to help maintain the orientation of the crystal at all loads. This precaution is particularly necessary and important in the case of uniaxial and biaxial crystals which are birefringent even in the undeformed state.

Figure 5.6 shows a compressing arrangement (of the direct loading type) for producing a linear stress on the crystal specimen. This method was employed by Pettersen [922] for his studies on sodium halides.

Employing the data obtained with a Babinet compensator, we proceed now to calculate the q_{ij} of a cubic crystal, say NaCl of the $m3m$ (O_h) class.

A rectangular prism with its edges parallel to the cube axes of the crystal is stressed along one cube axis, and light is propagated along another cube axis (Fig. 5.3).

Let f be the fringe width or constant of the Babinet compensator for the monochromatic light radiation used, and let S be the shift of a Babinet fringe when the crystal is stressed. If M is the additional mass attached to the lever, with a mechanical advantage y, then the path retardation is given by

$$\delta = \frac{(S/f)\lambda}{Pt} = \frac{S\lambda}{f\{(Mgy)/bt\}t} = \frac{S\lambda b}{fMgy} \tag{5.39}$$

Here b is the breadth and t is the thickness of the crystal specimen.

A correction for the change of thickness due to stress on the cubic crystal may be calculated thus:

$$\left\{\begin{matrix}\text{Change of thickness per unit} \\ \text{compressional stress}\end{matrix}\right\} = -s_{12} \text{ per centimeter thickness}$$

Therefore

$$\left\{\begin{matrix}\text{Additional path retardation introduced} \\ \text{per unit compressional stress}\end{matrix}\right\} = (-s_{12})(\Delta n)$$
$$= -s_{12}(n_0^3/2)(q_{11} - q_{12})$$

$$\{\text{Observed path retardation } \delta_0\} = (n_0^3/2)(q_{11} - q_{12})(1 - s_{12}) \qquad (5.40)$$

The stress birefringence measurements can be repeated on the NaCl crystal for two more orientations, namely, for a stress parallel to [1$\bar{1}$0] and observations along [001] and [110], and for stress parallel to [$\bar{1}$11] and observations along [01$\bar{1}$] and [$\bar{2}$11]. When we turn to Table 5.1 we notice that from the measurements so far taken using a Babinet compensator, we can calculate $(q_{11} - q_{12})$ and q_{44} with an internal check on either of the values; but we cannot separately calculate q_{11} and q_{12} using a Babinet compensator alone. Hence we are obliged to devise some other method to determine the individual values of q_{11} and q_{12}. This can be achieved by using an interferometer as Pockels did.

We next describe interferometric methods of obtaining q_{11} and q_{12}.

5.3.2. Measurement of Absolute Path Retardation by Interferometric Methods

5.3.2.1. Pockels' Method

To obtain the individual values of q_{11} and q_{12}, Pockels [949] used an interferometer of the type described below.

Monochromatic light from a source illuminates the slit S and is then made parallel by the collimating lens L_1 (see Fig. 5.7). One beam of light traverses a rectangular prism A of the crystal specimen under stress; another beam traverses a second similar crystal prism B which is unstressed. The two beams are brought nearer by being refracted inward by a compensator C which is an angle-shaped piece of glass with plane-parallel branches. The two beams of light then strike two slits S_1 and S_2 in front of the objective lens L_2 of a telescope, and interfere in the focal plane P of L_2, thereby producing the usual diffraction pattern of a double slit. This diffraction

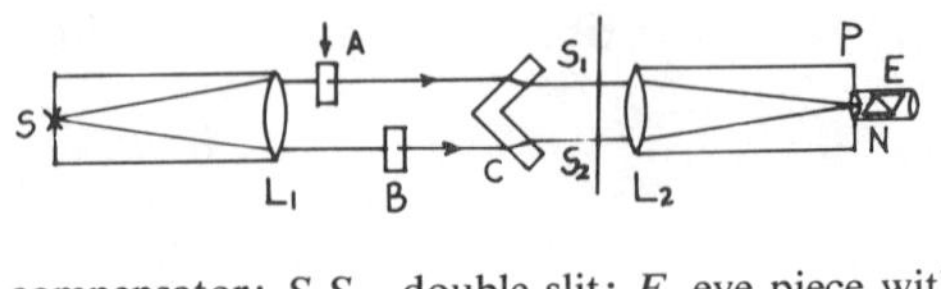

Fig. 5.7. Pockels' method for measuring q_{11} and q_{12} of solids. S, source of monochromatic light; A, B, two identical prisms of the solid, with A under stress; C, angle-shaped glass compensator; S_1S_2, double slit; E, eye piece with a Nicol N; P, focal plane of L_2.

pattern is viewed through an eyepiece E, which contains a Nicol N mounted in a circular scale.

When A is unstressed there is a diffraction pattern produced at P, but when a stress is applied, an additional retardation is experienced by the beam passing through A; consequently, the diffraction pattern is shifted, to one side or the other, from its original position, the shift depending upon the crystal, its orientation, the applied stress, wavelength of the light radiation, and temperature.

Since the retardation is different for the two differently plane-polarized waves, there are actually two patterns shifting differently but superposed one over the other. By setting the shorter diagonal of the Nicol N horizontal or vertical, each of these two patterns may be made separately visible in turn, and the two shifts can be measured. This was achieved by Pockels by a slight rotation of the compensating angle-shaped piece C in order to bring each pattern back to its original position.

If m fringes shift owing to a compressive stress P, then

$$m\lambda/Pt = \begin{Bmatrix}\text{path retardation per unit} \\ \text{stress per unit thickness}\end{Bmatrix} = \tfrac{1}{2}n^3 q_{11} - n_x s_{12} \tag{5.41}$$

when the Nicol N has its shorter diagonal parallel to the stress. Similarly, q_{12} can be determined keeping the Nicol N with its shorter diagonal normal to the stress.

Eppendahl [321] used a modified form of the Jamin interferometer to study the absolute path retardation in topaz.

5.3.2.2. Localized Interference Fringe Method

In recent years a method of using the crystal itself as an interferometer has come into vogue. Here localized interference fringes are formed in the crystal, as employed originally by Peters [917] for studying the variation of refractive index with temperature for optical glasses. More recently Ramachandran [993] has adopted this localized interference fringe method for studying dn/dT as well as the absolute values of the stress-optical coefficients of diamond. In Ramachandran's experimental setup a linear compressional stress is applied to the crystal specimen to determine the stress-optical coefficients. Bhagavantam and collaborators [119; 123; 128] have applied the same for a similar purpose in the case of alums and nitrates. α-Quartz and Rochelle salt have been studied by Narasimhamurty [850; 851] for their absolute values of q_{ij} using the same technique. Vedam and

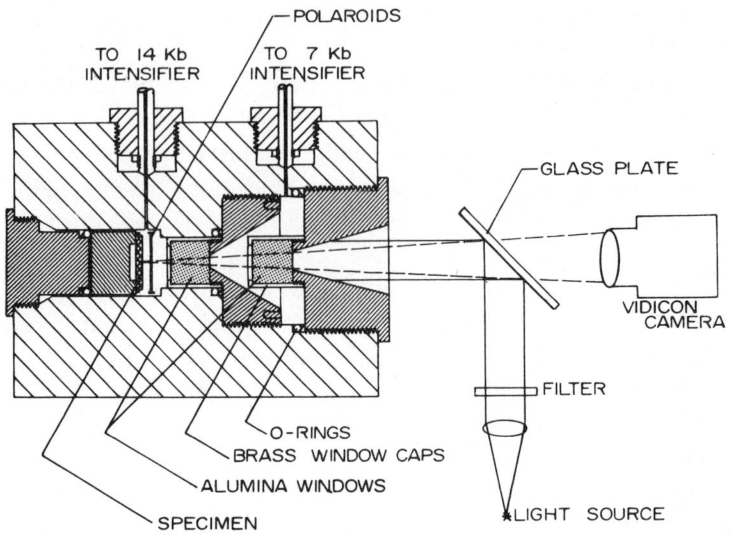

Fig. 5.8. Schematic drawing of an experimental arrangement for studying the effect of hydrostatic pressure on the index of refraction of crystals—Vedam's method. From [1301].

collaborators [1282–1301] have used the same technique (Fig. 5.8) with an additional provision to study the effect of hydrostatic pressure on the change of index of refraction of a number of crystals.

In Ramachandran's technique, the procedure adopted is to work with a pair of surfaces of the rectangular prism to get reasonably widely separated localized fringes (of the Newton's rings type), i.e., curves of equal thickness with reflected light. If the light is reflected at the usual 45° angle at a glass plate, the observation of the shift of the fringes is not possible when the light has its electric vector horizontal, because of the very low intensity. So light is reflected at an angle of about 15° on the pair of faces of the prism, and the localized fringes, formed by reflection of the beam at the two surfaces, are observed through a microscope. The microscope is focused on a reference mark on the front face of the crystal itself, and the movement of the fringes on loading is to be judged with reference to this mark. The shifting of the fringes is the combined result of the change in the refractive index and the change in the thickness of the crystal.

In general, complete shifts of one or two fringes are produced by stress of the same order as the one used for the Babinet observations. A Nicol is placed in the path of the light beam before the light is incident on the glass plate, and observations are taken for light beams having their electric vectors vertical and horizontal successively. The sign of the shift, namely whether the order of interference is decreasing or increasing at the point under consideration, settles the sign of the stress-optical constants unambiguously. In Vedam's experiments, hydrostatic pressures up to 14 kbars were applied to the crystal.

We can use Eq. (5.41) for calculating the relevant q_{ij}. Alternatively, the absolute values of the constants are calculated from the following equation, which is a modified form of Eq. (5.41):

$$\begin{aligned} n - n_x &= (\lambda/2t_2)[(2n/\lambda)(t_2 - t_1) - m] \\ q_{11} &= -(n - n_x) \cdot 2/(n^3 P_{xx}) = (n_x - n) \cdot 2/(n^3 P_{xx}) \end{aligned} \tag{5.42}$$

Here n is the refractive index of the unstrained crystal, n_x is the refractive index in the stressed crystal for light with vibrating direction vertical (i.e., parallel to the stress), $(t_2 - t_1)$ is the increase in the thickness of the crystal along the direction of observation for the corresponding load, λ is the wavelength of light, m is the increase in the order of interference at the point of observation, and P_{xx} is the stress producing the shift.

When the shift is away from the center of the fringe system, the order of the interference at the point of observation decreases; conversely, if the shift is toward the center, the order increases. Thus $m = -1$ when the shift is away from the center, and $m = +1$ if the shift is toward the center. It is well known that the accuracy attained is very low in these cases. Hence the values obtained by such interferometric measurements are combined with the measurements of differential path retardation using the Babinet compensator, giving due weight to both sets of observations, in order to arrive at the individual values of the relevant q_{ij}. Such a procedure was adopted by Pockels [948–956], Ramachandran [993], Bhagavantam and Suryanarayana [119], and Narasimhamurty [843; 850; 851].

5.3.2.3. *"High-Pressure Optics" Due to Vedam*

Vedam and collaborators [1282–1301] have suitably adapted the "localized interference fringe method" in order to study the effect of hydrostatic pressure on the change of refractive index of a number of cubic and noncubic crystals (Fig. 5.8).

Using an optical bomb capable of bearing hydrostatic pressure to 7 kbars, Vedam and Schmidt [1284] have been able to observe a shift of as many as 24 Newtonian-type interference fringes in the case of MgO crystal, in contrast to a shift of hardly a fringe under linear stresses on crystals.

A two-stage high-pressure optical vessel has been subsequently developed by Vedam and co-workers [1296], who have studied KBr and KI to a hydrostatic pressure of 14 kbars. For these crystals about 50–60 interference fringes are shifted.

The experimental technique and the results obtained from it by Vedam and co-workers on high-pressure optics constitute a major contribution to the studies of the photoelastic behavior of crystals.

First, Vedam has experimentally shown that there is a nonlinear variation of the refractive index with the applied hydrostatic pressure, whereas Pockels' assumption of a linear relation between Δn and stress is valid only at very low levels of linear stress. The magnitude of the hydrostatic pressure at which the nonlinearity commences depends on the experimental crystal. For example, for MgO the linear variation persists even to a pressure of 7 kbars, which can be justified by the fact that the elastic constants of MgO are fairly large and, consequently, the maximum volume strain is only about 0.13%. Similar is the case for diamond and sapphire. But in their studies on KBr and KI, Vedam and collaborators [1296] have found a pronounced nonlinear dependence of the refractive index above a pressure as low as 3 kbars.

However, by their studies on a very large number of cubic and noncubic crystals, they have established the fact that a perfect linear relationship exists between Δn, the change in the refractive index, and the Lagrangian volume strain, which is as high as 9% for KBr and KI, the Lagrangian volume strain having been computed using the nonlinear theory of elasticity. It is pertinent to caution at this stage that the validity of the statement that Δn bears a linear relation with the Lagrangian strain is limited in its scope and generally holds over the optical region far removed from the absorption edge in the crystal specimen. Thus, Vedam and co-workers have observed pronounced nonlinearity even at fairly low strain levels in α- and β-ZnS and CdS, etc. for 5893 Å, indicating the influence of the nearness of the absorption edge in these cases.

Yet another major significant point of interest in Vedam's investigations on high-pressure optics relates to an accurate evaluation of the individual values of q_{ij} and p_{ij} of crystals from the experimental data obtained by various experimental techniques.

For a complete description of the photoelastic behavior of crystals, whether cubic or noncubic, the techniques usually employed are (a) stress birefringence measurements, (b) the localized fringe method, and (c) ultrasonic studies on the elasto-optic ratios (for ultrasonic methods, see pages 220–278). Although the experimental method of determining the stress birefringence under a linear stress can be made sufficiently accurately using a Babinet compensator or a Soleil–Babinet compensator, the interferometric methods may yield q_{ij} values with errors as large as 50% in some cases. The ultrasonic methods due to Mueller [829] have their own limitations on the accuracy of the elasto-optic ratios, which may be in error by as much as 10% if not more, depending on their numerical values. Nevertheless, these have been the experimental methods applied to almost all the cubic crystals. But a major difficulty with Mueller's ultrasonic method arises when assigning the sign of the elasto-optic ratios; this difficulty becomes more prominent in the case of noncubic crystals, where some of the ratios are negative. It is at this juncture that Vedam's work on high-pressure optics plays a major role for computing the absolute values, in both magnitude and sign, of q_{ij} and p_{ij}. Thus Narasimhamurty, Veerabhadra Rao, and Pettersen [854] have been able to review the earlier work and obtain the individual values of q_{ij} and p_{ij} for ADP by combining the experimental data on stress birefringence under a linear stress, the elasto-optic ratios by Mueller's method C, and Vedam's data on ADP under hydrostatic pressure.

5.3.3. Photoelastic Studies of Optically Active Crystals

Crystals of $NaClO_3$ and $NaBrO_3$ belong to the crystal class T and exhibit optical activity. The usual methods of studying optically inactive substances cannot be employed in such cases for obvious reasons [998; 1001; 1017; 1018], and special techniques have been employed by Ramachandran and Chandrasekharan [998] to study $NaClO_3$. An optical method using the Rayleigh compensator has been employed by Narasimhamurty [842] to study $NaClO_3$ and $NaBrO_3$. It is very sensitive, being capable of measuring a birefringence of the order of $\lambda/500$. The sensitivity, which can be varied at will, is independent of the magnitude of the birefringence to be measured and remains uniform throughout the range of measurement. Such measurements give directly the values of $q_{11} - q_{12}$, $q_{11} - q_{13}$, and q_{44}.

The Rayleigh Compensator. A plane sheet of glass $ABCD$ (Fig. 5.9), bent into the form $A'B'C'D'$, constitutes the Rayleigh compensator [1037; 1015]. When the glass plate is strained as indicated, the region along NN' undergoes no strain and hence forms the neutral region. Portions of the plate above NN' experience a tensile stress and portions below NN' ex-

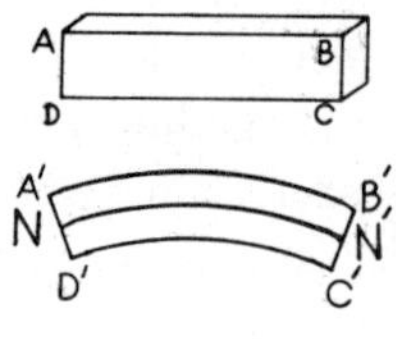

Fig. 5.9. Rayleigh compensator for studying artificial birefringence. *ABCD*, a rectangular glass strip under no strain; *A'B'C'D'*, the glass strip *ABCD* bent in its own plane; *NN'*, neutral axis. From [842].

perience a compressional stress. The result is that the upper half of the plate behaves as a positive uniaxial crystal, and the lower part as a negative uniaxial crystal, the two being separated by a neutral band. When such a plate is viewed through crossed Nicols, light is completely extinguished at the neutral region only, but appears over the rest of the field. Now if a birefringent medium *G* (e.g., a longitudinally compressed glass block (Fig. 5.10) is placed between the polarizer and the Rayleigh compensator, the dark band shifts upward or downward depending on the sign of birefringence of the medium *G* with respect to that of the Rayleigh compensator. The shift of the dark band can be detected by a low-power telescope. Further, if the telescope is provided with a finely divised scale in place of cross wires, the shift can be measured in terms of scale divisions and can be used as a measure of birefringence, after calibration. The sign of birefringence of the specimen can be determined unambiguously by comparing the direction of the shift for the specimen with that for glass.

If the specimen has optical activity but no birefringence, light is restored in the entire field of view, and the dark band *NN'* reappears in the field of view of the telescope when the analyzer is rotated through the appropriate angle. The band reappears now exactly at the former scale division. Thus the birefringence and optical activity could be studied.

Now to experiment with an optically active medium, the specimen is placed between the Rayleigh compensator and the analyzer; the analyzer is rotated until the neutral dark band reappears. The specimen is then compressed; consequently the band shifts upward or downward depending on its sign of birefringence. If in addition to the stress birefringence, the optical activity of the medium itself changes with stress, the dark band becomes diffuse and tends to disappear; it can be seen again by rotating the

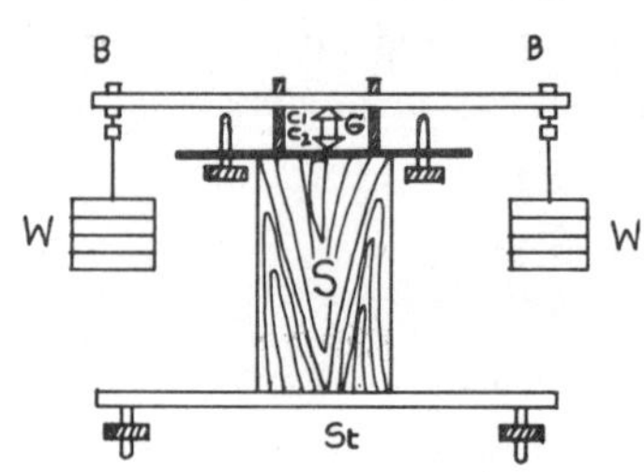

Fig. 5.10. Compressing arrangement for studying the photoelastic behavior of optically active crystals like $NaClO_3$. *G*, glass prism; C_1C_2, double knife edges; *BB*, loaded beam with loads *W*, *W*. From [842].

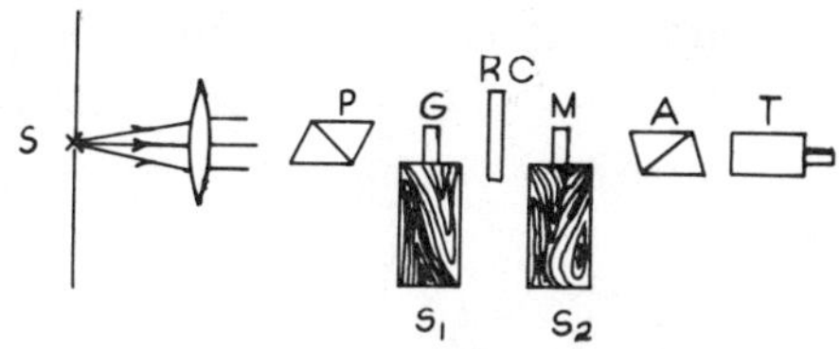

Fig. 5.11. Optical setup for studying the photoelastic behavior of crystals like $NaClO_3$ employing a Rayleigh compensator. *S*, source of monochromatic light; *M*, experimental solid material; *G*, glass prism for comparison; *RC*, Rayleigh compensator; *P*, polarizer; *A*, analyzer; *T*, telescope of a low power; S_1, S_2, compressing arrangements, taking *G* and *M*, respectively. From [842].

analyzer through the appropriate angle. Thus change in optical activity alone will render it diffuse but produces no shift, while birefringence alone will produce a shift but the band remains dark.

The Experimental Setup. The compressing arrangement is shown in Fig. 5.10. Since the Rayleigh compensator could detect birefringence of the order of $\lambda/500$, no lever arrangement would ordinarily be necessary. The beam *BB* carrying the loads *W* at either end rests on the specimen.

Two such compressing arrangements S_1, S_2 (Fig. 5.11) are set up, one on either side of the Rayleigh compensator RC. The polarizer *P* is set at 45° with the vertical.

A glass prism *G* ($1.00 \times 0.75 \times 0.45$ cm) is first stressed under a known load, and the consequent birefringence produced is determined with a Babinet compensator; the amount of birefringence (in terms of the light source used) is calculated for a load of 1000 g. The glass prism *G* is then used as the standard with which the photoelastic birefringence produced in the crystal specimen is compared.

The glass prism *G*, which is of isotropic medium, develops only birefringence under linear stress but no optical rotation, whereas the crystal prism *M*, which is initially optically active, becomes also birefringent under linear stress; hence there is an apparent rotation of the axes of birefringence [1017–1019]. For this reason the specimen *M* (i.e., of sodium chlorate or sodium bromate) is mounted on S_2 and the glass prism *G* on S_1. The dark band will now be seen only when the analyzer is rotated through an angle equal to the optical rotation by *M*. The specimen *M* is then stressed by a load of, say, 2000 g (i.e., 1000 g at either end), and the shift of the band is noted in the field of view of the telescope. The specimen *M* is then unloaded, and *G* is loaded suitably so that the dark band showed the same displacement as for the medium *M*. Hence the birefringence produced is the same for both. From this the corresponding piezo-optic constants of the medium could be calculated.

5.4. Dispersion of q_{ij} by Spectroscopic Methods

5.4.1. Birefringent Compensator for Studying Very Small Changes in Double Refraction

5.4.1.1. Experimental Setup

Although the Babinet compensator has been used successfully in routine measurements of double refraction, it is not sensitive enough to allow accurate measurement of the very small changes in double refraction occurring when the temperature of a birefringent crystal is altered, or for the study of dispersion of photoelasticity in glasses, cubic crystals, or birefringent crystals.

For studying the dispersion of photoelasticity of glasses, Filon [363] has developed a spectroscopic method. A modification of this method has been used by Bansigir and Iyengar [66] to study cubic crystals. But neither Filon's method nor its modification by Bansigir and Iyengar [67] is sensitive enough to study the photoelastic dispersion of uniaxial and biaxial crystals. For this reason, a birefringent-compensator method has been developed by Narasimhamurty and Ziauddin [844]. This method can be used for studying the photoelastic dispersion not only in glasses and cubic crystals but also in naturally birefringent crystals. Furthermore, the sensitivity of this method can be changed at will to facilitate measurement, whether the photoelastic dispersion is large or small.

The experimental arrangement is shown in Fig. 5.12. A parallel beam of white light passes through the polarizer P, the sample glass or crystal prism S, the birefringent compensator BC, the photoelastically calibrated glass prism M, and the analyzer A, and is incident on the slit of a constant-

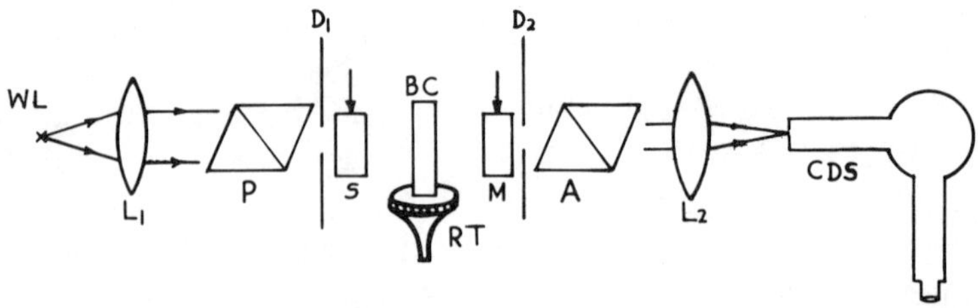

Fig. 5.12. Birefringent compensator method for studying photoelastic dispersion in solids. *WL*, source of white light; *M*, glass prism for comparison; *S*, sample glass or crystal prism; *BC*, birefringent compensator; D_1, D_2, diaphragms; *P*, polarizer; *A*, analyzer; *CDS*, constant deviation spectrometer; *RT*, rotating table; L_1, L_2, achromatic objectives. From [844].

deviation spectrometer. The prisms S and M can be stressed by adding weights to a system of levers. When the prism M and the birefringent compensator BC are removed, the remaining arrangement is that for Filon's method. Note that the positions of S and M can be interchanged.

The birefringent compensator BC is a rectangular prism of a birefringent crystal, say barite, cut parallel to its crystallographic axes. (On occasion the compensator may be of actually two such prisms). It is placed, with one of its axes vertical and the other two axes horizontal, on a prism table RT which can be rotated about its vertical axis. Because of the presence of the compensator, a number of bright and dark vertical fringes appears in the spectrometer telescope even without any load on the glass prisms. Such fringes are shown in Fig. 5.11. The number of fringes can be altered by varying the optical path in the compensator, which can be changed by suitably rotating it about a vertical axis. The change in the number of fringes is caused by a change in the difference between the refractive indices for the vertically and horizontally polarized beams of light. The compensator is rotated about the vertical axis until two or three fringes only are seen in the field of view of the constant deviation spectrometer.

5.4.1.2. *Theoretical Considerations*

If a uniform linear stress P is vertically applied to M, in the absence of S and the birefringent compensator BC a dark fringe appears in the otherwise continuous spectrum. The condition satisfied by the center of the dark fringe is

$$(m + \tfrac{1}{2})\lambda = C_\lambda Pt \qquad \text{(for parallel Nicols)}^\dagger \tag{5.43}$$

where m is the order of the fringe, λ is the wavelength of light at the center of the dark fringe, and C_λ is Brewster's constant at λ for the glass M.

When the second prism S under a uniform linear stress P' in the vertical direction is introduced, the dark fringe is shifted to λ', the direction of the shift depending on the sign of birefringence of S. Under these conditions the total birefringence is the sum of the birefringences produced in the individual prisms M and S. Hence

$$(m + \tfrac{1}{2})\lambda' = C_{\lambda'}Pt + C'_{\lambda'}P't' \tag{5.44}$$

where C', P', and t' refer to the prism S.

† Deduction along similar lines can also be made for crossed Nicols.

When the compensator BC is introduced between the two glass prisms M and S, the number of dark fringes increases. But by rotating the compensator about a vertical axis, it is possible to reduce the number of the fringes to about two or three in the entire field of view. Controlling the orientation of the compensator about the vertical axis, it is possible to record the readings on the same order of the fringe for any desired wavelength in the entire region of the spectrum. For parallel Nicols, the dark fringe now satisfies the following equation:

$$(m + \tfrac{1}{2})\lambda = C_\lambda Pt + C_\lambda' P't' + Q \tag{5.45}$$

Here Q is the birefringence produced by the compensator at λ. Suppose the stress on M is changed to P_1. Due to this, the dark fringe shifts to λ'. Equation (5.45) becomes

$$(m + \tfrac{1}{2})\lambda' = C_{\lambda'} P_1 t + C_{\lambda'}' P't' + Q' \tag{5.46}$$

Adjusting the stress on the second prism S to P_1', it is again possible to bring the dark fringe back to its initial position. The equation satisfying the above condition will be

$$(m + \tfrac{1}{2})\lambda = C_\lambda P_1 t + C_\lambda' P_1' t' + Q \tag{5.47}$$

By equating (5.45) and (5.47) we get

$$\begin{aligned} C_\lambda Pt + C_\lambda' P't' &= C_\lambda P_1 t + C_\lambda' P_1' t' \\ C_\lambda t(P - P_1) &= C_\lambda' t'(P_1' - P') \\ C_\lambda' &= C_\lambda t(P - P_1)/t'(P_1' - P') \end{aligned} \tag{5.48}$$

Hence C_λ', the stress-optical coefficient for the second prism S, can be calculated for the entire region of the spectrum if C_λ, the stress-optical coefficient of the first prism M, is known for the several wavelengths of light. For this purpose, the glass prism M which has already been photoelastically calibrated is used, and the graph of C_λ versus λ for M is used as a reference graph for the determination of the stress-optical coefficients of the prism S.

5.4.1.3. *Sensitivity of the Experimental Setup*

It has been mentioned earlier that the sensitivity of the method could be altered at will. In fact, it depends on the order of the fringe on which the observations are taken. The sensitivity decreases with increasing order of the fringe. This can be experimentally established as follows. The birefringent compensator is adjusted until a dark fringe appears near about 5890 Å. A suitable stress is applied on the prism *S*. The shift $\Delta\lambda$ is measured, and the load is removed. Then the compensator is slightly rotated so as to get the next lower order to the same value at about 5890 Å, and the shift of this dark fringe is determined for the same load on the prism *S*. The experiment is repeated for several orders. Figure 5.13 shows the appearance of the fringe shift with stress as the order of the fringe is altered by rotating the birefringent compensator. Figure 5.14 shows the linear variation of $1/\Delta\lambda$ with *m*, where *m* is the order of the fringe and $\Delta\lambda$ is the shift of the fringe, for the same stress on *S* [844]. The important fact to be noted here is that for higher orders of the fringe, the shift becomes smaller, though the actual value of stress birefringence produced in *S* remains the same. That is, the sensitivity of the method is less for large values of *m*, and vice versa.

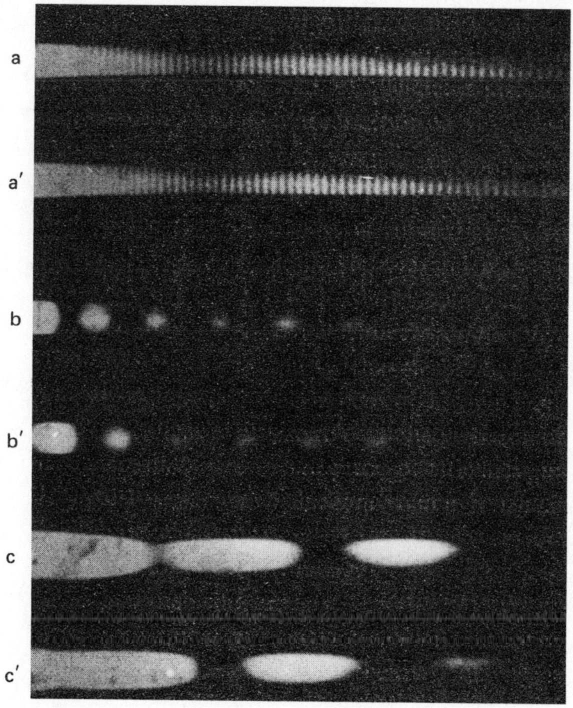

Fig. 5.13. Filon's fringe patterns in a crystal for different settings of the birefringent compensator. Fringe patterns (a), (b), and (c) are with the unstressed crystal. Patterns (a′), (b′) and (c′) are with the crystal under the same stress but with different settings of the birefringent compensator. Note the increasing shift with decreasing order of the fringe. From [844].

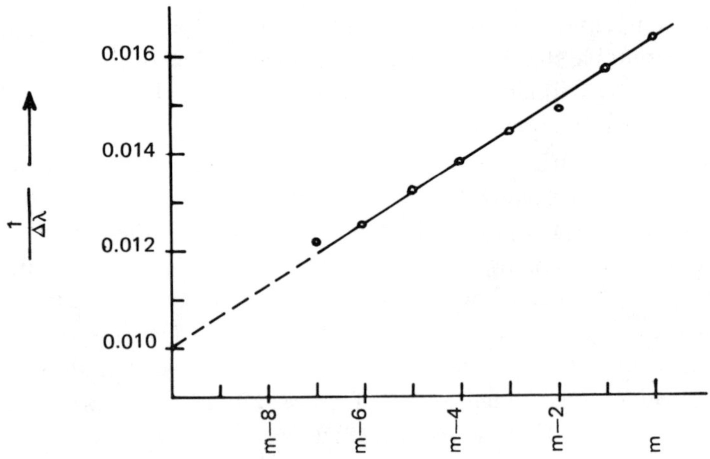

Fig. 5.14. Shift of a Filon fringe as a function of its order by the birefringent compensator method. $\Delta\lambda$ is the fringe shift in Å. From the graph $n = 29$. From [844].

The same result can be deduced thus: When a ray of plane-polarized light is passed through a uniaxial crystal of refractive indices n_v and n_h (for electric vectors along the vertical and horizontal directions), the condition for interference for a dark fringe to appear at a wavelength λ is given by the relations

$$(m + \tfrac{1}{2})\lambda = (n_v - n_h)t \qquad \text{(for parallel Nicols)} \tag{5.49}$$

$$m\lambda = (n_v - n_h)t \qquad \text{(for crossed Nicols)} \tag{5.50}$$

When the Nicols are parallel, the fringe shifts to λ' because an additional birefringence is caused by stress on the crystal. Then Eq. (5.49) becomes

$$(m + \tfrac{1}{2})\lambda' = (n_v' - n_h')t \tag{5.51}$$

From Eqs. (5.49) and (5.51) we get

$$[(n_v' - n_h') - (n_v - n_h)]t = (m + \tfrac{1}{2})(\lambda' - \lambda)$$

that is,

$$\Delta(n_v - n_h)\, t = (m + \tfrac{1}{2})\, \Delta\lambda \tag{5.52}$$

Dividing Eq. (5.52) by Eq. (5.49) we get

$$\frac{\Delta(n_v - n_h)}{(n_v - n_h)} = \frac{\Delta\lambda}{\lambda} \tag{5.53}$$

Thus, for the same λ and for the same stress birefringence $\Delta(n_v - n_h)$ in the specimen, when the initial birefringence $(n_v - n_h)$ is smaller, the value of $\Delta\lambda$ is larger. If the initial birefringence is small, the value of m is also small. Hence if m is small, the value of $\Delta\lambda$ is large. That is, the sensitivity of the method increases if the number of the order is reduced. A similar argument holds for crossed Nicols.

5.4.1.4. Determination of the Order of Fringe at Any Particular Wavelength

a. The X intercept of the straight-line graph (Fig. 5.14) gives the value of m.

b. The order of the fringe can also be computed thus. For parallel Nicols, whenever $(m + \frac{1}{2})\lambda_1 = (n_v' - n_h')t$, no light emerges from the optical system and hence the spectrum is crossed by a number of dark fringes. Taking into account a particular fringe of order m, which is at 5900 Å, the analyzer is rotated through 90°. The same dark fringe is shifted from $\lambda_1 = 5900$ Å to another wavelength λ_2, satisfying the equation $m\lambda_2 = (n_v'' - n_h'')t$. If the values of $(n_v - n_h)$ at the two neighboring wavelengths λ_1 and λ_2 are assumed to be the same, then, knowing the values of λ_1 and λ_2, the value of m can be calculated from the relation

$$m = \tfrac{1}{2}\lambda_1/(\lambda_2 - \lambda_1)$$

c. Alternatively, m can also be calculated thus: The wavelength of two successive fringes m and $m - 1$ are noted before applying the load. From Eq. (5.50) for crossed Nicols, $m\lambda = m_1\lambda_1 = (m - 1)\lambda_1$, which is nearly correct for two neighboring orders. Hence we can calculate m from the relation

$$m = \lambda_1/(\lambda_1 - \lambda).$$

5.4.1.5. Conclusion

Filon's spectroscopic method of studying photoelastic dispersion in glasses is more sensitive than the Babinet compensator, but it requires very large loads to shift the dark fringe from one end of the spectrum to the other end so as to scan the entire spectral region. This method fails for cubic crystals because most of these crystals cannot bear such large loads. The differential method by Iyengar and Bansigir removes the difficulty of using large stresses on cubic crystals, but both of these methods have the following common defects: (1) The dark fringe is oftentimes diffuse instead of being sharp. Consequently, there is a certain amount of error in locating the center of the dark fringe. (2) Neither of the methods is suitable for studying uniaxial and biaxial crystals because of the higher orders of the dark fringes or bands appearing in the field of view and the consequent decrease in sensitivity. (3) For the initially birefringent crystals the observations cannot be taken on the same order of the fringe. Since the sensitivity changes from order to order, it changes from one wavelength to another.

Compared with the difficulties enumerated above, the birefringent compensator has the following advantages: (1) It avoids large stresses on the experimental specimen. (2) The sharpness of the dark fringe is secured by changing the number of the order from first to second or third by rotating the birefringent compensator, and the consequent reduction in the sensitivity is more than compensated by the ease and reliability with which the center of the dark fringe can be located. This sharpness of the fringe cannot be secured easily in the former methods. (3) The same dark fringe can be changed to different wavelength regions by suitably rotating the birefringent compensator, and hence a constant load on the calibrated prism *M* and experimental prism *S* can be used for measurements over the entire spectral region. (4) Above all, the same sensitivity can be maintained throughout the entire range of wavelengths since the sensitivity depends on the order of the fringe, and the same order of fringe can be observed over the entire region by rotating the birefringent compensator. Thus it is possible to extend the measurements to crystals which show very small changes of birefringence caused by a mechanical stress, an electrical field, or a change of temperature. This method [844] has been successfully employed by Ziauddin [1424] and Ziauddin and Narasimhamurty [1422] to study the photoelastic dispersion in a number of cubic and noncubic crystals.

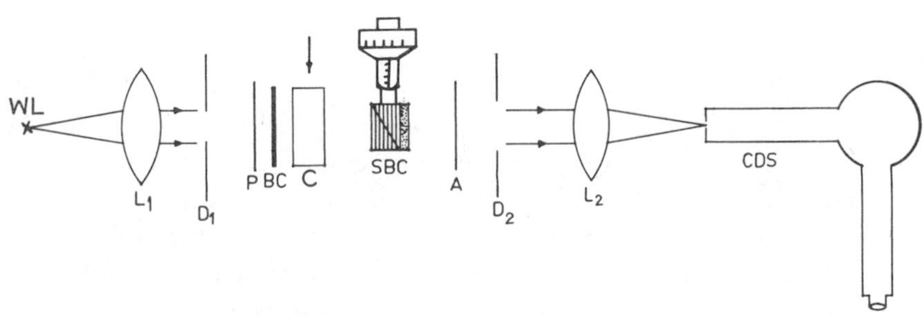

Fig. 5.15. Birefringent compensator method (modified form using a Soleil–Babinet compensator) for studying photoelastic dispersion in solids. *WL*, source of white light; *BC*, birefringent compensator; *C*, sample crystal prism; *SBC*, Soleil–Babinet compensator; *P*, polarizer; D_1, D_2, diaphragms; *A*, analyzer; *CDS*, constant deviation spectrometer; L_1, L_2, achromatic lenses. From [1303].

Veerabhadra Rao and Narasimhamurty [1303] have modified the above described birefringent compensator method [844] by substituting a Soleil–Babinet compensator for the standard glass prism M under stress in order to produce the desired artificial birefringence (Fig. 5.15), and they have applied it to study the dispersion of the photoelastic and Pockels' linear electro-optic effects in a number of noncubic crystals [1306]. The main advantage of this modified method is that one does not have to depend on the value of C_λ of the calibrated glass prism experimentally determined by Filon's method, which itself is not very accurate for studying the photoelastic dispersion of even cubic crystals.

5.4.2. *Dispersion of the Individual Stress-Optical Coefficients* q_{11} *and* q_{12} *of Vitreous Silica*

An experimental technique has been developed by Jog [586] for determining the dispersion of the individual q_{ij} of vitreous silica, and this technique has been applied over the range 2300–5500 Å. The method is important for two reasons: It is applicable for determining q_{11} and q_{12} separately, and also to study their dispersion extending over a wide range of the spectrum covering the visible and ultraviolet regions. The order of accuracy attainable in the measurement of q_{11} and q_{12} is comparable with that reached by Filon's method to obtain C_λ and $q_{11} - q_{12}$. It is, however, important to note that the principle of this method is entirely different from that of Filon's method. While Filon's method and its subsequent modifications are differential methods, Jog's method is not a differential method.

Jog combines the localized-fringe method with a quartz spectrograph. The experimental technique is schematically shown in Fig. 5.16. Here S_1 is a source of light, such as a mercury arc or a copper arc, capable of giving line spectra; S_2 is the fused quartz specimen under investigation. The light from S_1, collimated by a lens L_1, is reflected on the specimen by a plane-

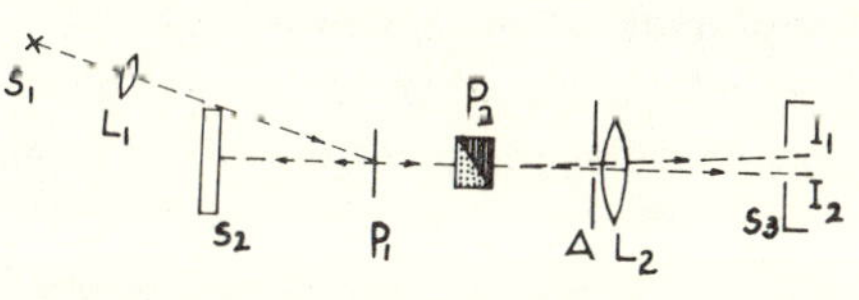

Fig. 5.16. Jog's experimental setup for determining q_{11} and q_{12} of vitreous silica. S_1, light source such as mercury and copper arcs giving line spectra; S_2, specimen of vitreous silica under stress; P_1, plane parallel plate of vitreous silica; P_2, double-image prism; A, aperture; S_3, slit of the quartz spectrograph; I_1, I_2, horizontally and vertically polarized images of the localized fringe systems, when the stress acts vertically on the solid; L_1, L_2, achromatic lenses. From [586].

parallel plate P_1 of fused quartz. The specimen S_2 has its broad faces slightly curved so that the plate itself acts as an interferometer, having developed localized fringes (of the type of Newton's rings). The converging lens L_2 throws an image of the localized fringes on the slit of a quartz spectrograph. The double-image prism P_2 separates out the two perpendicularly polarized images, and is so oriented that the two images I_1 and I_2 of the localized-fringe system falling on the slit have their electric vectors, respectively, parallel and perpendicular to the direction of stress acting on the specimen S_2. The fringe systems appearing in the field of view of the spectrograph are photographed, first with a dead load, then with a known additional load, and, finally, again with the original dead load.

A scratch made on the front face of the specimen S_2 which acts as a fiduciary point appears as a bright sharp line on the photographic negatives. From the measured diameters of the rings photographed with and without load, the fringe shift f as a fraction of the fringe width can be calculated using the relation

$$f = \frac{2(d_{m'}^2 - d_m{}^2)}{(d_{m+1}^2 - d_m{}^2) + (d_{m'}^2 - d_{m'-1}^2)} \tag{5.54}$$

where d_m and d_{m+1} are the diameters of the mth and $(m + 1)$st fringes respectively. The primed quantities refer to the corresponding diameters when the crystal is under a known stress.

The experiment can be repeated to study the fringe system formed by light transmitted by the specimen crystal. For this purpose the central part of the specimen crystal is semisilvered, and the light from the source S_1 is incident directly on S_2; the fringe system, seen under transmitted light, is focused on the slit of the quartz spectrograph, using a fused quartz lens or by means of a suitably oriented concave mirror. Using a double-image prism, the two polarized components are separated for taking the photographs for each component.

The value of $C_{\parallel}$ (the absolute path retardation, say, for the parallel component) at each wavelength can be calculated from the usual relation $f\lambda = \delta = C_{\parallel}Pt$, where P is the applied stress and t is the thickness of the specimen crystal. The quantity q_{11} can then be obtained from the relation

$$C_{\parallel} = n^3 q_{11} - 2ns_{12} \tag{5.55}$$

Similarly, q_{12} can be obtained from the observations taken on horizontally or perpendicularly vibrating light.

5.4.3. *Interference-Spectroscope Method of Studying the Absolute Photoelastic Coefficients of Glasses and Their Variation with Wavelength*

For the study of the photoelastic behavior of glasses and crystals, various methods have been applied by workers in this field. The Babinet compensator, Filon's method, the birefringent compensator method, and the ultrasonic method have been employed for this purpose. The Babinet compensator, Filon's method, and the birefringent compensator method give the differences of the changes of the refractive index for light polarized parallel and perpendicular to the stress, and the ultrasonic method gives their ratio. By the combination of these two measurements, absolute values of the photoelastic constants have been obtained. This also provides the technique for the study of the dispersion of the photoelastic constants for a number of glasses and crystals.

Direct determination of the absolute values of the stress-optical coefficients has also been made by Pockels by applying the interference method. But using an interferometer, the absolute values could not be obtained with sufficient accuracy. A Michelson interferometer in conjunction with a spectrometer was used by Braybon [188] in order to determine the absolute stress-optical coefficient of plastic (CR 39) taken as a thin lamina of the substance in the shape of a tension member. The spectrometer method for obtaining absolute values was adopted for diamond by Poindexter [961; 962].

A new technique has been developed by Narasimhamurty and Jeelani [847] to determine the absolute values of stress-optical coefficients for glass using an interference spectroscope. This has been successfully applied for the study of the photoelastic dispersion in glasses.

Figure 5.17 is a schematic representation of the familiar experimental setup to study the double-slit diffraction pattern of the Fraunhofer type. The length and width of each of the slits *E*, *F* are 15 and 2.5 mm, respectively, their centers being separated by 1 cm. The slit *Sp* of the collimator of

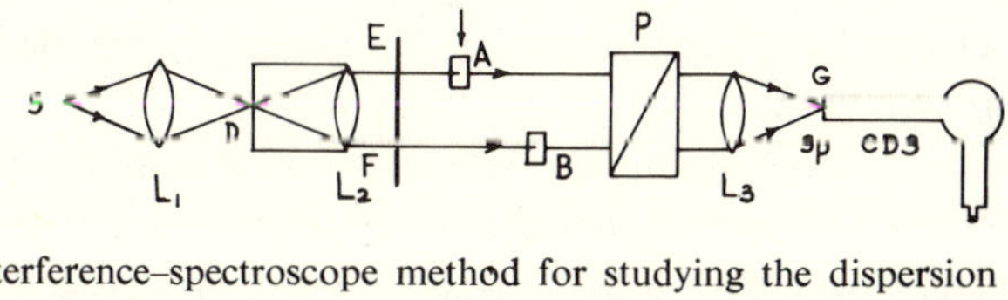

Fig. 5.17. Interference–spectroscope method for studying the dispersion of q_{11} and q_{12} of glass. *S*, source of white light; *E*, *F*, two slits to produce the double-slit diffraction pattern; *A*, *B*, rectangular glass prisms of approximately the same thickness: *A* is stressed; *Sp*, slit of the constant deviation spectrometer *CDS*; *G*, midpoint of the diffraction pattern of the double slit *E*, *F*; *P*, polarizer; L_1, L_2, L_3, achromatic lenses. From [847].

a constant-deviation spectrometer *CDS* is adjusted at *G*, the midpoint of the diffraction pattern of the double slit. The slit *Sp* thus receives the central bright diffraction band (together with some side bands, if the slit *Sp* is too wide). Looking through the eyepiece of the spectrometer, we see a bright, practically continuous, spectrum of white light given out by the source *S*. The structure of the central diffraction band (containing a number of interference fringes, which are the most important constituents for the Rayleigh refractometer) does not play any part in the present experimental setup.

In Fig. 5.17, *A* and *B* are two rectangular glass prisms cut from the same glass plate. A little variation in their relative dimensions is of little importance, as can be seen from the theory of the experiment. Prisms *A* and *B* are put in the paths of the two beams coming from the double slit *E* and *F*. Prism *B* is mounted on a slow-circular-motion device (not shown in the figure), and *A* is mounted on a stressing device. A slight rotation of *B*, if necessary, will cause the formation of dark and bright bands as in a channeled spectrum. These bands or fringes are obtained even without polarizers.

The slit of the collimator of the *CDS* is gradually narrowed until interference fringes are seen spread over the entire range of the spectrum. The sharpness of fringes is attained by narrowing down the width of the slit *D*. To verify that the fringes are due to interference between the two beams coming through the two rectangular glass prisms *A* and *B*, light from *E* or *F* is cut off. The fringes disappear and the spectrum becomes continuous. Next, the two beams are restored, and the block *B* is slowly rotated with the help of the rotating device. The fringes shift to one side.

By rotating the prism *B* in the proper direction, the number of fringes or the dark bands in the spectrum is reduced. The position of *B* is so adjusted that only three or four dark bands are seen in the field of view, so that the order of the fringes is very low, which is desirable for better sensitivity and accuracy.

Now a polarizer *P* is placed in the path of the rays coming from the double slit *E* and *F*. To determine $C_\lambda^{\parallel}$ of the prism *A*, *P* is placed with its light vibrations parallel to the stress direction of *A*.

The method of calculating the absolute values of the photoelastic constants of glass ($C_\lambda^{\parallel}$ and $C_\lambda^{\perp}$) is discussed below.

1. Let the thickness of prisms *A* and *B* be t_1 and t_2, respectively. For a dark band to appear at a wavelength λ, it is necessary that

$$(n - 1)t_2 - (n - 1)t_1 = (m + \tfrac{1}{2})\lambda \tag{5.56}$$

where n is the refractive index of the prisms A and B at the particular wavelength λ for light vibrations parallel to the stress, and m is the order of the dark band. Now the prism A is stressed by a stress P; then the dark band shifts from λ to λ', hence

$$(n' - 1)t_2 - (n' + \Delta n' - 1)(t_1 + \Delta t_1) = (m + \tfrac{1}{2})\lambda' \qquad (5.57)$$

where n' is the refractive index of the prisms A and B at the wavelength λ', $\Delta n'$ is the change of the refractive index, and Δt_1 is the change of thickness due to the applied stress on A.

Expanding Eq. (5.57), we obtain

$$(n' - 1)t_2 - [(n' - 1)(t_1 + \Delta t_1) + \Delta n'(t_1 + \Delta t_1)] = (m + \tfrac{1}{2})\lambda' \qquad (5.58)$$

Neglecting the small dispersion over the narrow range of λ to λ' and subtracting Eq. (5.58) from Eq. (5.56) we obtain

$$(n - 1)(\Delta t_1) + \Delta n t_1 = (m + \tfrac{1}{2})(\lambda - \lambda')$$

[ignoring the factor $(\Delta n)(\Delta t_1)$ in comparison with other terms]. Now substituting $C_\lambda^{\parallel}P$ for Δn, we obtain

$$C_\lambda^{\parallel} = \frac{(m + \frac{1}{2})(\lambda - \lambda') - (n - 1)\,\Delta t_1}{Pt_1}$$

or

$$C_\lambda^{\parallel} = \frac{(m + \frac{1}{2})(\lambda - \lambda')}{Pt_1} - (n - 1)\frac{\Delta t_1}{Pt_1}$$

$$= \frac{(m + \frac{1}{2})(\lambda - \lambda')}{Pt_1} - (n - 1)\frac{\sigma}{Y} \qquad (5.59)$$

where σ and Y are the Poisson's ratio and the Young's modulus of the glass. To get $C_\lambda^{\perp}$ the polarizer P is set normal to the stress.

2. The order of the dark band, m, which enters the calculations can be obtained as follows: Wavelengths of two adjacent dark bands are noted. The necessary condition for a dark band to appear at the wavelength λ is

$$(n - 1)t_2 - (n - 1)t_1 = (m + \tfrac{1}{2})\lambda$$

or

$$(n - 1)(t_2 - t_1) = (m + \tfrac{1}{2})\lambda \qquad (5.60)$$

Similarly for the next dark band to appear at the wavelength λ_1,

$$(n' - 1)(t_2 - t_1) = [(m + 1) + \tfrac{1}{2}]\lambda_1 \qquad (5.61)$$

Putting $n = n'$ within a short range of wavelengths (i.e., ignoring the dispersion within this narrow range) we can write

$$(m + \tfrac{1}{2})\lambda = (m + 1 + \tfrac{1}{2})\lambda_1 \tag{5.62}$$

from which m can be calculated.

5.5. Elliptic Vibrations and Elliptically Polarized Light

5.5.1. Composition of Two Rectangular Vibrations Giving an Ellipse: Use of the Sénarmont Compensator

5.5.1.1. Theoretical Considerations

a. Elliptic Vibration: Simple Case. Two linear vibrations of amplitudes a and b along any two mutually perpendicular directions when compounded with a phase difference $\hat{\delta}$ will yield, in general, an elliptic vibration.

Let A be the amplitude of a linear vibration OP making an angle β with OX (Fig. 5.18). Then A can be resolved into two coherent components, one along OX with the amplitude $a = A \cos \beta$ and the other along OY with the amplitude $b = A \sin \beta$ ($a \neq b$ except when $\beta = 45°$). Let these two vibrations combine with a phase difference $\hat{\delta}$. In the case of propagation of light through a biaxial crystal along its Z axis,

$$\hat{\delta} = 2\pi(n_x - n_y)d/\lambda \tag{5.63}$$

where d is the thickness. For a general discussion of elliptic vibrations, we do not need the relation between $\hat{\delta}$ and the optical parameters of the crystal. Now, let

$$\begin{aligned} x &= a \sin \omega t \\ y &= b \sin(\omega t + \hat{\delta}) \end{aligned} \tag{5.64}$$

The circular frequency ω is the same for both x and y. Both x and y have the same trigonometric function (namely sine), which implies that, if

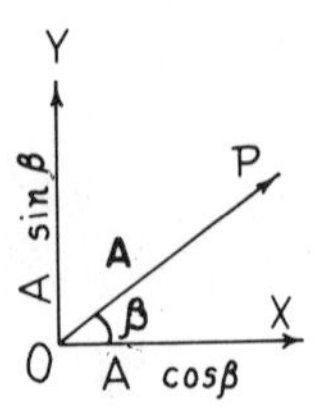

Fig. 5.18. Composition of two linear vibrations acting along two mutually perpendicular directions, and with a phase difference, yielding an elliptic vibration.

$\hat{\delta} = 0$, both the component vibrations are in phase initially at the point of incidence. From Eq. (5.64),

$$\frac{x}{a} = \sin \omega t \qquad \text{and} \qquad \frac{y}{b} = \sin(\omega t + \hat{\delta})$$

$$\begin{aligned} \frac{y}{b} &= \sin \omega t \cos \hat{\delta} + \cos \omega t \sin \hat{\delta} \\ &= \frac{x}{a} \cos \hat{\delta} + \left(1 - \frac{x^2}{a^2}\right)^{1/2} \sin \hat{\delta} \end{aligned}$$

Therefore

$$\left(\frac{y}{b} - \frac{x}{a} \cos \hat{\delta}\right)^2 = \sin^2 \hat{\delta}\left(1 - \frac{x^2}{a^2}\right)$$

$$\frac{y^2}{b^2} + \frac{x^2}{a^2} \cos^2 \hat{\delta} - \frac{2yx}{ba} \cos \hat{\delta} = \sin^2 \hat{\delta} - \sin^2 \hat{\delta}\, \frac{x^2}{a^2}$$

$$\frac{y^2}{b^2} + \frac{x^2}{a^2} - \frac{2yx}{ba} \cos \hat{\delta} = \sin^2 \hat{\delta} \tag{5.65}$$

Equation (5.65) shows that the resultant motion is, in general, an ellipse.

In Eq. (5.65) we shall study the nature of the resultant vibration for different values of $\hat{\delta}$ in some special cases.

1. Let $\hat{\delta} = 0$; then $(y/b - x/a)^2 = 0$. Hence $y/x = b/a$, thus describing a straight line.

2. Let $\hat{\delta} = \pi/2$; then $y^2/b^2 + x^2/a^2 = 1$, thus describing an ellipse of clockwise rotation with semiaxial lengths a and b, respectively, parallel to the X and Y axes.

3. Let $\hat{\delta} = \pi$; then $y/x = -b/a$, which describes again a straight line. Let $\tan \beta_1 = b/a$ in case 1 and $\tan \beta_2 = -b/a$ in case 3. Then $\beta_1 + \beta_2 = \pi$, and the two angles are supplementary.

4. Let $\hat{\delta} = 3\pi/2$. Then $y^2/b^2 + x^2/a^2 = 1$, which is an ellipse of anti-clockwise rotation with its semiaxial lengths a and b, respectively, parallel to the X and Y axes.

5. Proceeding along the lines indicated above, it is easily seen that the resultant vibration takes the following forms: (i) a straight line when $\hat{\delta} = 0, \pi, 2\pi, 3\pi, \ldots, \pm n\pi$; (ii) an ellipse with its major and minor axes parallel to the axes of the reference coordinate system when

$$\hat{\delta} = \frac{\pi}{2}, \frac{3\pi}{2}, \frac{5\pi}{2}, \ldots, \pm(2n + 1)\frac{\pi}{2} \tag{5.66}$$

An additional condition $a = b$ results in a circular vibration. (iii) For any value of $\hat{\delta}$ other than those indicated in (i) and (ii), the resultant vibration is an ellipse whose major and minor axes are not parallel to the axes of the reference coordinate system. It is an oblique ellipse whose axes are tilted with respect to the X and Y axes of the Cartesian system.

b. Elliptic Vibration—General Case; the Tilt and Ratio of Its Axes. Now we shall determine the tilt of the axes of the ellipse with respect to the coordinate axes OX and OY and also the ratio of the semiaxes a_1 and b_1 of the ellipse.

Let P be a point on the ellipse (Fig. 5.19a). It has x, y as coordinates with respect to the original axes OX, OY, and x_1, y_1 as coordinates with respect to the transformed coordinate axes OX_1, OY_1 inclined at an angle θ with respect to the original system OXY. Here the transformed axes are chosen parallel to the axes of the tilted ellipse.

From the geometry of Fig. 5.19b we note that

$$OA = x, \qquad OB = y, \qquad OA_1 = x_1, \qquad OB_1 = y_1$$

Now

$$\begin{aligned} x &= OA = OB' - AB' = x_1 \cos\theta - y_1 \sin\theta \\ y &= OB = AP = AB_2 + B_2P = x_1 \sin\theta + y_1 \cos\theta \end{aligned} \tag{5.67}$$

(Note that these values of x and y can also be obtained in terms of x_1 and y_1 and θ by treating OP as a vector (a tensor of rank one) and hence relating the components in one system to the components in the other system.)

We next transform Eq. (5.65) describing the ellipse in the original system of axes X and Y into an equation describing the same ellipse in the second system X_1 and Y_1 whose axes are parallel to the axes of the ellipse.

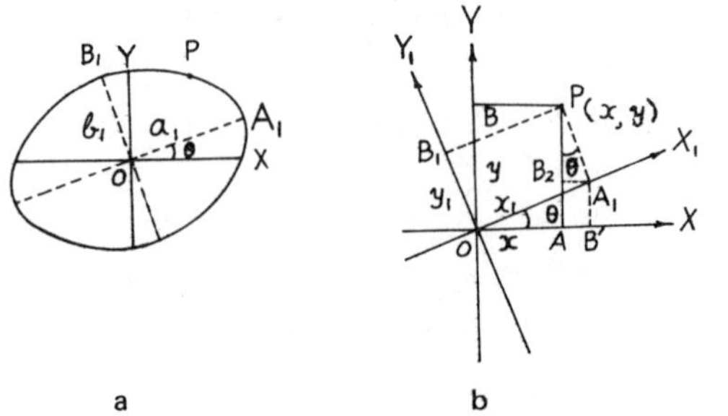

Fig. 5.19. Elliptic vibrations with oblique axes—general case (see text above).

Substituting the values of x and y obtained in terms of x_1 and y_1 and θ from Eq. (5.67), we have Eq. (5.65) (reproduced below)

$$\frac{y^2}{b^2} + \frac{x^2}{a^2} - \frac{2yx}{ba}\cos\hat{\delta} = \sin^2\hat{\delta} \tag{5.65}$$

yielding the following equation:

$$\frac{(y_1\cos\theta + x_1\sin\theta)^2}{b^2} + \frac{(x_1\cos\theta - y_1\sin\theta)^2}{a^2} - \frac{2(y_1\cos\theta + x_1\sin\theta)}{b}\,\frac{(x_1\cos\theta - y_1\sin\theta)}{a}\cos\hat{\delta} = \sin^2\hat{\delta}$$

Therefore

$$\frac{y_1^2\cos^2\theta + x_1^2\sin^2\theta + 2x_1y_1\sin\theta\cos\theta}{b^2} + \frac{x_1^2\cos^2\theta + y_1^2\sin^2\theta - 2x_1y_1\cos\theta\sin\theta}{a_1} - \frac{2(y_1x_1\cos^2\theta + x_1^2\sin\theta\cos\theta - y_1^2\sin\theta\cos\theta - x_1y_1\sin^2\theta)}{ba}\cos\hat{\delta} = \sin^2\hat{\delta}$$

Dividing both sides of the above equation by $\sin^2\hat{\delta}$ and rearranging the terms, we have

$$x_1^2\left\{\frac{\sin^2\theta}{b^2\sin^2\hat{\delta}} + \frac{\cos^2\theta}{a^2\sin^2\hat{\delta}} - \frac{2\sin\theta\cos\theta}{ba\sin^2\hat{\delta}}\cos\hat{\delta}\right\} + y_1^2\left\{\frac{\cos^2\theta}{b^2\sin^2\hat{\delta}} + \frac{\sin^2\theta}{a^2\sin^2\hat{\delta}} + \frac{2\sin\theta\cos\theta}{ba\sin^2\hat{\delta}}\cos\hat{\delta}\right\} + 2\frac{x_1y_1}{\sin^2\hat{\delta}}\left\{\frac{\sin\theta\cos\theta}{b^2} - \frac{\sin\theta\cos\theta}{a^2} - \frac{\cos^2\theta - \sin^2\theta}{ba}\cos\hat{\delta}\right\} = 1 \tag{5.68}$$

Let Eq. (5.68) be expressed as $x_1^2/a_1^2 + y_1^2/b_1^2 = 1$, an ellipse whose semiaxes a_1 and b_1 are, respectively, parallel to OX_1 and OY_1 of the transformed coordinate system. Hence

$$\frac{1}{a_1^2} = \frac{\sin^2\theta}{b^2\sin^2\hat{\delta}} + \frac{\cos^2\theta}{a^2\sin^2\hat{\delta}} - \frac{2\sin\theta\cos\theta}{ab\sin^2\hat{\delta}}\cos\hat{\delta}$$

$$\frac{1}{b_1^2} = \frac{\cos^2\theta}{b^2\sin^2\hat{\delta}} + \frac{\sin^2\theta}{a^2\sin^2\hat{\delta}} + \frac{2\sin\theta\cos\theta}{ab\sin^2\hat{\delta}}\cos\hat{\delta}$$

Also, the coefficient of $(x_1 y_1)$ is equal to zero. Hence

$$\frac{\sin 2\theta}{b^2 \sin^2 \hat{\delta}} - \frac{\sin 2\theta}{a^2 \sin^2 \hat{\delta}} - \frac{\cos \hat{\delta}}{ba \sin^2 \hat{\delta}} 2 \cos 2\theta = 0$$

from which the angle of tilt θ can be obtained thus:

$$\frac{\sin 2\theta}{\cos 2\theta} \left\{ \frac{1}{b^2} - \frac{1}{a^2} \right\} = \frac{2 \cos \hat{\delta}}{ab}$$

Therefore

$$\tan 2\theta = 2 \frac{ab}{a^2 - b^2} \cos \hat{\delta} \tag{5.69}$$

Corollary. Equation (5.65) can be put in the form

$$a_1 x^2 + b_1 y^2 + 2c_1 xy = 1 \tag{5.70}$$

where

$$a_1 = \frac{1}{(a \sin \hat{\delta})^2}, \qquad b_1 = \frac{1}{(b \sin \hat{\delta})^2}, \qquad c_1 = -\frac{\cos \hat{\delta}}{a \sin \hat{\delta}\, b \sin \hat{\delta}}$$

Substituting this into Eq. (5.69) we see that

$$\tan 2\theta = \frac{2c_1}{a_1 - b_1} \tag{5.71}$$

Thus Eq. (5.71) will yield θ in terms of a_1, b_1, c_1 when Eq. (5.70) represents the ellipse.

From Eq. (5.65) we can draw the following conclusions: If $\hat{\delta} = \pi/2$, $3\pi/2$, $(2n+1)\pi/2$, then $\cos \hat{\delta} = 0$. Therefore $\tan 2\theta = 0$ and hence $\theta = 0$. Thus there is no tilt of the axes of the ellipse with respect to the reference coordinate axes.

Let $\hat{\delta} = (2n + 1)(\pi/2)$ and $a = b$; then $\tan 2\theta$ is indeterminate. This is the case of circular vibration.

Let $a = b$ and let $\hat{\delta}$ have any value other than $(2n + 1)(\pi/2)$. Then $\tan 2\theta = \infty$, so that for all values of $\hat{\delta}$ other than $(2n + 1)(\pi/2)$ the value of θ is always 45°.

In Eq. (5.65) let $a = b$ (and hence $\theta = 45°$). Then

$$\frac{1}{{a_1}^2} = \frac{\sin^2 \theta}{b^2 \sin^2 \hat{\delta}} + \frac{\cos^2 \theta}{a^2 \sin^2 \hat{\delta}} - \frac{2 \sin \theta \cos \theta}{ab \sin^2 \hat{\delta}} \cos \hat{\delta}$$

reduces to

$$\frac{1}{a_1^2} = \frac{\sin^2\theta}{a^2\sin^2\hat{\delta}} + \frac{\cos^2\theta}{a^2\sin^2\hat{\delta}} - \frac{\sin 2\theta}{a^2\sin^2\hat{\delta}}\cos\hat{\delta} = \frac{(1-\cos\hat{\delta})}{a^2\sin^2\hat{\delta}}$$

and similarly,

$$\frac{1}{b_1^2} = \frac{\cos^2\theta}{a^2\sin^2\hat{\delta}} + \frac{\sin^2\theta}{a^2\sin^2\hat{\delta}} + \frac{2\sin\theta\cos\theta}{a^2\sin^2\hat{\delta}}\cos\hat{\delta}$$

reduces to

$$\frac{1}{b_1^2} = \frac{1+\cos\hat{\delta}}{a^2\sin^2\hat{\delta}}$$

Therefore

$$\frac{b_1^2}{a_1^2} = \frac{1-\cos\hat{\delta}}{1+\cos\hat{\delta}} = \tan^2\frac{\hat{\delta}}{2}$$

$$\frac{b_1}{a_1} = \tan\frac{\hat{\delta}}{2} \tag{5.72a}$$

If ϕ describes the ellipticity, then $\tan\phi = b_1/a_1$, the ratio of the semiaxes of the ellipse. Hence

$$\tan\phi = \tan\frac{\hat{\delta}}{2} \tag{5.72b}$$

or

$$2\phi = \hat{\delta} \tag{5.73}$$

This is a theoretical result of considerable importance and will be applied to the Sénarmont compensator and photometric methods. Here the phase retardation $\hat{\delta}$ is determined by measuring the ratio of the axes, or vice versa.

5.5.1.2. The Sénarmont Compensator

We describe the experimental procedure for determining birefringence of a specimen using the principle of the Sénarmont compensator. A polarizing microscope is used in the experiment. A plane wave front of monochromatic light *ML* is incident on the polarizer *P* (Fig. 5.20). The analyzer *A* is crossed with respect to *P*.

The specimen crystal plate *C* cut with its axes, say *X* and *Z*, lying in its face is placed on the microscope stage, and the stage along with the

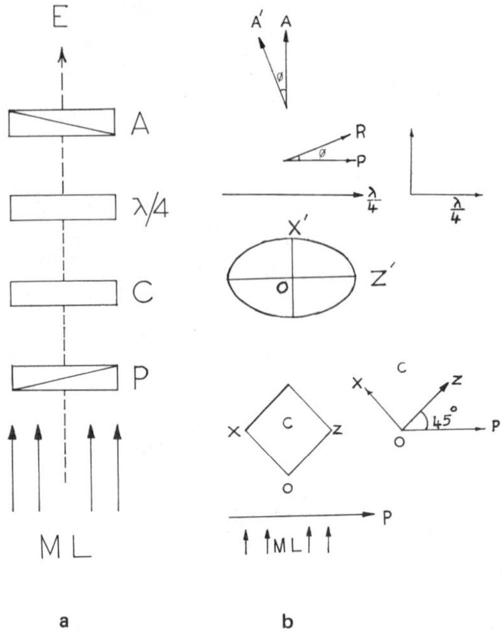

Fig. 5.20. Sénarmont compensator. (a) Optical components and (b) alignment of the relative axes. *P*, polarizer; *A*, analyzer; *C*, specimen crystal plate; *ML*, monochromatic light; $\lambda/4$, quarter-wave plate; *E*, eye or phototube.

crystal *C* is rotated in its own plane (i.e., about the line of sight along the axis of the polarizing microscope) until extinction occurs. The crystal *C* in this position has its vibration directions parallel to those of *P* and *A*. The position of the analyzer *A* is read.

The stage along with the crystal *C* is rotated through 45° in its own plane. Then the light vibrations of amplitudes *a* and *b*, respectively, parallel to the *X* and *Z* axes of the crystal *C* become equal; i.e., $a = b$. Consequently the emergent ellipse from the specimen crystal *C* is rotated through an angle of 45° with respect to the *X* and *Z* axes of the crystal *C*, according to the relation

$$\tan 2\theta = \frac{2ab}{a^2 - b^2} \cos \delta \tag{5.69}$$

thus bringing its axes parallel and perpendicular to those of the polarizer. Here $\theta = 45°$ whatever be the phase difference δ, which is given by the relation

$$\delta = 2\pi(n_z - n_x)d/\lambda$$

Now a $\lambda/4$ plate is placed with one of its axes parallel to that of the polarizer *P*, thereby making the axes of the $\lambda/4$ plate parallel to the axes of the ellipse emergent from *C*. Therefore the emergent ellipse is reduced to a straight line by the $\lambda/4$ plate, the straight line R making an angle ϕ with P.

Now the analyzer A is rotated until extinction occurs. Let the angle of rotation of the analyzer A be ϕ so as to produce a dark field. Then from Eq. (5.72b), $\tan\phi = \tan(\hat{\delta}/2)$, where $\hat{\delta}$ is the phase difference produced by the specimen. The birefringence $(n_z - n_x)$ of the crystal plate C is calculated from a knowledge of $\hat{\delta}$ and the thickness d of the plate.

The above experimental procedure using a polarizing microscope (Fig. 5.20) is usually adopted by mineralogists [477]. This optical method was followed by Rodgers [1062] to study the Pockels effect in hexamethylenetetramine, also called hexamine. However, the same experiment can be set up with equal facility on an optical bench; for example, Lee [721] used the latter method in order to study the linear electro-optic behavior of hexamethylenetetramine crystals.

5.5.2. *Photometric Method for the Measurement of Photoelastic Birefringence*

A photometric method has been developed by Robertson [1057] for the measurement of the photoelastic birefringence of a photoelastic model of CR-39 in the form of a circular disc.

Plane-polarized light is incident on a double-refracting medium (double refraction being caused by a mechanical stress acting on the photoelastic model). The amplitudes of vibration of the two linearly vibrating components are a and b, respectively, parallel to OU and OV, the directions of the principal stresses (Fig. 5.21). These two components are transmitted by the medium with a phase difference $\hat{\delta}$, and at the point of emergence from the crystal may be represented by $u = a\cos(\omega t + \hat{\delta})$ and $v = b\cos\omega t$, respectively; consequently they combine into an ellipse whose axes are in general tilted with respect to the principal stresses, or in the case of a naturally double-refracting crystal, inclined to the crystallographic axes.

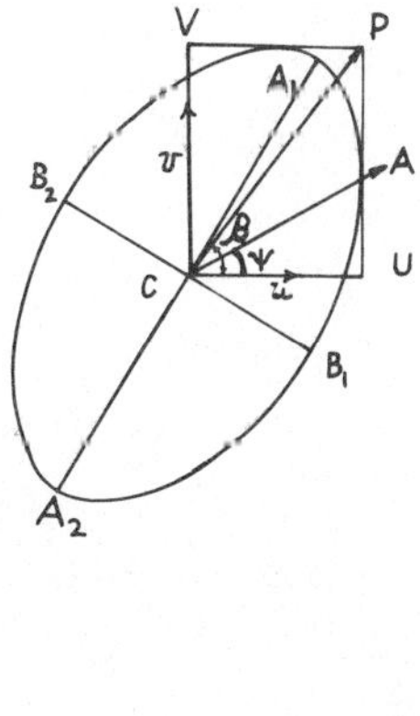

Fig. 5.21. Photometric method of determining stress birefringence: geometric representation of the principle (Robertson's method). Incident vibrations u along OU and v along OV. P, polarizer; A, analyzer; P, vibration direction of the polarizer; A, vibration direction of the analyzer. From [1057].

The elliptically polarized light is then viewed through an analyzer whose vibration direction makes an angle ψ with OU.

The resultant vibration of the light transmitted by the analyzer is given by

$$u_1 = u \cos\psi + v \sin\psi \tag{5.74}$$

$$= a\cos(\omega t + \hat{\delta})\cos\psi + b\cos\omega t\sin\psi$$

$$= (a\cos\psi\cos\hat{\delta} + b\sin\psi)\cos\omega t - a\cos\psi\sin\hat{\delta}\sin\omega t \tag{5.75}$$

Now we can express the above relation as

$$u_1 = a_1\cos(\omega t + \Delta) \tag{5.76}$$

where the amplitude a_1 can be obtained thus: Equating the like coefficients of $\cos\omega t$ in Eqs. (5.75) and (5.76), and similarly those of $\sin\omega t$, we obtain

$$a_1\cos\Delta = a\cos\psi\cos\hat{\delta} + b\sin\psi$$

$$a_1\sin\Delta = a\cos\psi\sin\hat{\delta}$$

$$a_1^2 = a^2\cos^2\psi\cos^2\hat{\delta} + b^2\sin^2\psi + 2ab\cos\psi\sin\psi\cos\hat{\delta} + a^2\cos^2\psi\sin^2\hat{\delta}$$

$$= a^2\cos^2\psi + b^2\sin^2\psi + ab\sin 2\psi\cos\hat{\delta} \tag{5.77}$$

Therefore

$$a_1 = \{a^2(1+\cos 2\psi) + b^2(1-\cos 2\psi) + 2ab\sin 2\psi\cos\hat{\delta}\}^{1/2}/2^{1/2} \tag{5.78}$$

Let

$$\frac{a^2 - b^2}{[(a^2-b^2)^2 + 4a^2b^2\cos^2\hat{\delta}]^{1/2}} = \sin 2\alpha$$

and

$$\frac{2ab\cos\hat{\delta}}{[(a^2-b^2)^2 + 4a^2b^2\cos^2\hat{\delta}]^{1/2}} = \cos 2\alpha$$

so that

$$\tan 2\alpha = \frac{a^2 - b^2}{2ab\cos\hat{\delta}} \tag{5.79}$$

Hence from Eq. (5.78)

$$a_1 = [(a^2+b^2) + (a^2-b^2)\cos 2\psi + 2ab\sin 2\psi\cos\hat{\delta}]^{1/2}/2^{1/2}$$

$$= \{(a^2+b^2) + [(a^2-b^2)^2 + 4a^2b^2\cos^2\hat{\delta}]^{1/2}\sin(2\alpha + 2\psi)\}^{1/2}/2^{1/2} \tag{5.80}$$

We note that a_1 is maximum if $\sin(2\alpha + 2\psi)$ has the maximum value ($= 1$) and a_1 is minimum if $\sin(2\alpha + 2\psi)$ has the minimum value ($= -1$). Thus

$$(a_1)_{\max} = \{(a^2 + b^2) + [(a^2 - b^2)^2 + 4a^2b^2 \cos^2 \delta]^{1/2}\}^{1/2}/2^{1/2} \tag{5.81a}$$

and

$$(a_1)_{\min} = \{(a^2 + b^2) - [(a^2 - b^2)^2 + 4a^2b^2 \cos^2 \hat{\delta}]^{1/2}\}^{1/2}/2^{1/2} \tag{5.81b}$$

Then $I_{\max}$ and $I_{\min}$ (the maximum and minimum intensities of the light transmitted when the analyzer is parallel to the major and minor axes of the ellipse, respectively) are proportional to $(a_1)^2_{\max}$ and $(a_1)^2_{\min}$. Now, if $\beta = 45°$, thus setting the polarizer with its vibration direction inclined at 45° to the direction of the principal stress (or to the principal axes of the crystal plate), then $a = b$. Hence

$$I_{\max} \propto \{(a^2 + b^2) + [(a^2 - b^2)^2 + 4a^2b^2 \cos^2 \hat{\delta}]^{1/2}\} \propto 2a^2(1 + \cos \hat{\delta}) \tag{5.82}$$

and

$$I_{\min} \propto \{(a^2 + b^2) - [(a^2 - b^2)^2 + 4a^2b^2 \cos^2 \hat{\delta}]^{1/2}\} \propto 2a^2(1 - \cos \hat{\delta}) \tag{5.83}$$

$$I_{\max} - I_{\min} \propto 4a^2 \cos \hat{\delta} \tag{5.84}$$

Now $4a^2$ represents the total intensity I (incident on the photoelastic model or birefringent crystal plate), being equal to the sum of $I_{\max}$ and $I_{\min}$ (which are themselves the intensities transmitted by the analyzer in the two positions perpendicular to each other). Therefore

$$\frac{I_{\max} - I_{\min}}{I} = \frac{4a^2 \cos \hat{\delta}}{4a^2} = \cos \hat{\delta} \tag{5.85}$$

In an actual experiment with the polarizer at 45° with OU the analyzer is set for the maximum and the minimum of transmitted intensities. Let the ratio of $I_{\max}$ to $I_{\min}$ be R. Then

$$\cos \hat{\delta} = \frac{I_{\max} - I_{\min}}{I_{\max} + I_{\min}} = \frac{R - 1}{R + 1} \tag{5.86}$$

from which $\cos \hat{\delta}$ and hence $\hat{\delta}$ are calculated.

Differentiating Eq. (5.78), we obtain the condition for a_1 to become a maximum or a minimum, which happens when $da_1/d\psi = 0$. Thus $a_{\max}$

or $a_{\min}$ occurs when

$$-2a^2 \sin\psi \cos\psi + 2b^2 \sin\psi \cos\psi + 2ab \cos 2\psi \cos\hat{\delta} = 0$$

i.e., when

$$\tan 2\psi = \frac{2ab}{a^2 - b^2} \cos\hat{\delta} \tag{5.87}$$

If $\beta = 45°$, then $a = b$, and consequently $\tan 2\psi = \infty$ and hence $\psi = 45°$. Thus for all values of $\hat{\delta}$ other than zero, the principal axes of the emerging ellipse are tilted at an angle of 45° to the directions of the principal stresses or the axes of the birefringent crystal plate under observation.

It may be noted that Eq. (5.87) has already been derived on page 250 in discussing the general case of elliptic vibrations and the theory of the Sénarmont compensator.

5.5.3. The Poincaré Sphere and Its Application to the Study of the Photoelastic Behavior of Optically Active Crystals

5.5.3.1. General Description and Theory

Light waves can be polarized in three different ways: (a) linear, (b) elliptic, and (c) circular. The most general case is elliptic polarization, the linear and circular ones being the two extreme cases.

An elliptically polarized beam of monochromatic light can be described by two quantities: (a) the orientation of the major axis of the ellipse, specified by the angle λ which the major axis of the ellipse makes with a given direction, and (b) the ratio of the semiaxes a, b of the ellipse, given by b/a, this ratio being defined in terms of an angle η, where $\tan\eta = b/a$. These two angles λ and η are called, respectively, the azimuthal and extinction angles of the elliptic vibration. Another important characteristic of an elliptic vibration is its direction of rotation—left- or right-handed, also known as clockwise or anticlockwise, describing the sense of rotation.

The concept of the optical-index ellipsoid was developed by Fresnel in order to explain the phenomenon of double refraction exhibited by crystals. Similarly, Poincaré developed the concept of a spherical surface in order to represent the state of polarization by a specific point on a sphere and thereby to study and predict the effect of optical retarders and polarizers on a beam of polarized light. The Poincaré sphere is a very simple and instructive device in which the surface of a sphere of unit radius acts as a map; each point on the surface or map uniquely signifies a definite state of polarization. Any problem involving the effect of an optical retarder or rotator on a monochromatic polarized beam of light is solved by navigating on the surface of the sphere. The navigation is always along the arc of a circle on the Poincaré sphere, thus making the process of computation very simple.

Let us consider the Poincaré sphere [960]. Let $HCVD$ be the equator and L_nHR_s the zero longitude of the sphere (Fig. 5.22a). Then the points

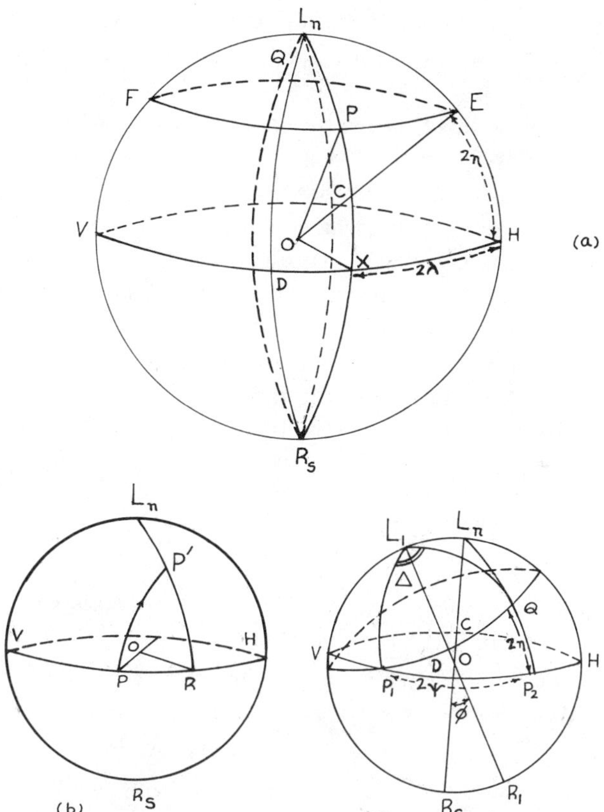

Fig. 5.22. Poincaré sphere. (a) The state of polarization of the point P is specified in terms of the azimuthal angle 2λ (measured clockwise), and the extinction angle 2η (measured upward from the equator); (b) use of the Poincaré sphere in predicting the effect of a linear retarder R (with its fast axis at $\lambda = \frac{1}{2}$ arc HR) on a linearly polarized incident beam P; the state of polarization of the emergent beam is indicated by the point P'; (c) Poincaré sphere method of computing the state of polarization of the light beam emerging from a crystal having both optical activity and birefringence (see text above). From [998].

H and V represent horizontally and vertically vibrating, and thus linearly polarized, beams of light; both are on the equator and diametrically opposite to each other and hence at an angle of π rad apart. Any two diametrically opposite points represent an orthogonal pair of polarization states. The north and south poles are L_n and R_s, respectively, and represent left and right circular vibrations.

Each point on the equator $HCVD$ represents a different state of linear polarization or plane of vibration, implying a different vibration direction; its vibration direction makes with the horizontal an angle equal to half of the angle indicated by the longitude at the point on the sphere. Thus the points C and D, which are $\pi/2$ from H and V, respectively (taken in the positive direction), correspond to linear vibrations at $+\pi/4$ and $-\pi/4$ with the horizontal. Points between the equator and the south pole represent right elliptical polarization and those above the equator represent left elliptical polarization. All points on the same longitude (say, L_nPR_s) correspond to elliptic vibrations having their major axes inclined at the same angle λ with the horizontal. All points on the same latitude 2η (say, $EQFP$) represent elliptic vibrations having the same axial ratio ($b/a = \tan \eta$). Thus a point P (of longitude 2λ and latitude 2η) represents an elliptic vibration of azimuth λ (i.e., the orientation of its major axis with the horizontal), and ellipticity η (i.e., the ratio of its axes $b/a = \tan \eta$). Thus the extinction angle η defines the ellipticity.

Application to Retarders. Using the device of the Poincaré sphere, we find the effect of a retarder on a polarized light beam as follows (Fig. 5.22b):

1. Mark on the sphere the point P that represents the polarization state of the incident light beam.
2. Then mark the point R that defines the fast axis of the retarder, and draw the appropriate arc PP'.
3. The axis of the arc PP' is the radius vector from the center of the sphere to the point R, and the starting point of the arc is point P.
4. The magnitude of the arc PP' in degrees gives the retardation $\hat{\delta}$ of the retarder in degrees.
5. The arc is always to be drawn in the clockwise direction as judged by an observer situated outside the sphere on an extension of the radius vector through R.
6. The final point P' of the arc defines the state of polarization of the light emerging from the retarder.

The method is universally applicable. The retarder may be of linear, circular, or elliptic type, and the polarization of the incident beam may have any form. In every case the problem is solved by an identical procedure. It may be noted that no attention need be paid to amplitudes, intensities, or phases.

When there are several retarders in series, the end point of one arc serves as the starting point for the next. This means that we do not have to interpret the end point of each arc resulting from each rotator or retarder in terms of its ellipticity, slope of the major axis, and the sense of rotation. Using the Poincaré sphere to deal with several retarders in series is like using a computer to carry out several mathematical operations in succession; the intermediate values may be completely ignored. It is only the end result that is of importance and hence to be recorded.

Let us now consider the case of plane-polarized light propagated through an optically active, birefringent crystal. Let the point P_1 on the Poincaré sphere (Fig. 5.22c) represent the plane-polarized light incident on the crystal of thickness t. For an optically active crystal with no birefringence, the emergent light will remain linearly polarized and be represented by the point P_2 obtained by a rotation about the polar axis L_nR_s along the equator $HCVD$ through an angle $2\varrho t$, where ϱ is the specific rotation per unit thickness of the crystal. If the crystal is birefringent but not optically active, the point P_1 is rotated about the axis HV along the longitude circle $L_nP_1R_s$. Under the combined influence of birefringence and optical activity, the state of polarization of the light emergent from the crystal would be represented by a point Q. Here the point P_1 can be considered to have been rotated through an angle Δ about the axis L_1R_1 of the Poincaré sphere, where Δ, the rotation for $t = 1$ cm, is given by

$$\Delta = [\hat{\delta}^2 + (2\varrho)^2]^{1/2} \tag{5.88}$$

The axis L_1R_1 makes an angle ϕ with L_nR_s, where $\tan\phi = \hat{\delta}/2\varrho$; here $\hat{\delta}$ is the birefringence of the crystal. The angle ψ made by the major axis of the emergent ellipse with the vibration direction of the incident plane polarized light and the ellipticity η are then determined using a Babinet compensator or any other usual analyzer of polarized light. Then ϕ and Δ can be calculated from a knowledge of ψ and η.

If the incident plane-polarized light has its plane of vibration inclined at an angle α with one of the principal vibration directions of the crystal, then ϕ and Δ, which relate $\hat{\delta}$ and 2ϱ, can be expressed in terms of the experimentally determinable quantities ψ and η as [998; 1001; 1003]

$$\tan \phi = \frac{\sin 2\eta}{\cos 2\alpha - \cos 2\eta \cos 2(\alpha + \psi)}$$
$$\cos \Delta = 1 - \frac{1 - \cos 2\eta \cos 2\psi}{1 - \sin^2 \phi \cos^2 2\alpha} \tag{5.89}$$

In the experimental investigations on photoelastic and electro-optic effects the orientation of the experimental crystal is known. So α can be given any chosen value. Thus if $2\alpha = 0$, Eqs. (5.89) reduce to

$$\tan \phi = \frac{\sin 2\eta}{1 - \cos 2\eta \cos 2\psi}$$
$$\sin \Delta = \frac{\cos 2\eta \sin 2\psi}{\cos \phi} \tag{5.90}$$

If $2\alpha = \pi/2$, we have

$$\tan \phi = \tan 2\eta / \sin 2\psi \tag{5.91}$$

and

$$\cos \Delta = \cos 2\eta \cos 2\psi \tag{5.92}$$

Equations (5.91) and (5.92) are easy to handle for the experimental determination of ψ and Δ and hence of $\hat{\delta}$ and 2ϱ.

5.5.3.2. *Measurement of ψ and η*

The polarizer P is set at an angle of $+45°$ with the stress direction of the crystal. The angle of rotation ψ of the major axis of the elliptically polarized light emergent from the crystal is determined using a half-shade.

The ratio of the axes of the ellipse can be obtained using a Babinet compensator by an indirect method [998]. From page 251 we see that when the polarizer is set at 45° with the vibration direction of the crystal, then $a = b$ and whatever be the phase retardation produced by the crystal, the emergent ellipse is tilted and its ellipticity η is equal to half the phase retardation. If a_1 and b_1 are the semimajor and semiminor axes of the tilted ellipse, then, by definition, $\tan \eta = b_1/a_1$. Hence in order to determine η, the ellipticity of the emergent ellipse, a Babinet compensator is inserted in the path of the emergent light, the axes of the Babinet compensator making 45° with the axes (major and minor) of the ellipse. The shift s of the fringes and their fringe width f are determined. Then 2η = phase retardation $= 2\pi(s/f)$.

The experiment can be repeated with the polarizer set at other convenient angles.

From a knowledge of ψ and η, one can calculate ϕ and Δ using Eqs. (5.91) and (5.92) and hence $\hat{\delta}$ and 2ϱ from the following relations:

$$\hat{\delta} = \Delta \sin \phi \qquad \text{and} \qquad 2\varrho = \Delta \cos \phi \tag{5.93}$$

Another experimental method applicable particularly to optically active cubic crystals for determining their photoelastic behavior is the elliptic-analyzer method [998]. It is essential here that the principal directions of vibration for the stressed crystal be known. The experimental crystal is placed between two crossed quarter-wave plates whose axes are oriented parallel to the axes of the stressed crystal specimen. A polarizing Nicol P allows plane-polarized light to be incident on the first quarter-wave plate, while the light emergent from the second quarter-wave plate is observed through an analyzing Nicol A; however, P and A are mounted together so that they are crossed and can be rotated together in the crossed position. For different settings of the crossed Nicols, the polarizer P and the first quarter-wave plate Q_1 together transmit elliptically polarized light of different axial ratios, and in the absence of the crystal specimen the light is completely stopped by the second $\lambda/4$ plate Q_2 and analyzer A. With the optically active, stressed crystal between the two quarter-wave plates, the elliptic vibrations incident on the crystal are, in general, altered, and hence the light is not extinguished. However, for a particular position of the polarizer P and analyzer A, light is completely stopped, and this happens when the incident light is represented by the points L_1 or R_1 of the Poincaré sphere in Fig. 5.22c. In this position of P, Q_1 the elliptically polarized light incident on the crystal specimen C is propagated unchanged through the crystal and hence stopped by the elliptic analyzer combination Q_2, A. For different positions of L_1 the ratio $\hat{\delta}/2\varrho$ changes (since $\hat{\delta}/2\varrho = \tan \phi$) and the ellipticity of the light from the P, Q_1 combination changes; but for all the positions of the P, Q_1 combination all the transmitted elliptically polarized beams will have their axes parallel to H and V. Hence in an actual experiment one determines the ratio of the axes b/a of the ellipse that is propagated without change through the crystal. Then ϕ can be calculated from the relation

$$\tan \frac{1}{2}\left(\frac{\pi}{2} - \phi\right) = \frac{b}{a} \tag{5.94}$$

and hence $\hat{\delta}$ is known from the relation $\tan \phi = \hat{\delta}/2\varrho$.

5.6. Ultrasonic Methods of Studying the Elasto-Optic Behavior of Crystals

5.6.1. Introduction

Studies on the photoelastic behavior of solids consist of determining the elasto-optic and piezo-optic constants (the p_{ij} and q_{kl} in Pockels' scheme) for the given solid, amorphous or crystalline. Experimentally one determines the very small changes in refractive index in different directions when a stress of known magnitude is applied on crystal specimens of known orientations. The bulk of the results obtained for p_{ij} and q_{kl} is based on the optical measurements made on crystals under static load; for the optical measurements one uses a Babinet compensator or its modification, a Soleil–Babinet compensator, in order to determine the relative path retardation, and an interferometer of the Jamin or Rayleigh type to determine the absolute path retardations. It is well known that the accuracy of the results obtained by the Babinet compensator method is usually 5–10%, whereas the interferometric methods often yield results differing on occasion by as much as even 50% from the Babinet measurements. The discovery of ultrasonics by Langevin in 1918 [716] during the First World War and the classical experiments of Debye and Sears [286] and of Lucas and Biquard [746; 747] in 1932 on the diffraction of light by ultrasonically excited transparent liquids and solids led Bergmann and Fues [108] in 1936 to make experimental investigations on the photoelastic behavior of glasses by observing the intensities of the diffraction spectra and their polarization characters.

In the case of an isotropic solid, Neumann [871] described its photoelastic behavior by means of two strain-optical constants p and q. In Pockels' scheme for an isotropic solid, the constants p_{ij} are reduced to only two, p_{11} and p_{12}. Neumann's constants p and q relate a mechanical strain to the changes which it produces in the index of refraction n for light polarized parallel and normal, respectively, to the direction of the strain. The relation between the change Δn in the index of refraction and a typical strain ε_{zz} is given by

$$\begin{aligned} \Delta n_z &= n_z - n = -n^2 q \varepsilon_{zz} \\ \Delta n_x &= n_x - n = -n^2 p \varepsilon_{zz} \end{aligned} \tag{5.95}$$

The Pockels constants p_{ij} and the Neumann constants p and q for an isotropic solid are related by

$$p_{11} = 2q/n \qquad \text{and} \qquad p_{12} = 2p/n \tag{5.96}$$

The photoelastic behavior represented by the elasto-optic ratio p/q was first studied by Bergmann and Fues (1936) using a point source of light. The situation improved greatly when Hiedemann and Hoesch [513–517] showed how easily line diffraction patterns could be obtained. The theory of diffraction of light in solids given by Nath and Mueller (1938) [861] was experimentally verified by Hiedemann (1938) [522]. Subsequently, Mueller (1938) [829] suggested in detail three different methods of evaluating the elasto-optic ratio p/q of glasses and p_{ij}/p_{kl} of cubic crystals by studying the intensities and polarization characters of the diffraction spectra. Mueller's ultrasonic methods in one form or another have been subsequently applied successfully to glasses and crystals (cubic and noncubic) by a number of workers. It may be pointed out that ordinarily the ultrasonic methods yield results to the same order of accuracy as the Babinet compensator method; hence by the combination of these two sets of measurements, optical and ultrasonic, the absolute values of p_{ij} and q_{kl} can be obtained more accurately than by using the data obtained by interferometric methods alone.

Such a combination of dynamic and static experiments in order to calculate the individual values of p_{ij} assumes that the values of p_{ij} which occur in the dynamic experiments are the same as those in static experiments. Schaefer and Dransfeld [1092] used a dynamic method to measure p/q of the same set of glasses numbering about 150, for which Schaefer and Nassenstein [1091] determined the values of p and q by static methods. Comparison of the values of p/q obtained by the two methods yield an average difference of 1.5%, with the dynamic values averaging slightly higher. Since the authors claim an overall accuracy of only 5%, this slight difference of 1.5% seems to be quite insignificant, and we may safely assume hereafter that the values obtained by the two methods, dynamic and static, remain unaffected.

5.6.2. *Mueller's Theory*

We describe below a simplified version of Mueller's theory for determining the elasto-optic ratios by ultrasonic methods.

Mueller's theory is an extension of that of Raman and Nath [1007–1011] which explains the diffraction of light by ultrasonic waves on the basis of the variation of the refractive index of the medium traversed by the sound waves.

If a mechanical disturbance is propagated in an arbitrary direction in a crystal, then, in general, it will be in the form of quasilongitudinal and quasitransverse waves. In our present investigations, we are interested only in the purely longitudinal waves and the diffraction patterns they produce. Therefore, we consider only waves propagated along certain specific directions of cubic crystals.

1. Consider a longitudinal wave propagated along the Z axis of a cubic crystal of $m3m$ (O_h) class such as NaCl. This wave produces longitudinal strains along the Z axis only. We shall deliberately avoid the formation of shear waves in the crystal; hence their influence on the optical ellipsoid is eliminated. For the particular direction of observation parallel to the Y axis, before the crystal is strained the cross section of the optical ellipsoid is represented by

$$B_{11}^0 x^2 + B_{33}^0 z^2 = 1 \tag{5.97}$$

Under the influence of the longitudinal waves (and of the consequent longitudinal strains), every volume element of the crystal becomes birefringent. Hence, in general, the cross section of the optical ellipsoid now becomes

$$B_{11}x^2 + B_{33}z^2 + 2B_{31}zx = 1 \tag{5.98}$$

Then according to the laws of photoelasticity due to Pockels the polarization constants B_{ij} for NaCl are given by

$$\begin{aligned} B_{11} - B_{11}^0 &= B_1 - B_1{}^0 = p_{11}\varepsilon_1 + p_{12}\varepsilon_2 + p_{12}\varepsilon_3 \\ B_{33} - B_{33}^0 &= B_3 - B_3{}^0 = p_{12}\varepsilon_1 + p_{12}\varepsilon_2 + p_{11}\varepsilon_3 \\ B_{31} &= B_5 = p_{44}\varepsilon_5 \end{aligned} \tag{5.99}$$

For a strain in an arbitrary direction the ellipse (5.98) has its axes in the directions which make angles θ and $\theta + 90$ with the Z axis, where $\tan 2\theta = 2B_{31}/(B_{11} - B_{33})$. These directions do not vary with time and are the same for every volume element.

For a longitudinal wave along the Z axis the only strain that remains is ε_3. Therefore

$$\begin{aligned} B_{11} - B_{11}^0 &= p_{12}\varepsilon_3 \\ B_{33} - B_{33}^0 &= p_{11}\varepsilon_3 \end{aligned} \tag{5.100}$$

thus avoiding the tilt of axes of the optical ellipsoid.

a. Let a longitudinal progressive wave propagated in the Z direction be represented by

$$\xi_z = A\cos(\omega t - k^* z) \tag{5.101}$$

Here ω and k^* refer to the sound field in the medium, ω being the cyclic frequency (equal to $2\pi\nu^*$, where ν^* is the frequency of the piezoelectric transducer) and k^* the acoustic wave vector (equal to $2\pi/\lambda^*$, where λ^* is the acoustic wavelength in the medium). Obviously $\omega = k^*V^*$, where V^* is the acoustic wave velocity in the medium.

Then the longitudinal strain is given by

$$\varepsilon_z = \varepsilon_{zz} = \frac{\partial \xi_z}{\partial z} = +k^*A \sin(\omega t - k^*z) \tag{5.102}$$

b. For a longitudinal standing wave along the Z axis, we can write

$$\xi_z = A \cos \omega t \cos k^*z \tag{5.103}$$

and the longitudinal strain is

$$\varepsilon_z = \varepsilon_{zz} = \frac{\partial \xi_z}{\partial z} = -k^*A \cos \omega t \sin k^*z \tag{5.104}$$

From Eqs. (5.100) and (5.102) we have, for a progressive wave,

$$\begin{aligned} B_{11} &= B_{11}^0 + p_{12}\varepsilon_3 = B_{11}^0 + p_{12}\varepsilon_z \\ &= 1/(n_{11}^0)^2 + p_{12}k^*A \sin(\omega t - k^*z) \\ &= 1/(n_{11}^0)^2 + i_{\mathrm{I}} \sin(\omega t - k^*z) \end{aligned} \tag{5.105}$$

and similarly,

$$\begin{aligned} B_{33} &= 1/(n_{33}^0)^2 + p_{11}\varepsilon_3 \\ &= 1/(n_{33}^0)^2 + p_{11}k^*A \sin(\omega t - k^*z) \\ &= 1/(n_{33}^0)^2 + i_{\mathrm{III}} \sin(\omega t - k^*z) \end{aligned} \tag{5.106a}$$

where

$$i_{\mathrm{I}} = p_{12}k^*A \qquad \text{and} \qquad i_{\mathrm{III}} = p_{11}k^*A \tag{5.106b}$$

For a standing wave, from Eqs. (5.100) and (5.104), we have

$$\begin{aligned} B_{11} &= 1/(n_{11}^0)^2 + p_{12}\varepsilon_z \\ &= 1/(n_{11}^0)^2 - p_{12}k^*A \cos \omega t \sin k^*z \\ &= 1/(n_{11}^0)^2 - i_{\mathrm{I}} \cos \omega t \sin k^*z \end{aligned} \tag{5.107}$$

and similarly,

$$\begin{aligned} B_{33} &= 1/(n_{33}^0)^2 + p_{11}\varepsilon_z \\ &= 1/(n_{33}^0)^2 - p_{11}k^*A \cos \omega t \sin k^*z \end{aligned} \tag{5.108}$$

Here, too,

$$i_{\mathrm{I}} = p_{12}k^*A \qquad \text{and} \qquad i_{\mathrm{III}} = p_{11}k^*A$$

Thus for both progressive and stationary longitudinal waves, the values of i_{I} are the same, and similarly for i_{III}; here i represents the maximum change in $1/n^2$ produced by the ultrasonic waves. In either case

$$i_{\mathrm{III}}/i_{\mathrm{I}} = p_{11}/p_{12} \tag{5.109}$$

Now if μ_{I} and μ_{III} are the amplitudes of the refractive-index variations along the X and Z directions, respectively, then (putting B's in both one- and two-suffix notations) the amplitude of ΔB_{ij} is

$$\Delta B_1 = \Delta B_{11} = \frac{1}{(n_{11}^0 + \mu_{\mathrm{I}})^2} - \left(\frac{1}{n_{11}^0}\right)^2 \approx \frac{2\mu_{\mathrm{I}}}{(n_{11}^0)^3} \tag{5.110a}$$

and similarly,

$$\Delta B_3 = \Delta B_{33} = \frac{1}{(n_{33}^0 + \mu_{\mathrm{III}})^2} - \left(\frac{1}{n_{33}^0}\right)^2 \approx \frac{2\mu_{\mathrm{III}}}{(n_{33}^0)^3} \tag{5.110b}$$

Comparing Eqs. (5.105)–(5.108) with (5.110a) and (5.110b), we obtain

$$\mu_{\mathrm{I}} = [(n_{11}^0)^3 i_{\mathrm{I}}]K = (n_x{}^3 i_{\mathrm{I}})K \tag{5.111a}$$

$$\mu_{\mathrm{III}} = [(n_{33}^0)^3 i_{\mathrm{III}}]K = (n_z{}^3 i_{\mathrm{III}})K \tag{5.111b}$$

where K is a constant at any particular time instant t and distance Z. Hence

$$\mu_{\mathrm{III}}/\mu_{\mathrm{I}} = (n_z{}^3/n_x{}^3)(i_{\mathrm{III}}/i_{\mathrm{I}}) = p_{11}/p_{12} = R \tag{5.111c}$$

for both progressive and stationary longitudinal waves.† We have assumed here that $n_x = n_z$, which holds for cubic crystals.

Furthermore, these amplitudes of the refractive indices for the electric vectors parallel and perpendicular to the acoustic wave normals cause corrugations of two different magnitudes on the optical wave front emerging out of the ultrasonically excited transparent solid; each of these corrugated wave fronts causes a diffraction pattern. There are then two diffraction patterns, one with its electric vector parallel to the sound wave normals and the other perpendicular to them. The intensity of each order of a particular pattern depends on the μ causing the pattern, and whether the sound field is due to progressive or stationary waves.

† In conformity with our convention (see page XXV) we define R by $R = \Delta n_{\parallel}/\Delta n_{\perp}$, whereas Mueller [829] has used $R = \Delta n_{\perp}/\Delta n_{\parallel}$.

It may be noted that Raman and Nath assume that there is no significant bending of light in the acoustic field. This assumption is valid provided L, the width of the acoustic field, is not too large nor the gradient of refractive index too large.

If the incident light vector of amplitude E is polarized in any arbitrary direction, its two components with their electric vectors normal and parallel to the acoustical wave normal, respectively (i.e., parallel to the principal axes), will each produce a diffraction pattern whose amplitude in the mth order is, according to the Raman–Nath theory of ultrasonic diffraction of light, given by

$$\begin{aligned} E_{m\mathrm{I}} &= E_{\mathrm{I}} J_m(v_{\mathrm{I}}) \\ E_{m\mathrm{III}} &= E_{\mathrm{III}} J_m(v_{\mathrm{III}}) \end{aligned} \tag{5.112}$$

where $v_{\mathrm{I}} = 2\pi\mu_{\mathrm{I}} L/\lambda$ and $v_{\mathrm{III}} = 2\pi\mu_{\mathrm{III}} L/\lambda$; J_m is the mth-order Bessel function, L the width of the sound field, and λ the wavelength of light in vacuum. $E_{m\mathrm{I}}$ and $E_{m\mathrm{III}}$ are, respectively, normal and parallel to the sound field. The two components are diffracted to the same extent and have the same frequency.

2. We shall next express the ratio of intensities of the parallel and perpendicular components in terms of $v_{\mathrm{III}}/v_{\mathrm{I}}$ or $\mu_{\mathrm{III}}/\mu_{\mathrm{I}}$.

a. For a progressive ultrasonic wave, the Raman–Nath theory gives for the intensity of the mth diffraction order

$$I_m = J_m{}^2(v) = \left[\sum_{r=0}^{\infty} \frac{(-1)^r v^{2r+m}}{2^{2r+m} r!(r+m)!} \right]^2 \tag{5.113}$$

Therefore for the zero order

$$I_0 = J_0{}^2(v) = \left[1 - \left(\frac{v}{2} \right)^2 + \cdots \right]^2 \tag{5.114}$$

and for the first order

$$I_1 = J_1{}^2(v) = \left[\frac{v}{2} + \cdots \right]^2 \tag{5.115}$$

Here $I_1 \to (v/2)^2$ as $v \to 0$. If R is the ratio of amplitudes of the two electric vectors $E_{\parallel}$ and $E_{\perp}$, then R^2, the ratio of intensities of the two mutually perpendicular components, is obtained as follows. For the zero order:

$$R_0{}^2 = \frac{(I_0)_{\mathrm{III}}}{(I_0)_{\mathrm{I}}} \approx \frac{[1 - (v_{\mathrm{III}}/2)^2]^2}{[1 - (v_{\mathrm{I}}/2)^2]^2}$$

which does not easily yield $v_{\mathrm{III}}/v_{\mathrm{I}}$. For the first order:

$$R_1{}^2 = \frac{(I_1)_{\mathrm{III}}}{(I_1)_{\mathrm{I}}} \approx \frac{[\frac{1}{2}v_{\mathrm{III}}]^2}{[\frac{1}{2}v_{\mathrm{I}}]^2} = \frac{v_{\mathrm{III}}^2}{v_{\mathrm{I}}{}^2} = \frac{\mu_{\mathrm{III}}^2}{\mu_{\mathrm{I}}{}^2}$$

$$= \left(\frac{n_z{}^3 i_{\mathrm{III}}}{n_x{}^3 i_{\mathrm{I}}}\right)^2 = \left(\frac{p_{11}}{p_{12}}\right)^2 \qquad (5.116)$$

since $n_x = n_z$ for cubic crystals.

Equation (5.116) is a simple relation and easy to handle in order to determine p_{11}/p_{12} or p_{ij}/p_{kl} in terms of the ratio of intensities. However, because of the series expansion for J_m this result is valid only for very small values of v. Therefore the experiment must be performed with very low acoustic intensity, and finally the value must be extrapolated to zero acoustic intensity.

b. For standing ultrasonic waves, each diffraction order is composed of subcomponents which are of slightly different frequencies and of different states of polarization. If ν and ν^* are the optical and acoustic frequencies, respectively, the rth components of even orders have the frequencies

$$\nu \pm 2r\nu^* \qquad (r = 0, 1, 2, \ldots)$$

and their intensities are given by

$$I_{m,r} = J_{s-r}^2 \frac{v}{2} J_{s+r}^2 \frac{v}{2} \qquad \text{where} \qquad s = \frac{m}{2} \qquad (5.117)$$

The total intensity for the mth order (even) is then given by

$$I_m = I_{m,0} + 2 \sum_{r=1}^{\infty} I_{m,r} \qquad (5.118)$$

For odd orders the subcomponents have frequencies

$$\nu \pm (2r + 1)\nu^* \qquad (r = 0, 1, 2, \ldots)$$

and intensities

$$I_{m,r} = J_{s-r}^2 \frac{v}{2} J_{s+r+1}^2 \frac{v}{2} \qquad (5.119)$$

where $s = (m - 1)/2$. Thus the total intensity for the mth order (odd) is

$$I_m = 2 \sum_{r=0}^{\infty} I_{m,r} \qquad (5.120)$$

The two expressions for I_m can be together written thus:

$$I_m = \sum_{r=-\infty}^{+\infty} J_r^{\,2} \frac{v}{2} J_{r-m}^2 \frac{v}{2} \tag{5.121}$$

Thus

$$\begin{aligned} I_0 &= J_0^{\,2} J_0^{\,2} + 2J_1^{\,2} J_1^{\,2} + \cdots \\ &= \left[1 - \left(\frac{v'}{2}\right)^2 + \cdots\right]^4 + \cdots = 1 - 4\left(\frac{v'}{2}\right)^2 + \cdots \end{aligned} \tag{5.122}$$

where $v' = v/2$ and

$$\begin{aligned} I_1 &= 2J_0^{\,2} J_1^{\,2} + 2J_1^{\,2} J_2^{\,2} + \cdots \\ &= 2\left\{\left[1 - \left(\frac{v'}{2}\right)^2 + \cdots\right]^2 \left[\frac{v'}{2} + \cdots\right]^2 + \cdots\right\} \\ &= \frac{1}{2} v'^2 \qquad \text{as} \qquad v' \to 0 \end{aligned} \tag{5.123}$$

(neglecting higher powers of v'). We note that as in the case of the progressive waves,

$$R_0^{\,2} = \frac{I_{0,\mathrm{III}}}{I_{0,\mathrm{I}}} = \frac{1 - v_{\mathrm{III}}'^2}{1 - v_{\mathrm{I}}'^2} \tag{5.124}$$

(which does not easily yield $v_{\mathrm{III}}'/v_{\mathrm{I}}'$);

$$R_1^{\,2} = \frac{I_{1,\mathrm{III}}}{I_{1,\mathrm{I}}} = \frac{v_{\mathrm{III}}'^2}{v_{\mathrm{I}}'^2} \tag{5.125}$$

a result which is the same as that for progressive waves. Therefore,

$$R_1^{\,2} = \left(\frac{v_{\mathrm{III}}}{v_{\mathrm{I}}}\right)^2 = \left(\frac{p_{11}}{p_{12}}\right)^2 \tag{5.126}$$

a relation which holds for both progressive and stationary waves in the limit of vanishingly small acoustic power. Gates and Hiedemann [413] have discussed in great detail the theoretical variation of $\tan\beta$ with v^2, for standing waves in the solid (Fig. 5.23). For the significance of $\tan\beta$ see Section 5.6.3 below.

5.6.3. Experimental Determination of p_{ij}/p_{kl} by Three Different Methods Due to Mueller

On the basis of the theory discussed in the previous section, Mueller [829] has suggested three possible methods of investigating the photoelastic behavior of solids.

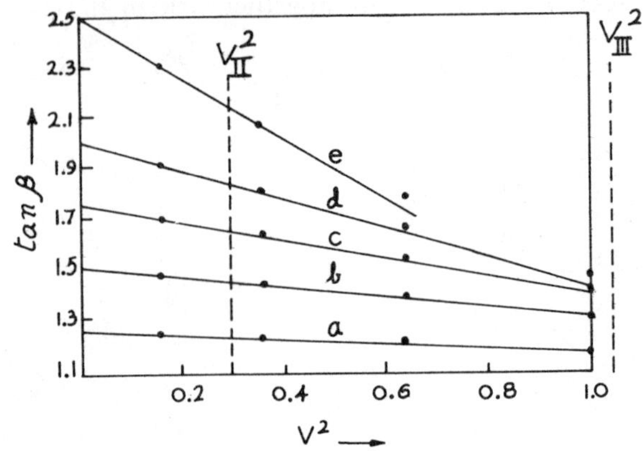

Fig. 5.23. Theoretical variation of tan β with v^2 in Mueller's ultrasonic methods of determining the ratios of elasto-optic constants. β, extinction angle; R, ratio of elasto-optic constants; V_{II}^2, value of v^2 when the second order appears; V_{III}^2, value of v^2 when the third order appears; v, Raman–Nath parameter; a, b, c, d, e: theoretical graphs for $R = 1.25$, 1.50, 1.75, 2.00, and 2.50, respectively. From [413].

5.6.3.1. *Method A*

In this method, the experiment consists in observing the Schaefer–Bergmann pattern of the crystal placed between crossed Nicols. Here a study is made of the diffraction orders produced by both longitudinal and shear waves (see page 165). As the crystal is rotated about the line of light propagation, it will be found that at certain positions of the crystal for which the X axis of the crystal makes an angle α with the plane of vibration of the polarizer, one diffraction spot in the direction of ϕ will disappear. For both longitudinal and transverse waves (i.e., for spots in the inner and outer rings) α is plotted against ϕ. For the same value of ϕ, let α_L and α_T be the corresponding values of α for the two types of waves. Then

$$\left[\frac{2p_{44}}{(p_{11} - p_{12})}\right]^2 = -\tan 2\alpha_L \tan 2\alpha_T \qquad (5.127)$$

This method is independent of the intensity of acoustic waves, and holds for both progressive and stationary waves; however it fails for glasses. This method does not seem to have been applied to any crystal so far. We may note here that this method gives no additional information than what we obtain by much simpler experiments using static load.

5.6.3.2. Method B

a. Experiments with Progressive Waves. We have seen in Section 5.6.2 [see Eq. (5.112)] that if the incident light is plane polarized in any arbitrary direction with respect to the acoustic wave propagation, the two components of light vibrations will each produce a diffraction pattern whose amplitude in the *m*th order is given by

$$\begin{aligned} E_{m\mathrm{I}} &= E_{\mathrm{I}} J_m(v_{\mathrm{I}}) \qquad \text{where} \qquad v_{\mathrm{I}} = (2\pi/\lambda)L\mu_{\mathrm{I}} \\ E_{m\mathrm{III}} &= E_{\mathrm{II}} J_m(v_{\mathrm{III}}) \qquad \text{where} \qquad v_{\mathrm{III}} = (2\pi/\lambda)L\mu_{\mathrm{III}} \end{aligned} \tag{5.128}$$

The two components are diffracted to the same extent and have the same frequency. Furthermore, Nath and Mueller [861] have pointed out that for glasses and cubic crystals they are in the same phase. Hence they combine to produce plane-polarized images in the various diffraction spectra.

Let the incident plane-polarized light of electric intensity E have its plane of vibration inclined at an angle α with the acoustic wave front. Then E_{I} and E_{III}, the components of E normal and parallel to the acoustic wave normal, are

$$E_{\mathrm{I}} = E \cos \alpha \qquad \text{and} \qquad E_{\mathrm{III}} = E \sin \alpha$$

Therefore

$$\begin{aligned} E_{m\mathrm{I}} &= E \cos \alpha \, J_m(v_{\mathrm{I}}) \\ E_{m\mathrm{III}} &= E \sin \alpha \, J_m(v_{\mathrm{III}}) \end{aligned} \tag{5.129}$$

Since there is no phase difference between the two diffracted beams at the point of emergence from the acoustic field, they combine to produce plane-polarized images in the various diffraction spectra; the resulting plane vibration is inclined at an angle β_m with the acoustic wave front, given by

$$\begin{aligned} \tan \beta_m &= E_{m\mathrm{III}}/E_{m\mathrm{I}} \\ &= \tan \alpha \, J_m(v_{\mathrm{III}})/J_m(v_{\mathrm{I}}) \end{aligned} \tag{5.130}$$

Hence the plane of polarization is different for different orders, thus undergoing, in effect, a rotation when the incident plane-polarized light is diffracted into various orders. The angle of rotation depends on both the acoustic power and the order of diffraction.

When the experiment is conducted with the solid between crossed Nicols, for progressive waves of sound, all diffraction orders are plane polarized, and any spot in the Schaefer–Bergmann pattern or any line in the Hiedemann pattern can be extinguished by suitably rotating the analyzer. In the experiment, it is most convenient to set the polarizer at $\alpha = +45°$, and hence the analyzer at $-45°$ with respect to the direction of propagation of longitudinal acoustic waves, and excite the Hiedemann pattern of the crystal, with light propagated perpendicular to the propagation of acoustic waves. Then the analyzer is rotated to extinguish the mth diffraction order.

If γ_m is the angle of rotation of the analyzer from the crossed position, for extinguishing the mth order, then

$$\tan(\gamma_m + 45°) = \frac{J_m(v_{\text{III}})}{J_m(v_{\text{I}})} = \frac{J_m(Rv_{\text{I}})}{J_m(v_{\text{I}})} \tag{5.131}$$

since $R = v_{\text{III}}/v_{\text{I}} = i_{\text{III}}/i_{\text{I}}$.

For small values of the acoustic power, $\Delta\mu$ will be very small, and hence v also will be very small. Thus for the first-order diffracted images, $J_1(v_{\text{I}}) = v_{\text{I}}/2$ and $J_1(v_{\text{III}}) = v_{\text{III}}/2$, and because $\alpha = 45°$, we have for the first-order images

$$\tan\beta_1 = \tan(\gamma_1 + 45°) = \frac{v_{\text{III}}}{v_{\text{I}}} = \frac{\Delta\mu_{\text{III}}}{\Delta\mu_{\text{I}}} = \frac{p_{11}}{p_{12}} \tag{5.132}$$

Thus a measurement of the angle β_1 for the first order gives the ratio p_{11}/p_{12}. Similarly one can take a measurement on the second or third order.

The ratio of intensities of the components of the diffracted light beams with their electric vectors parallel and perpendicular to the acoustic wave normal is given by

$$\frac{I_{\text{III}}}{I_{\text{I}}} = \left(\frac{E_{m\text{III}}}{E_{m\text{I}}}\right)^2 = \tan^2\alpha\,\frac{J_m^{\,2}(v_{\text{III}})}{J_m^{\,2}(v_{\text{I}})} \tag{5.133}$$

For small sound intensities, and at $\alpha = 45°$, for the first order, we have

$$\frac{I_{\text{III}}}{I_{\text{I}}} = \frac{(v_{\text{III}})^2}{(v_{\text{I}})^2} = \frac{(\Delta\mu_{\text{III}})^2}{(\Delta\mu_{\text{I}})^2} = \left(\frac{p_{11}}{p_{12}}\right)^2 \tag{5.134}$$

Evaluation of the strain optical ratios using the intensity ratios was first made by Bergmann and Fues [108] for glasses. For longitudinal progressive acoustic waves excited in solids, one obtains $R = q/p$, where p and q are Neumann's constants.

For cubic crystals of T_d, O, and O_h classes, Table 5.6 shows the expressions for R for various directions of sound and light propagation. Note that $R = \Delta n_{\parallel}/\Delta n_{\perp}$ in conformity with our convention mentioned on page XXV.

From Eq. (5.132) we note that for very small amplitudes of acoustic power (i.e., $v \ll 1$), γ_m is positive for $R > 1$ and negative for $R < 1$.

Table 5.6. *Elasto-Optic Ratio R for Cubic Crystals of Classes* $\bar{4}3m$ (T_d), 432 (O), *and* $m3m$ (O_h)

Direction of		R in terms of p_{ij} $(\Delta n_\parallel/\Delta n_\perp)$
sound propagation	observation	
[100]	[010]	p_{11}/p_{12}
[100]	[011]	p_{11}/p_{12}
[110]	[1$\bar{1}$0]	$(p_{11}+p_{12}+2p_{44})/2p_{12}$
[110]	[001]	$(p_{11}+p_{12}+2p_{44})/(p_{11}+p_{12}-2p_{44})$
[111]	Any direction perpendicular to [111]	$(p_{11}+2p_{12}+4p_{44})/(p_{11}+2p_{12}-2p_{44})$

Usually one investigates the behavior of the first-order diffraction pattern and for this purpose, the value of $\gamma_{m=1}$ (i.e., the angle of extinction for the first order) is determined as a function of the current or the voltage in the piezoelectric transducer and a graph is drawn. This method leads one to obtain both the magnitude and sign of R.

The theory outlined above offers a relatively simple method for determining R using progressive sound waves. However, in practice, it is very difficult to excite progressive longitudinal waves, and instead, it is the stationary waves that can be excited with very little effort, so we shall next see how R can be determined by exciting stationary waves in the solid.

b. Experiments with Stationary Waves. The structure and state of polarization of the diffraction spectra due to stationary waves in a solid are more complicated than those of the diffraction spectra due to progressive waves in the solid, because the various subcomponents of any particular diffraction order are polarized differently and hence any diffraction line cannot be completely extinguished by a setting of the analyzer. For weak sound waves the most intense subcomponents are those for which $r = 0$. Hence the intensity of each diffraction line will pass through a minimum if this subcomponent is extinguished by the analyzer. This minimum-intensity position of the analyzer occurs for an angle $\gamma_{m,0}$ of the analyzer, which is given by

$$\tan(\gamma_{m,0} + 45°) = \frac{J_s^2(v_{\mathrm{III}}/2)}{J_s^2(v_{\mathrm{I}}/2)} \qquad \text{if} \qquad m = 2s \tag{5.135a}$$

or

$$\tan(\gamma_{m,0} + 45°) = \frac{J_s(v_{\mathrm{III}}/2)J_{s+1}(v_{\mathrm{III}}/2)}{J_s(v_{\mathrm{I}}/2)J_{s+1}(v_{\mathrm{I}}/2)} \qquad \text{if } m = 2s + 1 \tag{5.135b}$$

In the limit $v \to 0$, a series development of these expressions shows that $\gamma_{m,0}$ tends to the angle of extinction as in the case of progressive waves. Hence Eq. (5.131) is valid also for standing waves. The theoretical variation of $\tan\beta$ with v^2 is shown in Fig. 5.23 [413].

5.6.3.3. Method C

Originally this method was used by Bergmann and Fues for glasses [108]. The method consists in passing natural (i.e., unpolarized) light through a double-image prism before the light enters the glass specimen. The double-image prism splits the light beam into two beams of equal intensity, their planes of polarization being mutually perpendicular. Each beam gives rise to a diffraction pattern and the ratio of intensities of the corresponding spectra in the two patterns is measured.

Measurements are very much facilitated when a slit source is used instead of a point source. The slit is oriented parallel to the purely longitudinal stationary acoustic wave front and hence normal to the sound field [Fig. 5.24(c)]. A double-image prism is suitably oriented so that the two beams emerging from it are polarized parallel and normal to the source slit. Then the ratio of the intensities of the two polarized beams for electric vectors parallel and perpendicular to the acoustic wave normal is given by

$$J_m^{\,2}(v_{\mathrm{III}})/J_m^{\,2}(v_{\mathrm{I}})$$

Extrapolation to zero sound amplitude, for both progressive and standing waves, yields

$$\lim_{v \to 0} \frac{J_m^{\,2}(v_{\mathrm{III}})}{J_m^{\,2}(v_{\mathrm{I}})} = (i_{\mathrm{III}}/i_{\mathrm{I}})^2 = R_m^{\,2} = \frac{f(p_{ij})}{g(p_{kl})} \tag{5.136}$$

Hence one obtains the ratio of the corresponding elasto-optic constants but cannot decide the sign of the ratio. In Eq. (5.136), the subscript m stands for the diffraction order.

Whenever the ratio of intensities is the quantity experimentally measured, the sign of R cannot be determined unambiguously from the ultrasonic measurements, and it will have to be assigned an appropriate sign so as to agree with the optical measurements for which the sign is known unambiguously; see [854; 926].

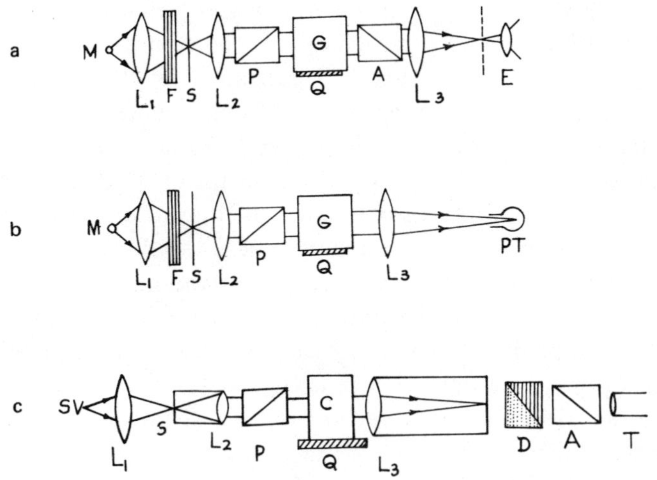

Fig. 5.24. (a) Hagelberg's optical system for Mueller's method B for determining the elasto-optic ratios of glasses and cubic crystals; (b) Hagelberg's optical system for Mueller's method C for determining the elasto-optic ratios of glasses and cubic crystals; (c) Narasimhamurty's optical system for Mueller's method C for studying the elasto-optic ratios of initially birefringent or nonbirefringent crystals and amorphous solids. *M*, mercury arc; L_1, L_2, L_3, lenses; *F*, green filter; *S*, narrow slit; *P*, polarizer; *A*, analyzer; *G*, glass sample or cubic crystal; *Q*, quartz transducer; *E*, eye piece; *PT*, photomultiplier tube; *SV*, sodium vapor lamp; *C*, crystal sample (cubic or noncubic) or amorphous solids; *D*, double-image prism; *T*, telescope. Parts (a) and (b) are from [464], part (c) is from [843].

In the ultrasonic determination of elasto-optic ratios by Mueller's method B, the technique usually employed consists in allowing a plane-polarized beam of light through the experimental crystal prism in a direction perpendicular to the direction of ultrasonic wave normals (the plane of the polarization of the beam of light making an angle of 45° with the direction of propagation of the ultrasonic waves) [Fig. 5.24(a)]. An analyzer is used to study the state of polarization of the diffracted orders in the Hiedemann pattern. This method is suitable only for glasses and cubic crystals, but not for uniaxial and biaxial crystals which are birefringent (even in the unexcited state).

Mueller's method C is universally applicable. Here the technique of passing natural light through a double-image prism and examining it with an analyzer, suggested by Mueller, may be employed. An analogous method was used by Bergmann and Fues (1936) [108], by Gates and Hiedemann (1956) [413], and Hagelberg (1961) [464] to study glasses [Fig. 5.24(b)]. In its modified form, this method was applied by Narasimhamurty

(1955) [843; 845] to study uniaxial and biaxial crystals. Figure 5.24(c) indicates the experimental arrangement. A plane parallel beam of monochromatic light is passed through a polarizing Nicol P and is polarized at 45° with the acoustic wave normal. This plane-polarized beam passes through the crystal specimen C. The lens L_3 forms a diffraction pattern which is viewed through the double-image prism D. The double-image prism is mounted with its axis parallel to the direction of observation and is so adjusted that the two images are seen one above the other. The analyzer A can be rotated about the direction of observation. It is initially set in such a way that the plane of vibration of light passing through it is horizontal. The double-image prism is then rotated through a right angle, if necessary, to make the upper image disappear. This means that the lower image has its plane of vibration horizontal and the upper one vertical. The analyzer is then rotated through 45°, the two images now having equal intensity. Now the ultrasonic power is applied to the experimental crystal along one of its axes and the Hiedemann pattern is observed. The ultrasonic power is next adjusted, so that only first orders are excited [845]. The images of the first orders for the two perpendicularly vibrating components differ in their intensity. The analyzer is then rotated to equalize the intensities of the first-order components. Then if β is the angle of rotation of the analyzer from the zero position, $\tan\beta$ gives directly $R = p_{ij}/p_{kl}$, the ratio of the corresponding elasto-optic constants.

In the case of noncubic crystals $\tan\beta$ gives a quantity R' which involves, besides p_{ij}/p_{kl}, the refractive indices of the crystal for the electric vector parallel and perpendicular to the direction of the acoustic wave propagation, as shown by Narasimhamurty [845]. For example, for the ultrasonic excitation along the Z axis and the observation along the Y axis of a quartz crystal belonging to the class 32 (D_3) of the trigonal system, Narasimhamurty [845] has shown that

$$\tan\beta = R' = \frac{p_{33}}{p_{13}} \cdot \frac{n_z^{\,3}}{n_x^{\,3}}$$

For cubic crystals, since $n_x = n_y = n_z$, $\tan\beta$ gives directly p_{ij}/p_{kl} [see also Eq. (5.116)]. For noncubic crystals, from the observed value of $\tan\beta = R'$, one can calculate the value of $R = p_{ij}/p_{kl}$ for the particular orientation of the noncubic crystal.

Table 5.7 gives the values of R in terms of the elasto-optic constants for specific directions of ultrasonic excitation and light propagation in the case of some uniaxial and biaxial crystals; Figure 5.24 is the schematic diagram of the optical systems employed by Hagelberg to study glasses,

Table 5.7. Elasto-Optic Ratios R for Some Noncubic Crystals

Direction of		Expression for the ratio R in terms of p_{ij}
sound propagation	observation	$(\Delta n_{\parallel}/\Delta n_{\perp})$
a. R for classes D_{2d} ($\bar{4}2m$), C_{4v} ($4mm$), D_4 (422), D_{4h} ($4/mmm$)		
X	Y	p_{11}/p_{31}
X	Z	p_{11}/p_{12}
Z	X	p_{33}/p_{13}
b. R for classes C_{3v} ($3m$), D_3 (32), D_{3d} ($\bar{3}m$)		
X	Y	p_{11}/p_{31}
X	Z	p_{11}/p_{12}
Z	X	p_{33}/p_{13}
c. R for all classes of the orthorhombic system		
X	Y	p_{11}/p_{31}
X	Z	p_{11}/p_{21}
Y	Z	p_{22}/p_{12}
Y	X	p_{22}/p_{32}
Z	X	p_{33}/p_{23}
Z	Y	p_{33}/p_{13}

and by Narasimhamurty to study initially birefringent and nonbirefringent crystals in regard to their photoelastic behavior. Figure 5.25 is the diagram of a typical crystal holder for ultrasonic studies on the elastic and elasto-optic behavior of glasses and transparent crystals.

It is seen from their respective experimental arrangements that though both Narasimhamurty [845] and Hagelberg [464] follow the same method C due to Mueller, the former author compares the amplitudes of the two states of vibration in the diffraction orders, whereas the latter compares their intensities using a photoelectric cell.

Furthermore, Mueller's method C is based upon the experimental procedure adopted by Bergmann and Fues [108] for the determination of p/q of glasses [Fig. 4.9(a)]. This method introduces discrepancies through the use of a photographic method for light intensity measurements and the use of a Wollaston double-image prism, as shown by Gates and Hiedemann

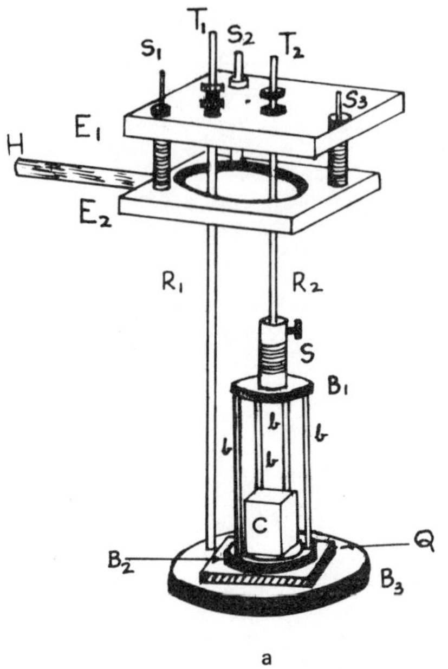

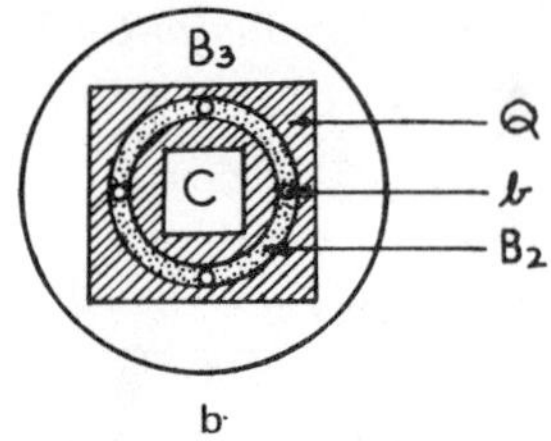

Fig. 5.25. Crystal holder for ultrasonic studies on the elasto-optic behavior of solids. (a) Q, quartz transducer, resting on brass plate B_3; T_1, T_2, rf voltage terminals of the connecting brass rods R_1, R_2; S_1, S_2, S_3, adjusting screws to keep B_3 horizontal; H, ebonite handle; E_2, E_2, ebonite plates; B_1, brass plate; B_2, brass annular ring resting on Q; b, b, b, b, brass rods rigidly connecting B_1 and B_2; C, crystal specimen standing directly on Q (through B_2); S, metal spring pressing against B_1 (for good electrical contact); (b) sectional view of the lower part of the crystal holder. From [1306].

[413]. Hagelberg modified the above experimental technique so as to eliminate both the discrepancies. Light intensity measurements are made by means of a photomultiplier–microphotometer. The Wollaston prism is eliminated by using a polarizer which may be rotated in order to permit readings of the light intensity for the two mutually perpendicular states of polarization.

5.6.4. Pettersen's Method of Determining p_{ij}/p_{kl}

This method [921–924] is based on the principle that R, the ratio of elasto-optic constants, could also be determined if the intensities of any particular order for the two different states of polarization were made equal and the rf voltages required in the two cases were compared.

Since the intensities are equal

$$(I_m)_\parallel = (I_m)_\perp \tag{5.137}$$

where $\parallel$ and $\perp$ refer to the components having electric vectors parallel and perpendicular to the acoustic wave normals. For progressive waves

$$J_m{}^2(v_\parallel) = J_m{}^2(v_\perp) \tag{5.138}$$

and for standing waves

$$\sum_{r=-\infty}^{+\infty} J_r^2 \frac{v_{\parallel}}{2} J_{r-m}^2 \frac{v_{\parallel}}{2} = \sum_{r=-\infty}^{+\infty} J_r^2 \frac{v_{\perp}}{2} J_{r-m}^2 \frac{v_{\perp}}{2} \tag{5.139}$$

Since the two series in Eq. (5.138) which are set equal to each other are identical, their arguments must be equal. Similarly, in Eq. (5.139) the arguments, namely, v, the Raman–Nath parameters, must be equal. Hence

$$v_{\parallel} = v_{\perp} \tag{5.140}$$

Therefore

$$2\pi\mu_{\parallel}L/\lambda = 2\pi\mu_{\perp}L/\lambda$$

$$\mu_{\parallel} = \mu_{\perp} \tag{5.141}$$

or from Eqs. (5.96), (5.106b), and (5.111) we obtain for glasses with Neumann's constants p and q

$$\tfrac{1}{2}n^3pA_{\perp}k^* = \tfrac{1}{2}n^3qA_{\parallel}k^* \tag{5.142}$$

Then

$$p/q = A_{\parallel}/A_{\perp} = V_{\parallel}/V_{\perp} \tag{5.143}$$

assuming that the sound amplitude A is proportional to the transducer voltage V.

For cubic crystals, with ultrasonic excitation along a cube axis, say, the X axis, and for observation along another cube axis, say, Y, we have

$$p_{11}/p_{12} = A_{\perp}/A_{\parallel} = V_{\perp}/V_{\parallel} \tag{5.144}$$

Pettersen's experimental setup is indicated in Fig. 5.26(a). In the actual experiment one studies the variation of I, the light intensity of the diffraction orders, for both parallel and perpendicular components, as a function of V, the voltage applied to the piezoelectric transducer, and graphs are drawn to indicate the variation of I with V. The voltages are then read from curves I and II for the same light intensity for parallel and perpendicular components, respectively. This is repeated at points corresponding to different values of light intensity, and also for as many orders as possible. Figure 5.26(b) shows the theoretical variation I_0 for the zero order of the light diffracted by stationary ultrasonic waves versus the Raman–Nath parameter v. It is easily seen that $I_0 = \sum_p J_p^4(v/2)$.

The main advantages gained in Pettersen's method over the earlier

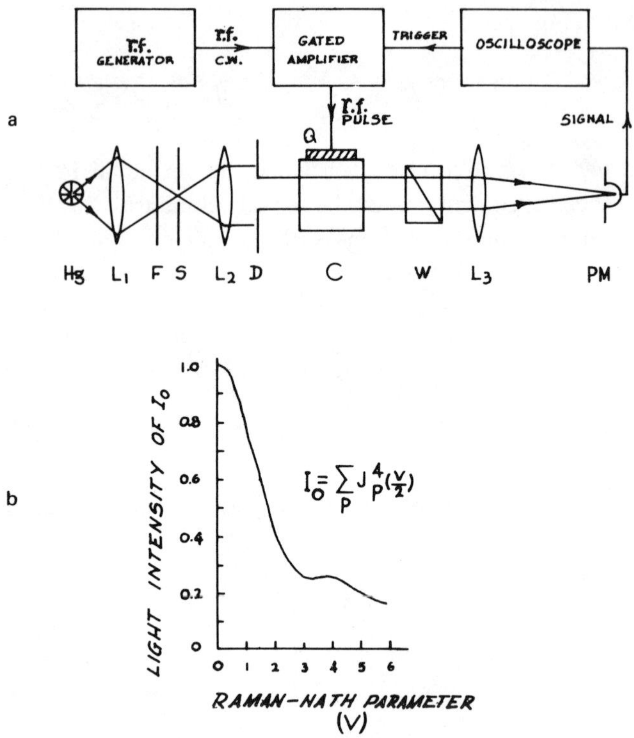

Fig. 5.26. Pettersen's method for determining the elasto-optic ratios of glasses and crystals (cubic and noncubic). (a) Schematic diagram of apparatus for using pulsed rf. Hg, mercury arc; L_1, L_2, L_3, lenses; *F*, green filter; *S*, narrow slit; *D*, diaphragm; *C*, crystal sample (cubic or noncubic); *Q*, quartz transducer; *W*, Wollaston prism; *PM*, photomultiplier tube; (b) theoretical variation of the intensity of the zeroth order of light diffracted by stationary ultrasonic waves. From [924].

methods are as follows: (i) Pettersen's method does not require extrapolation to zero acoustic power, which has been the most important condition stipulated in Mueller's methods followed earlier. (ii) Since Eq. (5.144) holds for all diffraction orders, one could with advantage take the experimental observations making use of the comparatively large intensities of the light in the zero order of the diffraction; in other methods one is forced to make use of the first order, which, as is well known, is of very low intensity compared with that of the zero order.

Pettersen's method is applicable to all solids, isotropic and crystalline (cubic and noncubic), with equal facility and to the same order of accuracy. From the above discussion on Mueller's and Pettersen's methods we conclude the following.

When we compare the intensities, at the same voltage, for the two mutually perpendicularly polarized diffraction orders

$$(p_{ij})^2 \propto \text{the intensity } I$$

or

$$p_{ij} \propto \text{the amplitude } A$$

On the other hand, if we compare the voltages at which the intensities are the same, then

$$p_{ij} \propto 1/V$$

5.6.5. Bragg Diffraction Method of Determining the Individual Values of p_{ij}

This method is of recent origin, and unlike Mueller's methods, where the ultrasonic frequency is of the order of 1–10 MHz, here the working frequency is very high, being 100–300 MHz; the acoustic grating is in this case said to be a "thick grating,"† and it produces only one order of diffraction pattern whose intensity can be controlled over a wide range with a theoretical possibility of 100%, the entire incident light intensity thus being diffracted into the first order alone. In fact, it is the case of Bragg reflection resulting only on one side of the zero order, thereby causing asymmetric distribution of intensity, as shown in Fig. 5.27. In the usual ultrasonic diffraction phenomenon of the Debye–Sears type, the grating is said to be a thin grating, with a larger number of diffraction orders, symmetrically distributed about the central or zero order for normal incidence of light.

The method here consists in measuring the intensity of the light beam diffracted by the solid specimen and comparing it with that diffracted by a solid or a liquid of known photoelastic property. The technique for this was first proposed and applied by Smith and Korpel [1152] and is shown in Fig. 5.28.

Pulses of very high-frequency ultrasonic waves are propagated by the transducer into the Bragg cell containing the solid specimen placed inside a standard reference liquid, usually water. The orientation of the crystalline solid is suitably chosen so as to give the desired strain-optical constant.

The intensity of the Bragg-diffracted light is measured at three regions *A*, *B*, and *C* of the Bragg cell; the source of light is a laser beam. *B* is in the center of the solid, and *A* and *C* are in the liquid, equidistant from *B*.

† The ultrasonic grating is said to be thick if the incident ray produced crosses a large number of grating lines for the same angle of incidence.

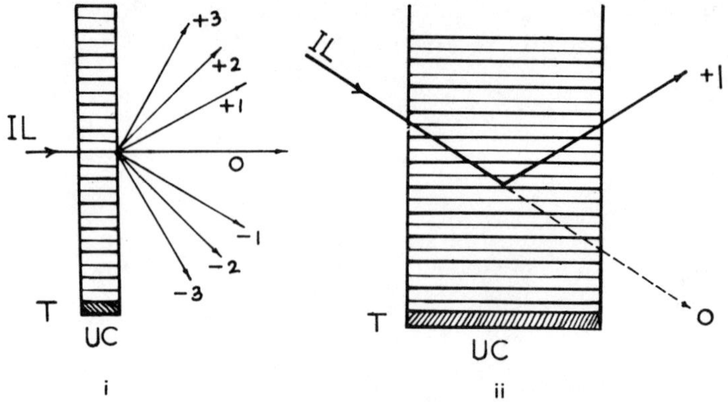

Fig. 5.27. Ultrasonic diffraction by (i) a thin grating (Raman–Nath region), and (ii) a thick grating (Bragg region). *UC*, ultrasonic cell; *T*, transducer; *IL*, incident light direction; 0, zero-order spectrum (central image); ± 1, ± 2, ± 3, different orders of the diffraction spectrum (note that the Bragg diffraction caused by a thick grating causes only one diffracted image). From [32].

The Bragg angle is different for the liquid from that for the solid (Fig. 5.28).

Let the incident acoustic power at the three points *A*, *B*, and *C* be, respectively, P_a, P_b, and P_c, and the intensities of the corresponding Bragg-diffracted beams be I_a, I_b, and I_c, and let I_0 be that of the incident laser light. Then

$$I_a \propto (M_l P_a) I_0$$

$$I_b \propto (M_s P_b) I_0$$

$$I_c \propto (M_l P_c) I_0$$

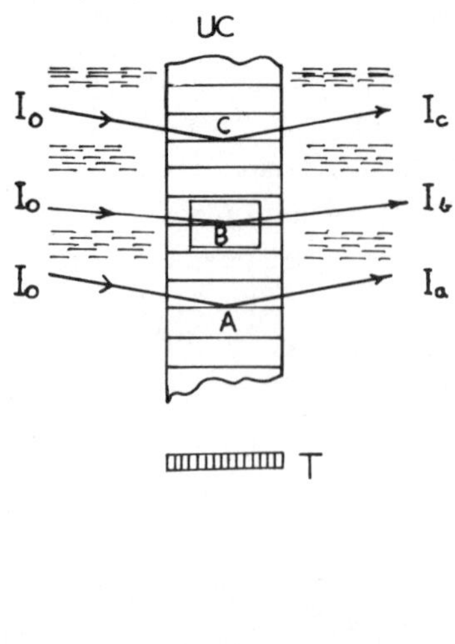

Fig. 5.28. Bragg diffraction method for studying the acousto-optic effect in solids: Smith and Korpel's technique. *UC*, ultrasonic cell; *T*, transducer; *B*, solid specimen, surrounded by a liquid of known figure of merit; I_0, intensity of the incident laser beam; I_a, I_b, and I_c, intensities of the Bragg-diffracted laser beams at the three points *A*, *B*, and *C*, respectively. From [1152].

If K is the constant of proportionality, then we have

$$I_a = KM_l P_a$$
$$I_b = KM_s P_b \qquad (5.145)$$
$$I_c = KM_l P_c$$

Here M is called the figure of merit of the photoelastic material (solid or liquid), depending upon the appropriate properties of the material, and can be defined as the fractional amount of incident light diffracted by the application of a given magnitude of acoustic power; thus M measures the elasto-optic performance of the material. For a crystalline solid, M depends also upon the orientation of the crystal specimen.

It is easily seen that

$$P_b = (P_a P_c)^{1/2} \qquad (5.146)$$

a relation which is independent of the sound alternation in the solid and liquid, and independent of the coefficient of transmission of sound at the solid–liquid interface, since the two boundaries are symmetrical.

$$\frac{I_b}{M_s} = \left(\frac{I_a}{M_l}\,\frac{I_c}{M_l}\right)^{1/2} \qquad (5.147a)$$

or

$$\frac{M_s}{M_l} = \frac{I_b}{(I_a I_c)^{1/2}} \qquad (5.147b)$$

The value of M_s can be expressed in terms of Pockels' strain-optical constants p_{ij} of the specimen. For small acoustic pressures, according to Raman–Nath theory, we have for the first order the amount of light diffracted

$$I' \propto \left(\frac{v}{2}\right)^2 \propto \left(\frac{2\pi\mu L}{\lambda}\right)^2 \propto (\Delta n)^2$$

letting $\mu = \Delta n$, the maximum change in the refractive index. But

$$\Delta n = -(n^3/2)(p_{ij}s_{ij}) \qquad (5.148)$$

per unit acoustic pressure, n being the appropriate refractive index of the specimen in the undisturbed state. Hence

$$I' \propto (\Delta n)^2 \propto (n^3 p_{ij} s_{ij})^2 \qquad (5.149)$$

where I' stands for I_a, I_b, or I_c, and the case for other symbols is similar.

The relation between s_{ij} and the acoustic power P_i per unit area of the transducer can be derived thus: The energy density in a strained medium is $\frac{1}{2}(s_{ij}^2 c_{ij})$, where c_{ij} is the elastic constant, equal to ϱV^2. Hence, the power P_i in the ultrasonic beam of unit cross section, being equal to the product of the energy density and the velocity of the sound wavefront, is given by

$$P_i = \tfrac{1}{2}s_{ij}^2(\varrho V^2)V = \tfrac{1}{2}s_{ij}^2\varrho V^3 \tag{5.150}$$

Combining Eqs. (5.149) and (5.150) we obtain M_s, the elasto-optic performance (also called the acousto-optic figure of merit) of the solid:

$$M_s = \frac{I'}{P_i}\Big/ I_0 \propto \frac{(n^3 p_{ij} s_{ij})^2}{\frac{1}{2}s_{ij}^2 \varrho V^3} \propto \frac{n^6 p_{ij}^2}{\varrho V^3} \tag{5.151}$$

From Eq. (5.147)

$$\left(\frac{n^6 p_{ij}^2}{\varrho V^3}\right)_{\text{solid}} \Big/ \left(\frac{n^6 p_{ij}^2}{\varrho V^3}\right)_{\text{liquid}} = \frac{I_b}{(I_a I_c)^{1/2}}$$

Knowing the value of p_{ij} of the reference liquid, we obtain the appropriate p_{ij} of the solid.

It should be noted that this method does not yield the sign of the photoelastic constant p_{ij} of the specimen solid.

The technique described above cannot be applied to the determination of photoelastic constants relating to acoustic shear waves because of the inability of liquids to transmit acoustic shear waves. This difficulty has been overcome by Dixon and Cohen [294]. They replaced the liquid medium by a solid, which naturally transmits shear waves also. A buffer rod of fused quartz whose photoelastic properties are well known is bonded to the specimen crystal as in Fig. 5.29.

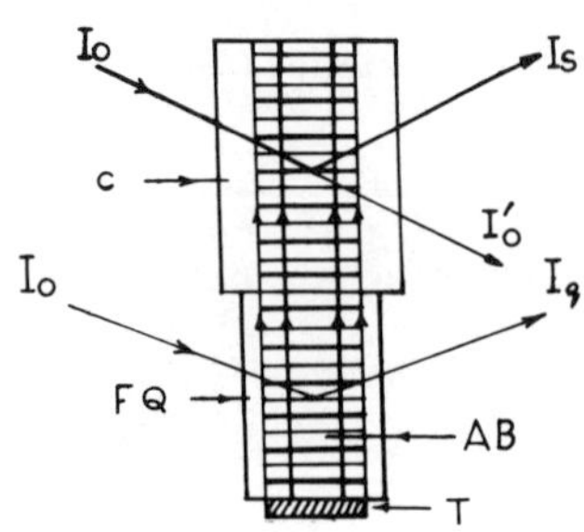

Fig. 5.29. Bragg diffraction method for studying the acousto-optic effect in solids due to acoustic longitudinal and shear waves: Dixon and Cohen's technique. *FQ*, fused-quartz rod; *T*, transducer; *C*, crystal prism of the desired orientation; *AB*, acoustic beam (field); I_q, Bragg-diffracted beam in the fused quartz; I_s, Bragg-diffracted beam in the crystal prism; I_0, incident light beam; I_0', zero-order beam (central image) due to the crystal prism. Note that only two of the five Bragg-diffracted light beams are shown in the figure (see text above). From [294].

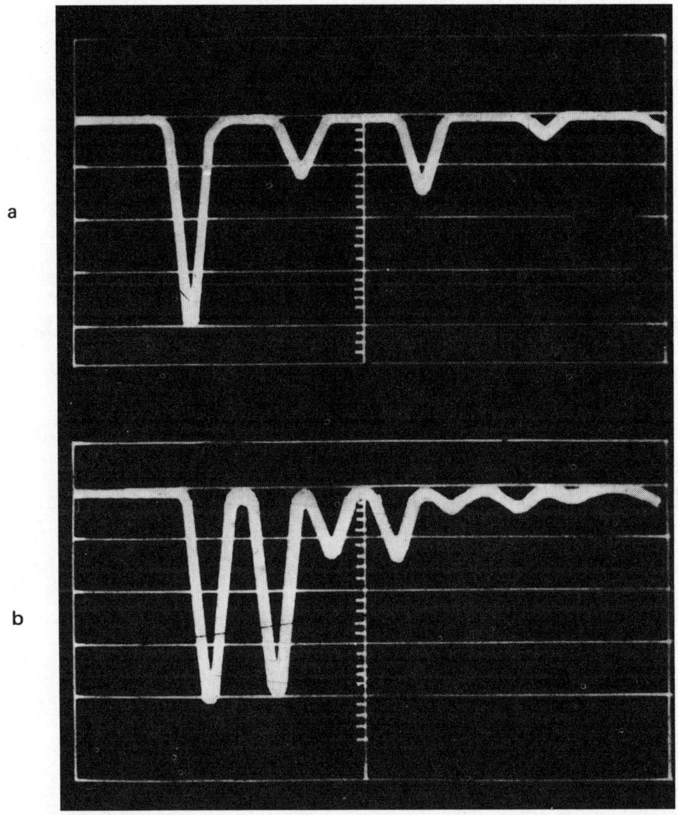

Fig. 5.30. Oscilloscope display of the photomultiplier output when a laser beam traverses (a) the fused quartz, and (b) the lithium niobate crystal (in the Bragg diffraction method of determining p_{ij} of solids—Dixon and Cohen's technique). From [294].

Dixon and Cohen's technique consists in diffracting a laser beam by pulsed ultrasound, first in quartz and then in the specimen crystal, which are bonded together. Two sets of observations are taken using a photomultiplier as a detector. The light signal is fed into a cathode ray oscilloscope. In each case the light beam is set at the appropriate Bragg angle for the fused quartz and the specimen crystal, respectively.

Figure 5.30(a) shows the light pulses diffracted in the fused quartz. The first light pulse I_1 is due to the fused quartz buffer rod with the acoustic wave advancing towards the specimen crystal from the transducer. The second pulse I_2 is due to the partial reflection of the acoustic wave from the bonding medium into quartz. The third pulse I_3 is due to that portion of the acoustic wave which travels through the bonding medium into the specimen crystal and then gets reflected from its free surface and back into quartz through the bonded medium.

Figure 5.30(b) shows the Bragg-diffracted light pulses when the laser beam traverses the crystal specimen. The first light pulse I_4 is due to the advancing acoustic pulse and the second light pulse I_5 is due to the pulse reflected at the free end of the specimen crystal. Then

$$\left(\frac{n_s^6 p_s^2}{\varrho_s V_s^3}\right) \Big/ \left(\frac{n_Q^6 p_Q^2}{\varrho_Q V_Q^3}\right) = \frac{M_s}{M_Q} = \frac{(I_4 I_5)_s}{(I_1 I_3)_Q} \tag{5.152}$$

We assume here that the above ratio does not depend upon the bonding medium between the reference rod of quartz and the specimen since the transmission coefficient of the bonding medium for the advancing acoustic wave is assumed to be reciprocal of that due to reflection. The acoustic loss in the material also has no influence on the ratio. In Eq. (5.152) p_s, n_s, ϱ_s, and V_s stand for the appropriate quantities for the sample, and similarly, p_Q, n_Q, ϱ_Q, and V_Q stand for the quartz buffer rod.

This experimental method also does not enable one to assign unambiguously the sign of p_{ij} of the specimen solid.

5.6.6. Borrelli and Miller's Method of Determining the p_{ij} of Glass

5.6.6.1. Determination of the Individual p_{ij} of Glass by an Ultrasonic Technique Due to Borrelli and Miller

Borrelli and Miller [176] have employed a composite piezoelectric resonator intercepting one of the two beams of light of an interferometer. The sample of glass forms one of the two components of the composite piezoelectric resonator, and so it is excited ultrasonically, thus giving rise to an interference pattern which varies with time. The individual strain-optical coefficients p_{ij} are deduced from the modulation characteristics of the interference pattern employing a photoelectric method. Using a Babinet compensator along with the same composite piezoelectric resonator one can also determine the relative strain-optical coefficient $(p_{11}-p_{12})$ of the glass sample.

Figure 5.31a shows the experimental setup for interferometric studies. The ultrasonic frequency applied to the resonator was 50 kHz. The light source consisted of a He–Ne gas laser giving light of $\lambda = 6328$ Å. A compensating plate C of the same thickness and material as the specimen glass G was placed in the path of the reference arm of a Mach–Zehnder interferometer. The introduction of the compensating plate C ensures zero

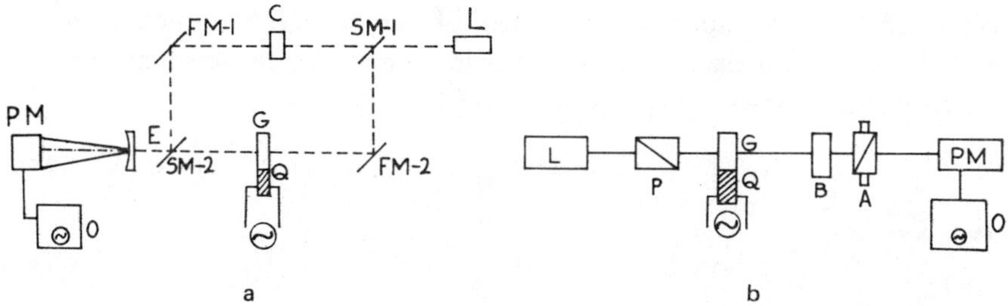

Fig. 5.31. Borrelli and Miller's experimental setup for determining (a) p_{11} and p_{12} and (b) $(p_{11} - p_{12})$ of glass by an ultrasonic method. *G*, glass sample bonded to a quartz driver *Q*; *C*, compensating plate of glass; *L*, laser; *SM*-1, *SM*-2, semisilvered glass plates; *FM*-1, *FM*-2, fully silvered mirrors; *PM*, photomultiplier; *O*, oscilloscope; *P*, polarizer; *A*, analyzer; *B*, Babinet compensator. From [176].

phase retardation between the two beams of light at the point of exit *E* of the interferometer when the experimental glass specimen is in the unstrained state, and also helps in equalizing the losses due to absorption of light by the experimental sample *G*.

In the unexcited state of *G* the light intensity distribution I_1 across an interference fringe is given by

$$I_1 = I_0 \cos^2(\hat{\delta}_0/2) \tag{5.153}$$

I_0 is the intensity of the incident light and $\hat{\delta}_0$ is the phase retardation.

In the ultrasonically excited state of *G*, at the time *t*,

$$I_2 = I_0 \cos^2\left[\tfrac{1}{2}\hat{\delta}_0 + \frac{\pi}{\lambda} d_s(\Delta n)_t + \frac{\pi}{\lambda}(n_0 - 1)(\Delta d)_t\right] \tag{5.154}$$

Here d_s is the thickness of sample *G*, and $(\Delta d)_t$ is the change in thickness of the sample *G* (along the direction of light propagation) due to Poisson's ratio $(\Delta d)_t = -\sigma\varepsilon_s l_s$, where σ is Poisson's ratio, l_s is the length, and ε_s is the maximum longitudinal strain of *G*.

For a composite resonator

$$\varepsilon_s l_s = \varepsilon_d l_d \tag{5.155}$$

where the subscripts *s* and *d* stand, respectively, for the sample (*G*) and the driver (quartz).

Marx [780] has given a method of calculating ε_d when the piezoelectric driver is supported at its middle and excited longitudinally, which is the case in the present experimental setup. Hence ε_s can be calculated from

Eq. (5.155). Using appropriate values of the several parameters in Eq. (5.154), $(\Delta n)_t$ can be computed, knowing which we can evaluate p_{11} and p_{12} from the following expressions for (Δn):

$$(\Delta n)_t^{\parallel} = -(n_o^3/2)(p_{11} - 2\sigma p_{12})\varepsilon_s \tag{5.156}$$

and

$$(\Delta n)_t^{\perp} = -(n_o^3/2)[p_{12} - \sigma(p_{11} + p_{12})]\varepsilon_s \tag{5.157}$$

for light with its electric vector parallel and perpendicular, respectively, to the acoustic field. The transmission versus strain is measured at the center of a bright fringe where $\frac{1}{2}\hat{\delta}_0 = 0$.

Borrelli and Miller claim an accuracy of about 7% in the reported values of p_{ij} of their glass samples using their technique described above; it does not seem to have so far been applied to any crystal, even cubic.

5.6.6.2. *Measurement of Birefringence and* $(p_{11} - p_{12})$ *by an Ultrasonic Method*

The experimental setup by Borrelli and Miller is shown in Fig. 5.31b for the determination of the relative strain-optical coefficient equal to $(n_o^3/2)(p_{11} - p_{12})$. The incident light beam is polarized at 45° to the ultrasonic wave front and passed through the sample at its center (where the maximum strain occurs). The modulated light beam is then received by a photomultiplier connected to a cathode ray oscilloscope.

As in the preceding section describing Borrelli and Miller's method of determining the individual p_{ij} of glass, here too the transmission is determined for the ultrasonically excited glass component of the composite piezoelectric resonator. The transmission is given by

$$T = \sin^2(\hat{\delta}_t/2 - \theta) \tag{5.158}$$

where θ is the angle through which the analyzer is rotated from the crossed position, and $\hat{\delta}_t$ is the phase retardation at the instant t. Here

$$\hat{\delta}_t = \frac{2\pi d_s}{\lambda}(\Delta n)_t = -\frac{2\pi d_s}{\lambda}\frac{n_0^3}{2}(p_{11} - p_{12})(1 + \sigma)\varepsilon_s \tag{5.159}$$

where d_s is the thickness of the sample along the direction of observation, σ is Poisson's ratio, and ε_s is the strain.

Equation (5.158) shows that in the initial crossed position of the Nicol with the unexcited glass sample, $\hat{\delta}_t = 0$ and $\theta = 0$, and hence $T = 0$.

When $(\hat{\delta}_t/2) = \theta$, again $T = 0$; i.e., in the excited state of the glass sample, if the analyzer is rotated through an appropriate angle, the transmission is reduced to zero. Thus a measure of the angle of rotation of the analyzer enables one to obtain $\hat{\delta}_t$, leading to $(\Delta n)_t$, and finally to $(p_{11} - p_{12})$.

The numerical values of p_{ij} for some cubic and noncubic crystals are given in Tables 5.8 and 5.9.

5.6.7. *Technological Applications of the Acousto-Optic Effect*

The phenomenon of photoelastic behavior of crystals, which was put on a solid foundation by Pockels as early as the 1880s, and the discovery of the phenomenon of diffraction of light by ultrasonically excited liquids and solids by Debye and Sears in America, and by Lucas and Biquard in France, respectively, in 1932, have remained subjects of only academic interest until the early 1960s. However, with the subsequent advent of the laser there has been a phenomenal growth of activity in crystal photoelasticity because of its ever growing number of applications in industry. For example, the laser and high-frequency acoustic techniques, along with the discovery of new piezoelectric materials, with a high coupling factor, have made it increasingly possible to apply the acousto-optic effect to devices such as (i) light deflectors, (ii) light modulators, and (iii) signal processors. Figure 5.32(a) shows the Debye–Sears type of operation of a light modulator. Bragg diffraction of light by ultrasonic cells has found applications for both modulation and deflection of a light beam. Figure 5.32(b) shows the Bragg type of operation.

Figure 5.33 shows how the Bragg-diffracted and modulated light signal can be made to scan over a range by controlling the acoustic frequency and the angle of incidence of the light beam on the Bragg cell.

5.7. *Brillouin Scattering and Photoelasticity of Crystals*

We have discussed earlier the application of "Brillouin's scattering of light" method for determining the elastic constants of a crystalline medium (see Chapter 4 on elasticity). In order to determine the elastic constants, one takes a measurement of the frequency shifts of the Brillouin components from the Rayleigh line. These frequency shifts are related to the acoustic velocities in the appropriate directions of the crystal, and the elastic constants are calculated from a knowledge of the acoustic velocities. We shall indicate in this section how a further study of the Brillouin spectra

Table 5.8. *Values of the Photoelastic*

Crystal	Strain-optical constants						
	p_{12}/p_{11}	p_{44}/p_{11}	$p_{11}-p_{12}$	p_{11}	p_{12}	p_{44}	Wave-length, Å
NaF	3.04	−0.42	—	—	—	—	5461
	2.35	−0.32	−0.101	0.075	0.176	−0.024	5893
	—	—	−0.098	—	—	−0.019	—
	—	—	—	0.050	0.161	—	—
	—	—	—	0.048	0.146	—	—
NaCl	1.36	−0.083	—	—	—	—	5461
	1.37	−0.084	−0.047	0.126	0.174	−0.0107	5893
	1.39	−0.091	−0.043	0.11	0.153	−0.010	—
	1.35	−0.099	—	—	—	—	—
	1.31	—	—	—	—	—	5890
NaBr	1.24	−0.030	—	—	—	—	5461
	—	—	−0.0356	—	—	−0.0036	5893
	1.24	—	—	—	—	—	5893
	—	—	—	0.148	0.184	—	—
KF	0.79	−0.11	—	—	—	—	5461
	—	—	0.061	—	—	−0.027	5893
	0.74	−0.11	—	—	—	—	—
	0.77	—	—	0.268	0.207	—	5893
	—	—	—	0.290	0.229	—	5893
	—	—	—	0.235	0.174	—	—
KCl	0.76	−0.099	—	—	—	—	5461
	0.72	−0.12	0.064	0.231	0.168	−0.027	5893
	0.73	—	0.046	0.17	0.124	—	—
	0.76	−0.12	—	—	—	—	—
	0.74	−0.11	0.055	0.211	0.156	−0.024	5461
	0.77	−0.12	0.0552	0.2371	0.1819	−0.0275	—
KBr	0.79	−0.086	—	—	—	—	5461
	0.75	−0.10	0.049	0.193	0.144	−0.020	5893
	0.78	−0.12	0.049	0.22	0.171	−0.026	—
	0.77	−0.12	—	—	—	—	—
	0.78	−0.11	0.046	0.208	0.162	−0.022	5461
	0.79	−0.098	0.0509	0.2422	0.1913	−0.0237	—
KI	0.83	−0.038	—	—	—	—	—
	0.81	−0.047	0.039	0.209	0.170	−0.0098	5893
	0.80	—	0.041	0.210	0.069	—	—
	0.87	—	—	—	—	—	5461

[a] The data on p_{ij} are from [925]. For exhaustive data on p_{ij} and q_{ij} of crystals, see [83; 84, 1526]

Constants of Some Alkali Halides[a]

Stress-optical constants[b]					
$q_{11} - q_{12}$	q_{11}	q_{12}	q_{44}	Wavelength, Å	Reference
−1.39	0.05	1.44	−0.85	5893	[724]
—	—	—	—	—	—
—	—	—	—	—	—
—	—	—	—	—	—
—	—	—	—	—	—
−1.31	1.27	2.58	−0.84	5893	[724]
−1.18	—	—	−0.83	—	[216]
—	—	—	—	—	—
—	—	—	—	—	—
—	—	—	—	—	—
−1.21	—	—	−0.36	5893	[724]
—	—	—	—	—	—
—	—	—	—	—	—
—	—	—	—	—	—
1.22	—	—	−2.16	5893	[724]
—	—	—	—	—	—
—	—	—	—	—	—
—	—	—	—	—	—
—	—	—	—	—	—
—	—	—	—	—	—
1.88	4.75	2.87	−4.32	5893	[724]
1.70	—	—	−4.30	5900	[950]
—	—	—	—	—	—
—	—	—	—	—	—
—	—	—	—	—	—
—	—	—	—	—	—
1.68	4.62	2.93	−3.94	5893	[724]
1.76	—	—	−4.42	5893	[66]
—	—	—	—	—	—
—	—	—	—	—	—
—	—	—	—	—	—
—	—	—	—	—	—
1.67	6.18	4.49	−2.66	5893	[724]
1.84	—	—	−3.04	5893	[66]
—	—	—	—	—	—
—	—	—	—	—	—

[b] The values of q_{ij} are in units of 10^{-13} cm^2/dyne.

Table 5.9. *Values of the Photoelastic Constants of Some Noncubic Crystals*[a]

System/class	Crystal	Photoelastic constants[b]												Ref.
Hexagonal system		p_{11}	p_{12}	p_{13}	p_{31}	p_{33}	p_{44}							
Class: 6/*mmm* (D_{6h})	Beryl	0.0099	0.175	0.191	0.313	0.023	−0.152							[951]
Tetragonal system		p_{11}	p_{12}	p_{13}	p_{31}	p_{33}	p_{44}	p_{66}						
Class: $\bar{4}2m$ (D_{2d})	ADP	0.319	0.277	0.169	0.197	0.167	−0.058	−0.091						[854]
$\bar{4}2m$ (D_{2d})	KDP	0.287	0.282	0.174	0.241	0.122	−0.019	−0.064						[1307]
Trigonal system		p_{11}	p_{12}	p_{13}	p_{31}	p_{33}	p_{44}	p_{14}	p_{41}					
Class: 32 (D_3)	α-Quartz	0.16	0.27	0.27	0.29	0.010	−0.079	−0.030	−0.047					[850]
$\bar{3}m$ (D_{3d})	Calcite	0.095	0.189	0.215	0.309	0.178	−0.090	−0.006	0.010					[955]
$\bar{3}m$ (D_{3d})	Ruby	−0.23	−0.03	0.02	−0.04	−0.20	−0.10	0.00	0.01					[1365]
Orthorhombic system		p_{11}	p_{12}	p_{13}	p_{21}	p_{22}	p_{23}	p_{31}	p_{32}	p_{33}	p_{44}	p_{55}	p_{66}	
Class: 222 (D_2)	Rochelle salt	0.35	0.41	0.42	0.37	0.28	0.34	0.36	0.35	0.36	−0.030	+0.005	−0.25	[851]
222 (D_2)	α-Iodic acid	0.302	0.496	0.339	0.263	0.412	0.304	0.251	0.345	0.336	0.084	−0.030	0.098	[492]
Class: *mmm* (D_{2h})	α-Sulfur	0.324	0.307	0.268	0.272	0.301	0.310	0.203	0.232	0.270	0.143	0.019	0.118	[492]
mmm (D_{2h})	Barite	0.21	0.25	0.16	0.34	0.24	0.19	0.28	0.22	0.31	0.002	−0.012	0.037	[1279]
Monoclinic system														
	Taurin	p_{11} = 0.313	p_{21} = 0.281	p_{31} = 0.362	p_{51} = −0.014	p_{44} = 0.0025								
		p_{12} = 0.251	p_{22} = 0.252	p_{32} = 0.275	p_{52} = 0.006	p_{46} = −0.0056								[489]
		p_{13} = 0.270	p_{23} = 0.272	p_{33} = 0.308	p_{53} = 0.0048	p_{64} = 0.0024								
		p_{15} = −0.010	p_{25} = −0.0025	p_{35} = −0.003	p_{55} = 0.047	p_{66} = 0.0028								

Hexagonal system		q_{11}	q_{12}	q_{13}	q_{31}	q_{33}	q_{44}	
Class: 6/*mmm* (D_{6h})	Beryl	−0.356	0.592	0.739	0.931	−0.426	−2.32	[951]

Tetragonal system		q_{11}	q_{12}	q_{13}	q_{31}	q_{33}	q_{44}	q_{66}	
Class: $\bar{4}2m$ (D_{2d})	ADP	4.40	3.71	0.34	2.02	2.65	−6.70	−15.25	[854]
$\bar{4}2m$ (D_{2d})	KDP	4.15	4.08	1.12	3.60	0.44	−1.47	−10.26	[1307]

Trigonal system		q_{11}	q_{12}	q_{13}	q_{31}	q_{33}	q_{44}	q_{14}	q_{41}	
Class: 32 (D_3)	α-Quartz	1.35	2.69	1.94	3.11	0.07	−1.13	−0.11	−0.34	[850]
$\bar{3}m$ (D_{3d})	Calcite	−0.61	0.92	2.53	1.58	0.45	0.35	−1.11	0.67	[955]
3*m* (D_{3d})	Ruby	−0.52	0.08	0.13	0.01	−0.41	−0.71	−0.07	−0.01	[1365]

Orthorhombic system		q_{11}	q_{12}	q_{13}	q_{21}	q_{22}	q_{23}	q_{31}	q_{32}	q_{33}	q_{44}	q_{55}	q_{66}	
Class: 222 (D_2)	Rochelle salt	3.1	3.4	3.2	5.4	0.76	2.1	5.0	2.9	1.6	−2.5	1.5	−2.5	[851]
222 (D_2)	α-Iodic acid	0.047	0.067	0.058	0.042	0.054	0.053	0.041	0.043	0.063	0.050	−0.015	0.063	[492]
Class: *mmm* (D_{2h})	α-Sulfur	0.189	0.182	0.042	0.148	0.163	0.080	0.107	0.112	0.085	0.173	0.044	0.272	[492]
mmm (D_{2h})	Barite	−0.079	3.81	1.64	1.49	1.77	1.23	0.44	1.23	2.56	0.20	−0.435	1.39	[1279]

Monoclinic system							
	Taurin	q_{11} = 0.0288	q_{21} = 0.0113	q_{31} = 0.0371	q_{44} = 0.0034	q_{53} = 0.0081	
		q_{12} = −0.0119	q_{22} = −0.0113	q_{32} = −0.0187	q_{46} = −0.0056	q_{55} = 0.0498	[489]
		q_{13} = 0.1844	q_{23} = 0.2017	q_{33} = 0.2117	q_{51} = −0.0037	q_{64} = 0.0034	
		q_{15} = −0.0188	q_{25} = −0.0155	q_{35} = −0.0104	q_{52} = −0.0042	q_{66} = 0.0029	

[a] For exhaustive data on the p_{ij} and q_{ij} of crystals, see [83, 84, 1526].
[b] The values of q_{ij} are in units of 10^{-13} cm²/dyne.

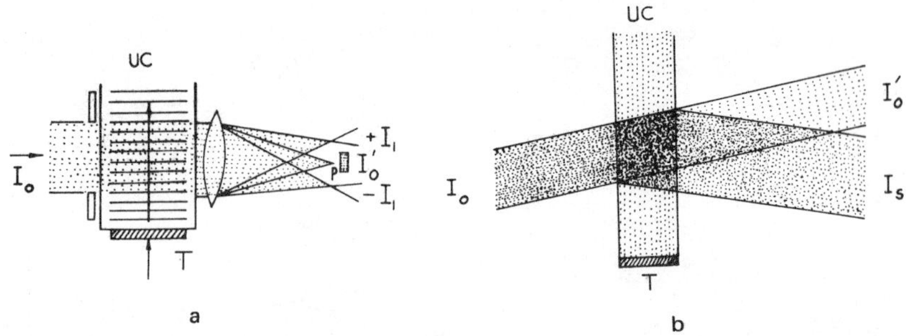

Fig. 5.32. Acousto-optic light modulators. (a) Debye–Sears operation; (b) Bragg operation. I_0, incident light beam of intensity I_0; $\pm I_1$, diffracted light beams of first orders of intensity I_1; I_0', zero order or central image of intensity I_0' obstructed by a plate P; UC, ultrasonic cell; T, transducer; I_s, Bragg-diffracted light beam of intensity I_s. Copyright © 1967 by the Institute of Electrical and Electronics Engineers, Inc. Reprinted, by permission, from *IEEE Spectrum*, Vol. 4, No. 5, May 1967, pp. 42–54.

leads us to the calculation of the photoelastic constants of a transparent crystal.

Owing to the propagation of thermally excited acoustic waves, elastic strains are developed in a crystal, and these strains fluctuate at each point, which in turn cause fluctuations in the optical dielectric constant tensor, since the latter depend on the former through the phenomenon of photoelasticity. Thus the intensity of the Brillouin components is related to the thermally excited acoustic waves.

By making an accurate study of the frequency shift, intensity, and polarization of each of the Brillouin components for different orientations of a crystal, one can obtain the directional variation of the photoelastic behavior of the crystal.

Leontowitsch and Mandelstamm [733] derived an expression for the intensity of the light transversely scattered by acoustic waves of thermal origin in a cubic crystal. Mueller [830] discussed in great detail the intensity of the Brillouin components in the light scattered by an isotropic solid, taking into consideration the contributions due to density scattering and

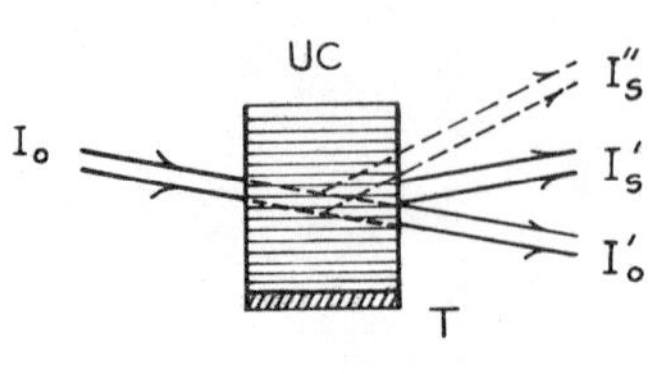

Fig. 5.33. Principle of acousto-optic modulation and deflection. I_0, incident light beam of intensity I_0; I_0', zero order or central image of intensity I_0'; I_s', I_s'', Bragg-diffracted light beams (being deflected at different angles) due to different acoustic frequencies; UC, ultrasonic cell; T, transducer. From [13].

to anisotropy or orientation scattering. Chandrasekharan [239; 240; 242] has given the theory of scattering of light in birefringent crystals and deduced expressions for the intensities of the 12 pairs of the Brillouin components that can appear in the most general case of a birefringent crystal; his theory is based on Mueller's theory of Brillouin scattering for isotropic solids. With the advent of lasers, studies on the Brillouin effect in crystals with a view to determining the photoelastic behavior of crystals have come into prominence, and the theory of Brillouin scattering has once again attracted several investigators, including Benedek and Fritsch [98], Hope [538], Vacher and co-workers [1246; 1247], and Humphreys and Maradudin [547]. Benedek and Fritsch applied their theory to cubic crystals KCl, RbCl, and KI, while Hope developed his theory to birefringent crystals, using the methods of physical optics.

We quote here, following Chandrasekharan [241], the final expressions for the intensity of light scattered in some special cases of the various Brillouin components for a cubic crystal belonging to any of the classes T_d, O, or O_h.

It is seen from Table 5.10 that for transverse scattering, the intensity due to the vertical electric component of the scattered radiation from a horizontally vibrating component of the incident light beam is equal to the intensity due to the horizontal component of the radiation scattered from a vertically vibrating incident light beam. Krishnan [682] discovered in his experiments on liquids such a reciprocity relation. Mueller [830] and Born and Huang [173] pointed out that the same reciprocity relation holds also for cubic crystals. However, Chandrasekharan [241] noted that, in general, V_h is not equal to H_v for cubic crystals; for example, for the acoustic wave normal along [111] and for transverse observation in the plane normal to [110], $H_v \neq V_h$.

Calculation of p_{ij}. The expressions in the last column in Table 5.10 relate the intensity of each one of the Brillouin components with the strain-optical constants p_{ij} and other optical and elastic constants of a crystal. Hence the individual values of p_{ij} can be determined by experimental studies on the Brillouin components.

From the standpoint of the photoelastic behavior of crystals, the early investigators were interested mainly in confirming Leontowitsch and Mandelstamm's theory regarding the relative intensities of the Brillouin components. Leontowitsch and Mandelstamm [733] have derived a formula for I, the sum of intensities of the three pairs of the Brillouin components in the transversely scattered light per unit volume of the scattering crystal,

Table 5.10. Transverse Brillouin Scattering in Cubic Crystals of Classes $\bar{4}3m\ (T_d)$, 432 (O), *and* $m3m\ (O_h)$[a]

Direction of the acoustic wave normal	Nature of the acoustic wave	Effective elastic stiffness constant	Plane of scattering T normal to	Intensity due to				Total intensity
				V_v	H_v	V_h	H_h	
[100]	L	c_{11}	[001]	p_{12}^2	0	0	$(p_{44}+\frac{1}{2}K)^2$	$\{p_{12}^2 + (p_{44} + \frac{1}{2}K)^2\}/c_{11}$
	T	c_{44} }		0	$\frac{1}{2}p_{44}^2$	$\frac{1}{2}p_{44}^2$	0	p_{44}^2/c_{44}
	T	c_{44} }						
[1$\bar{1}$0]	L	$c_{11}-\frac{1}{2}\gamma$	[001]	p_{12}^2	0	0	p_{44}^2	$(p_{12}^2 + p_{44}^2)/(c_{11} - \frac{1}{2}\gamma)$
	T	c_{44}		0	$\frac{1}{2}p_{44}^2$	$\frac{1}{2}p_{44}^2$	0	p_{44}^2/c_{44}
	T	$c_{44}+\frac{1}{2}\gamma$		0	0	0	0	0
[111]	L	$c_{11}-\frac{2}{3}\gamma$	[112]	$(p_{12}+\frac{1}{3}K)^2$	0	0	p_{44}^2	$\{(p_{12}+\frac{1}{3}K)^2+p_{44}^2\}/(c_{11}-\frac{2}{3}\gamma)$
	T	$c_{44}+\frac{1}{3}\gamma$ }		$K^2/18$	$\frac{1}{2}(p_{44}+\frac{1}{3}K)^2+K^2/36$	$\frac{1}{2}(p_{44}+\frac{1}{3}K)^2+K^2/36$	$K^2/72$	$\{(p_{44}+\frac{1}{3}K)^2+\frac{1}{8}K^2\}/(c_{44}+\frac{1}{3}\gamma)$
	T	$c_{44}+\frac{1}{3}\gamma$ }						

[a] The plane of scattering T is assumed to be horizontal. h denotes the horizontal component of the incident electric vector, so h is parallel to T. v denotes the vertical component of the incident electric vector, so v is perpendicular to T. H denotes the horizontal component of the scattered electric vector, so H is parallel to T. V denotes the vertical component of the scattered electric vector, so V is perpendicular to T. Thus V_v is the vertical component of the scattered electric vector for a unit vertical component of the incident electric vector. Similarly, we define H_v, V_h, and H_h [241]. The elastic anisotropy factor $\gamma = c_{11} - c_{12} - 2c_{44}$. The elasto-optic anisotropy factor $K = p_{11} - p_{12} - 2p_{44}$. From [241].

per unit solid angle. They have shown that

$$I = \frac{\pi^2 kTn^8}{2\lambda^4}\left\{\frac{2(p_{12}^2 + p_{44}^2)}{c_{11} + c_{12} + 2c_{44}} + \frac{p_{44}^2}{c_{44}}\right\} \tag{5.160}$$

where k is the Boltzmann constant and T the absolute temperature. Equation (5.160) holds only for a particular orientation of the crystal, namely, when the effective acoustic wave coincides with a dodecahedral direction and the plane of scattering is a cubic plane.

As pointed out by Mueller [830] the first term inside the brackets is the contribution due to longitudinal acoustic waves, while the second term represents that due to transverse waves. Computation of intensities of these two sets of Brillouin components, namely, longitudinal and transverse, on the basis of the strain-optical constants as determined by Ramachandran [997], leads to the conclusion that the transverse components are about 1.6 times more intense than the longitudinal components. This is in good agreement with an earlier experimental result of Krishnan [683], who has studied the light-scattering properties of diamond in great detail (see also [238]).

According to Eq. (5.160), deduced by Leontowitsch and Mandelstamm [733], the intensity of the Brillouin components is directly proportional to the absolute temperature of the scattering medium. This has been remarkably confirmed by Krishnan [683] by his investigations on diamond. He has shown by a series of photographs that the intensities of the Brillouin components are markedly influenced by temperature. One can notice from the photographs of the Brillouin spectra of diamond that at room temperature, the aggregate intensity of the components is less than one-fourth the intensity of the Raman line at 1332 cm^{-1}. Furthermore, in contrast with the 1332 cm^{-1} Raman line, the intensity of which is practically unaffected by the change of temperature of the experimental crystal specimen of diamond, the Brillouin components become rapidly weaker with falling temperature. This is to be expected on theoretical grounds, since for the Brillouin components, $h\nu_B \ll kT$, while for the Raman line $h\nu_R \gg kT$.

6

Atomistic Theory of Photoelasticity of Cubic Crystals

6.1. Introduction

A transparent solid, amorphous or crystalline, exhibits artificial double refraction under the influence of a mechanical stress. A cubic crystal belonging to any of the classes $\bar{4}3m$ (T_d), 432 (O), and $m3m$ (O_h) develops uniaxial birefringence under a stress along a cube normal or cube diagonal. The birefringence produced in the former case involves the differential strain-optical constant $(p_{11} - p_{12})$, while that in the latter case involves the strain-optical constant p_{44}. Under a stress in any other direction every cubic crystal becomes biaxial. In all cases the birefringence arises due to the changes brought about by a physical deformation of the crystal in the Lorentz–Lorenz field, the Coulomb field, and the intrinsic polarizabilities of the ions. The theory of photoelasticity has so far been attempted only for cubic crystals because the evaluation of the contributions due to the several changes mentioned above is extremely difficult for crystals of lower symmetry. We give below a brief review of the several theories of the photoelastic effect in cubic crystals.*

The earliest attempts to explain the natural birefringence of crystals are those of Ewald [332; 333] and Born [172]. According to them, the natural birefringence of crystals is due to the anisotropy of the Lorentz–Lorenz force. In ionic crystals the anisotropy of the Coulomb forces is yet another cause of birefringence. Taking into account both these effects, Hylleras [550; 551] calculated the optical properties of calomel, rutile, anatas, and quartz. Bragg [184; 185] developed a somewhat different theory by which he explained the natural double refraction of some carbonates and nitrates. This explanation is based primarily upon the optical anisotropy of the CO_3 and NO_3 groups in these crystals.

* Mueller's atomistic theory of photoelasticity of cubic crystals (1935) is based on Pockels' phenomenological theory, which does not distinguish T and T_h classes from T_d, O, and O_h classes. The distinction was made by Bhagavantam (1942).

As already mentioned, a cubic crystal subjected to a mechanical stress normal to a cube face gets deformed and becomes doubly refracting. Pockels assumed that except for optical properties the crystal remains cubic under low stresses.* Proceeding on the lines of the theories of natural double refraction formulated earlier, one may hope to explain the mechanical double refraction of cubic crystals. However, neither Ewald–Born's nor Bragg's theory can satisfactorily explain the photoelastic behavior of even cubic crystals.

We shall briefly outline some salient features of the earlier theories in general and Mueller's [826] theory in particular to explain the photoelastic effect in cubic crystals.

The attempts to explain photoelastic birefringence have been less successful compared with those for natural double refraction. Banerjee [65] considered the effect of physical deformation of the crystal on the Lorentz–Lorenz field. In order to evaluate the contribution by the Lorentz–Lorenz field to the mechanical birefringence produced in a cubic crystal, a spherical or cubical boundary is taken around the chosen atom or ion. The solid outside the boundary is considered to be a continuum, and the field of the dipoles within the boundary (or cavity) is superposed on the field due to the polarized continuum, the latter being $(4/3)\pi P$, where P is the polarization. While applying this method Banerjee considered a cubical boundary; however, he assumed that the boundary remained cubic in both the unstrained and strained states of the crystal; this assumption avoids any change in the field due to the continuum outside the boundary whether the crystal is free or strained. Mueller criticized such an assumption as wholly untenable. A cubical cavity in the unstrained state cannot remain cubical in the deformed state of the crystal, and it must become a rectangular parallelepiped. Allowing for such a change in the geometry of the cavity, Mueller computed the effect of the continuum and obtained the factors which determine the change of the Lorentz–Lorenz field caused by a uniaxial or linear strain and by a shear, and found that the results thus obtained agreed very well with those by the Ewald–Born method, thus proving that the Ewald–Born and the Lorentz–Bragg methods are equivalent. Banerjee compared his predictions with the observed behavior of only two substances, NaCl and KCl, and found satisfactory agreement in these two cases; he did not apply his theory to any other substance. Furthermore, he did not consider the Coulomb field.

Herzfeld [509] and Herzfeld and Lee [510], using the lattice sum

* Otherwise, for a cubic crystal under stress there should be seven constants instead of three.

method of Madelung, took into consideration the effect of both the Lorentz–Lorenz field and the Coulomb field in the deformed crystal, and obtained the resultant modification of the electronic energy levels which determine the polarizability of the ions; but they ignored the change in the polarizability of ions due to strain. They made several assumptions in their theory regarding the proper values of some constants. Furthermore, Mueller pointed out some errors in their computations.

Mueller [826] has shown that yet another cause of the photoelastic effect must be considered besides those considered by Banerjee [65], Herzfeld [509], and Herzfeld and Lee [510]. A deformation of the lattice changes the energy levels and transition probabilities of the optical electrons, and hence alters the refraction of the atoms. Following the results of Fajans and Joos [338], who showed that in crystals the refraction of ions is usually smaller than in solutions, the decrease being attributed to the Coulomb forces between the ions, Mueller assumes that a decrease of the lattice constant reduces the refraction still further. But, he does not attribute this effect to the Coulomb forces between the ions. The effect exists also in homopolar crystals and is caused by a change of the charge distribution or the binding of the optical electrons, and is not due to the Coulomb forces between the ions. It can be considered to have been produced by the repulsive forces between the atoms and depends primarily on the arrangement of, and the distance between, the nearest neighboring atoms. Heckmann [505] discussed the possibility that the repulsive forces may deform, and even polarize an atom, giving rise to negative polarizabilities. In photoelastic experiments, the lattice distances parallel to the stress are decreased, and the distances normal to the stress are increased; this gives rise to an anisotropy of the atomic refractions. Besides the Lorentz–Lorenz and Coulomb fields, this anisotropy of the atomic refractions is the third cause considered, and is investigated in detail by Mueller for simple cubic lattices.

By taking into account all these effects, namely, Lorentz–Lorenz field, Coulomb field, and optical anisotropy due to strain or deformation, Mueller was able to give a fairly satisfactory explanation of the observed photoelasticity of some simple cubic crystals, with *D* lattices. In developing his theory of photoelasticity, he seems to have been guided by the numerical values regarding the photoelastic behavior of just a few crystals [826], and that too only at one wavelength of light. Subsequent experimental work by Iyengar [565; 566], Srinivasan [1174–1176], and Bansigir and Iyengar [66; 67] on the dispersion of photoelasticity in some cubic crystals has warranted some modifications, particularly regarding λ_0, the strain polarizability of ions.

6.2. Mueller's Theory—A Brief Survey

Mueller [826] has taken into consideration the contribution due to the Lorentz–Lorenz field (being the sum total of the effect of the continuum outside the boundary and that of the dipoles within the cavity) and the Coulomb field due to the ions within the strained cavity. He found that the calculated value of the change in index of refraction was different from the observed value. In order to bring about reconciliation between theory and experiment, he had to introduce a correction factor, and he achieved this by assuming that the atoms (or ions) become optically anisotropic under a physical strain. Following Fajans and Joos [338], he assumed that a compression reduces the polarizabilities of the ions. In evaluating the contribution of the polarizability change due to strain, he assumed a linear proportionality between the molar polarizability of the ions and the strain, provided the strains are small. He thus introduced the concept of the strain polarizability constant.

When a crystal is subjected to a mechanical stress, the solid experiences a physical strain. In the experimental investigations on the photoelastic behavior of crystals, it is common to apply a linear stress on the crystal in an appropriate direction. Though the stress is linear, the resultant strain in the solid is not the same in all directions due to the anisotropy in the elastic behavior of crystals (including the cubic system). Therefore any physical property which depends upon the arrangement of the atoms and ions must reflect this anisotropic nature of the strain. Having realized this fact, Mueller assumed that the polarizability of ions of the strained crystal is different in different directions, i.e., the elastic deformations produce optical anisotropy of the atoms or ions, thus conforming to the anisotropic strain under a linear stress. Hence in the final analysis he combined the strain polarizability with the Lorentz–Lorenz field and the Coulomb field in order to explain the mechanical birefringence in cubic crystals with *D* lattices (e.g., NaCl, KCl, CaF_2, and CsCl).

That the polarizability of ions changes with strain follows readily from the classical electromagnetic theory as shown below.

Following the Lorentz–Lorenz theory, the interaction of light with crystalline matter under a mechanical stress can be represented by the influence of the electromagnetic field of the incident light wave on a charged oscillator, which in this case is an electron. The incident electric vector of light radiation causes the electron to get displaced from its equilibrium position. The extent to which the electron or the oscillator gets displaced depends upon the incident electric vector. However, since we are dealing with an ionic crystal, the oscillator is influenced not only by the applied field, which is due to the incident electric vector, but also by the Coulomb fields, due to other ions, and the Lorentz–Lorenz field, which itself is a consequence of the medium getting polarized due to the external electric field. The displacement of the electron (the charged particle) results in the production of a dipole moment of value

$$\boldsymbol{\mu}_j = \alpha_j \mathbf{E}_{\text{eff}} \tag{6.1}$$

where α_j is the polarizability of the ion and $\mathbf{E}_{\text{eff}}$ is the effective electric field.

From classical electromagnetic theory of light, we write down the equations

$$\mathbf{D} = k\mathbf{E} \tag{6.2}$$

and

$$\mathbf{D} = \mathbf{E} + 4\pi\mathbf{P} \tag{6.3}$$

Here $\mathbf{D}$ is the dielectric displacement caused by an external field $\mathbf{E}$ acting on a medium of dielectric constant k, and $\mathbf{P}$ is the dipole moment induced per unit volume. Maxwell's equations describing the propagation of electromagnetic waves through a nonabsorbing dielectric medium result in the relationship

$$k = n^2 \tag{6.4}$$

where n is the refractive index of the medium. In deriving this equation it is assumed that the magnetic permeability is equal to one. Combining Eqs. (6.2) and (6.3) with (6.4) we have

$$(n^2 - 1)\mathbf{E} = 4\pi\mathbf{P} \tag{6.5}$$

The quantity $\mathbf{P}$ is the sum total of the contributions by the different oscillators. If in a cubic centimeter N_j is the number of oscillators of type j, and $\boldsymbol{\mu}_j$ the electric moment per oscillator, then the total dipole moment per unit volume is given by

$$\mathbf{P} = \sum_j N_j \boldsymbol{\mu}_j = \sum_j N_j \alpha_j \mathbf{E}_{\text{eff}} \tag{6.6}$$

Hence

$$(n^2 - 1)\mathbf{E} = 4\pi\mathbf{P} = 4\pi \sum_j N_j \alpha_j \mathbf{E}_{\text{eff}} \tag{6.7}$$

While studying the photoelastic phenomenon, we are concerned with the change in the refractive index brought about by a mechanical strain developed in the medium. Such a relation is obtained by differentiating Eq. (6.7) with respect to the strain ε. Thus we have*

$$2n \frac{dn}{d\varepsilon} \mathbf{E} = 4\pi \sum_j \left[\frac{dN_j}{d\varepsilon} \alpha_j \mathbf{E}_{\text{eff}} + N_j \frac{d\alpha_j}{d\varepsilon} \mathbf{E}_{\text{eff}} + N_j \alpha_j \frac{d\mathbf{E}_{\text{eff}}}{d\varepsilon} \right] \tag{6.8}$$

The first term inside the brackets describes the effect due to the change in the number of oscillators of type j per cubic centimeter due to the strain. The third term describes the effect due to the change in the total effective electric field due to strain. Both these terms were known and considered by earlier workers. Mueller's contribution to the atomistic theory lies in his recognizing the role of the second term $d\alpha_j/d\varepsilon$ in the photoelastic phenomenon. Thus it can be readily seen that the polarizability of an ion depends upon the strain, and Mueller assumed that the strain polarizability $d\alpha_j/d\varepsilon$ is a constant for small strains, which means that the polarizability is a linear function of strain.

The fact that α, the polarizability of an ion, changes with strain can also be seen from thc following.

Let us take the Lorentz–Lorenz equation in the form

$$\alpha = \frac{3}{4\pi N} \frac{n^2 - 1}{n^2 + 2} \tag{6.9}$$

where N is the number of molecules per unit volume. The molar refractivity R can be expressed as

$$R = \frac{4\pi}{3} A\alpha = \frac{M}{\varrho} \frac{n^2 - 1}{n^2 + 2} \tag{6.10}$$

where A is the Avogadro's number, M the molecular weight, and ϱ the density.

Let us now study the effect of hydrostatic pressure on the refractive index of a cubic crystal. Differentiating Eq. (6.10), we obtain

$$\left(\frac{dn}{d\varrho} \right)_R = \frac{(n^2 - 1)(n^2 + 2)}{6n\varrho} \tag{6.11}$$

considering the molar refractivity R independent of changes in density.

From the available experimental data for $dn/d\varrho$, Mueller observed that $\varrho(dn/d\varrho)$ is less than $(n^2 - 1)(n^2 + 2)/6n$. Hence he concluded that the

* This interpretation of the term "strain polarizability" is due to Pettersen [922].

molar refractivity R is not independent of the density. Since $R = \frac{4}{3}\pi A\alpha$, we note that α is also not independent of density. Now taking the relation $\varrho V =$ mass, V being the volume,

$$\frac{d\varrho}{\varrho} = -\frac{dV}{V}$$

Thus α, which is not independent of density, is also not independent of strain. This is the information revealed to us by Eq. (6.8) also. Hence we see that $d\alpha/d\varepsilon$ describes the strain polarizability.

Introducing Mueller's Parameters p_x, p_z, and $p_{x'}$. The strain-optical constants p_{ij} in the phenomenological theory of Pockels can be related to terms in Eq. (6.8) by the following equation:

$$p_x = p_o + (p_x)_L + (p_x)_C + (p_x)_A \tag{6.12}$$

with similar equations for p_z and p_x'. Here p_o represents the contribution from the change in the oscillator density, $(p_x)_L$ that arising from the Lorentz–Lorenz part of the effective field, and $(p_x)_C$ that from the change in the Coulomb part of the effective field; hence the two $(p_x)_L$ and $(p_x)_C$ together arise out of $d\mathbf{E}_{\text{eff}}/d\varepsilon$. Finally, $(p_x)_A$ is the contribution from the change in polarizability $d\alpha_j/d\varepsilon$. The importance and success, to the extent achieved, of Mueller's theory is in the term $(p_x)_A$ arising out of $d\alpha_j/d\varepsilon$.

The lone term p_x on the left-hand side of Eq. (6.12) can be related to p_{ij} as shown below.

We know from Pockels' theory that

$$(B_i - B_i^0) = p_{ij}\varepsilon_j$$

Therefore

$$\Delta n_i = (-n_o^3/2)p_{ij}\varepsilon_j \qquad (i, j = 1\text{–}6) \tag{6.13}$$

On differentiating both sides of the above equation with respect to ε_j we obtain

$$p_{ij} = -\frac{2}{n_o^3}\frac{\partial(\Delta n_i)}{\partial\varepsilon_j} = -\frac{2}{n_o^3}\frac{\partial(n_i - n_o)}{\partial\varepsilon_j} = -\frac{2}{n_o^3}\frac{\partial n_i}{\partial\varepsilon_j} \tag{6.14}$$

since n_o is constant.

Comparing Eq. (6.14) with Eq. (6.8) we note that p_{ij} is related to $dN_i/d\varepsilon$, $d\alpha/d\varepsilon$, and $d\mathbf{E}_{\text{eff}}/d\varepsilon$. Under a linear stress along the Z axis, the constant p_x introduced by Mueller is related to p_{12} of Pockels' scheme thus:

$$\left.\begin{aligned} p_x &= \frac{n^4}{(n^2-1)^2} p_{12} = \frac{-2n}{(n^2-1)^2} \frac{\partial n_1}{\partial \varepsilon_3} \\ \text{Similarly,} \quad p_z &= \frac{n^4}{(n^2-1)^2} p_{11} = \frac{-2n}{(n^2-1)^2} \frac{\partial n_3}{\partial \varepsilon_3} \\ \text{and} \quad p_x' &= \frac{2n^4}{(n^2-1)^2} p_{44} = \frac{-2n}{(n^2-1)^2} \frac{\partial n_4}{\partial \varepsilon_3} \end{aligned}\right\} \tag{6.15}$$

Here n is the refractive index for the unstrained state and n_1, n_3, and n_4 are for the strained state of the crystal.

We see from Eq. (6.12) that the three constants p_x, p_z, and p_x' introduced by Mueller are all related to the strain-induced changes in the oscillator density, the polarizability, and the effective field.

6.3. *Effect of Hydrostatic Pressure on the Index of Refraction* n; *the Strain Polarizability Constant* λ_0

When a solid is subjected to a mechanical stress, its optical properties are changed due to a slight change in the distribution of its electron cloud. The calculation of the photoelastic constants of an ionic crystal is based on the evaluation of the changes in (i) the Lorentz–Lorenz field, (ii) the Coulomb field, and (iii) the intrinsic polarizability of the ions.

Because of the inherent symmetry of a cubic crystal in the undeformed state, the contribution of the Coulomb field to the refractive index is zero; however, since the cubic symmetry is slightly disturbed when the crystal undergoes a deformation anisotropically, the contribution of the Coulomb field becomes significant. On the other hand the changes in the Lorentz–Lorenz field are dependent on the changes in both density and geometrical symmetry. Thus when a cubic crystal is subjected to a hydrostatic stress, the density increases, and hence the Lorentz–Lorenz field is altered; the latter's contribution to the refractive index becomes significant, whereas the contribution due to the Coulomb field remains unaltered, and thus it is zero. But under a unidirectional or linear stress, say along a cube axis, there results an anisotropy in both the Coulomb field and Lorentz–Lorenz field.

Hydrostatic pressure on a crystal causes a change in the density, which in turn changes the refractive index of the crystal. But under a hydrostatic pressure the geometrical symmetry of the crystal is not altered, and hence the symmetry in the Lorentz–Lorenz and Coulomb forces is not changed; consequently, the two do not directly contribute to the polarizability of the ions. There is only a change in the polarizability of ions due to strain resulting in a change of the refractive index with density ϱ, which alone has to be considered now.

Let n be a function of density ϱ and ionic molar refraction R; thus let

$$n = f(\varrho, R)$$

Then

$$\frac{dn}{d\varrho} = \left(\frac{\partial n}{\partial \varrho}\right)_R + \left(\frac{\partial n}{\partial R}\right)_\varrho \frac{dR}{d\varrho} \tag{6.16}$$

We see that the previous discussion considered only the first term on the right. This expression can be evaluated by assuming either the Lorentz–Lorenz formula or the Drude formula for the crystal. Following Mueller [826] we shall use here the Lorentz–Lorenz formula for evaluating $(\partial n/\partial \varrho)_R$.

Then taking the Lorentz–Lorenz equation in the form

$$R = \frac{M}{\varrho} \frac{n^2 - 1}{n^2 + 2}$$

and differentiating it with respect to ϱ, we get

$$\frac{dn}{d\varrho} = \frac{(n^2 - 1)(n^2 + 2)}{6n\varrho} + \frac{(n^2 - 1)(n^2 + 2)}{6nR} \frac{\partial R}{\partial \varrho} \tag{6.17}$$

Assuming that R varies linearly with density and hence with strain, let

$$\begin{aligned} R &= R_0\left(1 + \lambda_0 \frac{\Delta V}{V}\right) \\ &= R_0\left(1 - \lambda_0 \frac{\partial \varrho}{\varrho}\right) \end{aligned} \tag{6.18}$$

so that

$$\frac{\partial R}{\partial \varrho} = -\lambda_0 \frac{R_0}{\varrho} \tag{6.19}$$

where R_0 is the molar refractivity of the unstrained crystal and λ_0 is a phenomenological constant, called the strain polarizability constant, with the subscript zero indicating the case of hydrostatic pressure. Hence Eq.

(6.17) becomes

$$\frac{dn}{d\varrho} = \frac{(n^2 - 1)(n^2 + 2)}{6n\varrho} + \frac{(n^2 - 1)(n^2 + 2)}{6nR_0}\left(-\frac{\lambda_0 R_0}{\varrho}\right) \tag{6.20}$$

$$= (1 - \lambda_0)\frac{(n^2 - 1)(n^2 + 2)}{6n\varrho} = (1 - \lambda_0)\left(\frac{\partial n}{\partial \varrho}\right)_R \tag{6.21}$$

Thus the observed values of $\varrho(dn/d\varrho)$ are less than the values predicted by the expression $(n^2 - 1)(n^2 + 2)/6n$, and from their difference λ_0 can be calculated. See under "Discussion" (p. 310) for calculation of λ_0 from photoelastic data.

6.4. Anisotropy of R_j and λ_j

The experimental observation that

$$\varrho\frac{dn}{d\varrho} \neq \frac{(n^2 - 1)(n^2 + 2)}{6n}$$

leads to our assumption that R is a linear function of strain as in Eqs. (6.18) and (6.19). Once we recognize this fact, we should note that the actual value of R depends upon both the nature and magnitude of strain; that is, R for a linear strain is different from that for a shearing strain. Hence for a linear strain ε_z along the Z direction, and for a light vector in the X direction,

$$R_{jx} = R_j(1 + \lambda_{jx}\varepsilon_z) \tag{6.22}$$

where R_j is the molar refractivity of the jth ion in the unstrained condition, R_{jx} that for the same jth ion under strain ε_z for light vibrations with their electric vector parallel to the X axis, and λ_{jx} is the new phenomenological strain polarizability constant of the crystal under strain for light vibrations parallel to the X axis.

Similarly,

$$R_{jz} = R_j(1 + \lambda_{jz}\varepsilon_z) \tag{6.23}$$

For a strain along a face diagonal, say the X' direction, and for light vibrations along the X' and Y' directions (the light being propagated along the Z axis) we have

$$R_{jx'} = R_j(1 + \lambda_{jx'}\varepsilon_{x'}) \tag{6.24}$$

and

$$R_{jy'} = R_j(1 + \lambda_{jy'}\varepsilon_{x'}) \tag{6.25}$$

The description of the phenomenon in Eqs. (6.24) and (6.25) by introducing two different symbols $\lambda_{jx'}$ and $\lambda_{jy'}$ [and similarly in Eqs. (6.22) and (6.23)] is in recognition of the fact that the strain polarizability is an anisotropic property, arising from the deformation of the electron cloud surrounding the ion when the crystal is strained.

Final Expression for p_{ij} in Terms of R_j. Molar refractivity R is a function of the ionic polarizability α, which is a function of ε, and ε governs, through p_{ij}, the changes in the refractive index n. Therefore the p_{ij} are related to the R's, the molecular refractivities of the different ions in an ionic cubic crystal.

Thus combining the several relations for crystals of the NaCl type, Mueller's equations for p and p', the birefringence produced for linear and shear stresses, respectively, are as follows*:

$$\begin{aligned} p &= \frac{n^4(p_{11} - p_{12})}{(n^2 - 1)^2} \\ &= \frac{1}{(R_1 + R_2)^2}\left[0.09(R_1{}^2 + R_2{}^2) + 5.82R_1R_2 + 2.82\left(\frac{R_1{}^2}{f_1} - \frac{R_2{}^2}{f_2}\right)\beta\right] \\ &\quad - \frac{\lambda}{3}\frac{n^2 + 2}{n^2 - 1} \end{aligned} \tag{6.26}$$

and

$$\begin{aligned} p' &= \frac{2n^4 p_{44}}{(n^2 - 1)^2} \\ &= \frac{1}{(R_1 + R_2)^2}\left[0.60(R_1{}^2 + R_2{}^2) - 2.52R_1R_2 - 1.86\left(\frac{R_1{}^2}{f_1} - \frac{R_2{}^2}{f_2}\right)\beta\right] \\ &\quad - \frac{\lambda'}{3}\frac{n^2 + 2}{n^2 - 1} \end{aligned} \tag{6.27}$$

where R_1 and R_2 are the ionic refractivities of the negative and positive ions, respectively, f_1 and f_2 their oscillator strengths, β is the valence of both the ions, which in the case of the alkali halides is unity, and λ and λ' are the strain polarizabilities.

* In this section λ stands for the strain polarizability but not for wavelength.

The linear-strain polarizability is

$$\lambda = (\lambda_z - \lambda_x) = \frac{\sum_j \lambda_{jz} N_j R_j - \sum_j \lambda_{jx} N_j R_j}{\sum_j N_j R_j} \tag{6.28a}$$

and the shear-strain polarizability is

$$\lambda' = \frac{\sum_j (\lambda_{jx'} - \lambda_{jy'}) N_j R_j}{2 \sum_j N_j R_j} \tag{6.28b}$$

The strain polarizability λ_0 under hydrostatic pressure [Eq. (6.18)] is related to λ_x and λ_z by the equation

$$\lambda_0 = \tfrac{1}{3}(\lambda_z + 2\lambda_x) \tag{6.29}$$

The first two terms in the brackets in Eqs. (6.26) and (6.27) are due to the Lorentz–Lorenz contribution to the effective field, while the third term is the Coulomb contribution to the effective field. The term outside the brackets is due to the optical anisotropy of the ions due to strain, and hence represents the contribution of the strain polarizability. The magnitude and the sign of birefringence, therefore, depend on the relative contributions of the above-mentioned factors.

Discussion

i. In order to calculate $(p_{11} - p_{12})$ and p_{44} from Mueller's theory, one requires a knowledge of R_i, f_i, λ, and λ' entering Eqs. (6.26) and (6.27). At least some approximate values of R_1, R_2, f_1, and f_2 are known for several ions, but λ and λ' are not known.

ii. On the other hand one can compute λ and λ' from the experimentally determined values of p and p' in Eqs. (6.26) and (6.27) and the calculated values of the Lorentz–Lorenz and Coulomb contributions using R_i and f_i determined otherwise. Mueller [826] found that λ_0 has roughly the same value for NaCl and KCl.

iii. From Eq. (6.21) we have

$$\frac{dn}{d\varrho} = (1 - \lambda_0)\left(\frac{\partial n}{\partial \varrho}\right)_R$$

Now $dn/d\varrho$, which describes the variation of refractive index with density under hydrostatic pressure, can be determined experimentally.

We can relate $dn/d\varrho$ with p_{ij} by the following equation. Under hydrostatic pressure the strains $\varepsilon_{xx} = \varepsilon_{yy} = \varepsilon_{zz}$. For cubic crystals of T_d, O, and O_h classes, according to Pockels,

$$-\frac{2\Delta n}{n^3} = (B_{11} - B_{11}^0) = (p_{11} + 2p_{12})\varepsilon_{xx}$$

and the strain

$$\varepsilon_{xx} = \frac{1}{3}\frac{dV}{V} = -\frac{1}{3}\frac{d\varrho}{\varrho}$$

Hence

$$\frac{dn}{d\varrho} = \frac{n^3}{6\varrho}(p_{11} + 2p_{12}) \qquad (6.30)$$

from which $dn/d\varrho$ can be calculated from the experimentally determined values of p_{11} and p_{12}. This procedure was first adopted by Pockels to obtain $\varrho(dn/d\varrho)$ for certain cubic crystals.

Since Δn, the observed increase* of n with density ϱ, is smaller than the value given by differentiating the Lorentz–Lorenz equation [Eq. (6.11)], Mueller introduced the factor λ_0 to account for the observations on the assumption that the compression reduces the polarizabilities of the atoms or ions, giving the equation

$$\varrho\frac{dn}{d\varrho} = \frac{(n^2 - 1)(n^2 + 2)}{6n}(1 - \lambda_0) \qquad (6.21')$$

Here λ_0 can be calculated from a knowledge of p_{ij} using Eqs. (6.30) and (6.21). But, from Eq. (6.29) $\lambda_0 = \frac{1}{3}(\lambda_z + 2\lambda_x)$, and from Eq. (6.28) $\lambda = \lambda_z - \lambda_x$. Thus from a knowledge of λ calculated from Eq. (6.26), and λ_0 from Eq. (6.21), we can obtain the individual values of λ_z and λ_x. Some typical values of λ_z and λ_x for sodium halides are shown in Table 6.1.

iv. Burstein and Smith [218] have given a physical meaning to the term λ_0. When matter changes from a gaseous to a solid state there follows a decrease in the molar polarizability owing to the overlap of the electronic cloud of the ions in the solid state. And now if the solid is compressed, thereby increasing further its density, there is an additional decrease in the polarizability of the ions. λ_0 represents this change and hence is a measure of the overlap of the electron cloud of the ions, when the crystal lattice is strained, and the magnitude of λ_0 will represent the homopolar bonding

* Δn is obtained from p_{11} and p_{12}.

Table 6.1. Strain Polarizability Constants[a]

	λ_0	$\lambda = \lambda_z - \lambda_x$	λ_z	λ_x	λ'
NaF	0.590	1.486	1.581	0.095	−0.096
NaCl	0.379	1.177	1.164	−0.013	0.058
NaBr	0.305	1.158	1.077	−0.081	0.087

[a] Taken from Pettersen's thesis [922].

present in the crystal. Mueller, who first calculated λ_0 for NaCl, KCl, and CaF_2, concluded that for all cubic crystals λ_0 has a value nearly equal to 0.5. But the subsequent experimental research on the photoelastic dispersion in some cubic crystals by Iyengar [565] and Bansigir and Iyengar [66; 67] contradicts Mueller's conclusion regarding the constancy of λ_0 (see Sections 6.6 and 6.7 for further comments).

6.5. Thermo-Optic Behavior of Crystals and Photoelastic Behavior

Pockels has noted that the thermal variation of the refractive index of a solid is due to two factors: (i) the physical deformation caused by temperature; here the physical deformation or strain causes a change in the refractive index through the photoelastic effect; and (ii) the direct influence of temperature on the refractive index.

Thus the thermo-optic behavior can be represented by

$$\frac{dn}{dt} = \left(\frac{dn}{dt}\right)_{\text{dil}} + \frac{\partial n}{\partial t} \tag{6.31}$$

Here dn/dt is the experimentally observed change in the refractive index due to temperature, $(dn/dt)_{\text{dil}}$ that due to the thermal dilatation, and $\partial n/\partial t$ the true temperature coefficient.

The physical significance of Eq. (6.31) can be described thus: When a crystal experiences a deformation due to either a mechanical stress or a temperature change, its refractive index changes. Although the lattice parameter is altered to the same extent by the application of a hydrostatic stress or by lowering the temperature, the changes in the refractive index are not the same in the two cases, thereby indicating that the refractive index is not only a function of the lattice parameter, but there is also a pure temperature effect. This is due to the fact that, even if the lattice parameters are the same, the ions in the crystal at a higher temperature vibrate with greater amplitudes. Consequently this increases the overlap of the electron clouds, which in turn affects the polarizability of the ions.

The contribution $(dn/dt)_{\text{dil}}$ due to thermal dilatation can be calculated from the known values of p_{ij}, the strain optical coefficients, and α_{ij}, the linear thermal expansion coefficients. Since the strain ε_{ij} in this case is equal to α_{ij}, the fractional change in length per 1°C rise of temperature, we have

$$B_1 - B_1^0 = (p_{11}\alpha_{xx} + p_{12}\alpha_{yy} + p_{13}\alpha_{zz}) \tag{6.32}$$

In Eq. (6.32) α_{ij} represent the coefficients of linear thermal expansions and not the direction cosines.

Now, for a rise of temperature of ΔT, we find

$$-\frac{2\Delta n}{n^3} = (p_{11} + 2p_{12})\alpha_{xx}\Delta T$$

for a cubic crystal of T_d, O, and O_h classes, because $p_{13} = p_{12}$ and $\alpha_{xx} = \alpha_{yy} = \alpha_{zz}$. Hence Δn per 1°C rise of temperature is given by

$$\left(\frac{dn}{dt}\right)_{\text{dil}} = -\frac{n^3}{2}(p_{11} + 2p_{12})\alpha_{xx} \tag{6.33}$$

As an example, the data [1302] for CaF_2 are as follows. At $\lambda = 5890$ Å: $dn/dt = -1.15 \times 10^{-5}$ at 20°C; $p_{11} = 0.0258$ and $p_{12} = 0.202$; $n = 1.434$; $\alpha_{xx} = 20.01 \times 10^{-6}$. Then $\partial n/\partial t = 0.118 \times 10^{-5}$ per °C.

$\partial n/\partial t$, the pure thermal coefficient of variation of refractive index, is always positive though it is very small for some crystals.

6.6. Pockels' Photoelastic Groups in Cubic Crystals and Mueller's Theory

Pockels [956] has divided the cubic crystals into four different groups,* according to the signs of their strain-optical constants $(p_{11} - p_{12})$ and p_{44} as shown in Table 6.2. Pockels could not give any example under group I.

* This is in accordance with Pockels' scheme of photoelastic matrices, wherein $p_{13} = p_{12}$ for all classes of cubic crystals.

Table 6.2. Photoelastic Groups in Cubic Crystals

Group number	Sign of $(p_{11} - p_{12})$	Sign of p_{44}	Examples given by Pockels
I	+	+	—
II	−	−	NaCl
III	−	+	CaF_2
IV	+	−	KCl

Having obtained the strain polarizability constant λ_0 under hydrostatic pressure from Eq. (21) for NaCl, KCl, and CaF_2, Mueller found it to be nearly 0.5 for all the three crystals. On the basis of these results and those obtained by him for glasses, he felt that λ_0 would be nearly the same for all cubic crystals. He did not discuss either the significance of λ_0 or its dependence on wavelength of light and temperature. However, having set arbitrary limits for the contribution of polarizability change due to strain in the case of linear and shear stresses, Mueller has predicted the photoelastic behavior of various cubic crystals. Following Pockels' classification of cubic crystals into four groups, Mueller [826] has predicted that crystals of the NaCl type such as K, Rb, and NH_4 halides and the bivalent salts of BaO, CaO, etc., would come under group IV, while Li and Na halides would come under group II, and CsCl and CaF_2 under group III. Crystals of ZnS and Na_2O should also come under group III irrespective of whether the binding is homopolar or ionic. There was no example available under group I.

Recent experimental investigations on a number of cubic crystals of the NaCl type have confirmed Mueller's predictions in regard to their photoelastic group, for example K and Rb halides. His prediction has also been found to be correct in the case of NH_4Cl at room temperature [841]. This crystal belongs to the CsCl type below 120°C, above which it changes its phase to the NaCl type. It will be interesting to investigate if p_{44} of NH_4Cl changes in sign from positive to negative as the crystal is heated to well beyond the phase transition.

On the other hand we note that departures from Mueller's predictions have been found. For example, thallium halides which have the crystal structure of the CsCl type should belong to group III according to Mueller, but West and Makas [1382] and Burstein, Smith, and Henvis [216] have experimentally found that the thallium halides come under group I. In fact, these crystals are the only examples under group I, for which examples were not available earlier. Similarly, ZnS, which according to Mueller should belong to group III, has for its p_{44} a negative sign, which also is contrary to Mueller's prediction [826].

6.7. *Photoelastic Dispersion in Cubic Crystals; λ_0 as a Function of Crystalline Material, Wavelength of Light, and Temperature*

Pockels' and Mueller's classification of cubic crystals into four groups breaks down when the photoelastic behavior of some cubic crystals is observed over a range of wavelengths instead of at one wavelength. See Fig. 6.1 for the dispersion of KCl, KBr, and KI by Srinivasan [1174; 1175], and also Figs. 6.2, 6.3, 6.8, and 6.9 for the dispersion of KCl, KBr, KI and RbCl, RbBr, RbI by Rahman and Iyengar [984; 985]. Dubenskii, Kaplyanskii, and Lozovskaya [307] studied the photoelastic dispersion in cubic ZnSe in the spectral region 5625–4610 Å, and have observed a reversal of sign of the effect in the spectral region around 5250 Å. Wardzyoski [1352] has noticed in cadmium telluride single crystals that the birefringence induced by uniaxial stress causes a reversal of sign near the absorption edge.

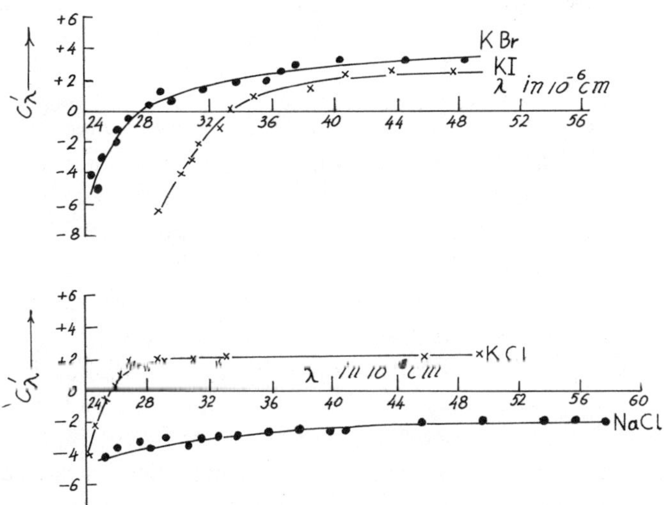

Fig. 6.1. Stress-optical dispersion of KBr, KI, KCl, and NaCl. λ is in units of 10^{-6} cm; $C_\lambda' = \frac{1}{2}n_0^3(q_{11} - q_{12})$, and is measured in brewsters. From [1175].

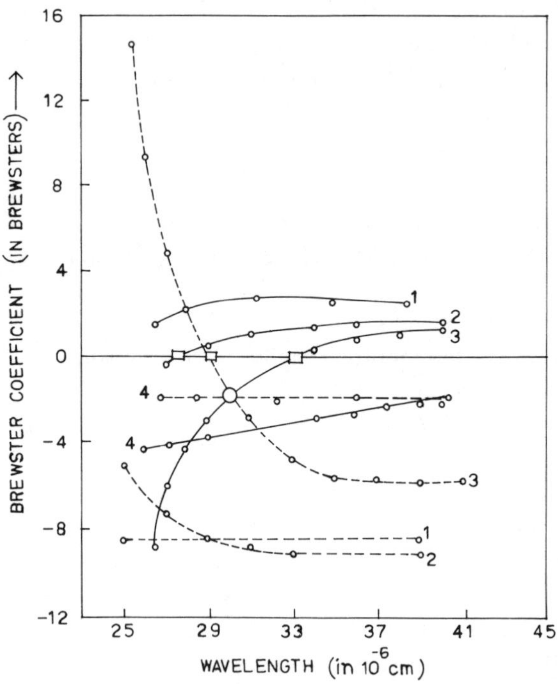

Fig. 6.2. Brewster coefficients C_λ' and C_λ'' versus wavelength for KCl, KBr, KI, and NaCl. $C_\lambda' = \frac{1}{2}n_0^3(q_{11} - q_{12})$ (○—○); $C_\lambda'' = \frac{1}{2}n_0^3 q_{44}$ (○– – –○); C_λ' and C_λ'' are measured in brewsters. (1) KCl; (2) KBr; (3) KI; (4) NaCl. (□) reversal; (○) isotropy. From [986].

Regarding the strain polarizability constant λ_0 under hydrostatic pressure, recent experimental work on a large number of cubic crystals proves unambiguously that λ_0 is not the same for all crystals, and what is more important it not only varies from crystal to crystal but also with the wavelength of light used and temperature. (See Tables 6.3, 6.4, and 6.5, and Figs. 6.4 and 6.5.)

For the first time Iyengar [565; 566] has noticed the variation of λ_0 from crystal to crystal and its dispersion for the same crystal; subsequently a similar variation of λ_0 has been observed by Srinivasan [1174; 1175] and Bansigir and Iyengar [66; 67]. This experimental observation regarding the behavior of λ_0 has prompted Bansigir and Iyengar [70; 71] to critically examine Mueller's theory, and they have successfully developed a theory of mechanical birefringence in crystals of NaCl structure for a linear stress normal to a cube face. They considered all three factors governing the photoelastic effect, namely, the Lorentz–Lorenz field, the Coulomb field, and the strain

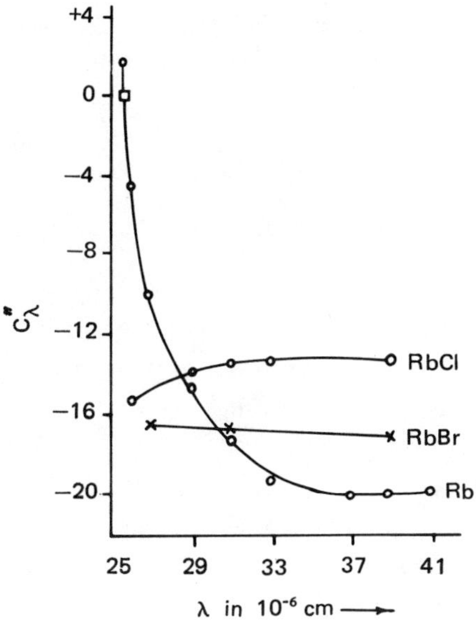

Fig. 6.3. Dispersion of C_λ'' for RbCl, RbBr, and RbI. $C_\lambda'' = \frac{1}{2}n_0^3 q_{44}$ and is measured in brewsters. (□) reversal. From [986].

Table 6.3. *Strain Polarizability Constants λ_0 of Some Cubic Crystals at Room Temperature and at λ = 5890 Å*[a]

Crystal	$\varrho \cdot \frac{dn}{d\varrho}$	$\varrho \cdot \left(\frac{\partial n}{\partial \varrho}\right)_R$	Strain polarizability constant λ_0			
			K[b]	M[c]	B and S[d]	I[e]
KCl	0.3477	0.5763	0.3966	0.48	0.61	0.489
KBr	0.3937	0.6773	0.4187	—	0.48	0.495
KI	0.4573	0.8477	0.4606	—	0.48	—
NaCl	0.2922	0.6543	0.5534	0.536	0.63	—
LiF	0.1040	0.4417	0.7645	—	0.7	0.715
MgO	−0.3887	0.9690	1.401	—	1.4	—
NH_4Cl	0.4877	0.8081	0.3966	—	—	0.389
$Ba(NO_3)_2$	2.298	0.7003	2.282	—	—	—
$Pb(NO_3)_2$	1.203	1.053	−0.142	—	—	−0.245

[a] From Ref. [681].
[b] Krishna Murty.
[c] Mueller.
[d] Burnstein and Smith.
[e] Iyengar.

***Table 6.4.** Strain Polarizability Constant λ_0 of Some Cubic Crystals at Room Temperature at Different Wavelengths*[a]

Crystal	Strain polarizability constant λ_0						
	4800 Å	5000 Å	5200 Å	5400 Å	5600 Å	5800 Å	6000 Å
NaCl	0.5752	0.5637	0.5567	0.5549	0.5528	0.5499	0.5447
KCl	0.4281	0.4239	0.4170	0.4093	0.4038	0.3932	0.3848
KBr	0.4633	0.4558	0.4471	0.4389	0.4284	0.4158	0.4068
KI	0.5401	0.5263	0.5112	0.4976	0.4823	0.4667	0.4501
LiF	0.7514	0.7544	0.7576	0.7598	0.7614	0.7647	0.7673
MgO	1.4117	1.4089	1.4071	1.4044	1.4026	1.4019	1.3994
NH_4Cl	0.4948[b]	0.4762	0.4436	0.4060	0.3862	0.3816	0.3957
$Ba(NO_3)_2$	−2.358[b]	−2.348	−2.333	−2.329	−2.324	−2.316	−2.292
$Pb(NO_3)_2$	−0.209	−0.190	−0.172	−0.163	−0.153	−0.148	−0.138

[a] From Ref. [681].
[b] At 4900 Å.

***Table 6.5.** Strain Polarizability Constant λ_0 of Some Cubic Crystals at $\lambda = 5890$ Å at Different Temperatures*[a]

Crystal	Strain polarizability constant λ_0					
	Room temperature	100°C	150°C	200°C	250°C	300°C
NaCl	0.5534	0.5774	0.5922	0.5805	0.5491	—
KCl	0.3966	0.4061	0.4170	0.4259	0.4271	—
KBr	0.4187	0.4444	0.4549	0.4596	0.4609	—
KI	0.4606	0.4802	0.5203	0.5470	0.5599	0.562
LiF	0.7645	0.7738	0.7739	0.7804	0.7849	—
MgO	1.4012	1.3957	1.3980	1.4036	1.4091	—
NH_4Cl[b]	0.3966	0.3647	0.3385	0.3136	—	—

[a] From Ref. [681].
[b] The values given in the table are determined at room temperature, 60, 80, and 100°C, respectively.

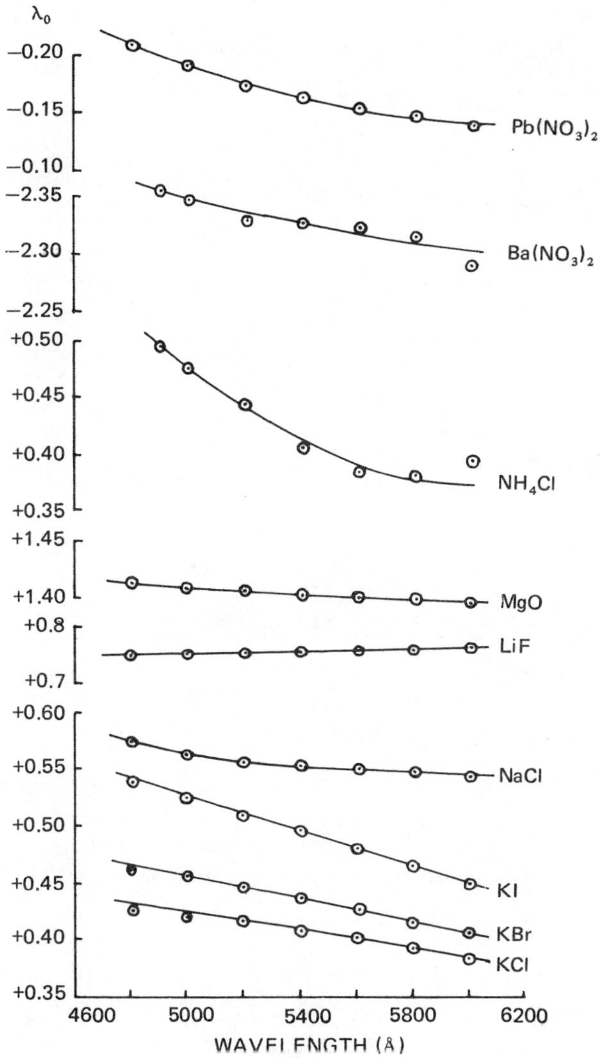

Fig. 6.4. Variation of λ_0 of some cubic crystals with wavelength at room temperature. From [681].

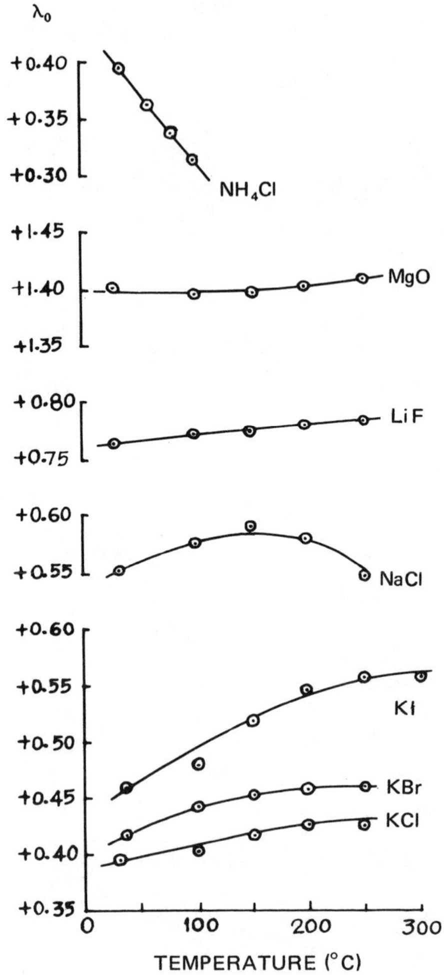

Fig. 6.5. Variation of λ_0 of some cubic crystals with temperature at $\lambda = 5890$ Å. From [681].

polarizability. These authors disagree with some of the assumptions in Mueller's theory. The expressions derived by Bansigir and Iyengar for the strain-optical constants and the strain polarizability constant λ_0 under hydrostatic pressure are given below:

$$\begin{aligned} \tfrac{1}{2}n^3(p_{11} - p_{12}) &= 3M - (1 + \sigma)A \\ \tfrac{1}{2}n^3(p_{11} + 2p_{12}) &= (3L - A)(1 - 2\sigma) \\ \tfrac{1}{6}n^3(p_{11} + 2p_{12}) &= L(1 - \lambda_0) \end{aligned} \tag{6.34}$$

where

$$\lambda_0 = 2\sigma + A(1 - 2\sigma)/3L$$

$$M = \frac{(n^2 - 1)^2(n^2 + 5)}{45n}$$

$$L = \frac{(n^2 - 1)(n^2 + 2)}{6n}$$

$$\begin{aligned} A = {} & \frac{(n^2 + 2)^2(n^2 + 5)}{54n}\left[1.458 + 2.9005\beta\left(\frac{1}{f_1} + \frac{1}{f_2}\right)\right]\theta^2 \\ & - \frac{(n^2 - 1)(n^2 + 2)(n^2 + 5)}{18n}\left[2.916 + \frac{4.2038}{f_1}\beta + \frac{7.5927}{f_2}\beta\right]\theta \\ & + \frac{(n^2 - 1)^2(n^2 + 5)}{3n}\left[0.2711 + \frac{1.6922}{f_2}\beta\right] \end{aligned}$$

In the above equations p_{11} and p_{12} are the strain-optical constants, n is the refractive index of the unstrained crystal, σ Poisson's ratio, β the number of valence electrons, f_1 and f_2 the oscillator strengths, and $\theta = \alpha_j/r^3$, where α_j is the polarizability of the jth ion and r the interionic distance.

An important conclusion that emerges from the theory above is that λ_0 can vary from crystal to crystal and also is not independent of wavelength of light and temperature.

6.8. *Effect of Elastic Deformation on the Oscillator Strengths and Dispersion Frequencies of Optical Electrons*

We have seen earlier that α, the polarizability of atoms and ions, changes with density. This change in polarizability may be considered as being due to a change in ν_r, the absorption frequencies, or f_r, the corresponding oscillator strengths, or both since α is related to both of them according to the quantum theory by Eq. (6.35):

$$\alpha = C \sum \frac{f_r}{\nu_r^2 - \nu^2} \tag{6.35}$$

where C is a constant and ν is the frequency of the incident light radiation. The magnitudes of the changes in f_r and ν_r may be evaluated as follows.

The change in the refractive index of a crystal under a hydrostatic pressure has been obtained earlier as

$$\varrho \frac{dn}{d\varrho} = \tfrac{1}{6}n^3(p_{11} + 2p_{12}) \tag{6.30'}$$

We can also obtain $\varrho(dn/d\varrho)$ by differentiating a dispersion formula either of the Drude type,

$$n^2 - 1 = \sum \frac{a_r}{\nu_r^2 - \nu^2} \tag{6.36}$$

or of the Lorentz–Lorenz type,

$$\frac{n^2 - 1}{n^2 + 2} = \sum \frac{a_r'^2}{\nu_r^2 - \nu^2} \tag{6.37}$$

where a_r and a_r' are quantities involving the density and strengths of oscillators, and the absorption frequencies; n is the refractive index at the frequency ν of the incident monochromatic light.

Ramachandran [994; 995] and Radhakrishnan [981–983] have fitted the absorption frequencies of a number of cubic crystals into the Drude formula, and obtained expressions for the refractive index over a wide range of wavelengths. The following discussion uses their approach. Differentiating the Drude formula, Eq. (6.36),

$$2n \frac{dn}{d\varrho} = \left\{ \sum \frac{a_r \lambda_r^2 \lambda^2}{\lambda^2 - \lambda_r^2} \frac{1}{a_r} \frac{da_r}{d\varrho} + 2 \sum \frac{a_r \lambda_r^2 \lambda^4}{(\lambda^2 - \lambda_r^2)^2} \frac{1}{\lambda_r} \cdot \frac{d\lambda_r}{d\varrho} \right\} \frac{1}{c^2}$$

Therefore,

$$\varrho \frac{dn}{d\varrho} = \frac{\varrho}{2n} \left[\sum \frac{a_r \lambda_r^2 \lambda^2}{\lambda^2 - \lambda_r^2} \eta_r + 2 \sum \frac{a_r \lambda_r^2 \lambda^4}{(\lambda^2 - \lambda_r^2)^2} \psi_r \right] \frac{1}{c^2} \tag{6.38}$$

where λ is the wavelength of the incident light, λ_r are the absorption wavelengths corresponding to the absorption frequencies ν_r, and c^2 is the velocity of light,

$$\eta_r = \frac{1}{a_r} \frac{da_r}{d\varrho} \quad \text{and} \quad \psi_r = \frac{1}{\lambda_r} \frac{d\lambda_r}{d\varrho}$$

r taking the values 1, 2, 3, and 4.

From Eqs. (6.30) and (6.38) and the experimental values of $(p_{11}+2p_{12})$ at different wavelengths, Krishna Murty [681] calculated the values of η_r and ψ_r for several cubic crystals including potassium halides, NaCl, NH_4Cl, LiF, and MgO, and concluded that both a_r and λ_r are affected by the changes in density. A negative sign before ψ or η indicates a shift in the absorption band towards shorter wavelength, or a decrease in the oscillator strength with increase in density. A positive sign has just the opposite meaning. There is no direct experimental evidence in support of the changes in a_r and λ_r; but in the case of KI Fesefeldt's work [360] on the temperature dependence of the absorption band at 2190 Å lends support to Krishna Murty's conclusions.

Furthermore, Krishna Murty has drawn the conclusion that the stress-optical dispersion arises on two counts: (1) a shift in the absorption frequency and (2) a change in the oscillator strength caused by a physical strain in the crystal. Such changes in the frequency generated by a mechanical stress have been conceived by earlier workers Burstein and Smith [218] and Iyengar [565; 566], and have actually been experimentally observed in Raman effect studies by Marie and Mathieu [774]. The stress-optical dispersion in glasses was explained earlier by Ramaseshan and Sivaramakrishnan [1021] on the basis of the variation of the oscillator strengths and the dispersion frequencies of the optical electrons. Krishna Murty's observations have, by and large, corroborated the findings previously made by Bansigir and Iyengar [67] on some alkali halides.

The applicability of Mueller's theory to a series of alkali halides has been thoroughly analyzed by Pakhnev, Shaskol'skay, and Gorbach [903].

6.9. *Temperature Dependence of Stress-Optical Dispersion*

Detailed studies on the effect of temperature on the stress-optical dispersion in a number of cubic crystals have been made by Krishna Murty [681]. His results on some of the cubic crystals (KCl, KBr, KI, and NaCl) are presented in Fig. 6.6; here C_λ' corresponds to the relative stress-optical constant $(q_{11}-q_{12})$. The increase in the dispersion of C_λ' with increasing temperature, according to him, is probably due to a large variation of the oscillator strengths and the absorption frequencies of the optical electrons with temperature.

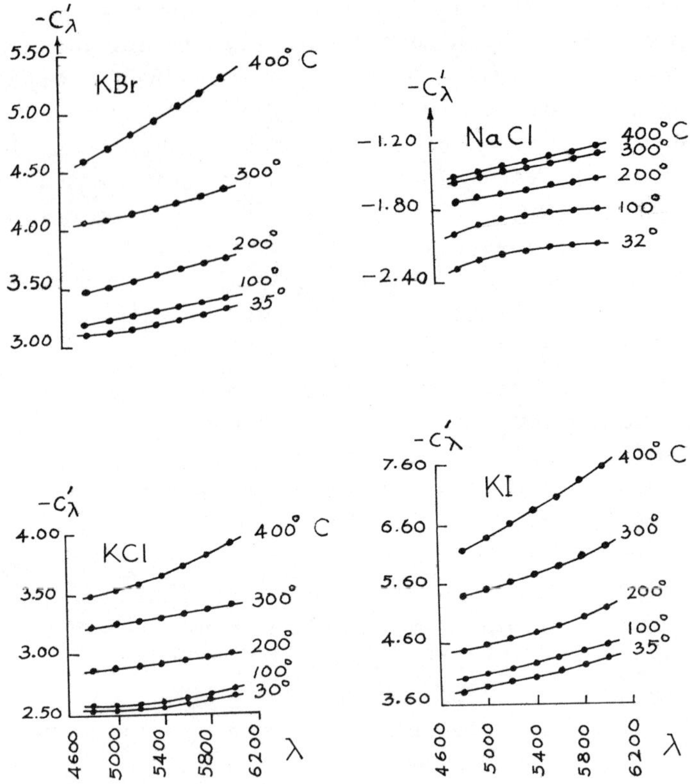

Fig. 6.6. Dispersion of C_λ' of some cubic crystals at different temperatures. $C_\lambda' = \frac{1}{2}n_0^3(q_{11} - q_{12})$, and is measured in brewsters; λ in angstroms. From [681].

6.10. Reversal of the Sign of Stress Birefringence in Pure and Mixed Crystals

6.10.1. Pure Crystals

It has been pointed out earlier (see page 315) that Srinivasan's [1174; 1175] studies on the photoelastic dispersion in some cubic crystals show that Pockels' and Mueller's classification of cubic crystals into four photoelastic groups breaks down. The reason, as can be seen from Fig. 6.1, is that the sign of $(q_{11} - q_{12})$ for KCl, KBr, and KI changes with λ, the wavelength of light. At room temperature Srinivasan has observed that a reversal of sign of $(q_{11} - q_{12})$, and hence of $(p_{11} - p_{12})$, occurs for these three crystals at the following wavelengths: for KCl at $\lambda = 2550$ Å, for KBr at $\lambda = 2760$ Å, and for KI at $\lambda = 3380$ Å.

Table 6.6. *Wavelengths of Sign Reversal of Stress Birefringence in Some Cubic Crystals*[a]

Crystal	Reversal wavelength (Å)		Reference
	$(q_{11} - q_{12})$	q_{44}	
KCl	2550	—	Srinivasan [1175]
	2480	—	Rahman [986]
KBr	2760	—	Srinivasan [1175]
	2750	2400	Rahman [986]
	2740	—	Laiho and Korpela [710]
KI	3380	—	Srinivasan [1175]
	3380	2750	Rahman [986]
RbBr	2020	2140	Laiho and Korpela [710]
	2450	2350	Rahman [986]
RbI	2420	2300	Laiho and Korpela [710]
	2680	2550	Rahman [986]
ZnSe	5250	—	Dubenskii *et al.* [307]
CsCl	2240	2650	Laiho and Korpela [710]
CsI	2590	2150	Laiho and Korpela [710]

[a] From [986].

Krishna Murty concludes from his studies on the temperature dependence of the strain polarizability (Fig. 6.5) that the wavelengths corresponding to the reversal of sign of $(p_{11} - p_{12})$ of these crystals may change with temperature.

Laiho and Korpela [710] have observed a sign reversal of $(q_{11} - q_{12})$ and q_{44} of RbBr and RbI, and also in CsCl and CsI. Dubenskii, Kaplyanskii, and Lozovskaya [307] have reported a reversal of sign of both $(q_{11} - q_{12})$ and q_{44} of ZnSe, a semiconductor.

Rahman [986] has made further studies on the sign reversal of $(q_{11} - q_{12})$ and q_{44} of a number of cubic crystals at room temperature. Table 6.6 summarizes the results of the various authors on the wavelengths at which a reversal of sign has been reported. There is general agreement on the reported values.

6.10.2. *Mixed Crystals of* KCl *and* KBr

The stress-optical behavior of cubic crystals KCl and KBr, pure and mixed, in various proportions, was investigated by Ethiraj [330] over the

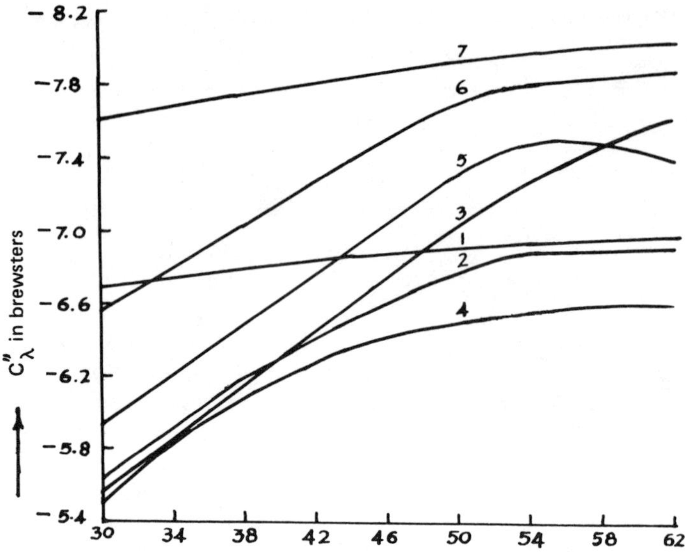

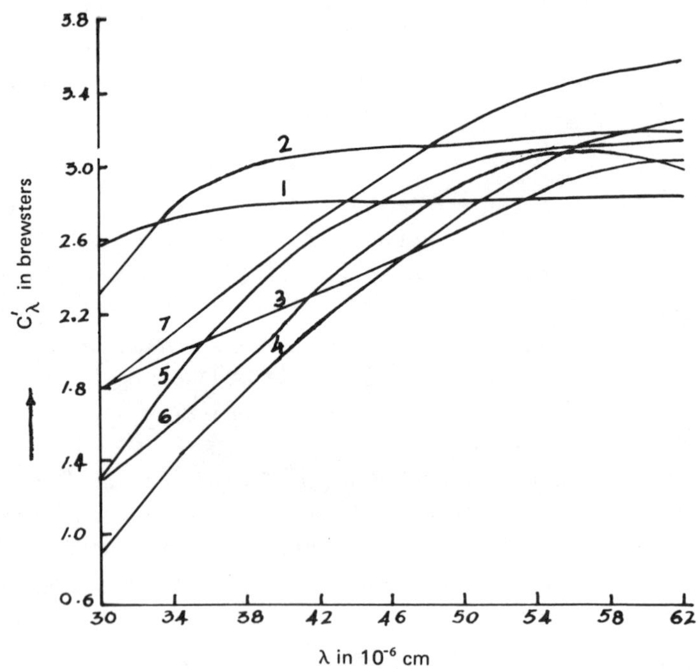

Fig. 6.7. Dispersion of C_λ' and C_λ'' of cubic crystals KCl and KBr, pure and mixed in various proportions. $C_\lambda' = \frac{1}{2}n_0^3(q_{11} - q_{12})$; $C_\lambda'' = \frac{1}{2}n_0^3 q_{44}$. (1) KCl; (2) $KCl_{0.86}KBr_{0.14}$; (3) $KCl_{0.69}KBr_{0.31}$; (4) $KCl_{0.59}KBr_{0.41}$; (5) $KCl_{0.43}KBr_{0.57}$; (6) $KCl_{0.21}KBr_{0.79}$; (7) KBr. From [330].

region 3000–6200 Å (Fig. 6.7). His observation is that in the ultraviolet region the slopes of the dispersion curves of $C_\lambda' = \frac{1}{2}n_0^3(q_{11} - q_{12})$ and $C_\lambda'' = \frac{1}{2}n_0^3 q_{44}$ tend to show that the reversal wavelengths of their mixed crystals lie at longer wavelengths compared to those of their unmixed crystals. The reversal wavelength changes as the proportion of their individual concentrations changes and attains the largest value around equimolar concentration. The actual reversal wavelengths have not yet been recorded.

The nonlinear variation of the piezo-optic behavior with the composition of the mixed crystal is in conformity with the dependence of certain other physical properties of the KCl–KBr system like microhardness, dielectric constants, and elastic constants on the composition of the mixed crystals.

While for KCl and KBr the stress-optical birefringence is of the same sign, it is well known that KCl and NaCl have opposite signs for their Brewster constant C_λ' over the entire visible region and down to 2800 Å, and hence it would be interesting to study the dispersion of stress birefringence in mixed crystals of NaCl and KCl with varying concentration.

6.11. Stress-Optical and Strain-Optical Isotropy in Cubic Crystals

Along the same lines as defining the elastic isotropy of cubic crystals which occurs when the relation $(c_{11} - c_{12} = 2c_{44})$ holds, one can define the condition for stress-optical isotropy for a cubic crystal as

$$q_{11} - q_{12} = q_{44} \tag{6.39}$$

(since for an isotropic solid $q_{44} = q_{11} - q_{12}$). Dubenskii, Kaplyanskii, and Lozovskaya [307] have observed such a condition at a certain wavelength near the absorption edge for ZnSe. At this wavelength, the stress birefringence is independent of the direction of uniaxial stress. The same effect was observed by Rahman [986] for some alkali halides at room temperature (Fig. 6.2).

Similarly, the condition for strain-optical isotropy for cubic crystals can be put in the form

$$p_{11} - p_{12} = 2p_{44} \tag{6.40}$$

[since for an isotropic solid $p_{44} = \frac{1}{2}(p_{11} - p_{12})$]. Hence the isotropy factors are related thus:

$$\frac{c_{11} \quad c_{12}}{2c_{44}} = \frac{q_{11} - q_{12}}{q_{44}} = \frac{p_{11} - p_{12}}{2p_{44}} \tag{6.41}$$

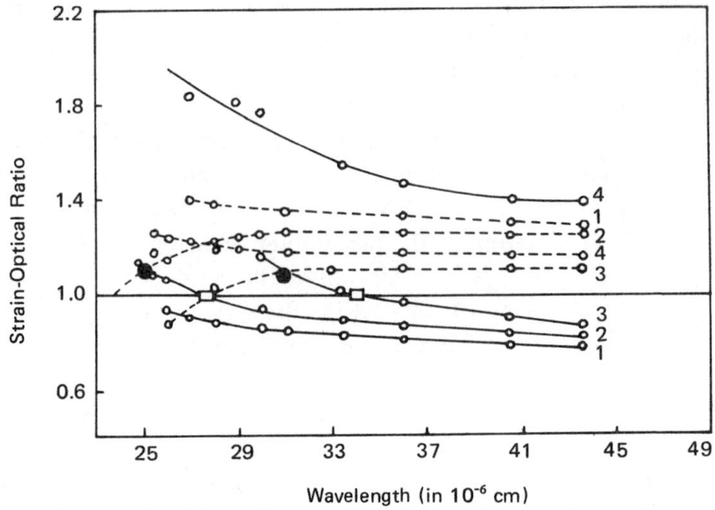

Fig. 6.8. Dispersion of the strain-optical ratio for KCl, KBr, KI, and NaCl. (○—○) p_{12}/p_{11}; (□) reversal; (○- - -○) $(p_{11} + p_{12} - 2p_{44})/(p_{11} + p_{12} + 2p_{44})$; (●) isotropy. (1) KCl; (2) KBr; (3) KI; (4) NaCl. From [988].

The stress-optical and strain-optical coefficients are related as follows:

$$p_{11} - p_{12} = (c_{11} - c_{12})(q_{11} - q_{12}) \qquad \text{and} \qquad p_{44} = c_{44}q_{44}$$

Therefore it is obvious that the two conditions of photoelastic isotropy may not be satisfied for the same wavelength of light unless the quantity

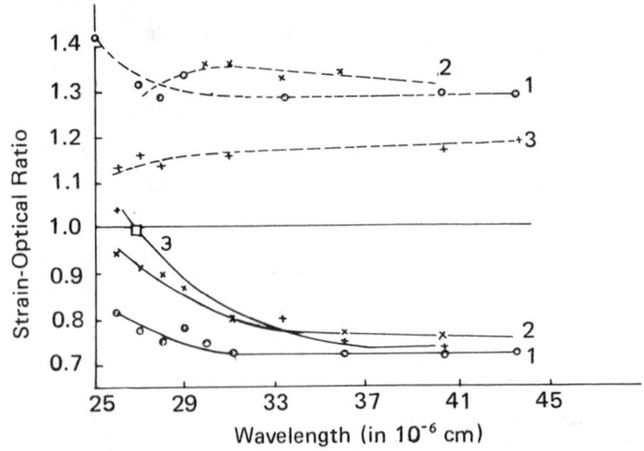

Fig. 6.9. Dispersion of the strain-optical ratio for RbCl, RbBr, and RbI. (○—○) p_{12}/p_{11}; (○- - -○) $(p_{11} + p_{12} - 2p_{44})/(p_{11} + p_{12} + 2p_{44})$; (□) reversal. (1) RbCl; (2) RbBr; (3) RbI. From [986].

Table 6.7. Wavelengths of Photoelastic Isotropy in Some Cubic Crystals[a]

Crystal	Wavelength	
	Stress-optical isotropy (Å)	Strain-optical isotropy (Å)
NaCl	4040	—
KBr	2420	2500
KI	3020	3100
RbBr	—	2400
RbI	—	2580

[a] From [986].

$(c_{11} - c_{12})/2c_{44}$, which is a measure of the elastic anisotropy, is also unity. For example, the elastic anisotropy factor is not unity for potassium and rubidium halides. Hence for these crystals, the wavelength of stress-optical isotropy is different from that of strain-optical isotropy. At the wavelength of strain-optical isotropy, we have the relation (see Figs. 6.8 and 6.9)

$$\frac{p_{12}}{p_{11}} = \frac{p_{11} + p_{12} - 2p_{44}}{p_{11} + p_{12} + 2p_{44}} \tag{6.42}$$

Table 6.7 shows the observed wavelengths of stress-optical and strain-optical isotropies of some alkali halides [986].

6.12. *Optic Axial Angle and Its Dispersion in Stressed Cubic Crystals of T and T_h Classes*

Cubic crystals of T_d, O, and O_h classes become optically uniaxial under a linear stress along any trigonal or tetragonal axis of symmetry, the optic axis coinciding with the direction of stress, whereas a linear stress applied along any other direction makes them optically biaxial. Although the stress birefringence is proportional to the applied stress, it is important to note that the optic axial angle $2V$ depends only on the direction of the linear stress, but not on its magnitude.

Cubic crystals of T and T_h classes become uniaxial only when the stress is along a cube diagonal (i.e., along a 3-fold axis of symmetry); for all other directions of a linear stress, the crystals become biaxial, and here also

$2V$ depends only on the direction of the linear stress but not on its magnitude.

All uniaxial crystals (tetragonal, hexagonal, and trigonal) become biaxial under any linear stress unless the stress direction is parallel to the optic axis of the unstressed crystal.

We shall examine now the optical behavior of cubic crystals of 23 (T) and $m3$ (T_h) classes under a linear stress parallel to a cube edge. Cubic crystals belonging to these two classes become optically biaxial even when stressed parallel to a cube edge, the principal axes of the optical index ellipsoid being parallel to the three cube edges. When a linear compressional stress is applied along [001], from Pockels' stress-optical relations, we have, for crystals of these two classes,

$$\Delta B_1 = -q_{13}P_3, \qquad \Delta B_2 = -q_{12}P_3, \qquad \Delta B_3 = -q_{11}P_3 \tag{6.43}$$

If the three principal refractive indices of the deformed crystal are n_s, n_m, and n_g, where $n_s < n_m < n_g$, the half-optic-axial angle V is given by

$$\cos^2 V = \left(\frac{1}{n_m^2} - \frac{1}{n_g^2}\right)\Big/\left(\frac{1}{n_s^2} - \frac{1}{n_g^2}\right) \tag{6.44}$$

and the optic axes lie in the plane defined by n_s and n_g.

From Eq. (6.43) one obtains

$$\Delta n_1 = \tfrac{1}{2}n^3q_{13}, \qquad \Delta n_2 = \tfrac{1}{2}n^3q_{12}, \qquad \Delta n_3 = \tfrac{1}{2}n^3q_{11}$$

Let us assume that for a given crystal $q_{11} < q_{13} < q_{12}$. Then $n_3 < n_1 < n_2$, which correspond, respectively, to $n_s < n_m < n_g$. Therefore, from Eq. (6.44),

$$\cos^2 V = (q_{12} - q_{13})/(q_{12} - q_{11}) \tag{6.45}$$

Equation (6.45) can be thrown into a more convenient form in terms of the experimentally observed stress birefringences,

$$C_1 = \tfrac{1}{2}n^3(q_{11} - q_{12}) \qquad \text{and} \qquad C_2 = \tfrac{1}{2}n^3(q_{11} - q_{13})$$

Thus

$$\cos^2 V = (C_1 - C_2)/C_1 \tag{6.46}$$

Rahman [986; 987] has studied lead nitrate and strontium nitrate of the T_h class in great detail and obtained the dispersion of the half-optic-axial angle V of these crystals in the spectral region 4100–3300 Å. For both

Table 6.8. Dispersion of V, the Half-Optic-Axial Angle[a,b]

Å	C_1 (brewsters)	C_2 (brewsters)	$C_1 - C_2$ (brewsters)	V (degrees)
		$Pb(NO_3)_2$		
4100	−50.0	−38.1	−11.9	60.8
3900	−50.7	−39.0	−11.7	61.0
3700	−53.1	−39.9	−13.2	60.1
3500	−56.0	−40.9	−15.1	58.7
3300	−59.2	−41.6	−17.6	57.0
		$Sr(NO_3)_2$		
4100	−29.3	−22.1	− 7.2	60.3
3900	−30.2	−22.1	− 8.1	58.8
3700	−31.1	−22.1	− 9.0	57.4
3500	−32.2	−22.1	−10.1	56.0
3300	−33.8	−22.1	−11.7	54.0

[a] From [986].
[b] $C_1 = \frac{1}{2}n^3(q_{11} - q_{12})$; $C_2 = \frac{1}{2}n^3(q_{11} - q_{13})$.

the crystals he found that both C_1 and C_2 are negative and that the magnitude of C_1 is greater than that of C_2. Hence using Eq. (6.46) he has evaluated V at different wavelengths. His results are given in Table 6.8. For both crystals V decreases towards the shorter wavelengths, the dispersion being about 10 and 6.5% in the region studied for $Sr(NO_3)_2$ and $Pb(NO_3)_2$, respectively.

7

Piezoelectricity

7.1. Introduction

In the year 1880 the brothers Jacques and Pierre Curie [278] discovered the phenomenon of piezoelectricity. They found that when a weight is placed on the surface of an *X*-cut quartz plate electric charges are developed on either surface of the quartz plate. They measured the charge thus developed and found it to be linearly proportional to the weight placed. The electromechanical interaction in crystals is called piezoelectricity. Their experiment demonstrates the direct piezoelectric effect. Several crystals including quartz, tourmaline, and Rochelle salt were found to exhibit the piezoelectric property.

Following the discovery of the direct piezoelectric effect by the Curie brothers, Lippmann [739] in 1881 predicted theoretically the converse piezoelectric effect, according to which a voltage applied to a piezoelectric crystal produces a physical strain in the crystal. Lippmann's prediction was confirmed by the Curies. Pockels [953] made a thorough investigation of the piezoelectric effect in several crystals, and contributed especially to the phenomenological theory of the linear electro-optic effect in crystals, which goes by his name today. In his celebrated work, "Lehrbuch der Kristallphysik," Voigt [1334] gave an exhaustive account of piezoelectricity in crystals.

As happens to several of the scientific discoveries, the piezoelectric effect also remained just a scientific curiosity for about a third of a century. The First World War became a blessing in disguise to this phenomenon and spurred interest in the practical applications of the phenomenon of piezoelectricity. During the war the French government posed a problem to Langevin of Paris; he was asked to devise a method of detecting the presence of enemy submarines. After experimenting with several devices, he finally concluded that piezoelectric quartz plates could be used as underwater detectors of submarines. This is a landmark in the history of piezoelectricity and it aroused interest the world over on the possible technological applications of piezoelectric crystals as ultrasonic sources. In 1921 Cady [225] showed that quartz crystals could be used to control and stabilize electronic oscillators.

7.2. Direct and Converse Piezoelectric Effects

We have already stated that piezoelectricity is the phenomenon in which a crystal under the action of a mechanical stress becomes electrically charged with opposite charges appearing at the two ends of the crystal. This effect occurs only with some special classes of crystals which have polar axes, positive and negative charges being developed at the ends of the polar axes. The phenomenon may be explained as follows. Matter is electrical in nature, and a crystal is made up of a large number of positive and negative ions which ordinarily neutralize each other resulting in no dipole moment. But the stressed crystal undergoes a slight geometrical deformation, thereby disturbing and deforming the charge distribution; this causes a net resultant dipole moment, which can be detected by an electrometer or some other suitable electrical devices. This phenomenon is different from pyroelectricity, where a crystal when uniformly heated develops electrical charges on a pair of opposite faces, with one end of the crystal becoming positively charged and the other end negatively charged. This effect also occurs with only some special classes of crystals.

The converse piezoelectric phenomenon is also possible as predicted by Lippmann [739]; that is, under the action of an electric field elastic strains are developed in a crystal. Both the direct and converse piezoelectric effects are exhibited by the same classes of crystals. All crystals in a ferroelectric state are also piezoelectric. A crystal may be piezoelectric without being ferroelectric.

The nature of charges developed on the opposite faces of a piezoelectric crystal depends upon the nature of the applied stress. Thus if a compressional stress on a crystal produces a positive dipole moment, then a tensile stress on the same crystal specimen produces in it a negative dipole moment; that is, a reversal of the nature of stress reverses the nature of the accumulated charges on a pair of its opposite faces, say the end faces. Similarly if a positive electric field produces an elongation strain in a piezoelectric crystal specimen, then the reversal of the electric field produces a compressional strain. It is in this respect that the converse piezoelectric effect differs from what is called electrostriction. Owing to the phenomenon of electrostriction, the physical or geometrical strain produced in a crystal (piezoelectric or otherwise) is proportional to the square of the applied electric field, and consequently independent of the field direction. On the other hand, owing to piezoelectricity, the strain is linearly proportional to the applied electric field and hence the nature of strain changes its sign with the reversal of the electric field.

Electrostriction is a universal phenomenon in the sense that it is exhibited by crystals of all classes (both centrosymmetric and noncentrosymmetric) whereas the converse piezoelectric effect is shown only by noncentrosymmetric crystals. Electrostriction is responsible for the Kerr effect shown by gases, liquids, and all solids.

7.3. *Mathematical Formulation, Piezoelectric Constants d_{ijk} in Tensor Notation, and d_{ij} in Two-Suffix Notation; Relation between d_{ijk} and d_{ij}*

The piezoelectric effect is a vector–tensor effect and thus a third-rank tensor property of a crystal. The applied stress, being a second-rank tensor, has nine components in general, and the resulting electric polarization (the dipole moment, which is equal to the charge multiplied by the separation of the charges), being a vector, has three components. The three components P_i of electric polarization are linearly related to all the nine stress components σ_{ij} in the most general case, given by the equation

$$P_i = d_{ijk}\sigma_{jk} \tag{7.1}$$

There are 27 d_{ijk} piezoelectric stress components as shown below:

$$\begin{aligned}
P_x &= d_{xxx}\sigma_{xx} + d_{xyy}\sigma_{yy} + d_{xzz}\sigma_{zz} + d_{xyz}\sigma_{yz} + d_{xzy}\sigma_{zy} + d_{xzx}\sigma_{zx} \\
&\quad + d_{xxz}\sigma_{xz} + d_{xxy}\sigma_{xy} + d_{xyx}\sigma_{yx} \\
P_y &= d_{yxx}\sigma_{xx} + d_{yyy}\sigma_{yy} + d_{yzz}\sigma_{zz} + d_{yyz}\sigma_{yz} + d_{yzy}\sigma_{zy} + d_{yzx}\sigma_{zx} \\
&\quad + d_{yxz}\sigma_{xz} + d_{yxy}\sigma_{xy} + d_{yyx}\sigma_{yx} \\
P_z &= d_{zxx}\sigma_{xx} + d_{zyy}\sigma_{yy} + d_{zzz}\sigma_{zz} + d_{zyz}\sigma_{yz} + d_{zzy}\sigma_{zy} + d_{zzx}\sigma_{zx} \\
&\quad + d_{zxz}\sigma_{xz} + d_{zxy}\sigma_{xy} + d_{zyx}\sigma_{yx}
\end{aligned} \tag{7.2}$$

Since the elastic strain ε_{kl} developed in a crystal is proportional to the stress acting, we note that $P_i \propto \varepsilon_{kl}$. Hence let $P_i = e_{ikl}\varepsilon_{kl}$. The 27 e_{ikl} are called piezoelectric strain components.

It is customary, however, to use the contracted notation of the stress components, and since the stress is a second-rank symmetric tensor we have

$$P_i = d_{il}\sigma_l \tag{7.3}$$

which in the expanded form becomes

$$\begin{aligned}
P_1 &= d_{11}\sigma_1 + d_{12}\sigma_2 + d_{13}\sigma_3 + d_{14}\sigma_4 + d_{15}\sigma_5 + d_{16}\sigma_6 \\
P_2 &= d_{21}\sigma_1 + d_{22}\sigma_2 + d_{23}\sigma_3 + d_{24}\sigma_4 + d_{25}\sigma_5 + d_{26}\sigma_6 \\
P_3 &= d_{31}\sigma_1 + d_{32}\sigma_2 + d_{33}\sigma_3 + d_{34}\sigma_4 + d_{35}\sigma_5 + d_{36}\sigma_6
\end{aligned} \tag{7.4}$$

There are thus 18 d_{il} for the most anisotropic crystal, which can be related to the 27 d_{ijk} of Eq. (7.2). For example, equating the like coefficients of P_x (= P_1) in Eqs. (7.2) and (7.4), we note

$$\begin{aligned}
d_{11} &= d_{111}, & d_{14} &= 2d_{123} = 2d_{132} \\
d_{12} &= d_{122}, & d_{15} &= 2d_{113} = 2d_{131} \\
d_{13} &= d_{133}, & d_{16} &= 2d_{112} = 2d_{121}
\end{aligned} \tag{7.5a}$$

In a similar manner, the relationships between the other d_{ijk} and d_{il} could be deduced. The results are collected below:

$$\begin{aligned}
d_{21} &= d_{211}, & d_{24} &= 2d_{223} = 2d_{232} \\
d_{22} &= d_{222}, & d_{25} &= 2d_{213} = 2d_{231} \\
d_{23} &= d_{233}, & d_{26} &= 2d_{212} = 2d_{221} \\
d_{31} &= d_{311}, & d_{34} &= 2d_{323} = 2d_{332} \\
d_{32} &= d_{322}, & d_{35} &= 2d_{313} = 2d_{331} \\
d_{33} &= d_{333}, & d_{36} &= 2d_{312} = 2d_{321}
\end{aligned} \tag{7.5b}$$

The mathematical formulation of the piezoelectric effect can alternatively be put in a form relating the deformation of a crystal when placed in an electric field. Thus

$$\varepsilon_{jk} = d_{ijk}E_i \tag{7.6}$$

Note that the coefficients d_{ijk} connecting the field E_i and the strain ε_{jk}

in the converse effect are just the same as those connecting the stress σ_{jk} and the polarization P_i in the direct effect.

The electric field has three components and the deformation being a strain is a second-rank symmetric tensor having six components. Hence there are 18 components in the two-suffix notation of the piezoelectric tensor of third rank. The equations (7.6) when referred to the conventional Cartesian system of coordinates reduce to

$$\begin{aligned} \varepsilon_{xx} &= d_{11}E_x + d_{21}E_y + d_{31}E_z \\ \varepsilon_{yy} &= d_{12}E_x + d_{22}E_y + d_{32}E_z \\ \varepsilon_{zz} &= d_{13}E_x + d_{23}E_y + d_{33}E_z \\ \varepsilon_{yz} &= d_{14}E_x + d_{24}E_y + d_{34}E_z \\ \varepsilon_{zx} &= d_{15}E_x + d_{25}E_y + d_{35}E_z \\ \varepsilon_{xy} &= d_{16}E_x + d_{26}E_y + d_{36}E_z \end{aligned} \tag{7.7}$$

The d_{kl} in the two-suffix notation relate ε_1 to E_k thus: $\varepsilon_l = d_{kl}E_k$.

The two sets of values of d_{ij} are the same in both the cases owing to the reciprocity relations [1430]. The d_{kl} matrix in Eq. (7.7) is the transpose of the d_{ij} matrix in Eq. (7.4).

d_{ij} may be expressed in centimeters per volt or in Coulombs per dyne.

Because of the various symmetry elements associated with each crystal class, some constants vanish and some others become equal among themselves. Thus, e.g., α quartz has only five piezoelectric constants that are not zero and of these only two are independent, namely,

$$d_{12} = -d_{11}, \qquad d_{25} = -d_{14}, \qquad d_{26} = -2d_{11}$$

For tourmaline of class $3m$ (C_{3v}) there are four constants, namely,

$$d_{11},\ d_{15},\ d_{31},\ \text{and}\ d_{33}$$

and for Rochelle salt only three constants d_{14}, d_{25}, and d_{36} exist independently.

Another classical example of a third-rank tensor property of crystals is piezomagnetism, i.e., the production of a magnetic moment in a crystal under an applied stress in an appropriate direction. But the piezomagnetic effects are very much smaller compared with piezoelectric effects in any crystal.

7.4. Deduction of the Surviving d_{ijk} for Some Crystal Classes by Tensor Method, and the d_{ij} Matrices for the 21 Noncentrosymmetric Classes

Using the transformation matrix of each of the symmetry elements of any particular point group and applying Neumann's principle as discussed in Chapter 2, we show how some of the coefficients become zero and others equal among themselves.

Case (i) Point Group $\bar{1}$ (C_i). This has only a center of symmetry. The transformation matrix of this symmetry operation is

$$\begin{bmatrix} -1 & 0 & 0 \\ 0 & -1 & 0 \\ 0 & 0 & -1 \end{bmatrix}$$

For tensors of rank three, the general transformation law from the old to the new system is

$$T'_{ijk} = \alpha_{il}\alpha_{jm}\alpha_{kn}T_{lmn} \tag{7.8}$$

and by Neumann's principle

$$T'_{ijk} = T_{ijk} \tag{7.9}$$

Now the direction cosines α_{ij} are zero except when

$$i = l, \qquad j = m, \qquad \text{and} \qquad k = n$$

in which case each nonvanishing $\alpha_{ij} = -1$. Consequently Eq. (7.8) reduces to

$$T'_{ijk} = -T_{ijk} \tag{7.10}$$

and, from Neumann's principle T'_{ijk} should be equal to T_{ijk} since any physical property should be invariant under a symmetry operation. Combining Eqs. (7.9) and (7.10) we get

$$T_{ijk} = T'_{ijk} = -T_{ijk}$$

Hence

$$T_{ijk} = 0$$

Hence, crystals having a center of symmetry cannot have any of the components of a third-rank tensor property surviving.

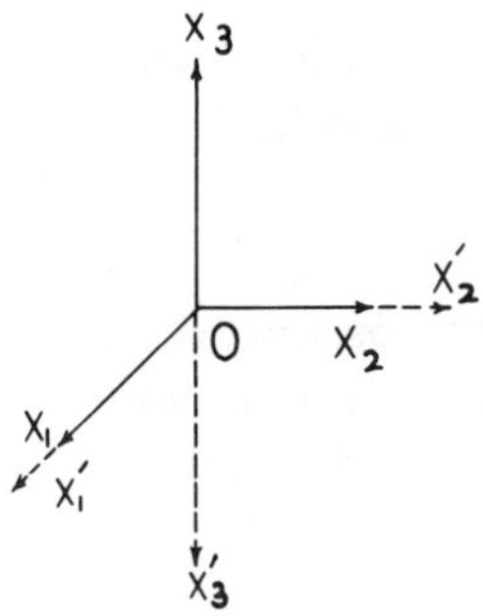

Fig. 7.1. Transformation of axes: X_1OX_2 is a plane of reflection.

Case (ii) Point Group m (C_s). This has only a plane of symmetry which is assumed to be normal to the X_3 axis (Fig. 7.1). Its transformation matrix is

$$\begin{bmatrix} 1 & 0 & 0 \\ 0 & 1 & 0 \\ 0 & 0 & -1 \end{bmatrix}$$

With this transformation matrix, any third-rank tensor component must vanish if it has one 3 or three 3's as its subscripts. Thus the vanishing coefficients are those that are canceled in the following matrix:

$$\begin{bmatrix} d_{111} & d_{122} & d_{133} & \cancel{d_{123}} & \cancel{d_{131}} & d_{112} & \cancel{d_{132}} & \cancel{d_{113}} & d_{121} \\ d_{211} & d_{222} & d_{233} & \cancel{d_{223}} & \cancel{d_{231}} & d_{212} & \cancel{d_{232}} & \cancel{d_{213}} & d_{221} \\ \cancel{d_{311}} & \cancel{d_{322}} & \cancel{d_{333}} & d_{323} & d_{331} & \cancel{d_{312}} & d_{332} & d_{313} & \cancel{d_{321}} \end{bmatrix}$$

Hence the resulting matrix is

$$\begin{bmatrix} d_{111} & d_{122} & d_{133} & 0 & 0 & 2d_{112} \\ d_{211} & d_{222} & d_{233} & 0 & 0 & 2d_{212} \\ 0 & 0 & 0 & 2d_{323} & 2d_{331} & 0 \end{bmatrix} = \begin{bmatrix} d_{11} & d_{12} & d_{13} & 0 & 0 & d_{16} \\ d_{21} & d_{22} & d_{33} & 0 & 0 & d_{26} \\ 0 & 0 & 0 & d_{34} & d_{35} & 0 \end{bmatrix}$$

For the monoclinic system, class m with the reflection plane perpendicular to the X_2 axis (standard orientation) the resulting d_{ij} matrix is

$$\begin{bmatrix} d_{11} & d_{12} & d_{13} & 0 & d_{15} & 0 \\ 0 & 0 & 0 & d_{24} & 0 & d_{26} \\ d_{31} & d_{32} & d_{33} & 0 & d_{35} & 0 \end{bmatrix}$$

Case (iii) Point Group 2 (C_2). This has only a diad axis parallel to X_3 (i.e., Z axis). Here the transformation matrix is

$$\begin{bmatrix} -1 & 0 & 0 \\ 0 & -1 & 0 \\ 0 & 0 & 1 \end{bmatrix}$$

As in the previous case we notice that the direction cosines α_{ij} are zero except when

$$i = l, \qquad j = m, \qquad k = n$$

The product $\alpha_{il}\alpha_{jm}\alpha_{kn}$ is equal to $+1$ whenever the subscripts consist of a single 3 or all three 3's. Any combination other than the above will necessarily be zero. Hence the surviving third-rank tensor coefficients are

$$\begin{matrix} d_{123} & d_{213} & d_{312} & d_{311} & d_{132} & d_{231} & d_{321} \\ d_{113} & d_{223} & d_{333} & d_{322} & d_{131} & d_{232} & \end{matrix}$$

We note that $d_{ijk} = d_{ikj}$. Hence in the contracted or two-suffix notation, the d_{ij} matrix for class 2 with diad parallel to X_3 is then given by

$$\begin{bmatrix} 0 & 0 & 0 & d_{14} & d_{15} & 0 \\ 0 & 0 & 0 & d_{24} & d_{25} & 0 \\ d_{31} & d_{32} & d_{33} & 0 & 0 & d_{36} \end{bmatrix}$$

The contraction is made only after deducing the surviving d_{ijk}.

With diad axis parallel to X_2 (standard orientation) the d_{ij} matrix is given by

$$\begin{bmatrix} 0 & 0 & 0 & d_{14} & 0 & d_{16} \\ d_{21} & d_{22} & d_{23} & 0 & d_{25} & 0 \\ 0 & 0 & 0 & d_{34} & 0 & d_{36} \end{bmatrix}$$

Case (*iv*). Proceeding on the lines above we can easily deduce the d_{ij} matrix for the class 222 (D_2) of the orthorhombic system.

This point group has three mutually perpendicular diads which are coincident with the three coordinate axes X_1, X_2, and X_3. The transformation matrices corresponding to the three diads are, respectively,

$$\begin{bmatrix} 1 & 0 & 0 \\ 0 & -1 & 0 \\ 0 & 0 & -1 \end{bmatrix}, \quad \begin{bmatrix} -1 & 0 & 0 \\ 0 & 1 & 0 \\ 0 & 0 & -1 \end{bmatrix}, \quad \begin{bmatrix} -1 & 0 & 0 \\ 0 & -1 & 0 \\ 0 & 0 & 1 \end{bmatrix}$$

Taking X_2 as the diad axis, we note as before that the direction cosines α_{ij} are zero unless

$$i = l, \qquad j = m, \qquad k = n$$

Hence the product $\alpha_{ij}\alpha_{jm}\alpha_{kn}$ will be equal to $+1$ for any combination of subscripts involving a single 2 or three 2's. Consequently the components with the combinations of subscripts other than those listed above vanish. The vanishing constants are thus

$$d_{113} \quad d_{131} \quad d_{221} \quad d_{212} \quad d_{331} \quad d_{313} \quad d_{223} \quad d_{232} \quad d_{111}$$
$$d_{311} \quad d_{122} \quad d_{133} \quad d_{322} \quad d_{333}$$

or

$$d_{15} \quad d_{26} \quad d_{35} \quad d_{24} \quad d_{11}$$
$$d_{31} \quad d_{12} \quad d_{13} \quad d_{32} \quad d_{33}$$

Now we can find out the result of superposition of the diad axis X_2 over that of X_3. The resultant of the matrix of the tensor components in 2-suffix notation is given by

$$\begin{bmatrix} 0 & 0 & 0 & d_{14} & 0 & 0 \\ 0 & 0 & 0 & 0 & d_{25} & 0 \\ 0 & 0 & 0 & 0 & 0 & d_{36} \end{bmatrix}$$

Proceeding in the same manner for the remaining diad axis X_1 and transforming with the matrix operator

$$\begin{bmatrix} 1 & 0 & 0 \\ 0 & -1 & 0 \\ 0 & 0 & -1 \end{bmatrix}$$

it will be seen that no more terms become zero, and hence the nonvanishing components in the tensor notation are

$$d_{123} = d_{132}, \qquad d_{213} = d_{231}, \qquad d_{312} = d_{321}$$

and in two-suffix notation there are only three, d_{14}, d_{25}, and d_{36}, which completely describe the piezoelectric constants for crystals like Rochelle salt. In passing, it may be noted that the point group 222 ($= D_2$) has the symmetry elements E, C_2, C_2', C_2'', of which any two of the three mutually perpendicular diad axes X_1, X_2, X_3 form the generating elements of the point group. It is well known that in deducing the nonvanishing independent coefficients of any tensor property of a crystal it will be quite sufficient if we study the influence of the generating elements of the crystal. This fact has been completely proved above in the deduction of the third-rank tensor components of the point group 222, where we have seen that the imposition of X_2 over X_3 gives all the required information since they form a set of

Table 7.1. d_{ij} *Matrices of the 21 Noncentrosymmetric Point Groups*

Class 1 (C_1): 18 constants

d_{11}	d_{12}	d_{13}	d_{14}	d_{15}	d_{16}
d_{21}	d_{22}	d_{23}	d_{24}	d_{25}	d_{26}
d_{31}	d_{32}	d_{33}	d_{34}	d_{35}	d_{36}

Class 2 (C_2): 8 constants

$2 \parallel X_3$

0	0	0	d_{14}	d_{15}	0
0	0	0	d_{24}	d_{25}	0
d_{31}	d_{32}	d_{33}	0	0	d_{36}

$2 \parallel X_2$ (standard orientation)

0	0	0	d_{14}	0	d_{16}
d_{21}	d_{22}	d_{23}	0	d_{25}	0
0	0	0	d_{34}	0	d_{36}

Class m (C_{1h}): 10 constants

$m \perp X_3$

d_{11}	d_{12}	d_{13}	0	0	d_{16}
d_{21}	d_{22}	d_{23}	0	0	d_{26}
0	0	0	d_{34}	d_{35}	0

$m \perp X_2$ (standard orientation)

d_{11}	d_{12}	d_{13}	0	d_{15}	0
0	0	0	d_{24}	0	d_{26}
d_{31}	d_{32}	d_{33}	0	d_{35}	0

Class $mm2$ (C_{2v}): 5 constants

0	0	0	0	d_{15}	0
0	0	0	d_{24}	0	0
d_{31}	d_{32}	d_{33}	0	0	0

Class 222 (D_2): 3 constants

0	0	0	d_{14}	0	0
0	0	0	0	d_{25}	0
0	0	0	0	0	d_{36}

Classes 4 (C_4) and 6 (C_6): 4 constants

0	0	0	d_{14}	d_{15}	0
0	0	0	d_{15}	$-d_{14}$	0
d_{31}	d_{31}	d_{33}	0	0	0

Class $\bar{4}$ (S_4): 4 constants

0	0	0	d_{14}	d_{15}	0
0	0	0	$-d_{15}$	d_{14}	0
d_{31}	$-d_{31}$	0	0	0	d_{36}

Classes $4mm$ (C_{4v}) and $6mm$ (C_{6v}): 3 constants

0	0	0	0	d_{15}	0
0	0	0	d_{15}	0	0
d_{31}	d_{31}	d_{33}	0	0	0

Class $\bar{4}2m$ (D_{2d}): 2 constants
$2 \parallel X_1$

0	0	0	d_{14}	0	0
0	0	0	0	d_{14}	0
0	0	0	0	0	d_{36}

Table 7.1 *(continued)*

Classes 422 (D_4) and 622 (D_6): 1 constant

0	0	0	d_{14}	0	0
0	0	0	0	$-d_{14}$	0
0	0	0	0	0	0

Class 3 (C_3): 6 constants

d_{11}	$-d_{11}$	0	d_{14}	d_{15}	$-2d_{22}$
$-d_{22}$	d_{22}	0	d_{15}	$-d_{14}$	$-2d_{11}$
d_{31}	d_{31}	d_{33}	0	0	0

Class 32 (D_3): 2 constants

d_{11}	$-d_{11}$	0	d_{14}	0	0
0	0	0	0	$-d_{14}$	$-2d_{11}$
0	0	0	0	0	0

Class $3m$ (C_{3v}): 4 constants
$m \perp X_2$

d_{11}	$-d_{11}$	0	0	d_{15}	0
0	0	0	d_{15}	0	$-2d_{11}$
d_{31}	d_{31}	d_{33}	0	0	0

Class $3m$ (C_{3v}): 4 constants
$m \perp X_1$ (standard orientation)

0	0	0	0	d_{15}	$-2d_{22}$
$-d_{22}$	d_{22}	0	d_{15}	0	0
d_{31}	d_{31}	d_{33}	0	0	0

Class $\bar{6}m2$ (D_{3h}): 1 constant
$m \perp X_2$

d_{11}	$-d_{11}$	0	0	0	0
0	0	0	0	0	$-2d_{11}$
0	0	0	0	0	0

Class $\bar{6}m2$ (D_{3h}): 1 constant
$m \perp X_1$ (standard orientation)

0	0	0	0	0	$-2d_{22}$
$-d_{22}$	d_{22}	0	0	0	0
0	0	0	0	0	0

Class $\bar{6}$ (C_{3h}): 2 constants

d_{11}	$-d_{11}$	0	0	0	$-2d_{22}$
$-d_{22}$	d_{22}	0	0	0	$-2d_{11}$
0	0	0	0	0	0

Classes 23 (T) and $\bar{4}3m$ (T_d): 1 constant

0	0	0	d_{14}	0	0
0	0	0	0	d_{14}	0
0	0	0	0	0	d_{14}

Class 432 (O) all constants vanish

0	0	0	0	0	0
0	0	0	0	0	0
0	0	0	0	0	0

generating elements and the application of the remaining diad axis X_1 gives no additional information.

The d_{ij} matrix for each of the 21 noncentrosymmetric classes can be deduced on the lines indicated above taking into consideration the effect of each one of the symmetry elements of a class on the several d_{ij} coefficients; in fact, it is sufficient if the generating elements alone of each point group are considered to arrive at the nonvanishing independent d_{ij} coefficients. The final d_{ij} matrices are collected in Table 7.1.

7.5. Concluding Remarks

We have discussed so far the phenomenon of piezoelectricity. This has been warranted because the converse piezoelectric effect has an important bearing on the measurements on the linear electro-optic effect. The piezoelectric strains cause a secondary effect producing optical birefringence through the phenomenon of photoelasticity and this secondary effect will have to be computed before arriving at the true or primary electro-optic effect.

To this end, one has to know the piezoelectric strains caused by the application of an electric field. Hence the set of piezoelectric equations (7.1) and (7.6) are of utmost importance while discussing the linear electro-optic effect.

The experimental methods of determining the piezoelectric constants d_{ijk} are described in great detail in several books, among which may be mentioned W. G. Cady's *Piezoelectricity* [225] and W. P. Mason's *Piezoelectric Crystals and Their Applications to Ultrasonics* [783], and hence these experimental methods are not described in this book.

8

Electro-Optic Effects in Crystals: Pockels Linear Electro-Optic and Kerr Quadratic Electro-Optic Effects

8.1. Introduction

The development of the laser has largely been responsible for the tremendous growth of interest during the past decade and a half in the study of the electro-optic and nonlinear optical properties of solids. These optical phenomena are applied in fabricating a large number of optical devices such as tunable narrow band interference polarizing monochromators, light modulators, light beam deflectors, frequency shifters, and second-harmonic generators. The electro-optic phenomena as described in this book deal entirely with the effect of an electric field on the index of refraction of materials in the solid state.

A piece of glass that is optically isotropic becomes birefringent when it is subjected to a mechanical stress; this phenomenon, as we already know, is called photoelasticity. The same piece of glass becomes birefringent when subjected to an electric field, this phenomenon being known as the Kerr electro-optic effect. While the photoelastic birefringence is linearly proportional to stress, the Kerr electro-optic birefringence is proportional to the square of the electric field. Therefore, the reversal of sign of mechanical stress, say from compression to tension, changes the sign of birefringence, but the reversal of sign of the electrical field in the Kerr effect experiments has no influence on the sign of birefringence. These two phenomena, photoelastic effect and Kerr effect, are universal in the sense that they are exhibited by every transparent solid, and in fact, both of them are of the same tensor rank, namely, fourth. Furthermore, the matrices for both of them are the same for every crystal class. The Kerr effect is also called the quadratic electro-optic effect. Besides the above two phenomena, there exists yet another electro-optic effect which is shown only by piezoelectric crystals, such as $NaClO_3$, quartz, ADP, KDP, and Rochelle salt; this effect is called the Pockels effect, though the phenomenon was discovered by Röntgen. In this case the electro-optic birefringence is linearly proportional to the applied electric field. This linear electro-optic effect was thoroughly investigated in several crystals before the turn of the last century by Pockels.

8.2. Demonstration of the Electro-Optic Effects, Linear and Quadratic

A rectangular prism of quartz is cut with its length along the X axis, breadth along the Z axis, and thickness along the Y axis. A parallel beam of monochromatic plane-polarized light is passed through the rectangular quartz prism along the Y axis and observed through a Babinet compensator, the electric vector of the incident light vibrations making an angle of 45° with the X and Z axes.

Now the prism is subjected to an electric field parallel to the X axis. Since the quartz prism is piezoelectric, it experiences a piezoelectric strain parallel to its X axis. This strain is accompanied by strain birefringence, which can be measured by a Babinet compensator. The piezoelectric strain birefringence is to be distinguished from the photoelastic birefringence. It is interesting to note that the birefringence and the refractive indices of a crystal deformed by an electric field are different from those of a crystal deformed to the same extent by a mechanical stress.

The actual situation may be described as follows.

a. Owing to the applied electric field, there is the universal Kerr effect (quadratic electro-optic effect), causing a certain amount of optical path difference δ_K.

b. Owing to the piezoelectric nature of the quartz crystal, it experiences a physical strain, which in its turn causes a photoelastic path difference δ_p.

c. Apart from the above two effects, the external electric field may alter the dipole moment of the medium and hence produce some path difference δ_l. This last effect is called the pure linear electro-optic effect.

Thus in general the observed artificial path difference δ_0 is given by

$$\delta_0 = \delta_l + \delta_K + \delta_p \tag{8.1}$$

It should be noted that the pure electro-optic effect is exhibited only by piezoelectric crystals, and that too only for some specific orientations.

8.3. Historical Survey

8.3.1. Earlier Work

The linear electro-optic effect (also called the Pockels effect) was discovered for the first time in quartz and tourmaline crystals independently by Röntgen [1065; 1066] and by Kundt [698] in 1883, about eight years after Kerr discovered the quadratic electro-optic effect in glass and liquids. Further investigations of this effect in quartz were carried out by Pockels [953; 956], Tsi-Ze [1220], Günther [460], and Grant [447]. Pockels [953; 956] has investigated the electro-optic effect in Rochelle salt, tourmaline, and sodium chlorate and worked out a phenomenological theory of the linear electro-optic effect. Mueller [825; 831] and Valasek [1249] carried out thorough investigations on the quadratic and linear electro-optic effects in Rochelle salt. Schramm [1103] has studied this effect in zinc blende. Jacquerod [570] studied the Kerr effect in potassium dihydrogen phosphate and potassium dihydrogen arsenate. Zwicker and Scherrer [1428; 1429] reported the linear electro-optic effect in KDP and the salt resulting after the replacement of hydrogen by deuterium.

After World War II, the interest in investigating the electro-optic effect has grown beyond one's imagination, owing to the fact that the large magnitude of the effect in the artificially grown crystals of ADP and KDP has found practical applications in a variety of situations.

In 1947 Billings [144] indicated the use of electro-optic crystals like ADP and KDP for modulation of light and as optical shutters. This is the beginning of a new era in the history of the Pockels effect and its overwhelmingly large variety of applications in science and technology. Subsequently, innumerable workers have been attracted into this field, and particularly the last one and a half decades have witnessed a phenomenal growth in both the number and variety of crystals studied, which has been warranted by a systematic search for new crystals surpassing the known ones, and in the ever-increasing multidirectional applications.

8.3.2. More Recent Work

8.3.2.1. Amorphous Solids

It is only in the recent years that work has been initiated on glass ceramics. Gires and Paillette [424] reported the electro-optic effect in some colored glasses. Borrelli and Layton [177] and Borrelli [178] studied the electro-optic properties of glass ceramic systems. Others to study ceramics with regard to their electro-optic behavior include Land and collaborators [712; 713]. Land and Haertling [713] have reviewed the work on the electro-optic ceramic materials and their applications.

8.3.2.2. Cubic Crystals

Investigations on cubic crystals have come into prominence of late because they facilitate the design of wide-aperture electro-optic modulators owing to their optical isotropy in the unstrained state. Hexamethylenetetramine was studied with respect to its electro-optic behavior by several workers including McQuaid [796], Belyaev, Vlokh, Gilvarg, Dobrzhanskii, Netesov, Shamburov, and Shuvalov [89], Heilmeier [506], Lee [721], Rodgers [1062], and Jones and Forno [597]. Shamburov and Blokh [1118] gave an account of the phenomenological theory of the linear electro-optic effect in cubic crystals. Belyaev, Dobrzhanskii, Pisarevskii, Chernyshev, and Shaldin [90] studied the electro-optic properties of copper chloride and copper bromide. Vassel and Conwell [1271] reported the electro-optic constant of ammonium chloride, treating it as one lacking a center of symmetry. Warner, Robertson, and Parfit [1358] studied the electro-optic effect in sodium uranyl acetate. Kojima, Shikama, Kuninobu, Kawabata, and Tanaka [660] determined the electro-optic constant of zinc selenide. Shamburov, Kuznetsov, Lobachev, Kharikonova, and Soshnikov [1120] studied the electro-optic effect in zinc sulfide. Tada, Aoki, Washiyama, and Murai [1194] have studied the linear electro-optic effect in ZnTe.

The electro-optic effect in cubic crystals for light propagation along a cube diagonal has been discussed by Smirnov [1148]. Low-frequency linear electro-optic effect in $PbTe_2O_6$ has been studied by Bruton and White [206]. The dispersion of the electro-optic coefficient of the cubic ZnSe has been studied by Baillou and Lemercier [62] and Baillou, Lemercier, Rivaud, and Pavis [63]. The electro-optic and Raman coefficients in cubic semiconductors have been investigated by Flytzanis [377]. The temperature dependence of Raman scattering and the electro-optic properties of CuCl have been studied by Kaminow and Turner [621].

Haussühl and Hesse [488] studied the quadratic electro-optic effect in 15 alkali halides of sodium chloride type. The absolute quadratic electro-optic effects in LiF and α-$TlAl(SO_4)_2 \cdot 12H_2O$ have been measured by Haussühl and Walda [493]. Geusic, Kurtz, Van Uitert, and Wemple [416] investigated the Kerr effect in a number of crystals including KTN, $SrTiO_3$, $BaTiO_3$, and $KTaO_3$. Narasimhamurty [852] has pointed out the applicability of the Kerr effect methods to verify indirectly the validity of Bhagavantam's scheme of photoelastic constants of crystals as against Pockels' scheme.

8.3.2.3. Noncubic Crystals

There is an enormous wealth of information available now on the electro-optic properties of noncubic crystals, and the growth of the subject is so rapid these days that we can only barely touch upon a few topics here, leaving the Bibliography at the end of this book to describe a better picture of the subject.

Billings [145; 146] studied in detail the electro-optic effect of uniaxial crystals of some dihydrogen phosphates. Carpenter [228–230] measured the electro-optic coefficients of ADP and KDP. Vlokh, Zheludev, and Shamburov [1318] studied the electro-optic effect in crystals of pentaerythritol. Ho and Buhrer [530] determined the electro-optic constants of gallium arsenide. Gainon [409] studied cadmium sulfide. Buhrer, Ho, and Zucker [213] studied sodium thioantimonate. ADP and KDP and their isomorphs have attracted the attention of innumerable workers including, among others, Ott and Sliker [899], Vasilevskaya [1257], Rosner, Turner, and Kaminow [1068], Sidnenko and Zheludev [1132], Adhav [8–11], Veerabhadra Rao [1306], and Veerabhadra Rao and Narasimhamurty [1308]. Barium titanate has been studied by Johnston and Weingart [592] and Johnston [591; 594; 595], and calcium pyroniobate by Holmes, Spencer Ballman, and Lenzo [533]. Van Raalte [1251] studied the linear electro-optic effect in potassium tantaloniobate. Triglycine sulfate is another substance which has been widely studied; the authors include Vasilevskaya, Sonin, and Slepkov [1256] and Vlokh and Lutsiv-Shumskii [1319]. Smakula and Claspy [1147] studied lithium niobate and KTN. Rochelle salt has been studied by Anistratov and Aleksandrov [43]. Nitsche and Wild [877] grew crystals of copper tantalum-selenide and studied the electro-optic properties. Rez [1043] reviewed the work on the electro-optic crystals up to 1968. Emmensegger, Nitsche, and Miller [317] grew a number of double sulfates and determined their electro-optic constants. Hirano, Takei, and Koide [529] determined the linear electro-optic effect of $(Ba_{1-x}Sr_x)_2NaNb_5O_{15}$ crystals. Fujii and Sakudo [397] determined the electro-optic constants of strontium titanate. The electro-optic behavior of silver thiogallate was studied by Cound, Davies, Hulme, and Robertson [274]. The electro-optic effect in the crystals of iodic acid has been investigated by, among others, Vlokh, Kutnyi, and Nesterenko [1325] and Volkova, Dianova, Zuev, Izrail-enko, Lipatov, Parygin, Rashkovich, and Chirkov [1337]. Potassium tantaloniobate has been studied by Fox [384].

Adhav [8; 10] investigated the dispersion of the half-wave voltages of a number of tetragonal dihydrogen phosphates and arsenates. Johnston [595] studied the wavelength dependence of the electro-optic effect in barium titanate. Veerabhadra Rao [1306] and Veerabhadra Rao and Narasimhamurty [1308] have studied the dispersion of electro-optic constants of ADP, KDP, α-quartz, and Rochelle salt at constant electric fields.

Lenzo, Spencer, and Ballman [732] studied the temperature dependence of the electro-optic constants of strontium barium niobate. Vasilevskaya and Sonin [1269] have studied the temperature dependence of the electro-optic coefficients of the KDP-type crystals, which included RbDP, KDP, deuterated KDP, as well as potassium, rubidium, and caesium dihydrogen arsenates. Vlokh, Lutsiv-Shumskii, and Pylypyshin [1326] have studied the electro-optic and photoelastic properties of ADP and KDP at high temperatures.

The dependence of the linear electro-optic effect and dielectric constant on the melt composition in lithium niobate has been investigated by Turner, Nash, and Bridenbaugh [1230]. The influence of orientation of crystals in utilizing the linear electro-optic effect has been studied by a number of workers including Gisin [425; 426], Hulme [546], and Hookabe and Matsuo [535–537] with particular reference to light modulation. The influence of different degrees of deuteration on the electro-optic properties of KDP has been studied by a number of workers and recently by Volkova and Velichko [1339]. Both clamped and unclamped electro-optic behaviors of a large number of crystals have been studied by several investigators.

It is of special interest in the experimental investigations of the linear electro-optic effect in crystals to measure the electro-optic coefficients and their dependence upon (i) the wavelength of light that is transmitted through the crystal (dispersion), (ii) the electric field intensity in ferroelectrics (electro-optic hysteresis), (iii) the field frequency (frequency characteristics), and (iv) the temperature (especially for ferroelectrics).

The linear electro-optic effect in crystals has completely replaced the hitherto indispensable quadratic electro-optic effect in polar liquids for several reasons; the more outstanding ones are: (i) the Pockels effect in crystals is much larger than the Kerr effect in polar liquids; (ii) crystals are more convenient to work with; (iii) crystals can be used at very high frequencies, and what is still more important is that at their resonant frequencies the effect is very much enhanced owing to the indirect photoelastic effect generated by the converse piezoelectric effect in the crystals.

For technological applications some of the main considerations are the influence of the material and orientation of the transmitting electro-optic element upon the intensity and wavelength of light, the field strength, and the crystal thickness or half-wave voltage. A brief review of some of the applications of the Pockels effect is given in the last section of this chapter.

8.4. Pockels' Phenomenological Theory of the Linear Electro-Optic Effect in Three- and Two-Suffix Notations, R_{ijk} and r_{ij}

Under the action of an electric field, the optical properties of crystals change. Generally speaking the electric field can exert an influence (i) on the radiation processes (thereby causing the splitting of spectral lines, i.e., the Stark effect, or electroluminescence), (ii) on the change of absorption spectra, and (iii) on the refractive indices of the crystals. By electro-optic effect (as it is commonly understood in the scientific literature) we mean the phenomenon of the change in the indices of refraction or other related optical properties of the crystalline medium. Thus the action of the electric field results in a crystal, originally isotropic for light, becoming optically uniaxial or biaxial. An originally uniaxial crystal may become biaxial, and a biaxial crystal may have its refractive indices altered, or the axes of the optical index ellipsoid changed due to an impressed electric field. That is, in general, the shape, the size, and the orientation of the optical indicatrix may be changed.

The optical properties of crystals can be represented by an optical index ellipsoid (as we have already seen in Chapter 3, "Phenomenological Theory of Photoelasticity"). The optical index ellipsoid is a triaxial ellipsoid for biaxial crystals of triclinic, monoclinic, and orthorhombic systems, an ellipsoid of revolution for uniaxial crystals comprising trigonal, hexagonal, and tetragonal systems, and a sphere for anaxial crystals of the cubic system.

In its most general form the indicatrix or the optical index ellipsoid for an electrically unstrained crystal can be described by the equation below:

$$B^0_{11}x^2 + B^0_{22}y^2 + B^0_{33}z^2 + 2B^0_{23}yz + 2B^0_{31}zx + 2B^0_{12}xy - 1 \qquad (8.2)$$

This is the same as Eq. (3.8a), which was discussed in detail in Chapter 3 in connection with the phenomenological theory of photoelasticity due to Pockels. When the crystal is under the influence of an electric field, there will be in general a change of both the magnitude and direction of the principal axes defining the second-rank symmetrical tensor of optical polarization constants or the optical index ellipsoid. The difference between the original and the deformed tensors is again a symmetrical tensor. The electro-optic effects describe the relation between the above-mentioned second-rank symmetrical tensor (being the difference between the unstrained and strained ellipsoids) and the impressed electric field.

In order to explain the changes in the optical properties of a crystal under the influence of an applied electric field, Pockels assumes, as in the case of the photoelastic behavior of crystals, that the index ellipsoid defining the optical behavior of the crystal gets deformed owing to the applied electric field; the resulting deformation of the index ellipsoid can be described in terms of the changes ΔB_{ij} in the optical parameters in the indicatrix, the ΔB_{ij} being linearly related to the components E_k of the electric field. Thus the phenomenological tensor property relating a second-rank tensor ΔB_{ij} to a first-rank tensor E_k defines a third-rank tensor, described by the coefficients R_{ijk} called the linear electro-optic coefficients (also called the Pockels electro-optic constants). In order to bring out the full significance of R_{ijk} as a third-rank tensor, we represent the optical index ellipsoid in the form

$$B_{11}^0x^2 + B_{22}^0y^2 + B_{33}^0z^2 + B_{23}^0yz + B_{31}^0zx + B_{12}^0xy + B_{32}^0zy + B_{13}^0xz + B_{21}^0yx = 1 \tag{8.3}$$

for the unstrained crystal.

The deformed optical index ellipsoid will then be represented by

$$B_{11}x^2 + B_{22}y^2 + B_{33}z^2 + B_{23}yz + B_{31}zx + B_{12}xy + B_{32}zy + B_{13}xz + B_{21}yx = 1 \tag{8.4}$$

Now according to Pockels

$$B_{ij} - B_{ij}^0 = \Delta\left(\frac{1}{n_{ij}^2}\right) = \sum_k R_{ijk}E_k \tag{8.5}$$

Here E_k is the kth Cartesian component of the externally applied electric field.

Alternatively, the changes in the optical parameters B_{ij} can be linearly related to the electric-field-induced electric polarization P_k, in which case

$$B_{ij} - B_{ij}^0 = \Delta\left(\frac{1}{n_{ij}^2}\right) = \sum_k \tau_{ijk}P_k \tag{8.6a}$$

In Eqs. (8.5) and (8.6a) R_{ijk} and τ_{ijk} are interrelated thus:

$$\tau_{ijk} = R_{ijk}4\pi/(\varepsilon_k - 1) \tag{8.7}$$

The concepts developed in the previous paragraphs can be elaborated as follows (as in the case of photoelasticity of crystals).

The indicatrix or the optical index ellipsoid of a crystalline solid in the absence of an electric field can be written as

$$\begin{aligned} &B_{11}^0x^2 + B_{22}^0y^2 + B_{33}^0z^2 + B_{23}^0yz + B_{31}^0zx + B_{12}^0xy + B_{32}^0zy \\ &\qquad + B_{13}^0xz + B_{21}^0yx = 1 \end{aligned} \tag{8.3}$$

and under the influence of an electric field

$$\begin{aligned} &B_{11}x^2 + B_{22}y^2 + B_{33}z^2 + B_{23}yz + B_{31}zx + B_{12}xy + B_{32}zy + B_{13}xz \\ &\qquad + B_{21}yx = 1 \end{aligned} \tag{8.4}$$

Pockels assumes that the changes in the optical parameters B_{ij} can be expressed as homogeneous linear functions of the components of the applied electric field. Thus the relation in the three-suffix notation is written

$$\varDelta B_{ij} = B_{ij} - B_{ij}^0 = \sum_k R_{ijk}E_k \qquad (i, j = 1, 2, 3;\ k = 1, 2, 3) \tag{8.5}$$

or in the expanded form thus:

$$\begin{aligned}
B_{11} - B_{11}^0 &= \varDelta B_{11} = R_{111}E_x + R_{112}E_y + R_{113}E_z \\
B_{22} - B_{22}^0 &= \varDelta B_{22} = R_{221}E_x + R_{222}E_y + R_{223}E_z \\
B_{33} - B_{33}^0 &= \varDelta B_{33} = R_{331}E_x + R_{332}E_y + R_{333}E_z \\
B_{23} - B_{23}^0 &= \varDelta B_{23} = R_{231}E_x + R_{232}E_y + R_{233}E_z \\
B_{31} - B_{31}^0 &= \varDelta B_{31} = R_{311}E_x + R_{312}E_y + R_{313}E_z \\
B_{12} - B_{12}^0 &= \varDelta B_{12} = R_{121}E_x + R_{122}E_y + R_{123}E_z \\
B_{32} - B_{32}^0 &= \varDelta B_{32} = R_{321}E_x + R_{322}E_y + R_{323}E_z \\
B_{13} - B_{13}^0 &= \varDelta B_{13} = R_{131}E_x + R_{132}E_y + R_{133}E_z \\
B_{21} - B_{21}^0 &= \varDelta B_{21} = R_{211}E_x + R_{212}E_y + R_{213}E_z
\end{aligned} \tag{8.8}$$

The coefficients R_{ijk} are called the linear electro-optic constants in the three-suffix or tensor notation. The tensor nature of R as of third rank is obvious.

When the optical parameters are expressed in one-suffix notation, namely, B_i, then we get the linear electro-optic coefficients expressed in the matrix or two-suffix notation as given below:

$$\Delta B_i = B_i - B_i^0 = \sum_j r_{ij}E_j \qquad (i = 1, \ldots, 6;\ j = 1, 2, 3) \tag{8.9}$$

which when expanded yields the following:

$$\begin{aligned}
B_1 - B_1^0 &= \Delta B_1 = r_{11}E_x + r_{12}E_y + r_{13}E_z \\
B_2 - B_2^0 &= \Delta B_2 = r_{21}E_x + r_{22}E_y + r_{23}E_z \\
B_3 - B_3^0 &= \Delta B_3 = r_{31}E_x + r_{32}E_y + r_{33}E_z \\
B_4 - B_4^0 &= \Delta B_4 = r_{41}E_x + r_{42}E_y + r_{43}E_z \\
B_5 - B_5^0 &= \Delta B_5 = r_{51}E_x + r_{52}E_y + r_{53}E_z \\
B_6 - B_6^0 &= \Delta B_6 = r_{61}E_x + r_{62}E_y + r_{63}E_z
\end{aligned} \tag{8.10}$$

Note that, R_{ijk} being a third-rank tensor, there should be 27 coefficients of R_{ijk}. However, in Eq. (8.8) the terms on the left-hand side of each of the nine separate relations are the elements of a second-rank symmetric tensor; hence there are only six independent B_{ij} instead of nine. Since $B_{ij} = B_{ji}$ we get $R_{ijk} = R_{jik}$. Hence the last three relations of Eq. (8.8) are overlapped by the immediately preceding three relations. There are therefore only 18 independent linear electro-optic constants R_{ijk} defined by the first six relations in the set of 9 relations of Eq. (8.8).

Comparing Eqs. (8.8) and (8.10) we see that

$$R_{111} = r_{11}, \qquad R_{112} = r_{12}, \qquad R_{113} = r_{13}$$

and

$$R_{121} = r_{61}, \qquad R_{122} = r_{62}, \qquad R_{123} = r_{63}, \qquad \text{etc.}$$

Alternatively, ΔB_{ij} can be expressed in terms of the components of electric polarization P produced in the crystal consequent upon the application of the electric field E_i; here

$$P_i = \frac{1}{4\pi}(\varepsilon_i - 1)E_i \tag{8.11}$$

where ε_i is the dielectric constant of the crystalline solid. Thus, as indicated earlier, we can write

$$\Delta B_{ij} = B_{ij} - B_{ij}^0 = \sum_k \tau_{ijk}P_k \qquad (i, j = 1, 2, 3;\ k = 1, 2, 3) \tag{8.6b}$$

in three-suffix notation of τ_{ijk} and

$$\Delta B_i = B_i - B_i{}^0 = \sum_j \tau_{ij} P_j \qquad (i = 1 \cdots 6;\ j = 1, 2, 3) \tag{8.6c}$$

in two-suffix notation of τ_{ij}.

Equation (8.6c) can be written explicitly in terms of the components of electric polarization P thus:

$$\begin{aligned}
B_{11} - B_{11}^0 &= \Delta B_1 = \tau_{11} P_1 + \tau_{12} P_2 + \tau_{13} P_3 \\
B_{22} - B_{22}^0 &= \Delta B_2 = \tau_{21} P_1 + \tau_{22} P_2 + \tau_{23} P_3 \\
B_{33} - B_{33}^0 &= \Delta B_3 = \tau_{31} P_1 + \tau_{32} P_2 + \tau_{33} P_3 \\
B_{23} - B_{23}^0 &= \Delta B_4 = \tau_{41} P_1 + \tau_{42} P_2 + \tau_{43} P_3 \\
B_{31} - B_{31}^0 &= \Delta B_5 = \tau_{51} P_1 + \tau_{52} P_2 + \tau_{53} P_3 \\
B_{12} - B_{12}^0 &= \Delta B_6 = \tau_{61} P_1 + \tau_{62} P_2 + \tau_{63} P_3
\end{aligned} \tag{8.6d}$$

The two sets of equations (8.10) and (8.6d) are not independent, and τ_{ij} can be expressed in terms of r_{ij} thus:

$$\tau_{ij} = r_{ij} 4\pi/\varepsilon_j - 1 \tag{8.12}$$

It may be noted that it is the constants r_{ij} that are determined experimentally and thence τ_{ij} are computed because the latter are of greater importance to develop any atomistic (or microscopic) theory.

In what follows we confine our attention to r_{ij} only.

8.5. Derivation of the Relation between the Linear Electro-Optic Constants of a Crystal: Free and Clamped Constants

We derive below a relation between the free and clamped Pockels constants in the matrix or two-suffix notation (as a matter of convenience).

While discussing the linear electro-optic effect, we did not mention hitherto the mechanical state of the crystal. The crystal is either free (stress is zero) or clamped (deformation is zero, i.e., strain is zero). However, the mechanical state of the crystal cannot be considered as insignificant. If the crystal is free, then owing to the converse piezoelectric effect, a static electric field will give rise to deformations which in turn will result in changing the refractive indices and polarization constants as well. In this case, the total change of polarization constants in the matrix notation is deduced as follows.

In the most general case, the optical index ellipsoid is given by the following equation in single-suffix notation for B's under the influence of $\mathbf{E}$:

$$B_1x^2 + B_2y^2 + B_3z^2 + 2B_4yz + 2B_5zx + 2B_6xy = 1 \qquad (8.13)$$

The changes ΔB_i in the coefficients of the index ellipsoid due to both electro-optic and photoelastic effects can be expressed in terms of the electric field and either the stress or the strain in the crystal. Thus

$$\Delta B_i = \sum_{j=1}^{3} r_{ij}E_j + \sum_{k=1}^{6} q_{ik}\sigma_k \qquad (8.14)$$

$$\Delta B_i = \sum_{j=1}^{3} r'_{ij}E_j + \sum_{k=1}^{6} p_{ik}\varepsilon_k \qquad (8.15)$$

where σ_k and ε_k are the components of stress and strain, respectively, in one-suffix notation, q_{ik} and p_{ik} are the stress-optical and strain-optical constants, respectively, and r_{ij} and r'_{ij} are, respectively, the free and clamped linear electro-optic constants of the crystal. These relations are not independent of each other because of the piezoelectric relations:

$$\varepsilon_k = \sum_{l=1}^{6} s_{kl}\sigma_l + \sum_{j=1}^{3} d_{jk}E_j \qquad (k = 1 \cdots 6) \qquad (8.16)$$

where s_{kl} and d_{jk} are the elastic compliance and piezoelectric constants, respectively. Substituting Eq. (8.16) in Eq. (8.15), we obtain

$$\Delta B_i = \sum_{j=1}^{3} \left(r'_{ij} + \sum_{k=1}^{6} p_{ik}d_{jk}\right)E_j + \sum_{l=1}^{6} \left(\sum_{k=1}^{6} p_{ik}s_{kl}\right)\sigma_l \qquad (8.17)$$

On comparing Eq. (8.17) with Eq. (8.14), it is evident that

$$q_{ij} = \sum_{k=1}^{6} p_{ik}s_{kj} \qquad (8.18)$$

and

$$r_{ij} = r'_{ij} + \sum_{k=1}^{6} p_{ik}d_{jk} \qquad (8.19)$$

Therefore the electro-optic coefficient r_{ij} measured at zero stress (free crystal) differs from r'_{ij} measured at zero strain (clamped crystal).* From Eq. (8.19) we note, for ADP, for example that

$$r_{63} = r'_{63} + p_{66}d_{36} \tag{8.20}$$

For α-quartz

$$r_{11} = r'_{11} + (p_{11} - p_{12})d_{11} + p_{14}d_{14} \tag{8.21}$$

8.5.1. Discussion: Primary and Secondary Electro-Optic Effects, and Clamped and Unclamped Electro-Optic Coefficients

We can now distinguish between the primary and secondary electro-optic effects. The electro-optic effect that would arise if the crystal were not allowed to strain is called the primary (or true) effect. If the experimental prism is free, an electric field will cause a strain by the converse piezoelectric effect, and this in turn will cause a change in the index of refraction through the phenomenon of photoelasticity. The effect due to the piezoelectric strain (and hence due to photoelasticity) is then called the secondary (or false) effect. The primary effect is observed in a crystal which does not undergo deformation when an electric field is applied and the total effect is observed in a crystal which is deformed by a field (due to the converse piezoelectric effect). Hence we see that the secondary effect is equal to the difference between the total and primary effects.

It has already been pointed out that the linear electro-optic effect is possible with all but one class 432 of noncentrosymmetric crystals (which are also piezoelectric), and with them only. Because of the close relationship between this effect and the converse piezoelectricity, Röntgen [1065], who in 1883 was the first to make a thorough study of the linear electro-optic effect, believed that it was only a secondary consequence of the converse piezoelectric deformation, and he already knew the photoelastic phenomenon, which is the effect of a physical deformation on the refractive indices. Kundt [698] was also of the same opinion. It was Pockels [953] who through his classical investigations in 1894 conclusively proved the existence of a direct influence of the electric field on the optical parameters; that is, the refractive indices of a crystal deformed by an electric field are different from those of the crystal deformed to the same extent by a mechanical stress. The credit thus goes to Pockels, in whose honor this effect is called the Pockels effect.

* Equation (8.19), when expressed in tensor notation, takes the form $R^u_{ijk} = R^c_{ijk} + p_{ijmn}d_{kmn}$ (the superscripts u and c indicate the unclamped and clamped states of the crystal).

8.5.2. Methods of Obtaining the Primary and Secondary Linear Electro-Optic Effects

The primary effect can be obtained by suppressing the secondary effect. One method of achieving this is to use mechanical constraints for clamping the crystal. Another method is to produce the desired clamped or free effect by a proper choice of the frequency of the applied electric field.

When the crystal is under the influence of an external electric field at low frequencies, starting from zero frequency (i.e., dc field) up to frequencies low enough compared with the first mechanical resonance, which, depending upon the crystal dimensions, is usually of the order of 1 MHz, the crystal oscillates freely and it is the combined or overall effect that is observed. Hence, in this case the unclamped coefficient R^u, being the sum of the primary and secondary coefficients, is obtained.

If the electric field frequency is set very much higher than the elastic resonance frequency of the crystal prism, i.e., beyond the acoustic range of about 10^6 MHz, the crystal prism cannot deform macroscopically, since the mechanical deformation is excluded by the inertia of the crystal prism, thus the crystal behaves as though it were clamped. Hence, at such very high frequencies of the applied electric field the true or clamped coefficient R^c is obtained.

On the other hand, starting with a frequency equal to the fundamental mechanical resonance, which is usually of the order of 1 MHz and up to about 100 MHz, resonant vibrations occur at one or another upper harmonics of the crystal (depending upon its geometrical dimensions, elastic properties, and mode of vibration). In this region of electric field frequency, the secondary effect is greatly enhanced.

We see thus that the magnitude of R^c is different from that of R^u due to an indirect strain-optical contribution which always accompanies the piezoelectric strain induced by the applied electric field or polarization. It has been shown already that $R^u_{ijk} - R^c_{ijk} = \sum_{m,n} p_{ijmn} d_{kmn}$.

The magnitude of R^u_{ijk} may be either greater or less than R^c_{ijk} since the strain-optical contribution can be positive or negative.

The order of magnitude of the secondary effect (due to the converse piezoelectric effect, and hence due to the photoelastic effect) may be found as follows. A field E produces a piezoelectric strain $\varepsilon = d \cdot E$. This produces in turn a change in the optical parameter B_{ij} (of the optical index ellipsoid), where $\Delta B_{ij} = p\varepsilon = p(d \cdot E) = (pd)E$. Here p is the strain-optical constant. Since p is of the order of 10^{-1} and $(d \cdot E)$ of the order of 10^{-12}, we have ΔB_{ij} of the order of 10^{-13}.

It should be noted that the observed change in birefringence in a free crystal (for constant or low-frequency electric fields) is the sum of the primary and secondary effects. Also the universal Kerr effect is superposed in either case above.

Experiments [150] have shown that the contribution of the true linear electro-optic effect involving r_{63} (arising out of a longitudinal effect along the Z axis) for the crystal ADP is about 60% of the total effect, and for KDP it is about 90% of the total effect. For α-quartz calculations [1419] show that the unclamped r_{11} has 80% of it contributed by the true or clamped effect and only 20% is due to the secondary or false effect.

At frequencies far higher than 10^6 MHz of the applied electric field, and up to the interband absorption edge, only the electrons contribute to the linear electro-optic effect, which at such high frequencies is related to Brillouin scattering, Raman scattering, and optical second-harmonic generation.

8.6. *Kerr Quadratic Electro-Optic Effect: Pockels' Phenomenological Theory*

We have seen in Sec. 8.1 that the Kerr effect, also called the quadratic electro-optic effect, is a universal effect, and that it is shown by all crystals—centrosymmetric and noncentrosymmetric. Here ΔB_{ij}, which are the changes in the polarization constants, are linear functions of all the nine terms E_kE_l, where k and l run through all the values 1, 2, 3. The coefficients of E_kE_l form a symmetrical tensor of fourth rank. These coefficients in the two-suffix notation can straightaway be written in the following form:

$$\begin{bmatrix} B_1 - B_1^0 \\ B_2 - B_2^0 \\ B_3 - B_3^0 \\ B_4 - B_4^0 \\ B_5 - B_5^0 \\ B_6 - B_6^0 \end{bmatrix} = \begin{bmatrix} K_{11} & K_{12} & K_{13} & K_{14} & K_{15} & K_{16} \\ K_{21} & K_{22} & K_{23} & K_{24} & K_{25} & K_{26} \\ K_{31} & K_{32} & K_{33} & K_{34} & K_{35} & K_{36} \\ K_{41} & K_{42} & K_{43} & K_{44} & K_{45} & K_{46} \\ K_{51} & K_{52} & K_{53} & K_{54} & K_{55} & K_{56} \\ K_{61} & K_{62} & K_{63} & K_{64} & K_{65} & K_{66} \end{bmatrix} \begin{bmatrix} E_1^2 \\ E_2^2 \\ E_3^2 \\ E_2 \; E_3 \\ E_3 \; E_1 \\ E_1 \; E_2 \end{bmatrix} \tag{8.22}$$

The coefficients K_{ij} should not be mistaken for the components of a second-rank tensor; actually, for example, by measuring K_{14} we obtain K_{1123} (a fourth-rank tensor component) and a measurement of K_{22} yields K_{2222}. Here K_{ijkl}, a fourth-rank tensor, linking a symmetrical polarization tensor of the changes of polarization constants with a square of the polar vector (electric field) is analogous to the electrostriction tensor. The Kerr electro-optic effect, accompanying electrostriction, takes place in all dielectrics—solids, liquids, and gases. In fact, in crystals lacking a center of symmetry, the Pockels effect is accompanied by the Kerr effect, although its magnitude may be considerally less than that of the Pockels effect. In polar liquids (like nitrobenzene, alcohol, and turpentine) and in ferroelectric crystals [723] it is rather too large. The quadratic electro-optic effect in cubic crystals (centrosymmetric) and in polar liquids is made use of in the construction of Kerr cells for light modulation.

All centrosymmetric crystals fail to show the linear electro-optic effect (Pockels effect); however, they exhibit the universal quadratic (Kerr) electro-optic effect. As in the case of the Pockels effect, the quadratic (Kerr) electro-optic effect can be put in a mathematical formulation thus:

$$B_{ij} - B_{ij}^0 = \Delta B_{ij} = \Delta\left(\frac{1}{n_{ij}^2}\right) = \sum_{k,l=1,2,3} K_{ijkl} E_k E_l \tag{8.23}$$

Alternatively the phenomenological quadratic polarization optic (PO) coefficients g_{ijkl} can be defined by the relation below:

$$\Delta\left(\frac{1}{n_{ij}^2}\right) = \sum_{k,l=1,2,3} g_{ijkl} P_k P_l \tag{8.24}$$

where P_k and P_l are the electric polarizations.

Here again, as in the case of the linear electro-optic effect, the coefficient g_{ijkl} can show two values, clamped and unclamped, depending upon the experimental conditions. The tensor of electrostrictive strain ε_{kl} can be put in the form

$$\varepsilon_{kl} = \sum_{m,n} Q_{klmn} P_m P_n$$

Now, the difference between the unclamped g^u and clamped g^c values of g_{ijkl} can be obtained in a manner similar to that used to obtain Eq. (8.19). Thus we have

$$g_{ijkl}^u - g_{ijkl}^c = \sum_{m,n} p_{ijmn} Q_{mnkl} \tag{8.25}$$

which is the product of the strain-optical tensor p_{ijmn} and the quadratic electrostrictive tensor Q_{mnkl}.

Methods of deriving expressions for the Kerr electro-optic path retardations and the experimental techniques are exactly the same as those for determining the photoelastic constants. Hence they will not be described here.

The interrelation between the changes of polarization constants ΔB_{ij} and the electric field E_k can be analytically described by the following equation:

$$\Delta B_{ij} = aE_k + bE_kE_l + \cdots \tag{8.26}$$

Ordinarily the changes of the polarization constants are small. Now let the crystal on which E_k is acting have a center of symmetry. Let the crystal be reversed now with respect to **E**; this is equivalent to saying that **E** has been reversed, keeping the crystal undisturbed. Consequently the physical situation, and therefore the change of optical polarization constants, remain unaltered, but the signs of E_k and E_l in Eq. (8.26) must be reversed. We then have

$$\Delta B_{ij} = -aE_k + bE_kE_l \tag{8.27}$$

ignoring the third- and higher-order terms. Comparing Eqs. (8.26) and (8.27), we find that a is equal to zero, and hence

$$\Delta B_{ij} = bE_kE_l \tag{8.28}$$

which defines the Kerr effect. If, on the other hand, there were no center of symmetry present which made the two directions of the field equivalent from the point of view of the crystal, then there is no reason why a reversal of the electric field should not be accompanied by a change in the term ΔB_{ij} and hence a would not vanish, that is, the linear electro-optic effect can exist. Thus we conclude that the linear electro-optic effect (Pockels effect) cannot be exhibited by centrosymmetric crystals. On the other hand, in the crystals with a center of symmetry and in liquids and gases, the term bE_k^2 can exist, though the effect is small for several substances for moderate fields. Thus the Kerr effect is a universal effect.

The linear electro-optic effect has the same relation to the Kerr effect (change in the refractive index being a quadratic function of the applied electric field) as the converse piezoelectricity has to electrostriction (geometric deformation being a quadratic function of electric field). The double refraction produced at room temperature with paraelectric fields (up to 20 kV/cm) is only of the order of 10^{-4}. Although this is too small to appreciably change refractive angles, it is sufficient to produce optical path retardations of the order of a wavelength, and hence lead to interference phenomena making it possible to fabricate quarter-, half-, and full-wave retardation plates. Furthermore, these interference phenomena are of immense use to modulate polarized light. See Table 8.5 for the numerical values of g_{ijkl} (put in two-suffix notation) of cubic crystals.

Morphic Effects. Under the influence of an electric field the crystal symmetry may or may not lower. When the crystal symmetry is lowered, new electro-optic coefficients (which are zero in the earlier symmetry of the crystal) can arise corresponding to the new symmetry of the crystal under the action of the electric field. These new coefficients are proportional to the applied electric field, and consequently, they determine the changes of polarization constants which are proportional to the square of the field strength. This quadratic type of effect is called morphic, a term introduced by Mueller [831] in his studies on the electro-optic behavior of Rochelle salt. Such morphic effects can also arise in photoelasticity and thus the deformation-induced coefficients are not constant in their magnitude, the latter being proportional to the stress. This is tantamount to saying that a crystal belongs to its assigned class only so long as it is free from stress of any sort. But, in the strained state, crystals—particularly of higher symmetry—may get reduced to a lower symmetry. In general, morphic effects are not large, hence they escape notice. In our study of electro-optics and photoelasticity we avoid the morphic effects by working in the region where they do not manifest themselves.

8.7. Crystal Symmetry and the Number of Surviving Linear Electro-Optic Coefficients R_{ijk} and r_{ij} and Their Deduction by Tensor Method: r_{ij} Matrices for the 21 Noncentrosymmetric Classes

8.7.1. Crystal Symmetry and the Surviving Linear Electro-Optic Constants

Since R_{ijk} is a third-rank tensor, the linear electro-optic effect is exhibited only by noncentrosymmetric crystals by virtue of Neumann's principle (see Chapter 7 for d_{ijk}). There are 21 point groups or crystal classes which are noncentrosymmetric, among which there is one point group 432 (O) class which on detailed examination shows no surviving R_{ijk}. Crystals that belong to any of the remaining 20 classes can show a linear electro-optic effect, that is, a birefringence directly proportional to the applied voltage. The symmetry conditions for the occurrence of this effect are exactly the same as for the occurrence of the piezoelectric effect. Thus there is an exact analogy in regard to the symmetry conditions between the linear electro-optic effect (double refraction as a linear function of electric field) and the converse piezoelectric effect (geometrical or physical strain as a linear function of the electric field).

We have seen earlier that the maximum number of nonvanishing, independent linear electro-optic constants r_{ij} that can occur is 18 and this number occurs for the most anisotropic crystal class C_1 (1). This number, however, gets reduced with the increase of symmetry of the crystal. As in the case of q_{ij} and p_{ij} in photoelasticity, the number of nonvanishing independent constants of R_{ijk} for each crystal class can be deduced by (i) classical, (ii) tensor, and (iii) group theoretical methods. The matrices of the surviving piezoelectric constants for a few of the 20 crystal classes that exhibit piezoelectricity have been derived by tensor methods in Chapter 7. On similar lines we deduce the matrices for the Pockels effect (the linear electro-optic effect) which are given in Table 8.1.

8.7.2. Tensor Method of Deducing the Nonvanishing Independent R_{ijk}

First we examine the influence of a center of symmetry or inversion on R_{ijk}. Figure 8.1 shows the effect of the symmetry operation of inversion on the coordinate axes; let its transformation matrix be given by

$$\begin{bmatrix} \alpha_{11} & \alpha_{12} & \alpha_{13} \\ \alpha_{21} & \alpha_{22} & \alpha_{23} \\ \alpha_{31} & \alpha_{32} & \alpha_{33} \end{bmatrix}$$

For tensor coefficients of third rank, the law of transformation from the old to the new system of coordinate axes is

$$R'_{ijk} = \alpha_{il}\alpha_{jm}\alpha_{kn}R_{lmn} \tag{8.29}$$

Now the direction cosines α_{ij} are zero except when $i = l$, $j = m$, and $k = n$; in all these three cases, each nonvanishing $\alpha_{il} = -1$. Consequently Eq. (8.29) reduces to

$$R'_{ijk} = -R_{ijk} \tag{8.30}$$

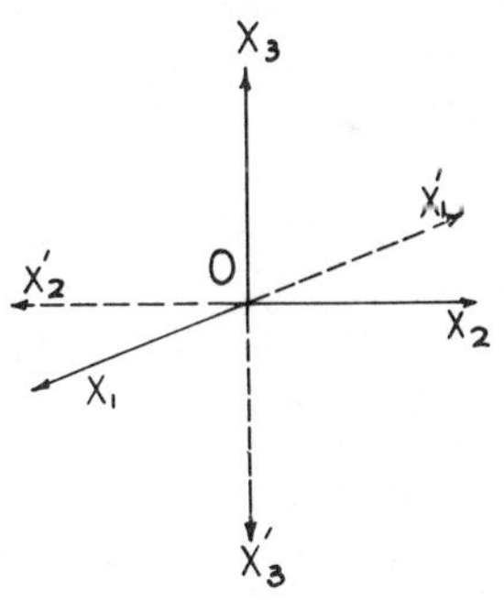

Fig. 8.1. Transformation of axes: the origin O is a center of symmetry.

From Neumann's principle R'_{ijk} should be equal to R_{ijk} since a symmetry operation must leave any physical property invariant; that is,

$$R'_{ijk} = R_{ijk} \tag{8.31}$$

Combining Eqs. (8.30) and (8.31) we obtain

$$R'_{ijk} = -R_{ijk} = R_{ijk} \qquad \therefore R_{ijk} = 0 \tag{8.32}$$

Hence all crystal classes that are centrosymmetric cannot have any of their components R_{ijk} surviving. Thus none of the centrosymmetric classes can exhibit the Pockels effect.

Next, we examine each one of the 21 noncentrosymmetric classes in regard to the surviving R_{ijk}.

8.7.2.1. Triclinic System: Class 1 (C_1)

This has only an identity operation E which leaves all the coordinate axes undisturbed. Figure 8.2 shows the disposition of the new and old coordinate systems, and the matrix of their direction cosines is

$$\begin{bmatrix} 1 & 0 & 0 \\ 0 & 1 & 0 \\ 0 & 0 & 1 \end{bmatrix}$$

As indicated earlier

$$R'_{ijk} = \alpha_{il}\alpha_{jm}\alpha_{kn}R_{lmn} \tag{8.29}$$

and the direction cosines α_{ij} are zero except when $i = l$, $j = m$, and $k = n$; in all these three cases each $\alpha_{il} = +1$. Hence for all i, j, k,

$$R'_{ijk} = R_{ijk} \tag{8.33}$$

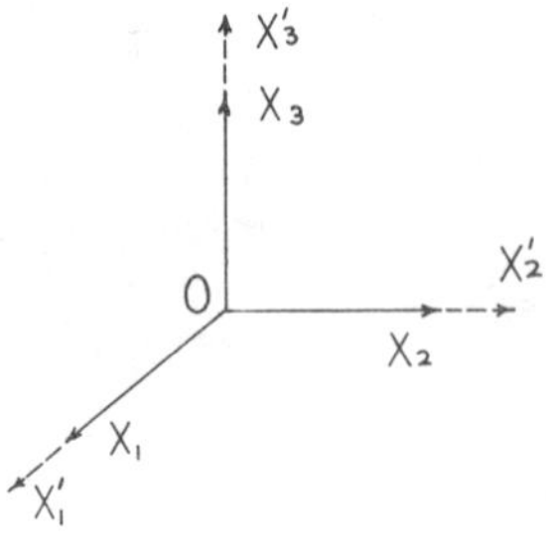

Fig. 8.2. Transformation of axes due to the symmetry element, identity E, or rotation by 360° about any axis.

From Neumann's principle, too, R'_{ijk} should be equal to R_{ijk} since the change in coordinate system resulting from the symmetry operation E must leave the physical property unaltered.

Hence all the R_{ijk} survive. Thus for this crystal class 1 (C_1), there are 18 independent nonvanishing R_{ijk} (see page 366) and in the two-suffix notation, there exist all the 18 r_{ij} whose matrix is given in Table 8.1.

8.7.2.2. *Monoclinic System*: *Class m* (C_{1h})

This has only a plane of symmetry, and according to the standard convention (IEEE Standards on Piezoelectric Crystals, 1949) this plane is perpendicular to the X_2 axis. The new coordinate system is shown in Fig. 8.3, and its transformation matrix is

$$\begin{bmatrix} 1 & 0 & 0 \\ 0 & -1 & 0 \\ 0 & 0 & 1 \end{bmatrix}$$

We give below the full procedure in this case.

The general formula for the transformation of the R_{ijk} from the old to the new axes is

$$R'_{ijk} = \alpha_{il}\alpha_{jm}\alpha_{kn}R_{lmn} \tag{8.29}$$

Now let us examine R_{111}. Inserting the appropriate values of i, j, k, l, m, n, and the direction cosines

$$\begin{aligned} R'_{111} &= \alpha_{1l}\alpha_{1m}\alpha_{1n}R_{lmn} \\ &= \alpha_{1m}\alpha_{1n}[\alpha_{11}R_{1mn} + \alpha_{12}R_{2mn} + \alpha_{13}R_{3mn}] \\ &= \alpha_{1n}\{\alpha_{11}[\alpha_{11}R_{11n} + \alpha_{12}R_{12n} + \alpha_{13}R_{13n}]\} \\ &= \alpha_{11}\alpha_{11}(\alpha_{11}R_{111} + \alpha_{12}R_{112} + \alpha_{13}R_{113}) \\ &= \alpha_{11}^3R_{111} = R_{111} \end{aligned} \tag{8.34}$$

By Neumann's principle, R'_{111} must be equal to R_{111}. Thus by both the symmetry operation and Neumann's principle $R'_{111} = R_{111}$. Hence R_{111} survives.

The above procedure is repeated for each one of the remaining R_{ijk}, and we note that, for example, R_{122} and R_{133} survive, whereas R_{123} vanishes; thus a few R_{ijk} survive and the remaining vanish. But it is a tedious procedure to examine each R_{ijk} individually. We can very much simplify the algebra by the method of inspection.

Table 8.1. Pockels' Linear Electro-Optic Constants r_{ij}[a]

Class 1 (C_1)
18 constants

$$\begin{bmatrix} B_{11}-B_{11}^0 \\ B_{22}-B_{22}^0 \\ B_{33}-B_{33}^0 \\ B_{23}-B_{23}^0 \\ B_{31}-B_{31}^0 \\ B_{12}-B_{12}^0 \end{bmatrix} = \begin{bmatrix} r_{11} & r_{12} & r_{13} \\ r_{21} & r_{22} & r_{23} \\ r_{31} & r_{32} & r_{33} \\ r_{41} & r_{42} & r_{43} \\ r_{51} & r_{52} & r_{53} \\ r_{61} & r_{62} & r_{63} \end{bmatrix} \begin{bmatrix} E_1 \\ E_2 \\ E_3 \end{bmatrix}$$

Class 2 (C_2), 8 constants, $2 \parallel X_3$

$$\begin{matrix} 0 & 0 & r_{13} \\ 0 & 0 & r_{23} \\ 0 & 0 & r_{33} \\ r_{41} & r_{42} & 0 \\ r_{51} & r_{52} & 0 \\ 0 & 0 & r_{63} \end{matrix}$$

Class 2 (C_2), 8 constants, $2 \parallel X_2$ standard orientation

$$\begin{matrix} 0 & r_{12} & 0 \\ 0 & r_{22} & 0 \\ 0 & r_{32} & 0 \\ r_{41} & 0 & r_{43} \\ 0 & r_{52} & 0 \\ r_{61} & 0 & r_{63} \end{matrix}$$

Class m (C_{1h}), 10 constants, $m \perp X_3$

$$\begin{matrix} r_{11} & r_{12} & 0 \\ r_{21} & r_{22} & 0 \\ r_{31} & r_{32} & 0 \\ 0 & 0 & r_{43} \\ 0 & 0 & r_{53} \\ r_{61} & r_{62} & 0 \end{matrix}$$

Class m (C_{1h}), 10 constants, $m \perp X_2$ standard orientation

$$\begin{matrix} r_{11} & 0 & r_{13} \\ r_{21} & 0 & r_{23} \\ r_{31} & 0 & r_{33} \\ 0 & r_{42} & 0 \\ r_{51} & 0 & r_{53} \\ 0 & r_{62} & 0 \end{matrix}$$

Class 222 (D_2), 3 constants

$$\begin{matrix} 0 & 0 & 0 \\ 0 & 0 & 0 \\ 0 & 0 & 0 \\ r_{41} & 0 & 0 \\ 0 & r_{52} & 0 \\ 0 & 0 & r_{63} \end{matrix}$$

Class $mm2$ (C_{2v}), 5 constants

$$\begin{matrix} 0 & 0 & r_{13} \\ 0 & 0 & r_{23} \\ 0 & 0 & r_{33} \\ 0 & r_{42} & 0 \\ r_{51} & 0 & 0 \\ 0 & 0 & 0 \end{matrix}$$

Classes 4 (C_4) and 6 (C_6), 4 constants

$$\begin{matrix} 0 & 0 & r_{13} \\ 0 & 0 & r_{13} \\ 0 & 0 & r_{33} \\ r_{41} & r_{51} & 0 \\ r_{51} & -r_{41} & 0 \\ 0 & 0 & 0 \end{matrix}$$

Class $\bar{4}$ (S_4), 4 constants

$$\begin{matrix} 0 & 0 & r_{13} \\ 0 & 0 & -r_{13} \\ 0 & 0 & 0 \\ r_{41} & -r_{51} & 0 \\ r_{51} & r_{41} & 0 \\ 0 & 0 & r_{63} \end{matrix}$$

Classes $4mm$ (C_{4v}) and $6mm$ (C_{6v}), 3 constants

$$\begin{matrix} 0 & 0 & r_{13} \\ 0 & 0 & r_{13} \\ 0 & 0 & r_{33} \\ 0 & r_{51} & 0 \\ r_{51} & 0 & 0 \\ 0 & 0 & 0 \end{matrix}$$

Class $\bar{4}2m$ (D_{2d}), 2 constants, $2 \parallel X_1$

$$\begin{matrix} 0 & 0 & 0 \\ 0 & 0 & 0 \\ 0 & 0 & 0 \\ r_{41} & 0 & 0 \\ 0 & r_{41} & 0 \\ 0 & 0 & r_{63} \end{matrix}$$

Table 8.1 (continued)

Classes 422 (D_4) and 622 (D_6)
1 constant

0	0	0
0	0	0
0	0	0
r_{41}	0	0
0	$-r_{41}$	0
0	0	0

Class 3 (C_3)
6 constants

r_{11}	$-r_{22}$	r_{13}
$-r_{11}$	r_{22}	r_{13}
0	0	r_{33}
r_{41}	r_{51}	0
r_{51}	$-r_{41}$	0
$-r_{22}$	$-r_{11}$	0

Class 32 (D_3)
2 constants

r_{11}	0	0
$-r_{11}$	0	0
0	0	0
r_{41}	0	0
0	$-r_{41}$	0
0	$-r_{11}$	0

Class $3m$ (C_{3v})
4 constants

$m \perp X_2$

r_{11}	0	r_{13}
$-r_{11}$	0	r_{13}
0	0	r_{33}
0	r_{51}	0
r_{51}	0	0
0	$-r_{11}$	0

Class $3m$ (C_{3v})
4 constants

$m \perp X_1$
standard
orientation

0	$-r_{22}$	r_{13}
0	r_{22}	r_{13}
0	0	r_{33}
0	r_{51}	0
r_{51}	0	0
$-r_{22}$	0	0

Class $\bar{6}m2$ (D_{3h})
1 constant

$m \perp X_2$

r_{11}	0	0
$-r_{11}$	0	0
0	0	0
0	0	0
0	0	0
0	$-r_{11}$	0

Class $\bar{6}m2$ (D_{3h})
1 constant

$m \perp X_1$
standard
orientation

0	$-r_{22}$	0
0	r_{22}	0
0	0	0
0	0	0
0	0	0
$-r_{22}$	0	0

Class $\bar{6}$ (C_{3h})
2 constants

r_{11}	$-r_{22}$	0
$-r_{11}$	r_{22}	0
0	0	0
0	0	0
0	0	0
$-r_{22}$	$-r_{11}$	0

Classes 23 (T) and $\bar{4}3m$ (T_d)
1 constant

0	0	0
0	0	0
0	0	0
r_{41}	0	0
0	r_{41}	0
0	0	r_{41}

Class 432 (O)

0	0	0
0	0	0
0	0	0
0	0	0
0	0	0
0	0	0

all constants vanish

[a] For the purpose of numerical computations the r_{ij} electro-optic matrix and the d_{ij} piezoelectric matrix should be consistent, whenever alternative conventions are followed.

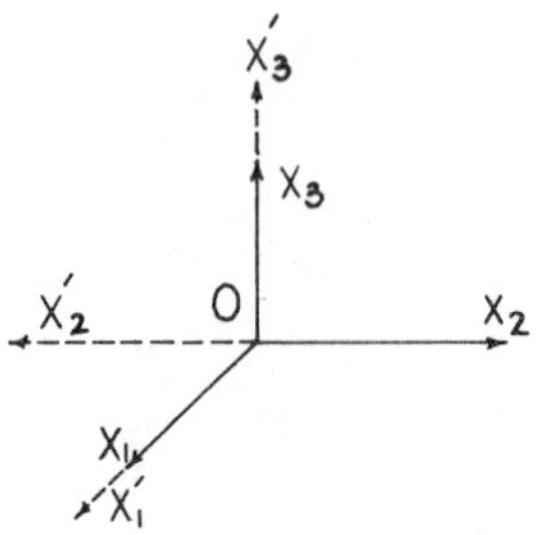

Fig. 8.3. Transformation of axes: the plane X_1OX_3 is a reflection plane.

With the transformation matrix in question, we note that if α_{22} appears once or thrice in Eq. (8.29) the corresponding R_{ijk} vanishes. That is, any R_{ijk} having one 2 or three 2's as its subscripts should vanish. Thus the vanishing R_{ijk} are those that are canceled as shown below:

$$\begin{array}{ccccccccc}
R_{111} & \cancel{R_{112}} & R_{113} & R_{221} & \cancel{R_{222}} & R_{223} & R_{331} & \cancel{R_{332}} & R_{333} \\
\cancel{R_{231}} & R_{232} & \cancel{R_{233}} & R_{311} & \cancel{R_{312}} & R_{313} & \cancel{R_{121}} & R_{122} & \cancel{R_{123}} \\
\cancel{R_{321}} & R_{322} & \cancel{R_{323}} & R_{131} & \cancel{R_{132}} & R_{133} & \cancel{R_{211}} & R_{212} & \cancel{R_{213}}
\end{array}$$

Hence the nonvanishing independent r_{ij} in the two-suffix notation are

$$r_{11} \quad r_{13} \quad r_{21} \quad r_{23} \quad r_{31} \quad r_{33} \quad r_{42} \quad r_{51} \quad r_{53} \quad r_{62}$$

and their matrix is given in Table 8.1.

For the class m (C_{1h}) of the monoclinic system with its reflection plane perpendicular to the X_3 axis, we have

$$\alpha_{11} = 1, \qquad \alpha_{22} = 1, \qquad \alpha_{33} = -1$$

and all other α_{ij} are zero. With this transformation all those R_{ijk} should vanish that have one 3 or three 3's as the subscripts. Thus the vanishing coefficients are those that are canceled as shown below:

$$\begin{array}{ccccccccc}
R_{111} & R_{112} & \cancel{R_{113}} & R_{221} & R_{222} & \cancel{R_{223}} & R_{331} & R_{332} & \cancel{R_{333}} \\
\cancel{R_{231}} & \cancel{R_{232}} & R_{233} & \cancel{R_{311}} & \cancel{R_{312}} & R_{313} & R_{121} & R_{122} & \cancel{R_{123}} \\
\cancel{R_{321}} & \cancel{R_{322}} & R_{323} & \cancel{R_{131}} & \cancel{R_{132}} & R_{133} & R_{211} & R_{212} & \cancel{R_{213}}
\end{array}$$

Hence the nonvanishing independent r_{ij} in the two-suffix notation are

$$r_{11} \quad r_{12} \quad r_{21} \quad r_{22} \quad r_{31} \quad r_{32} \quad r_{43} \quad r_{53} \quad r_{61} \quad r_{62}$$

and their matrix is given in Table 8.1.

8.7.2.3. *Monoclinic System*: *Class 2* (C_2)

This has only a diad axis as its element of symmetry (besides, of course, E), and according to the standard convention this twofold axis of symmetry is parallel to the X_2 axis. The new coordinate system is shown in Fig. 8.4, and its transformation matrix is

$$\begin{bmatrix} -1 & 0 & 0 \\ 0 & 1 & 0 \\ 0 & 0 & -1 \end{bmatrix}$$

As in the previous case we notice that the direction cosines are zero except when

$$i = l, \qquad j = m, \qquad k = n$$

The product $\alpha_{il}\alpha_{jm}\alpha_{kn}$ is equal to $+1$ whenever the subscripts of R_{ijk} consist of a single 2 or all three 2's. Any combination other than these will necessarily be zero. Hence the surviving R_{ijk} are as shown below:

$\cancel{R_{111}}$	R_{112}	$\cancel{R_{113}}$	$\cancel{R_{221}}$	R_{222}	$\cancel{R_{223}}$	$\cancel{R_{331}}$	R_{332}	$\cancel{R_{333}}$
R_{231}	$\cancel{R_{232}}$	R_{233}	$\cancel{R_{311}}$	R_{312}	$\cancel{R_{313}}$	R_{121}	$\cancel{R_{122}}$	R_{123}
R_{321}	$\cancel{R_{322}}$	R_{323}	$\cancel{R_{131}}$	R_{132}	$\cancel{R_{133}}$	R_{211}	$\cancel{R_{212}}$	R_{213}

Hence the nonvanishing independent r_{ij} are

$$r_{12} \quad r_{22} \quad r_{32} \quad r_{41} \quad r_{43} \quad r_{52} \quad r_{61} \quad r_{63}$$

and their matrix is given in Table 8.1.

For the same class 2 (C_2) of the monoclinic system if the diad axis is parallel to X_3 we have

$$\alpha_{11} = -1, \qquad \alpha_{22} = -1, \qquad \alpha_{33} = 1$$

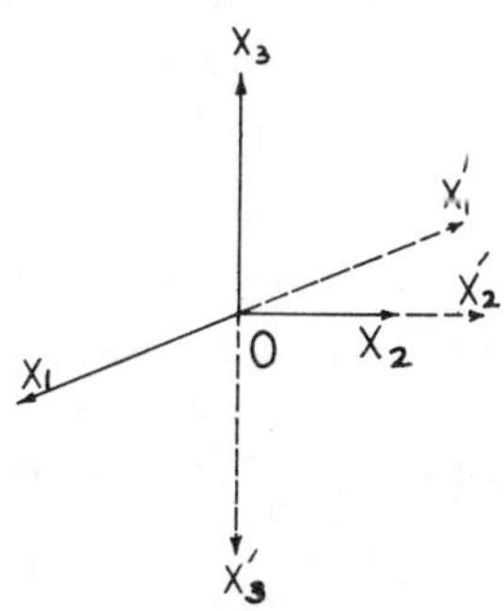

Fig. 8.4. Transformation of axes: OX_2' is a diad or twofold axis of rotation; here $OX_2' = OX_2$.

and all others are equal to zero. The product $\alpha_{il}\alpha_{jm}\alpha_{kn}$ is equal to $+1$ whenever the subscripts of R_{ijk} consist of a single 3 or all three 3's. Any combination other than the above must be zero. Hence the surviving coefficients are other than those that are canceled as indicated below.

$$\begin{array}{ccccccccc} \cancel{R_{111}} & \cancel{R_{112}} & R_{113} & \cancel{R_{221}} & \cancel{R_{222}} & R_{223} & \cancel{R_{331}} & \cancel{R_{332}} & R_{333} \\ R_{231} & R_{232} & \cancel{R_{233}} & R_{311} & R_{312} & \cancel{R_{313}} & \cancel{R_{121}} & \cancel{R_{122}} & R_{123} \\ R_{321} & R_{322} & \cancel{R_{323}} & R_{131} & R_{132} & \cancel{R_{133}} & \cancel{R_{211}} & \cancel{R_{212}} & R_{213} \end{array}$$

Hence the nonvanishing independent r_{ij} are

$$r_{13} \quad r_{23} \quad r_{33} \quad r_{41} \quad r_{42} \quad r_{51} \quad r_{52} \quad r_{63}$$

Their matrix is given in Table 8.1.

8.7.2.4. Orthorhombic System: Class 222 (D_2); Symmetry Elements: $E\ C_2\ C_2'\ C_2''$; Generating Elements: ($C_2\ C_2'$)

Proceeding on the lines indicated in Sec. 8.7.2.3 above we can easily deduce the surviving R_{ijk} for this class.

This class has three mutually perpendicular diads which are coincident with the three coordinate axes X_1, X_2, and X_3.

With X_2 as the diad axis, we find, as shown in Sec. 8.7.2.3 above, that the surviving independent R_{ijk} are

$$R_{112} \quad R_{222} \quad R_{332} \quad R_{231} \quad R_{233} \quad R_{312} \quad R_{121} \quad R_{123}$$

With X_3 as the diad axis, we have seen, the surviving independent R_{ijk} are

$$R_{113} \quad R_{223} \quad R_{333} \quad R_{231} \quad R_{232} \quad R_{311} \quad R_{312} \quad R_{123}$$

The superposition of a diad axis C_2 parallel to X_3 on another diad axis C_2'' parallel to X_2 will permit the survival of only three R_{ijk}, namely,

$$R_{231} \quad R_{312} \quad R_{123}$$

Proceeding in the same manner with the superposition of the remaining diad axis C_2' parallel to the X_1 axis, it will be seen that no more R_{ijk} coefficients become zero, and hence owing to all the three diads C_2'', C_2', and C_2 the nonvanishing independent R_{ijk} are

$$R_{231} \quad R_{312} \quad R_{123}$$

and in the two-suffix notation there are only three, namely,

$$r_{41} \quad r_{52} \quad r_{63}$$

which completely describe the linear electro-optic effect in crystals like Rochelle salt.

At this stage we may point out that the class 222 (D_2) has the symmetry elements $E\ C_2\ C_2'\ C_2''$, of which any two of the three mutually perpendicular diads form the generating elements of the crystal class. It is well known that in deducing the nonvanishing independent components of any tensor property of a crystal, it will be quite sufficient if we study the influence of only the generating elements of the crystal. This fact has been completely proved above in the deduction of the R_{ijk} components of the class 222 (D_2), where we have seen that the imposition of C_2 over C_2'' gives all the required information since they form a set of generating elements, and the application of the remaining diad C_2' gives no additional information, but serves as a useful check on the results obtained otherwise.

8.7.2.5. *Orthorhombic System*: *Class mm2* (C_{2v}); *Symmetry Elements*: $E\ C_2\ \sigma_v'\ \sigma_v''$; *Generating Elements*: $C_2\ \sigma_v'$

In this crystal class there are two vertical planes of symmetry intersecting along a diad axis parallel to X_3. Since the generating elements are C_2 and σ_v' we get the surviving coefficients by combining a vertical plane perpendicular to X_1 with a diad parallel to X_3.

We have already seen in Section 8.7.2.3 that with X_3 as a diad, the surviving coefficients of R_{ijk} are those with subscripts having one 3 and three 3's; hence they are

$$R_{113} \quad R_{223} \quad R_{333} \quad R_{231} \quad R_{232} \quad R_{311} \quad R_{312} \quad R_{123}$$

and their equivalents. Hence the nonvanishing independent r_{ij} are

$$r_{13} \quad r_{23} \quad r_{33} \quad r_{41} \quad r_{42} \quad r_{51} \quad r_{52} \quad r_{63}$$

Owing to the additional symmetry of a plane perpendicular to X_1, among the eight components mentioned above, all those that have their subscripts containing one 1 or all three 1's vanish. Thus the following three R_{ijk} vanish:

$$R_{123} \quad R_{231} \quad R_{312}$$

The imposition of the remaining element of symmetry σ_v'' perpendicular to X_2 will not result in any further reduction of R_{ijk}; this is what is to be expected since we have already exhausted the generating elements C_2, σ_r'. The nonvanishing independent R_{ijk} are therefore

$$R_{113} \quad R_{223} \quad R_{333} \quad R_{232} \quad R_{311}$$

and in the two-suffix notation they are

$$r_{13} \quad r_{23} \quad r_{33} \quad r_{42} \quad r_{51}$$

8.7.2.6. *Tetragonal System: Class 4* (C_4); *Symmetry Elements*: $E\ C_2\ 2C_4$; *Generating Element*: C_4

This class of crystals has a tetrad axis parallel to X_3 (Fig. 8.5); but a tetrad axis parallel to X_3 implies a diad axis parallel to the same X_3 (since $C_4 \cdot C_4 = C_2$). We have found earlier that the surviving constants for a diad parallel to X_3 are such that each combination of subscripts of R_{ijk} should contain either one 3 or three 3's. Thus the survivals, along with their equivalents, are

$$R_{113} \quad R_{223} \quad R_{333} \quad R_{231} \quad R_{232} \quad R_{311} \quad R_{312} \quad R_{123}$$

Therefore only these eight R_{ijk} need be considered for survival and equivalence when the C_4 symmetry element is imposed on them.

The transformation matrix for the C_4 about X_3 (which forms the only generating element for the crystal class C_4) is given below:

$$\begin{bmatrix} 0 & 1 & 0 \\ -1 & 0 & 0 \\ 0 & 0 & 1 \end{bmatrix}$$

All the 8 R_{ijk} will now have to be considered individually:

$$R'_{113} = \alpha_{12}\alpha_{12}\alpha_{33}R_{223} = R_{223}$$

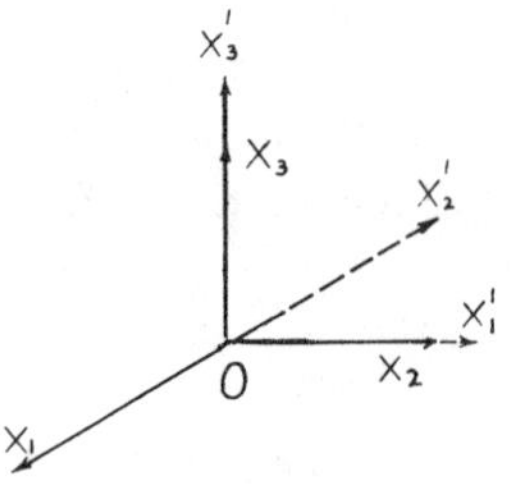

Fig. 8.5. Transformation of axes: the axis OX_3 is a tetrad or a fourfold axis of rotation; here $OX_3' = OX_3$.

But by Neumann's principle $R'_{113} = R_{113}$. Hence $R_{113} = R_{223}$.

$$R'_{223} = \alpha_{21}\alpha_{21}\alpha_{33}R_{113} = R_{113} = R_{223}$$

$$R'_{333} = \alpha_{33}\alpha_{33}\alpha_{33}R_{333} = R_{333}$$

$$R'_{231} = \alpha_{21}\alpha_{33}\alpha_{12}R_{132} = -R_{132} = R_{231}$$

$$R'_{232} = \alpha_{21}\alpha_{33}\alpha_{21}R_{131} = R_{131}$$

$$R'_{311} = \alpha_{33}\alpha_{12}\alpha_{12}R_{322} = R_{322} = R_{311}$$

$$R'_{312} = \alpha_{33}\alpha_{12}\alpha_{21}R_{321} = -R_{321} = R_{312}$$

$$R'_{123} = \alpha_{12}\alpha_{21}\alpha_{33}R_{213} = -R_{213} = R_{123}$$

But $R_{123} = R_{213}$ since $B_{ij} = B_{ji}$. Hence $R_{123} = 0 = R_{213}$. Thus we find the following relationships among the 8 R_{ijk}:

$$R_{113} = R_{223}, \qquad R_{123} = R_{213} = 0, \qquad R_{231} = -R_{132}, \qquad R_{232} = R_{131}$$

$$R_{311} = R_{322}, \qquad R_{312} = -R_{321}, \qquad R_{333} \text{ survives}$$

In the two-suffix notation, the surviving constants are

$$r_{13}\,(= r_{23}), \qquad r_{33}, \qquad r_{41}\,(= -r_{52}), \qquad r_{42}\,(= r_{51})$$

Note that the contraction is made only after deducing the surviving R_{ijk}.

8.7.2.7. *Tetragonal System*: *Class* $\bar{4}$ ($\tilde{S}_4$); *Symmetry Elements*: $E\ C_2\ 2\tilde{S}_4$; *Generating Element*: $\tilde{S}_4$

$\tilde{S}_4$ is a composite symmetry element and is defined by crystallographers as C_4 followed by *i*. These two cannot be separated. The axis of $\tilde{S}_4$ is parallel to X_3. The transformation matrix for the symmetry element $\tilde{S}_4$ (which forms the generating element for the crystal class) is (Fig. 8.6)

$$\begin{bmatrix} 0 & -1 & 0 \\ 1 & 0 & 0 \\ 0 & 0 & -1 \end{bmatrix}$$

We know that the $\bar{4}$ axis parallel to X_3 implies a diad axis parallel to X_3 (since $\tilde{S}_4 \cdot \tilde{S}_4 = C_2$), and we have found earlier that the surviving constants for a diad parallel to X_3 are such that each combination of subscripts of R_{ijk} contains either one 3 or three 3's. Thus the surviving R_{ijk} for the

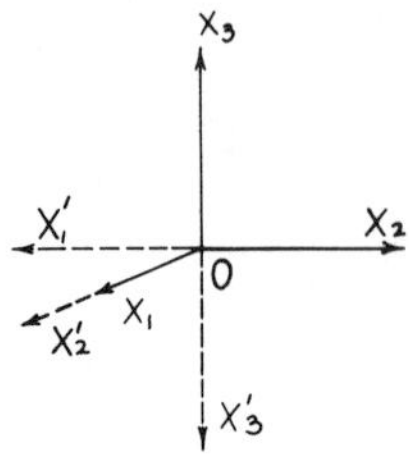

Fig. 8.6. Transformation of axes: the axis OX_3 is an axis of composite symmetry of a fourfold rotation followed by an inversion.

class 2 (C_2) are

$$R_{113} \quad R_{223} \quad R_{333} \quad R_{231} \quad R_{232} \quad R_{311} \quad R_{312} \quad R_{123}$$

and their equivalents.

Therefore only these 8 R_{ijk} need be considered for their survival and equivalence when $\bar{4}$ symmetry is imposed on them:

$$\begin{array}{llll}
\text{(a)} & R'_{113} = \alpha_{12}\alpha_{-2}\alpha_{33}R_{223} = -R_{223} = R_{113} & & \\
\text{(b)} & R'_{223} = \alpha_{21}\alpha_{21}\alpha_{33}R_{113} = -R_{113} = R_{223} & & \\
\text{(c)} & R'_{333} = \alpha_{33}\alpha_{33}\alpha_{33}R_{333} = -R_{333} = R_{333} & & \therefore R_{333} = 0 \\
\text{(d)} & R'_{231} = \alpha_{21}\alpha_{33}\alpha_{12}R_{132} = R_{132} = R_{231} & & \\
\text{(e)} & R'_{232} = \alpha_{21}\alpha_{33}\alpha_{21}R_{131} = -R_{131} = R_{232} & & \\
\text{(f)} & R'_{311} = \alpha_{33}\alpha_{12}\alpha_{12}R_{322} = -R_{322} = R_{311} & & \\
\text{(g)} & R'_{312} = \alpha_{33}\alpha_{12}\alpha_{21}R_{321} = R_{321} = R_{312} & & \\
\text{(h)} & R'_{123} = \alpha_{12}\alpha_{21}\alpha_{33}R_{213} = R_{213} = R_{123} & &
\end{array}$$

Thus the following relationships hold good among the R_{ijk}: $R_{113} = -R_{223}$; $R_{132} = R_{231}$; $R_{131} = -R_{232}$; R_{123} stands alone. Hence the nonvanishing independent r_{ij} are

$$r_{13}\ (= -r_{23}), \qquad r_{52}\ (= r_{41}), \qquad r_{51}\ (= -r_{42}), \qquad r_{63}$$

8.7.2.8. *Tetragonal System*: *Class* 4*mm* (C_{4v}); *Symmetry Elements*: $E\ 2C_4\ C_2\ 2\sigma_v'\ 2\sigma_v''$; *Generating Elements*: $C_4\ \sigma_v'$

Owing to the symmetry element C_4 parallel to X_3 the surviving R_{ijk} are

$$R_{113}\ (= R_{223}), \qquad R_{231}\ (= -R_{132}), \qquad R_{232}\ (= R_{131}), \qquad R_{333}$$

Owing to the additional vertical plane of symmetry perpendicular to X_1 all the R_{ijk} possessing a single 1 or three 1's as their subscripts vanish. Thus the surviving coefficients are

$$R_{113}\ (= R_{223}), \qquad R_{333}, \qquad R_{232}\ (= R_{311})$$

and their equivalents.

Hence the nonvanishing independent r_{ij} are

$$r_{13}\ (= r_{23}), \qquad r_{33}, \qquad r_{42}\ (= r_{51})$$

8.7.2.9. Tetragonal System: Class $\bar{4}2m$ (D_{2d}); Symmetry Elements: $E\ C_2\ C_2'\ C_2''\ 2\tilde{S}_4\ \sigma_v'\ \sigma_v''$; Generating Elements: $\tilde{S}_4\ C_2'$

This class can be discussed taking the crystal class $\bar{4}$ ($\tilde{S}_4$) as the basis. For the class $\bar{4}2m$ (D_{2d}) we have also to consider an additional symmetry element, a diad axis C_2' parallel to X_1 because the two elements $\tilde{S}_4$ parallel to X_3 and C_2' parallel to X_1 form the generating elements for the class $\bar{4}2m$ (D_{2d}).

Similar to the cases in Secs. 8.7.2.3 and 8.7.2.4, a diad parallel to X_1 removes all coefficients having no 1's or two 1's. In other words, only such coefficients survive which have in their subscripts one 1 and three 1's. We have therefore to examine the superposition of this additional condition over the surviving seven constants of the class $\bar{4}$ ($\tilde{S}_4$), which are

$$R_{113} \quad R_{223} \quad R_{132} \quad R_{231} \quad R_{131} \quad R_{232} \quad R_{123}$$

Out of the above seven R_{ijk}, the following four R_{ijk} vanish owing to the imposition of the symmetry C_2':

$$R_{113} \quad R_{223} \quad R_{131} \quad R_{232}$$

Hence the survivors are R_{132}, R_{231}, and R_{123} among which $R_{132} = R_{231}$ owing to the $\bar{4}$ symmetry. Thus the nonvanishing independent R_{ijk} are

$$R_{231}\ (= R_{132}), \qquad R_{123}$$

and the surviving r_{ij} are

$$r_{41}\ (= r_{52}), \qquad r_{63}$$

8.7.2.10. *Tetragonal System: Class* 422 (D_4); *Symmetry Elements:* $E\ 2C_4\ C_2\ 2C_2'\ 2C_2''$; *Generating Elements:* $C_4\ C_2'$

The superposition of the additional symmetry element C_2' (a diad parallel to X_1) on the class C_4 (having a tetrad parallel to X_3) results in the crystal class 422 (D_4).

The surviving R_{ijk} for class 4 (C_4) are

$$R_{113}\ (= R_{223}), \qquad R_{333}, \qquad R_{231} = -R_{132}, \qquad R_{232} = R_{131}$$

Owing to the additional diad axis parallel to X_1, it is only the R_{ijk} possessing a single 1 and three 1's that survive. Hence the surviving coefficients are

$$R_{231}, \qquad R_{132}\ (= -R_{231})$$

Hence the surviving r_{ij} are

$$r_{41}, \qquad r_{52}\ (= -r_{41})$$

8.7.2.11. *Trigonal System: Class 3* (C_3); *Symmetry Elements:* $E\ 2C_3$; *Generating Element:* C_3

Here the deduction is more involved, although the principle of the method is the same as before. We transform the axes by a symmetry operation of the crystal and equate the coefficients in the old system of axes with those in the new system.

The transformation matrix for a triad parallel to X_3 (Fig. 8.7) is

$$\begin{bmatrix} -1/2 & 3^{1/2}/2 & 0 \\ -3^{1/2}/2 & -1/2 & 0 \\ 0 & 0 & 1 \end{bmatrix}$$

We have

$$\begin{aligned} R'_{111} &= \alpha_{1l}\alpha_{1m}\alpha_{1n}R_{lmn} \\ &= \alpha_{1m}\alpha_{1n}[\alpha_{11}R_{1mn} + \alpha_{12}R_{2mn}] \\ &= \alpha_{11}\alpha_{1n}[\alpha_{11}R_{11n} + \alpha_{12}R_{12n}] + \alpha_{12}\alpha_{1n}[\alpha_{11}R_{21n} + \alpha_{12}R_{22n}] \\ &= \alpha_{11}^2[\alpha_{11}R_{111} + \alpha_{12}R_{112}] + \alpha_{11}\alpha_{12}[\alpha_{11}R_{121} + \alpha_{12}R_{122}] \\ &\quad + \alpha_{12}\alpha_{11}[\alpha_{11}R_{211} + \alpha_{12}R_{212}] + \alpha_{12}^2[\alpha_{11}R_{221} + \alpha_{12}R_{222}] \\ &= \alpha_{11}^3R_{111} + \alpha_{11}^2\alpha_{12}R_{112} + 2\alpha_{11}^2\alpha_{12}R_{121} + 2\alpha_{11}\alpha_{12}^2R_{122} \\ &\quad + \alpha_{11}\alpha_{12}^2R_{221} + \alpha_{12}^3R_{222} \end{aligned} \tag{8.35}$$

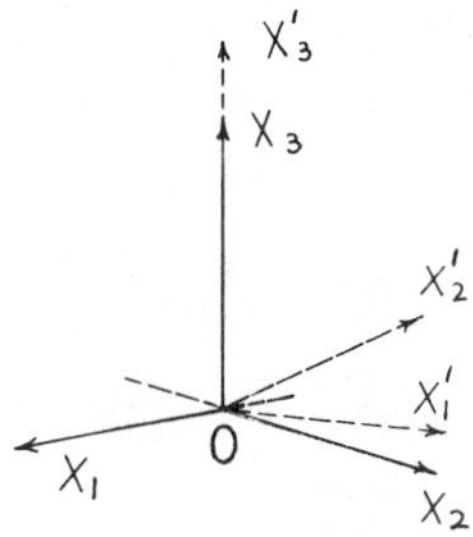

Fig. 8.7. Transformation of axes: the axis OX_3 is a triad or a threefold axis of rotation; here $OX_3' = OX_3$.

By Neumann's principle $R'_{111} = R_{111}$. Therefore substituting R_{111} for R'_{111} on the left-hand side, and transposing, we find

$$(1 - \alpha_{11}^3)R_{111} - \alpha_{11}^2\alpha_{12}R_{112} - 2\alpha_{11}^2\alpha_{12}R_{121} - 2\alpha_{11}\alpha_{12}^2R_{122} - \alpha_{11}\alpha_{12}^2R_{221} - \alpha_{12}^3R_{222} = 0 \qquad (8.36a)$$

Equation (8.36a) is a general relation which holds among the several R_{ijk} under a symmetry operation of an n-fold axis of rotation. Now, for the class 3 (C_3), we have $n = 3$, for which the transformation matrix is given on the previous page. Then Eq. (8.36a) yields

$$\frac{9}{8}R_{111} - \frac{3^{1/2}}{8}R_{112} - \frac{3^{1/2}}{4}R_{121} + \frac{3}{4}R_{122} + \frac{3}{8}R_{221} - \frac{3(3)^{1/2}}{8}R_{222} = 0 \qquad (8.36b)$$

Similarly, taking R_{112}, R_{121}, R_{122}, R_{221}, and R_{222} which appear in Eq. (8.36b), we obtain five more equations. All the six equations are together given below:

$$\frac{9}{8}R_{111} - \frac{3^{1/2}}{8}R_{112} - \frac{3^{1/2}}{4}R_{121} + \frac{3}{4}R_{122} + \frac{3}{8}R_{221} - \frac{3(3)^{1/2}}{8}R_{222} = 0 \qquad (8.36b)$$

$$\frac{3^{1/2}}{8}R_{111} - \frac{3}{8}R_{112} + \frac{3}{4}R_{121} + \frac{3^{1/2}}{4}R_{122} - \frac{3^{1/2}}{8}R_{221} + \frac{3}{8}R_{222} = 0 \qquad (8.37)$$

$$\frac{3}{8}R_{111} - \frac{3(3)^{1/2}}{8}R_{112} + \frac{3^{1/2}}{4}R_{121} - \frac{3}{4}R_{122} + \frac{9}{8}R_{221} - \frac{3^{1/2}}{8}R_{222} = 0 \qquad (8.38)$$

$$\frac{3^{1/2}}{8}R_{111} + \frac{9}{8}R_{112} - \frac{3}{4}R_{121} - \frac{3^{1/2}}{4}R_{122} + \frac{3(3)^{1/2}}{8}R_{221} + \frac{3}{8}R_{222} = 0 \qquad (8.39)$$

$$\frac{3}{8}R_{111} + \frac{3^{1/2}}{8}R_{112} - \frac{3^{1/2}}{4}R_{121} + \frac{3}{4}R_{122} - \frac{3}{8}R_{221} - \frac{3^{1/2}}{8}R_{222} = 0 \qquad (8.40)$$

$$\frac{3(3)^{1/2}}{8}R_{111} + \frac{3}{8}R_{112} + \frac{3}{4}R_{121} + \frac{3^{1/2}}{4}R_{122} + \frac{3^{1/2}}{8}R_{221} + \frac{9}{8}R_{222} = 0 \qquad (8.41)$$

Adding Eqs. (8.36b) and (8.38) we obtain

$$\frac{3}{2}(R_{111}+R_{221})-\frac{3^{1/2}}{2}(R_{112}+R_{222})=0 \tag{8.42}$$

Adding Eqs. (8.39) and (8.41) we obtain

$$\frac{3^{1/2}}{2}(R_{111}+R_{221})+\frac{3}{2}(R_{112}+R_{222})=0 \tag{8.43}$$

From Eqs. (8.42) and (8.43) we find

$$R_{111}=-R_{221}, \qquad R_{112}=-R_{222} \tag{8.44}$$

Subtracting Eq. (8.37) from Eq. (8.39) we obtain

$$\frac{3}{2}(R_{112}-R_{121})-\frac{3^{1/2}}{2}(R_{122}-R_{221})=0 \tag{8.45}$$

Subtracting Eq. (8.38) from Eq. (8.40) we obtain

$$\frac{3^{1/2}}{2}(R_{112}-R_{121})+\frac{3}{2}(R_{122}-R_{221})=0 \tag{8.46}$$

From Eqs. (8.45) and (8.46) we find

$$R_{112}=R_{121}, \qquad R_{122}=R_{221} \tag{8.47}$$

Thus combining Eqs. (8.44) and (8.47) we find

$$R_{111}=-R_{221}=-R_{122}, \qquad R_{222}=-R_{112}=-R_{121}$$

We continue the procedure for the remaining R_{ijk} also.

$$R'_{113}=\alpha_{1l}\alpha_{1m}\alpha_{3n}R_{lmn}=\frac{1}{4}R_{113}-\frac{3^{1/2}}{2}R_{213}+\frac{3}{4}R_{223}$$

$$\therefore \frac{3}{4}(R_{113}-R_{223})+\frac{3^{1/2}}{2}R_{213}=0 \tag{8.48}$$

$$R'_{223}=\alpha_{2l}\alpha_{2m}\alpha_{3n}R_{lmn}=\frac{3}{4}R_{113}+\frac{3^{1/2}}{2}R_{213}+\frac{1}{4}R_{223}$$

$$\therefore \frac{3}{4}(R_{113}-R_{223})+\frac{3^{1/2}}{2}R_{213}=0 \tag{8.49}$$

$$R'_{213}=\alpha_{2l}\alpha_{1m}\alpha_{3n}R_{lmn}=\frac{3^{1/2}}{4}R_{113}+\frac{1}{4}R_{213}-\frac{3}{4}R_{123}-\frac{3^{1/2}}{4}R_{223}$$

$$\therefore \frac{3^{1/2}}{4}(R_{113}-R_{223})-\frac{3}{2}R_{213}=0 \tag{8.50}$$

From Eqs. (8.49) and (8.50) we obtain

$$R_{113} = R_{223}, \qquad R_{213} = 0$$

$$R'_{311} = \alpha_{3l}\alpha_{1m}\alpha_{1n}R_{lmn} = \frac{1}{4}R_{311} - \frac{3^{1/2}}{4}R_{312} - \frac{3^{1/2}}{4}R_{321} + \frac{3}{4}R_{322}$$

$$\therefore \frac{3}{4}(R_{311} - R_{322}) + \frac{3^{1/2}}{4}(R_{312} + R_{321}) = 0 \tag{8.51}$$

$$R'_{322} = \alpha_{3l}\alpha_{2m}\alpha_{2n}R_{lmn} = \frac{3}{4}R_{311} + \frac{3^{1/2}}{4}R_{312} + \frac{3^{1/2}}{4}R_{321} + \frac{1}{4}R_{322}$$

$$\therefore \frac{3}{4}(R_{311} - R_{322}) + \frac{3^{1/2}}{4}(R_{312} + R_{321}) = 0 \tag{8.52}$$

$$R'_{312} = \alpha_{3l}\alpha_{1m}\alpha_{2n}R_{lmn} = \frac{3^{1/2}}{4}R_{311} + \frac{1}{4}R_{312} - \frac{3}{4}R_{321} - \frac{3^{1/2}}{4}R_{322}$$

$$\therefore \frac{3^{1/2}}{4}(R_{311} - R_{322}) - \frac{3}{4}(R_{312} + R_{321}) = 0 \tag{8.53}$$

$$R'_{321} = \alpha_{3l}\alpha_{2m}\alpha_{1n}R_{lmn} = \frac{3^{1/2}}{4}R_{311} - \frac{3}{4}R_{312} + \frac{1}{4}R_{321} - \frac{3^{1/2}}{4}R_{322}$$

$$\therefore \frac{3^{1/2}}{4}(R_{311} - R_{322}) - \frac{3}{4}(R_{312} + R_{321}) = 0 \tag{8.54}$$

Combining (8.51) and (8.53) we conclude that

$$R_{312} = -R_{321}, \qquad R_{311} = R_{322}$$

$$R'_{333} = \alpha_{3l}\alpha_{3m}\alpha_{3n}R_{lmn} = R_{333}$$

$\therefore R_{333}$ survives

$$R'_{331} = \alpha_{3l}\alpha_{3m}\alpha_{1n}R_{lmn} = -\frac{1}{2}R_{331} + \frac{3^{1/2}}{2}R_{332}$$

$$\therefore -\frac{3}{2}R_{331} + \frac{3^{1/2}}{2}R_{332} = 0 \tag{8.55}$$

$$R'_{332} = \alpha_{33}\alpha_{33}\alpha_{2n}R_{33n} = \alpha_{21}R_{331} + \alpha_{22}R_{332} = -\frac{3^{1/2}}{2}R_{331} - \frac{1}{2}R_{332}$$

$$\therefore -\frac{3^{1/2}}{2}R_{331} - \frac{3}{2}R_{332} = 0 \tag{8.56}$$

Combining Eqs. (8.55) and (8.56) we get

$$R_{331} = 0, \qquad R_{332} = 0$$

$$R'_{233} = \alpha_{2l}\alpha_{33}\alpha_{33}R_{l33} = -\frac{3^{1/2}}{2}R_{133} - \frac{1}{2}R_{233}$$

$$\therefore -\frac{3^{1/2}}{2}R_{133} - \frac{3}{2}R_{233} = 0 \tag{8.57}$$

$$R'_{133} = \alpha_{1l}\alpha_{33}\alpha_{33}R_{l33} = -\frac{1}{2}R_{133} + \frac{3^{1/2}}{2}R_{233}$$

$$\therefore -\frac{3}{2}R_{133} + \frac{3^{1/2}}{2}R_{233} = 0 \tag{8.58}$$

From Eqs. (8.57) and (8.58) we obtain

$$R_{233} = 0, \qquad R_{133} = 0$$

Thus we observe the following R_{ijk} and their equivalents surviving for a trigonal crystal of class 3 (C_3):

$$R_{111} = -R_{221} = -R_{122}, \qquad R_{222} = -R_{112} = -R_{121},$$

$$R_{113} = R_{223}, \qquad R_{312} = -R_{321}, \qquad R_{311} = R_{322}, \qquad R_{333}$$

Hence the nonvanishing independent r_{ij} are

$$r_{11} = -r_{21} = -r_{62}, \qquad r_{22} = -r_{12} = -r_{61},$$

$$r_{13} = r_{23}, \qquad r_{52} = -r_{41}, \qquad r_{51} = r_{42}, \qquad r_{33}$$

Having deduced the nonvanishing independent R_{ijk} for the class 3 (C_3) of the trigonal system, it becomes very simple to examine the remaining classes of the trigonal system.

8.7.2.12. Trigonal System: Class 3m (C_{3v}); Symmetry Elements: $E\ 2C_3\ 3\sigma_v$; Generating Elements: $C_3\ \sigma_v$

We consider the case of a plane of reflection σ_v parallel to X_2OX_3, of the class $3m$, in conformity with the convention of the IEEE Standards on Piezoelectric Crystals (1949). We arrive at the final matrix of R_{ijk} by superposing this plane of reflection σ_v (in the plane X_2OX_3) on the matrix

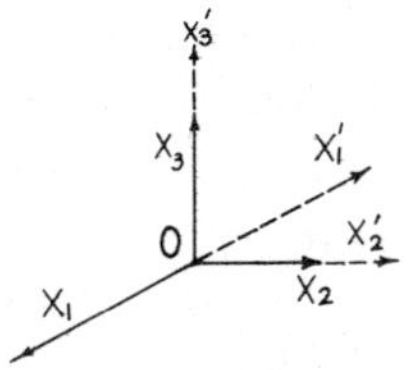

Fig. 8.8. Transformation of axes: for the trigonal class $3m$ (C_{3v}), with the axis OX_3 as a triad, the plane X_2OX_3 is a plane of reflection.

for the class 3 (C_3). The transformation matrix of the operation σ_v is (Fig. 8.8)

$$\begin{bmatrix} -1 & 0 & 0 \\ 0 & 1 & 0 \\ 0 & 0 & 1 \end{bmatrix}$$

The surviving R_{ijk} for class 3 (C_3) are

$$R_{111} = -R_{221} = -R_{122}, \qquad R_{222} = -R_{112} = -R_{121},$$
$$R_{113} = R_{223}, \qquad R_{312} = -R_{321}, \qquad R_{311} = R_{322}, \qquad R_{333}$$

With the transformation matrix given above, the coefficients R_{ijk} having one 1 and three 1's vanish. Hence R_{111} and R_{312} and their equivalents vanish. Therefore the resulting nonvanishing R_{ijk} are

$$R_{222} = -R_{112} = -R_{121}, \qquad R_{113} = R_{223}, \qquad R_{311} = R_{322}, \qquad R_{333}$$

And the surviving r_{ij} are

$$r_{22} = -r_{12} = -r_{61}, \qquad r_{13} = r_{23}, \qquad r_{51} = r_{42}, \qquad r_{33}$$

For the class $3m$ (C_{3v}) if the plane of symmetry σ_v is taken parallel to the plane X_1OX_3, then R_{222} and R_{312} vanish from among the 6 R_{ijk} of the class 3 (C_3). Hence the surviving r_{ij} are

$$r_{11} = -r_{21} = -r_{62}, \qquad r_{13} = r_{23}, \qquad r_{51} = r_{42}, \qquad r_{33}$$

8.7.2.13 *Trigonal System: Class 32 (D_3); Symmetry Elements: $E\ 2C_3\ 3C_2'$; Generating Elements: $C_3\ C_2'$*

To the crystal class 3 (C_3) are added three diad axes perpendicular to the triad axis X_3 in order to obtain the class 32 (D_3); the three diads make angles of 120° with one another, and any one of them can be taken parallel to the X_1 axis of the Cartesian system (IEEE Standards on Piezoelectric Crystals, 1949) for discussing the tensor properties in this class. Here the generating elements are C_3 parallel to X_3 and C_2' parallel to X_1. Hence the surviving Pockels linear electro-optic coefficients R_{ijk} can be obtained by taking the matrix for the class 3 (C_3) and imposing an additional symmetry of a diad C_2' parallel to X_1.

Among the six independent R_{ijk} of class 3 (C_3), we have

$$R_{111} = -R_{221} = -R_{122}, \qquad R_{222} = -R_{112} = -R_{121},$$
$$R_{113} = R_{223}, \qquad R_{312} = -R_{321}, \qquad R_{311} = R_{322}, \qquad R_{333}$$

Owing to the additional symmetry of C_2' parallel to X_1, those R_{ijk} whose subscripts contain one 1 and three 1's above survive. Thus the surviving R_{ijk} are

$$R_{111} = -R_{221} = -R_{122}, \qquad R_{312} = -R_{321}$$

and the surviving r_{ij} are

$$r_{11} = -r_{21} = -r_{62}, \qquad r_{52} = -r_{41}$$

8.7.2.14. Hexagonal System: Class 6 (C_6); Symmetry Elements: $E\ 2C_6\ 2C_3\ C_2$; Generating Elements: (a) $C_3\ C_2$ or (b) C_6

a. Considering C_3 and C_2 (both parallel to the X_3 axis) as the generating elements of the crystal class 6 (C_6) the surviving R_{ijk} can be deduced in a very simple way.

For the class 3 the surviving R_{ijk} are

$$R_{111} = -R_{221} = -R_{122}, \qquad R_{222} = -R_{112} = -R_{121},$$
$$R_{113} = R_{223}, \qquad R_{312} = -R_{321}, \qquad R_{311} = R_{322}, \qquad R_{333}$$

We have seen earlier that the presence of a diad axis parallel to the X_3 axis will result in the survival of only those R_{ijk} whose subscripts consist of only a single 3 or all three 3's. Any combination other than these must vanish.

The surviving R_{ijk} for the class 6 (C_6) are therefore

$$R_{113} = R_{223}, \qquad R_{312} = -R_{321}, \qquad R_{311} = R_{322}, \qquad R_{333}$$

and the r_{ij} are

$$r_{13} = r_{23}, \qquad r_{52} = -r_{41}, \qquad r_{51} = r_{42}, \qquad r_{33}$$

b. Without assuming the results for the class 3 (C_3), we can proceed here also exactly on the lines followed for the class 3 (C_3).

We have deduced earlier for the class 3 (C_3) that the following general

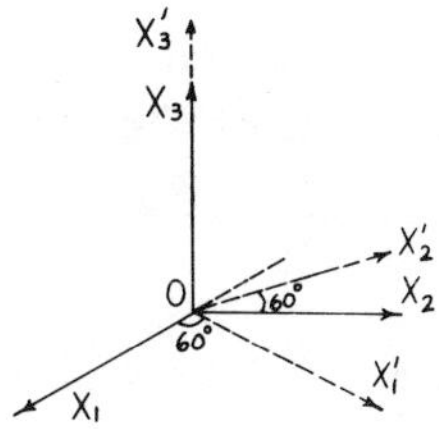

Fig. 8.9. Transformation of axes: OX_3 is a hexad or a sixfold axis of rotation; here $OX_3' = OX_3$.

relation holds good among the several R_{ijk} under a symmetry operation of an n-fold axis of rotation:

$$(1 - \alpha_{11}^3)R_{111} - \alpha_{11}^2\alpha_{12}R_{112} - 2\alpha_{11}^2\alpha_{12}R_{121} - 2\alpha_{11}\alpha_{12}^2R_{122} - \alpha_{11}\alpha_{12}^2R_{221} - \alpha_{12}^3R_{222} = 0 \tag{8.36a}$$

The transformation matrix for a hexad parallel to the X_3 axis is (Fig. 8.9)

$$\begin{bmatrix} 1/2 & 3^{1/2}/2 & 0 \\ -3^{1/2}/2 & 1/2 & 0 \\ 0 & 0 & 1 \end{bmatrix}$$

Then

$$\frac{7}{8}R_{111} - \frac{3^{1/2}}{8}R_{112} - \frac{3^{1/2}}{4}R_{121} - \frac{3}{4}R_{122} - \frac{3}{8}R_{221} - \frac{3(3)^{1/2}}{8}R_{222} = 0 \tag{8.59}$$

As in the former case of the class 3 taking R_{112}, R_{121}, R_{122}, R_{221}, and R_{222} which appear in Eq. (8.59), we obtain five more equations. All the six equations are together given below:

$$\frac{7}{8}R_{111} - \frac{3^{1/2}}{8}R_{112} - \frac{3^{1/2}}{4}R_{121} - \frac{3}{4}R_{122} - \frac{3}{8}R_{221} - \frac{3(3)^{1/2}}{8}R_{222} = 0 \tag{8.59}$$

$$\frac{3^{1/2}}{8}R_{111} + \frac{3}{8}R_{112} + \frac{5}{4}R_{121} + \frac{3^{1/2}}{4}R_{122} - \frac{3^{1/2}}{8}R_{221} - \frac{3}{8}R_{222} = 0 \tag{8.60}$$

$$\frac{3}{8}R_{111} + \frac{3(3)^{1/2}}{8}R_{112} - \frac{3^{1/2}}{4}R_{121} - \frac{3}{4}R_{122} - \frac{7}{8}R_{221} + \frac{3^{1/2}}{8}R_{222} = 0 \tag{8.61}$$

$$-\frac{3^{1/2}}{8}R_{111} + \frac{7}{8}R_{112} + \frac{3}{4}R_{121} - \frac{3^{1/2}}{4}R_{122} + \frac{3(3)^{1/2}}{8}R_{221} - \frac{3}{8}R_{222} = 0 \tag{8.62}$$

$$\frac{3}{8}R_{111} - \frac{3^{1/2}}{8}R_{112} + \frac{3^{1/2}}{4}R_{121} - \frac{5}{4}R_{122} - \frac{3}{8}R_{221} + \frac{3^{1/2}}{8}R_{222} = 0 \tag{8.63}$$

$$\frac{3(3)^{1/2}}{8}R_{111} - \frac{3}{8}R_{112} - \frac{3}{4}R_{121} + \frac{3^{1/2}}{4}R_{122} + \frac{3^{1/2}}{8}R_{221} + \frac{7}{8}R_{222} = 0 \tag{8.64}$$

Following the procedure, under the class 3 (C_3), we observe that the solution of the six simultaneous equations yields the following:

$$R_{111} = R_{221} = R_{122} = R_{222} = R_{112} = R_{121} = 0$$

The conditions for the remaining R_{ijk} coefficients are obtained by a procedure analogous to that for the class 3 (C_3), yielding finally the following:

$$R_{113} = R_{223}, \qquad R_{312} = -R_{321}, \qquad R_{311} = R_{322}, \qquad R_{333},$$
$$R_{213} = R_{331} = R_{332} = R_{233} = R_{133} = 0$$

Hence the surviving r_{ij} are

$$r_{13} = r_{23}, \qquad r_{52} = -r_{41}, \qquad r_{51} = r_{42}, \qquad r_{33}$$

8.7.2.15. Hexagonal System: Class $\bar{6}$ (C_{3h}); Symmetry Elements: $E\,2C_3\,\sigma_h\,2\tilde{S}_6$; Generating Elements: $C_3\,\sigma_h$

A plane of symmetry σ_h perpendicular to the X_3 axis as a triad is the symmetry element that added to the class 3 gives us the class $\bar{6}$. Hence, from among the following surviving R_{ijk} of the class 3

$$R_{111} = -R_{221} = -R_{122}, \qquad R_{222} = -R_{112} = -R_{121},$$
$$R_{113} = R_{223}, \qquad R_{312} = -R_{321}, \qquad R_{311} = R_{322}, \qquad R_{333}$$

we have to drop out all those R_{ijk} that have one 3 and three 3's as their subscripts. Hence the surviving R_{ijk} are

$$R_{111} = -R_{221} = -R_{122}, \qquad R_{222} = -R_{112} = -R_{121}$$

and the nonvanishing independent r_{ij} are

$$r_{11} = -r_{21} = -r_{62}, \qquad r_{22} = -r_{12} = -r_{61}$$

8.7.2.16. Hexagonal System: Class $\bar{6}m2$ (D_{3h}); Symmetry Elements: $E\,2C_3\,3C_2'\,\sigma_h\,2\tilde{S}_6\,3\sigma_v$; Generating Elements: $C_3\,\sigma_h\,\sigma_v$

To the symmetry of class $\bar{6}$ (C_{3h}) if we add a plane of symmetry parallel to X_3 (a triad), the result is the class $\bar{6}m2$. In conformity with the IEEE Standards on Piezoelectric Crystals (1949), let the X_2OX_3 plane be the symmetry plane. Then the R_{ijk} of the class $\bar{6}$ having two 1's or none alone survive; these are given below:

$$R_{222} = -R_{112} = -R_{121}$$

The surviving r_{ij} are

$$r_{22} = -r_{12} = -r_{61}$$

If the plane of reflection is perpendicular to the X_2 axis, then the R_{ijk} of the class $\bar{6}$ having two 2's or none alone survive, and others vanish. Hence the surviving R_{ijk} are

$$R_{111} = -R_{221} = -R_{122}$$

and the surviving r_{ij} are

$$r_{11} = -r_{21} = -r_{62}$$

8.7.2.17. *Hexagonal System: Class 6mm (C_{6v}); Symmetry Elements: $E\ 2C_6\ 2C_3\ C_2\ 3\sigma_v'\ 3\sigma_v''$; Generating Elements: $C_6\ \sigma_v'$*

When a plane of reflection parallel to the X_2OX_3 plane is added to the class 6, the result is the class 6*mm*. The addition of such a symmetry plane will result in the vanishing of all those R_{ijk} of the class 6 that have one 1 and three 1's in their subscripts. Thus among the four nonzero components of the class 6, namely,

$$R_{113} = R_{223}, \qquad R_{312} = -R_{321}, \qquad R_{311} = R_{322}, \qquad R_{333}$$

we note that R_{312} $(= -R_{321})$ disappears in the matrix of the class 6*mm*. Hence the surviving R_{ijk} are

$$R_{113} = R_{223}, \qquad R_{311} = R_{322}, \qquad R_{333}$$

and the corresponding r_{ij} are

$$r_{13} = r_{23}, \qquad r_{51} = r_{42}, \qquad r_{33}$$

8.7.2.18. *Hexagonal System: Class 622 (D_6); Symmetry Elements: $E\ 2C_6\ 2C_3\ C_2\ 3C_2'\ 3C_2''$; Generating Elements: $C_6\ C_2'$*

The addition of a diad axis parallel to X_1 of the class 6 results in the class 622; this consequently eliminates the R_{ijk} having two 1's or none in their subscripts. Thus among the four independent nonzero R_{ijk} of the class 6, the surviving coefficient is only

$$R_{312}\ (= -R_{321})$$

and hence only

$$r_{52}\ (= -r_{41})$$

8.7.2.19. Cubic System: Class 23 (T); Symmetry Elements: $E\,3C_2\,8C_3{}'$; Generating Elements: $C_2\,C_3{}'$

We can obtain the R_{ijk} matrix for this class from that of the class 222 (D_2) by applying an additional symmetry element of a triad equally inclined to the crystallographic axes of the orthorhombic system, class 222.

The transformation matrix for this triad axis which makes $X_1 \rightharpoonup X_2$, $X_2 \rightharpoonup X_3$, and $X_3 \rightharpoonup X_1$ (Fig. 8.10) is

$$\begin{bmatrix} 0 & 1 & 0 \\ 0 & 0 & 1 \\ 1 & 0 & 0 \end{bmatrix}$$

The surviving R_{ijk} for the class 222 are, as shown earlier,

$$R_{231}, \qquad R_{312}, \qquad R_{123}$$

Now for the class 23 (*T*)

$$\begin{aligned} R'_{231} &= \alpha_{23}\alpha_{31}\alpha_{12}R_{312} = R_{312} = R_{231} \\ R'_{312} &= \alpha_{31}\alpha_{12}\alpha_{23}R_{123} = R_{123} = R_{312} \\ R'_{123} &= \alpha_{12}\alpha_{23}\alpha_{31}R_{231} = R_{231} = R_{123} \end{aligned}$$

$$R_{231} = R_{312} = R_{123}$$

and hence the surviving r_{ij} are

$$r_{41} = r_{52} = r_{63}$$

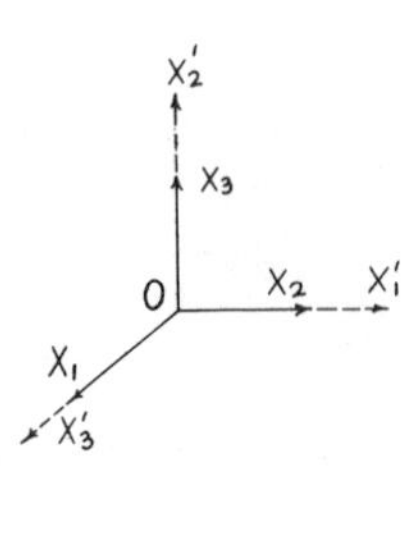

Fig. 8.10. Transformation of the coordinate axes: for all classes of the cubic system, any cube diagonal is a triad or a threefold axis of rotation.

8.7.2.20. *Cubic System: Class $\bar{4}3m$ (T_d); Symmetry Elements: $E\ 3C_2\ 8C_3'\ 6\sigma_d\ 6\tilde{S}_4$; Generating Elements: $\tilde{S}_4\ C_3'$*

Starting with the class $\bar{4}$ $(\tilde{S}_4)$ of the tetragonal system, we add a triad axis equally inclined to the X_1, X_2, X_3 axes.

The surviving R_{ijk} for the class $\bar{4}$ are

$$R_{113} = -R_{223}, \qquad R_{132} = R_{231}, \qquad R_{131} = -R_{232}, \qquad R_{123}$$

Applying the transformation matrix for the triad axis which makes $X_1 \rightharpoonup X_2$, $X_2 \rightharpoonup X_3$, $X_3 \rightharpoonup X_1$ (Fig. 8.10), we obtain the following results:

$$R'_{113} = \alpha_{12}\alpha_{12}\alpha_{31}R_{221} = R_{221} = 0$$

$$R'_{131} = \alpha_{12}\alpha_{31}\alpha_{12}R_{212} = R_{212} = 0$$

since for the class $\bar{4}$ both R_{221} and R_{212} vanish. Hence $R_{113} = 0$ and $R_{131} = 0$.

$$R'_{132} = \alpha_{12}\alpha_{31}\alpha_{23}R_{213} = R_{213} = R_{132}$$

$$R'_{123} = \alpha_{12}\alpha_{23}\alpha_{31}R_{231} = R_{231} = R_{123}$$

Bearing in mind that $R_{ijk} = R_{jik}$, we obtain

$$R_{123} = R_{312} = R_{231}$$

Thus the nonvanishing r_{ij} are as follows:

$$r_{41} = r_{52} = r_{63}$$

8.7.2.21. *Cubic System: Class 432 (O); Symmetry Elements: $E\ 3C_2\ 8C_3'\ 6C_2''\ 6C_4$; Generating Elements: $C_4\ C_3'$*

This crystal class is of special interest because it lacks a center of symmetry and nevertheless it does not exhibit piezoelectricity and Pockels (linear electro-optic) effect.

It has three tetrad axes X_3, X_1, X_2 and also four triad axes parallel to each one of the cube diagonals, besides a few other elements of symmetry as listed earlier. The generating elements are C_4 and C_3'.

First, taking X_3 as a tetrad axis, we note that for the class 4 the surviving R_{ijk} are

$$R_{113} = R_{223}, \qquad R_{333}, \qquad R_{232} = R_{131} = R_{311} = R_{322},$$

$$R_{312} = -R_{321} = -R_{231} = R_{132}$$

We examine only the four independent R_{ijk} under a triad axis parallel to a cube diagonal, a direction equally inclined to X_1, X_2, X_3 (Fig. 8.10), whose transformation matrix has been given earlier. Now, we have

$$R'_{113} = \alpha_{12}\alpha_{12}\alpha_{31}R_{221} = R_{221} = R_{113}$$

$$R'_{333} = \alpha_{31}\alpha_{31}\alpha_{31}R_{111} = R_{111} = R_{333}$$

$$R'_{232} = \alpha_{23}\alpha_{31}\alpha_{23}R_{313} = R_{313} = R_{232}$$

$$R'_{312} = \alpha_{31}\alpha_{12}\alpha_{23}R_{123} = R_{123} = R_{312}$$

However, for the crystal class 4 (C_4)

$$R_{221} = R_{111} = R_{313} = R_{123} = 0$$

Thus

$$R_{113} = R_{333} = R_{232} = R_{312} = 0$$

Hence all the R_{ijk} for class 432 become equal to zero, and thus all the r_{ij} are also zero.

8.8. Derivation of the Expressions for $\delta = f(r_{ij})$ for Some Typical Crystal Classes and Orientations

Proceeding on the lines indicated in Chapter 5 we can derive expressions for the linear electro-optic birefringence produced in any of the piezoelectric crystals for appropriate directions of observation, when an electric field acts in a specified direction. Such expressions for the birefringence are necessary in the experimental determination of the linear electro-optic constants, and also to estimate the figure of merit of a Pockels cell used in electro-optic devices.

We give below the derivations in some typical cases of a few crystal classes.

8.8.1. *Cubic System: Classes* 23 (T) *and* $\bar{4}3m$ (T_d)

8.8.1.1. *Case* (i): *To Determine* r_{41} *by a Transverse Effect*

We give the derivation in full for this orientation.

In the absence of an electric field, the indicatrix is given by

$$B_{11}^0x^2 + B_{22}^0y^2 + B_{33}^0z^2 + 2B_{23}^0yz + 2B_{31}^0zx + 2B_{12}^0xy = 1 \qquad (8.65)$$

and in the presence of an electric field it becomes

$$B_{11}x^2 + B_{22}y^2 + B_{33}z^2 + 2B_{23}yz + 2B_{31}zx + 2B_{12}xy = 1 \qquad (8.66)$$

From the definition of r_{ij} in the two-suffix notation

$$B_i - B_i^0 = \frac{1}{n_i^2} - \frac{1}{(n_i^0)^2} = \Delta\left(\frac{1}{n_i^2}\right) = \sum r_{ij}E_j \qquad (8.67)$$

For these two crystal classes of cubic system r_{41} alone exists (see page 367 for the r_{ij} matrix).

Let an electric field E_z be applied externally along the Z axis, and the light beam be propagated along the direction designated N, which is at 45° to the X and Y axes and perpendicular to the Z axis (Fig. 8.11).

From the electro-optic scheme due to Pockels for the two classes 23 and $\bar{4}3m$, we have

$$\begin{aligned}
B_{11} - B_{11}^0 &= B_1 - B_1^0 = 0 \cdot E_x + 0 \cdot E_y + 0 \cdot E_z \\
B_{22} - B_{22}^0 &= B_2 - B_2^0 = 0 \cdot E_x + 0 \cdot E_y + 0 \cdot E_z \\
B_{33} - B_{33}^0 &= B_3 - B_3^0 = 0 \cdot E_x + 0 \cdot E_y + 0 \cdot E_z \\
B_{23} - B_{23}^0 &= B_4 - B_4^0 = r_{41}E_x + 0 \cdot E_y + 0 \cdot E_z \\
B_{31} - B_{31}^0 &= B_5 - B_5^0 = 0 \cdot E_x + r_{41}E_y + 0 \cdot E_z \\
B_{12} - B_{12}^0 &= B_6 - B_6^0 = 0 \cdot E_x + 0 \cdot E_y + r_{41}E_z
\end{aligned} \qquad (8.68)$$

Since the direction of light propagation (and hence of observation) is not parallel to the crystallographic axes, let the indicatrix in general

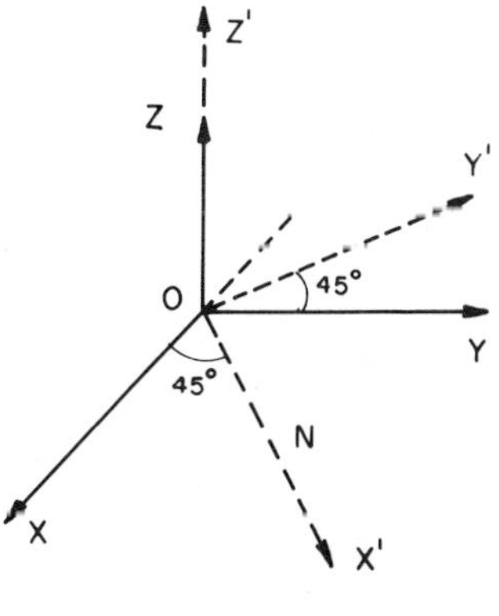

Fig. 8.11. Transformation of axes: a rotation of the X' system through an angle of 45° about the Z axis; here $OZ' = OZ$.

be given by

$$A_{11}^0x'^2 + A_{22}^0y'^2 + A_{33}^0z'^2 + 2A_{23}^0y'z' + 2A_{31}^0z'x' + 2A_{12}^0x'y' = 1 \quad (8.69)$$

For observation along X' (equally inclined to X and Y axes), the cross section of the optical index ellipsoid is

$$A_{22}^0y'^2 + A_{33}^0z'^2 + 2A_{23}^0y'z' = 1 \quad (8.70)$$

The matrix of coordinate transformation (Fig. 8.11) is

$$\begin{bmatrix} 1/2^{1/2} & 1/2^{1/2} & 0 \\ -1/2^{1/2} & 1/2^{1/2} & 0 \\ 0 & 0 & 1 \end{bmatrix}$$

Here

$$A_{22}^0 = \alpha_{2k}\alpha_{2l}B_{kl}^0 = \tfrac{1}{2}[B_{11}^0 + B_{22}^0 - 2B_{12}^0] \quad (8.71)$$

$$A_{33}^0 = \alpha_{3k}\alpha_{3l}B_{kl}^0 = B_{33}^0 \quad (8.72)$$

and

$$A_{23}^0 = \alpha_{2k}\alpha_{3l}B_{kl}^0 = -2^{-1/2}B_{13}^0 + 2^{-1/2}B_{23}^0 \quad (8.73)$$

Under the influence of an electric field E_z, the cross section of the optical index ellipsoid becomes, for observation in the direction X',

$$A_{22}y'^2 + A_{33}z'^2 + 2A_{23}y'z' = 1 \quad (8.74)$$

where

$$A_{22} = \tfrac{1}{2}[B_{11} + B_{22} - 2B_{12}]$$

$$A_{33} = B_{33} \qquad \text{and} \qquad A_{23} = -2^{-1/2}B_{13} + 2^{-1/2}B_{23}$$

Now

$$A_{23} - A_{23}^0 = -2^{-1/2}(B_{13} - B_{13}^0) + 2^{-1/2}(B_{23} - B_{23}^0) = 0$$

from Eq. (8.68); but $A_{23}^0 = 0$ since $B_{23}^0 = 0$ and $B_{13}^0 = 0$ for a cubic crystal.* Therefore the cross section of the ellipse in the strained state is given by

$$A_{22}y'^2 + A_{33}z'^2 = 1$$

Hence there is no tilt of axes of the ellipse. The electro-optic birefringence expressed in terms of r_{ij} will now have to be related to the optical path retardation δ which can be experimentally determined employing a suitable

* For all crystal systems except monoclinic and triclinic $B_{23}^0 = B_{31}^0 = B_{12}^0 = 0$.

optical technique. Now

$$A_{33} - A_{33}^0 = \frac{1}{n_{z'}^2} - \frac{1}{(n_{z'}^0)^2} = \frac{(n_{z'}^0)^2 - n_{z'}^2}{n_{z'}^2(n_{z'}^0)^2} = \frac{-2n_{z'}(\Delta n_{z'})}{n_{z'}^4}$$

$$\therefore \Delta n_{z'} = -\tfrac{1}{2}n_{z'}^3(A_{33} - A_{33}^0) = -\tfrac{1}{2}n_{z'}^3(B_{33} - B_{33}^0) = 0$$

Similarly

$$\begin{aligned}\Delta n_{y'} &= -\tfrac{1}{2}n_{y'}^3(A_{22} - A_{22}^0)\\ &= -\tfrac{1}{2}n_{y'}^3[(\tfrac{1}{2}B_{11} + \tfrac{1}{2}B_{22} - B_{12}) - (\tfrac{1}{2}B_{11}^0 + \tfrac{1}{2}B_{22}^0 - B_{12}^0)]\\ &= -\tfrac{1}{2}n_{y'}^3[\tfrac{1}{2}(B_{11} - B_{11}^0) + \tfrac{1}{2}(B_{22} - B_{22}^0) - (B_{12} - B_{12}^0)]\\ &= \tfrac{1}{2}n_{y'}^3 r_{41} E_z\end{aligned}$$

$$\Delta n = (\Delta n_z' - \Delta n_y') = -\tfrac{1}{2}n_o{}^3 r_{41} E_z \tag{8.75}$$

This value of Δn is the ordinary birefringence without any tilt of the axes of the ellipse for the particular direction of observation. Hence a Babinet compensator can be used. Furthermore, the birefringent compensator technique [844] or the Soleil–Babinet form of the birefringent compensator technique [1303] can be used for studying the dispersion of r_{41} and also its temperature variation.

Corollary: On the basis of the above derivation when applied to tetragonal crystals of class $\bar{4}2m$ (for example, ADP) and orthorhombic crystals of class 222 (for example, Rochelle salt) for an electric field E_z and observation along a direction equally inclined between the $+X$ and $+Y$ axes, we get

$$\Delta n = -\tfrac{1}{2}n_{xy}^3 r_{63} E_z$$

8.8.1.2. Case (ii): To Determine r_{41} of Cubic Crystals by a Longitudinal Effect

The problem of determining r_{41} by a longitudinal effect in cubic crystals can be treated exactly as that in the class $\bar{4}2m$ (D_{2d}) of the tetragonal system. Hence for the derivation and experiment, vide the latter.

8.8.2. Tetragonal System: Class $\bar{4}2m$ (D_{2d})

8.8.2.1. Case (i): To Determine r_{41} by a Longitudinal Effect

Crystals of this point group (for example, ADP and KDP) have two linear electro-optic constants r_{41} ($= r_{52}$) and r_{63}. According to Pockels' scheme

$$B_1 - B_1^0 = 0 \cdot E_x + 0 \cdot E_y + 0 \cdot E_z \qquad B_4 - B_4^0 = r_{41}E_x + 0 \cdot E_y + 0 \cdot E_z$$
$$B_2 - B_2^0 = 0 \cdot E_x + 0 \cdot E_y + 0 \cdot E_z \qquad B_5 - B_5^0 = 0 \cdot E_x + r_{41}E_y + 0 \cdot E_z$$
$$B_3 - B_3^0 = 0 \cdot E_x + 0 \cdot E_y + 0 \cdot E_z \qquad B_6 - B_6^0 = 0 \cdot E_x + 0 \cdot E_y + r_{63}E_z$$

Let the electric field be applied along X, hence only E_x exists, and let light be propagated along the same axis, namely, X. The crystal shows natural birefringence for this direction of light propagation. Then the equation of the elliptic cross section of the optical index ellipsoid perpendicular to X is given by

$$B_2^0y^2 + B_3^0z^2 + 2B_4^0yz = 1$$

for the unstrained crystal and

$$B_2y^2 + B_3z^2 + 2B_4yz = 1$$

for the electrically strained crystal.

Since $B_4 - B_4^0 = r_{41}E_x$ from Pockels' scheme and $B_4^0 = 0$ for the $\bar{4}2m$ class, we note that B_4 exists. Therefore the cross section of the optical index ellipsoid is rotated about X by an angle θ given by

$$\tan 2\theta = \frac{2B_4}{B_2 - B_3}$$

(as shown on page 204). Now

$$B_2 - B_2^0 = 0, \qquad \therefore B_2 = B_2^0 = \frac{1}{n_y^2} = \frac{1}{n_o^2}$$

$$B_3 - B_3^0 = 0, \qquad \therefore B_3 = B_3^0 = \frac{1}{n_z^2} = \frac{1}{n_e^2}$$

and

$$B_4 - B_4^0 = r_{41}E_x$$

but

$$B_4^0 = 0 \qquad \therefore B_4 = r_{41}E_x$$

Hence

$$\tan 2\theta = \frac{2r_{41}E_x}{1/n_o^2 - 1/n_e^2} \tag{8.76}$$

A Sénarmont compensator can be used to determine θ and hence r_{41} can be computed.

8.8.2.2. *Case (ii): To Determine r_{63} by a Longitudinal Effect*

The electric field is applied along Z and the light propagation also is along Z; hence it is a longitudinal effect that is studied in this situation. In the unstrained state the crystal does not show natural birefringence for this direction of light propagation. The axes of the electrically induced birefringence are inclined at 45° with respect to the crystallographic axes X and Y of the tetragonal crystal. The two refractive indices along the two principal axes of birefringence are given by the relations $1/n_+^2 = 1/n_0^2 + r_{63}E_z$ and $1/n_-^2 = 1/n_0^2 - r_{63}E_z$. Hence the field-induced birefringence is given by

$$\Delta n = +n_o^3 r_{63} E_z \tag{8.77}$$

A Sénarmont compensator can be used to determine Δn and hence r_{63} in the present case.

Alternatively, intensity measurements [145; 146] also permit the evaluation of r_{63} thus: A Z-cut plate of thickness t of ADP or KDP is placed between crossed plane polarizers and a photomultiplier receives the transmitted light whose intensity becomes maximum at an applied half-wave voltage $V_{\lambda/2} = V'$ say. Then from Eq. (8.77),

$$\lambda/2 = t(\Delta n) = n_o^3 r_{63} t E_z = n_o^3 r_{63} V'$$

$$\therefore r_{63} = \frac{\lambda}{2n_o^3 V'} \tag{8.78}$$

In case the transmitted intensity does not reach maximum within the maximum electric field applied to the crystal, still one can obtain the ratio of the transmitted to the incident intensity I/I_0, which yields r_{63} from the relation

$$\frac{I}{I_0} = \sin^2\left(\frac{\pi V n_o^3 r_{63}}{\lambda}\right) \tag{8.79}$$

where the applied voltage $V = E_z t$.

8.8.2.3. *Case (iii): Tetragonal System, Class 4mm (C_{4v})—To Determine r_{33} and r_{13} Using a Transverse Electro-Optic Effect*

Let the electric field be applied along the Z axis of the tetragonal crystal specimen and light be propagated along the Y axis of the crystal. For observation along the Y axis even in the absence of the electric field, owing to the uniaxial nature of the crystal, there exists a certain magnitude of birefringence. However, under the influence of an electric field E_z along the Z axis, the index of refraction changes for electric vectors parallel to the X and Z axes. The change Δn_3 for the electric vector parallel to the Z axis is given by

$$\Delta n_3 = -\tfrac{1}{2} n_e^3 r_{33} E_z$$

and for the electric vector parallel to the X axis, the change Δn_1 is given by

$$\Delta n_1 = -\tfrac{1}{2} n_o^3 r_{13} E_z$$

Therefore the linear electro-optic birefringence Δn is given by

$$\Delta n = \Delta n_3 - \Delta n_1 = -\tfrac{1}{2}(n_e^3 r_{33} - n_o^3 r_{13}) E_z \qquad (8.80)$$

The individual values of r_{33} and r_{13} can be obtained only by directly determining Δn_3 and Δn_1 by interferometric or other such methods.

8.8.3. *Trigonal System: Class 32* (D_3)

α-Quartz belongs to this class and has two independent r_{ij} in Pockels' scheme, namely, r_{11} and r_{41}. Thus

$$\begin{aligned}
B_1 - B_1^0 &= r_{11}E_x + 0 \cdot E_y + 0 \cdot E_z & \quad B_4 - B_4^0 &= r_{41}E_x + 0 \cdot E_y + 0 \cdot E_z \\
B_2 - B_2^0 &= -r_{11}E_x + 0 \cdot E_y + 0 \cdot E_z & \quad B_5 - B_5^0 &= 0 \cdot E_x - r_{41}E_y + 0 \cdot E_z \\
B_3 - B_3^0 &= 0 \cdot E_x + 0 \cdot E_y + 0 \cdot E_z & \quad B_6 - B_6^0 &= 0 \cdot E_x - r_{11}E_y + 0 \cdot E_z
\end{aligned} \qquad (8.81a)$$

It is thus seen that it is only the component of the electric field perpendicular to the optic axis that is effective.

8.8.3.1. *Case (i): To Determine* r_{11} *by a Transverse Effect*

Let an electric field E_x along the X axis exist, and let the observation be along the Y axis. Then we observe a transverse effect:

$$B_1 - B_1^0 = r_{11}E_x, \qquad \therefore \Delta n_x = -\tfrac{1}{2} n_x^3 r_{11} E_x$$

$$B_3 - B_3^0 = 0 \cdot E_x, \qquad \therefore \Delta n_z = 0$$

$B_5 - B_5^0 = 0 \cdot E_x$ and $B_5^0 = 0$ for this crystal class and hence $B_5 = 0$. Therefore there is no tilt of axes of the elliptical cross section perpendicular to the Y axis, which is the direction of observation. Therefore we have

$$\begin{aligned}\delta = \Delta n &= (\Delta n_x - \Delta n_z)\\ &= -\tfrac{1}{2}n_x{}^3 r_{11} = -\tfrac{1}{2}n_0{}^3 r_{11}\end{aligned} \tag{8.81b}$$

where δ is the path retardation per unit electric intensity per unit dimension along the direction of observation. A Babinet compensator can be used to observe this transverse effect and thus to calculate r_{11}.

8.8.3.2. *Case (ii): To Determine r_{11} and r_{41} by a Longitudinal Effect along the X Axis*

Let E_x be the electric field applied along the X axis, and the observation be also along the same axis, thus giving rise to a longitudinal effect.

The equation of the elliptic cross section normal to the X axis when the crystal is electrically strained is given by

$$B_2 y^2 + B_3 z^2 + 2B_4 yz = 1 \tag{8.82a}$$

Since $B_4 - B_4{}^0 = r_{41}E_x$ and here $B_4{}^0 = 0$, we note that B_{41} exists under the influence of E_x resulting in a tilt of the axes of the ellipse. Combining Eqs. (8.81a) and (8.82a), we get

$$(B_2{}^0 - r_{11}E_x)y^2 + B_3{}^0 z^2 + 2r_{41}E_x yz = 1$$

This is an ellipse rotated about the X axis through an angle θ depending on E_x, where

$$\tan 2\theta = \frac{2r_{41}E_x}{(B_2{}^0 - r_{11}E_x) - B_3{}^0} = \frac{2r_{41}E_x}{(1/n_o{}^2 - 1/n_e{}^2) - r_{11}E_x} \tag{8.82b}$$

Knowing r_{11} from the experiment under case (i) above, and with θ the angle of rotation of the Y', Z' axes with respect to the position of the Y, Z axes under an electric field E_x, we can calculate r_{41}.

Alternatively the value of θ may be determined at two different electric field strengths using a Sénarmont compensator, and hence the individual values of r_{11} and r_{41} are calculated.

8.8.3.3. *Case (iii): To Determine r_{41} by a Longitudinal Effect along the Y Axis*

Let E_y be parallel to Y and the observation also be along Y; here again we observe a longitudinal effect.

The deformed ellipse is given by

$$B_1x^2 + B_3z^2 + 2B_5xz = 1$$

From Eq. (8.81a) we see that $B_5 - B_5{}^0 = -r_{41}E_y$; hence B_5 exists, and there is a tilt of the axes of the ellipse with respect to the X and Z axes, the ellipse having rotated about the Y axis through an angle θ given by

$$\tan 2\theta = \frac{2B_5}{B_1 - B_3} = \frac{-2r_{41}E_5}{1/n_o{}^2 - 1/n_e{}^2} = -2r_{41}E_y \frac{n_o{}^2 n_e{}^2}{n_e{}^2 - n_o{}^2} \tag{8.83}$$

Here θ can be determined using a Sénarmont compensator.

8.8.3.4. *Case (iv): To Determine r_{11} and r_{41} by a Transverse Effect*

Let an electric field E_x be applied along the X axis, and the light beam be propagated at 45° to Y and Z and perpendicular to X; this direction of light propagation is designated M for convenience (see p. 211). Consequently we study here a transverse effect. The direction cosine scheme of the two systems of axes (Fig. 8.12) is

$$\begin{bmatrix} 1 & 0 & 0 \\ 0 & 1/2^{1/2} & 1/2^{1/2} \\ 0 & -1/2^{1/2} & 1/2^{1/2} \end{bmatrix}$$

The equation of the index ellipsoid in the unstrained state of the crystal is

$$A_{11}^0 x'^2 + A_{22}^0 y'^2 + A_{33}^0 z'^2 + 2A_{23}^0 y'z' + 2A_{31}^0 z'x' + 2A_{12}^0 x'y' = 1$$

Under the influence of an applied electric field, this changes to

$$A_{11}x'^2 + A_{22}y'^2 + A_{33}z'^2 + 2A_{23}y'z' + 2A_{31}z'x' + 2A_{12}x'y' = 1$$

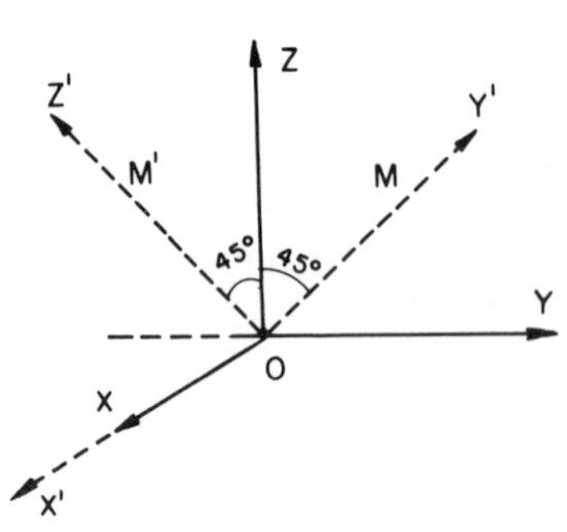

Fig. 8.12. Transformation of axes: a rotation of the X' system through an angle of 45° about the X axis; here $OX' = OX$.

When the direction of observation is along M, the cross sections of the ellipsoid in the unstrained and strained states are given, respectively, by

$$A_{11}^0 x'^2 + A_{33}^0 z'^2 + 2A_{31}^0 z'x' = 1$$

and

$$A_{11} x'^2 + A_{33} z'^2 + 2A_{31} z'x' = 1$$

The optical parameters A_{ij} and B_{ij} are related thus:

$$\begin{aligned} A_{11}^0 &= \alpha_{1k}\alpha_{1l}B_{kl}^0 = B_{11}^0 \\ A_{33}^0 &= \alpha_{3k}\alpha_{3l}B_{kl}^0 = \tfrac{1}{2}B_{22}^0 + \tfrac{1}{2}B_{33}^0 - B_{23}^0 \\ A_{31}^0 &= \alpha_{3k}\alpha_{1l}B_{kl}^0 = 2^{-1/2}B_{31}^0 - 2^{-1/2}B_{12}^0 \end{aligned}$$

Similarly the A_{ij} are related to the B_{ij}. We find here that

$$\begin{aligned} A_{31} - A_{31}^0 &= 2^{-1/2}(B_{31} - B_{12}) - 2^{-1/2}(B_{31}^0 - B_{12}^0) \\ &= 2^{-1/2}(B_{31} - B_{31}^0) - 2^{-1/2}(B_{12} - B_{12}^0) \\ &= 0 \text{ for an electric field } E_{x'} = E_x \end{aligned}$$

But

$$A_{31}^0 = 0 \qquad \text{since } B_{31}^0 = B_{12}^0 = 0$$

Hence there is no tilt of axes of the ellipse when the crystal is strained under an applied electric field along the X axis.

As in the earlier cases r_{ij} will now have to be related to

$$\Delta n = \Delta n_{x'} - \Delta n_{z'}$$

Now

$$A_{11} - A_{11}^0 = B_{11} - B_{11}^0 = r_{11}E_x = -2(\Delta n_{x'})/n_{x'}^3$$

and

$$\begin{aligned} A_{33} - A_{33}^0 &= \tfrac{1}{2}[(B_{22} - B_{22}^0) + (B_{33} - B_{33}^0)] - (B_{23} - B_{23}^0) \\ &= \tfrac{1}{2}[-r_{11}E_x] - r_{41}E_x = -2\Delta n_{z'}/n_z'^3 \end{aligned}$$

So the path difference δ introduced per unit electric field per unit dimension along the direction of observation is given by

$$\begin{aligned} \delta &= (\Delta n_{x'} - \Delta n_{z'}) = [-\tfrac{1}{2}n_{x'}^3 r_{11}] - \tfrac{1}{2}n_{z'}^3[\tfrac{1}{2}r_{11} + r_{41}] \\ &= (\Delta n_x - \Delta n_{yz}) = -\tfrac{1}{2}r_{11}[n_x^3 + \tfrac{1}{2}n_{yz}^3] - \tfrac{1}{2}n_{yz}^3 r_{41} \end{aligned} \qquad (8.84)$$

where

$$n_x = n_o \qquad \text{and} \qquad \frac{1}{n_{yz}^2} = \tfrac{1}{2}\left(\frac{1}{n_o^2} + \frac{1}{n_e^2}\right)$$

Table 8.2. *Expressions for the Linear Electro-Optic Path Retardation for Some Crystal Classes (Traditional Method Due to Pockels)*

No.	Point group	Electric field parallel to	Observation parallel to	Mode of observation[a]	Expression[b] for path retardation δ per unit electric field
	Cubic				
1	$\bar{4}3m$ (T_d) 23 (T)	[001]	N^c	T	$-\frac{1}{2}n_o^3 r_{41}$
2	$\bar{4}3m$ (T_d)	N	N'	T	$n_o^3 r_{41}$
3	$\bar{4}3m$ (T_d)	[001]	[001]	L	$n_o^3 r_{41}$
	Tetragonal				
4	$\bar{4}2m$ (D_{2d})	[001]	[001]	L	$+n_o^3 r_{63}$
5	$\bar{4}2m$ (D_{2d})	[001]	N	T	$-\frac{1}{2}n_o^3 r_{63}$
6	$\bar{4}2m$ (D_{2d})	[100]	M	T	$-\frac{1}{2}n_{yz}^3 r_{41}$
7	$\bar{4}2m$ (D_{2d})	[100]	[100]	L	$\tan 2\theta = 2r_{41}/[(1/n_o^2)-(1/n_e^2)]$
8	$4mm$ (C_{4v})	[001]	[010]	T	$-\frac{1}{2}(n_e^3 r_{33} - n_o^3 r_{13})$
	Trigonal				
9	$3m$ (C_{3v})	[001]	[100]	T	$\frac{1}{2}(n_e^3 r_{33} - n_o^3 r_{13})$
10	$3m$ (C_{3v})	[100] [010]	[001]	T	$n_o^3 r_{22}$
11	32 (D_3)	[100]	[010]	T	$-\frac{1}{2}n_x^3 r_{11}$
12	32 (D_3)	[100]	[001]	T	$-n_x^3 r_{11}$
13	32 (D_3)	[100]	M	T	$-\frac{1}{2}r_{11}(n_x^3 + \frac{1}{2}n_{yz}^3) - \frac{1}{2}n_{yz}^3 r_{41}$
14	32 (D_3)	[100]	M'	T	$-\frac{1}{2}r_{11}(n_x^3 + \frac{1}{2}n_{yz}^3) + \frac{1}{2}n_{yz}^3 r_{41}$
15	32 (D_3)	[010]	[010]	L	$\tan 2\theta = -2r_{41}/[(1/n_o^2) - (1/n_e^2)]$
	Orthorhombic				
16	222 (D_2)	[100]	M	T	$-\frac{1}{2}n_{yz}^3 r_{41}$
17	222 (D_2)	[100]	M'	T	$\frac{1}{2}n_{yz}^3 r_{41}$
18	222 (D_2)	[010]	L	T	$-\frac{1}{2}n_{zx}^3 r_{52}$
19	222 (D_2)	[010]	L'	T	$\frac{1}{2}n_{zx}^3 r_{52}$
20	222 (D_2)	[001]	N	T	$-\frac{1}{2}n_{xy}^3 r_{63}$
21	222 (D_2)	[001]	N'	T	$\frac{1}{2}n_{xy}^3 r_{63}$

[a] T = transverse, L = longitudinal.
[b] δ is obtained in terms of $(\Delta n_{\parallel} - \Delta n_{\perp})$, for a transverse mode of vibration.
[c] L, M, N, etc. have the same significance as in Table 5.2.

***Table 8.3.** Expressions for the Linear Electro-Optic Path Retardation for Some Trigonal Crystals (Expressions for δ in Terms of r_{ij} by Mandarino's Method)*

No.	Point group	Direction of electric field	Direction of observation	Expression for path retardation δ per unit electric field ($\Delta n = \Delta n_\parallel - \Delta n_\perp$)
1	$3m$ (C_{3v})	[001]	[100] [010]	$\{(n_x^2 n_z^2)/(n_x + n_z)\}(r_{33} - r_{13})$
2	32 (D_3)	[100]	M^a	$-\{(n_x^2 n_{yz}^2)/(n_x + n_{yz})\}(\frac{3}{2}r_{11} + r_{41})$
3	32 (D_3)	[100]	M'	$-\{(n_x^2 n_{yz}^2)/(n_x + n_{yz})\}(\frac{3}{2}r_{11} - r_{41})$

[a] M and M' have the same significance as in Table 5.3.

Corollary: To Study r_{41} of the Classes 222, $\bar{4}2m$, and $\bar{4}3m$ by a Transverse Effect. The method of derivation given above can wholly be applied to obtain r_{41} of orthorhombic [D_2 (222) class, example—Rochelle salt], tetragonal ($\bar{4}2m$ class, example—ADP), and cubic ($\bar{4}3m$, example—ZnS) crystals by putting $r_{11} = 0$ for each of them. In each case then, for an electric field E_x (along the X axis) and for observation along M (equally inclined between Y and Z axes in the YZ plane), the path difference per unit electric field per unit thickness is given by $\delta = \Delta n = -(n_{yz}^3/2)r_{41}$.

In Table 8.2 are given the expressions for the linear electro-optic path retardations for some crystal classes.

Mandarino's method of relating δ to q_{ij} (see pages 212 and 213) can also be applied with equal facility to derive expressions relating δ, the optical path retardation, to r_{ij}, the linear electro-optic constants in the Pockels effect.

Thus, for example, we mention here two typical cases for the crystal classes C_{3v} ($3m$) and D_3 (32) of the trigonal system. See Table 8.3.

8.9. Experimental Methods of Determining r_{ij}

8.9.1. General Description and Application to Some Typical Crystal Classes

Electro-optic constants r_{ij} can be determined by measuring the electro-optic birefringence and hence the changes of polarization constants ΔB_{ij} in the crystal corresponding to a given electrical field strength, similar to the measurement of photoelastic constants.

Let a specimen of the experimental crystal be cut in the form of a rectangular parallelepiped (prism) with its edges parallel to a new coordinate system X_1', X_2', X_3' having its axes rotated with reference to the original crystallographic axes of the crystal (IEEE convention). The polarization constants B'_{ij} for the transformed coordinate system can be easily calculated, since B'_{ij} form the components of a second-rank symmetric tensor.

Under the influence of an electric field (which may be either dc or ac) on the experimental crystal prism, the optical index ellipsoid may undergo any of the following changes.

a. Axes of the optical index ellipsoid retain their positions, but their lengths vary, the increase in lengths being proportional to **E**. A Babinet compensator can be used to determine $\hat{\delta}$, the phase difference.

b. Axes tilt through 45° exactly and then their axial lengths may change. The tilt is independent of **E**. A Sénarmont compensator may be used here to determine $\hat{\delta}$.

c. Axial tilt depends upon **E**; that is, the tilt varies from one value of **E** to another. Here also a Sénarmont compensator may be used to determine $\hat{\delta}$.

We now have to determine experimentally the influence of an applied electric field on the angle of rotation θ of the axes, and the changes in the lengths of the axes of the appropriate elliptical section of the optical index ellipsoid, employing a suitable optical technique using plane or circularly polarized light.

8.9.2. Some Experimental Methods

Wherever appropriate, one uses ordinarily either a Babinet compensator or a Sénarmont compensator. Other methods include the following.

a. The crystal specimen is placed between two crossed Nicols or parallel Nicols, along with a half-wave plate immediately after the crystal. The electric field on the crystal is raised until the electro-optic path retardation induced in the crystal is another half-wave. This position can be precisely ascertained either by using a half-shade device or a photoelectric cell [229]. A knowledge of the half-wave voltage $V_{\lambda/2}$ is thus gained.

b. Measurement of the optic axial angle allows one to study the electro-optic behavior of crystals [145; 146].

c. When the natural birefringence of the crystal is very large for the particular direction of observation, the crystal specimen is shaped into a wedge of a very small angle (a few minutes of an arc) and is placed between two polaroids, crossed or parallel, in order to obtain the usual Babinet type of fringes. The crystal is then subjected to an electric field in the appropriate direction and the induced path retardation is calculated from a knowledge of the measurement of the shift of the Babinet fringes [1428].

d. Conoscopic method is applicable where the initial circular fringes become oval on the application of an electric field. Here the crystal is placed between two crossed Nicols and linearly or circularly polarized light is observed through the crystal. When the direction of propagation of light is along the optic axis of the crystal then one can observe the usual circular fringes. On the application of an electric field the circular fringes become oval. The measurement of the change of the lengths of the major and minor axes of the ovals brought about by a known change in the applied electric field affords a method of studying the electro-optic behavior of the crystal [229; 1419].

e. The displacement of interference bands of a channeled spectrum obtained when the plate is placed between crossed polaroids and the transmitted light is viewed by a spectroscope (i.e., Filon's spectroscopic method or one of its modifications [844; 1303]).

Figure 8.13 represents some types of experimental setup to study the linear electro-optic effect in crystals and to modulate light [1419; 1118].

8.9.3. Methods of Applying the Electric Field to the Crystal Prism

a. Nontransparent electrodes are generally utilized if light is propagated through the crystal across the field.

b. Transparent or lattice electrodes are utilized when the light is passed in the direction of the electric field. Transparent electrodes are placed either directly on the crystal surface or on glass plates which are placed on the crystal surfaces in the form of thick films with a very small resistance.

8.9.4. Experimental Determination of r_{ij} in Some Specific Cases of Crystals

8.9.4.1. Cubic Crystals of Class T_d ($\bar{4}3m$). Examples: ZnS, CuCl

r_{41} of this class of crystals can be determined relatively easily using the transverse effect. Here the field is applied along the [001] direction and a parallel beam of light is propagated along the [110] direction. The double

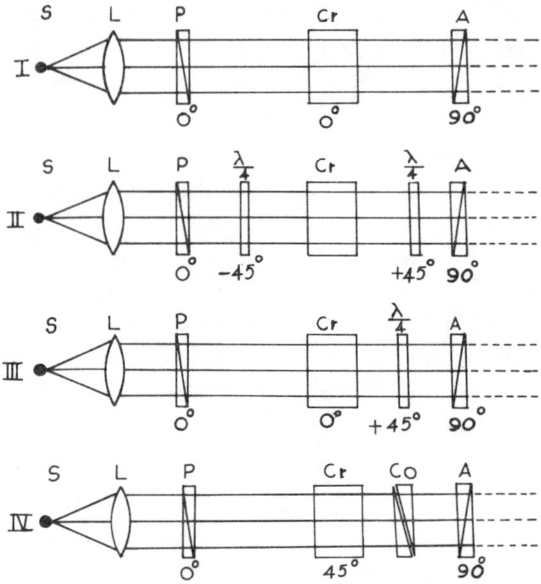

Fig. 8.13. Different types of experimental setups for studying the linear and quadratic electro-optic effects in crystals, and to modulate light. *S*, source of monochromatic light; *L*, converging lens; *P*, polarizer; *Cr*, crystal sample; $\lambda/4$, quarter-wave plate; *A*, analyzer; *Co*, compensator. The orientations of the various optical components are specified in degrees. From [1419].

refraction produced for this orientation is (from Table 8.2) $-\frac{1}{2}n_0{}^3r_{41}$ per unit electric field per unit dimension along the direction of light propagation. Then the path retardation δ is given by

$$\delta = -\tfrac{1}{2}n_o{}^3r_{41}E_zd$$

where d is the thickness along light propagation. Here δ can be determined using a Babinet or Soleil–Babinet compensator and hence r_{41} at one λ. When the dispersion of r_{41} is required, it is advantageous to use the birefringent compensator [844] or its modification [1303].

8.9.4.2. *Tetragonal Crystals of Class $\bar{4}2m$. Examples: ADP, KDP*

Under the action of the electric field E_z (along the optic axis), the circular Z section of the optical indicatrix is stretched into an ellipse oriented at $\pm 45°$ angle to the crystallographic axes of X and Y. Here the tilt of the electro-optic axes X' and Y' with reference to the crystallographic axes X and Y is equal to 45° and independent of E_z. When the crystal is placed between crossed polaroids, light passing along the Z axis, then the fraction of the intensity of light transmitted by the analyzer is expressed by the following formula [145; 146]:

$$\frac{I}{I_0} = \sin^2\left(\frac{\pi V n_o^3 r_{63}}{\lambda}\right) \tag{8.85}$$

where λ is the wavelength, n_o is the ordinary refractive index, and $V = E_z d$ is the voltage on the crystal. It follows that the intensity will be a maximum of one unit ($I/I_0 = 1$) at the voltage V', which produces a path retardation equal to $\lambda/2$, i.e., on the condition that $\sin^2(\pi V' n_o^3 r_{63}/\lambda) = 1$, which is satisfied when

$$(\pi V' n_o^3 r_{63}/\lambda) = \pi/2 \tag{8.86}$$

$$\therefore r_{63} = \lambda/2V' n_o^3$$

V' is thus the half-wave voltage represented by $V_{\lambda/2}$. Using the above formula, one can determine the dispersion of the electro-optic coefficient r_{63}. This can be achieved by transmitting monochromatic light of different wavelengths and measuring the respective voltages V' required for the path difference of $\lambda/2$. Figure 8.14 shows the dispersion of $V_{\lambda/2}$ for some ADP-type crystals [10].

The coefficient r_{63} can also be determined by measuring the compensation angle ϕ using a Sénarmont compensator. The compensation angle ϕ is the angle through which the analyzer should be rotated for compensating or reducing to minimum the light intensity which is transmitted at the given voltage on the crystal. The compensation angle ϕ depends linearly upon the value of the voltage on the crystal: $\phi = r_{63} n_o^3 \pi V/\lambda$. This method lends itself to determine r_{63} at any voltage on the crystal [145; 146].

8.9.4.3. *Trigonal Crystals, Class D_3 (32). Example: α-Quartz*

Here two Y-cut plates of equal thickness of quartz are placed together so as to compensate their double refraction (that is, the X axis of one plate is parallel to the Z axis of the other plate and conversely) and the combination is placed between two crossed Nicols whose vibration directions are at $\pm 45°$ with the X axis of one of the plates.

An electric field E_x is applied along the X axis of one of the plates and light is propagated along its Y axis. Under the influence of the electric field, the elliptically vibrating light emerging from the pair of quartz plates

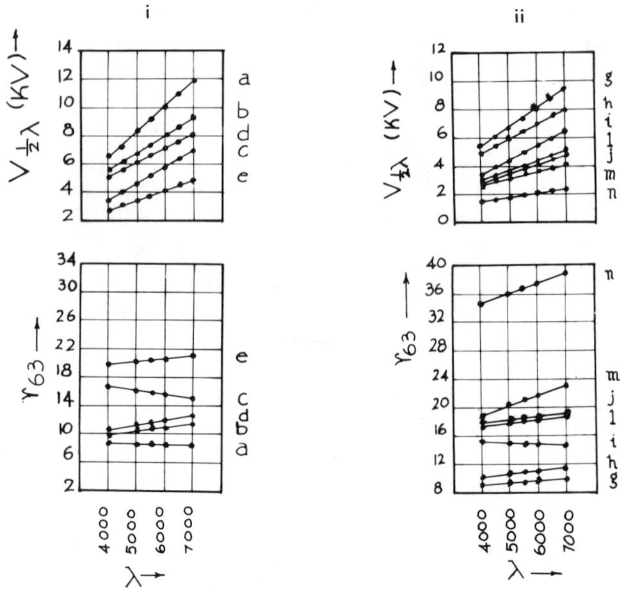

Fig. 8.14. Dispersion of $V_{\lambda/2}$ and r_{63} for some tetragonal phosphates and arsenates of ADP type. $V_{\lambda/2}$ in kilovolts; r_{63} in units of 10^{-10} cm/V; λ in angstroms. *i*: (a) ADP; (b) KDP; (c) RbDP; (d) AD*P; (e) KD*P. *ii*: (g) ADA; (h) KDA; (i) RbDA; (j) CsDA; (l) KD*A; (m) RbD*A; (n) CsD*A. From [10].

has its axes parallel to the X and Z axes of the quartz. If the applied voltage corresponds to the half-wave voltage of the quartz specimen, then the intensity of the transmitted light is maximum, and we have

$$r_{11} = \lambda/n_o^3 d_y E'$$

where E' is the applied electric intensity producing a half-wave retardation and d_y is the thickness of the crystal plate excited by E'.

Instead of measuring the transmitted intensity, one can use a Babinet compensator and determine the path difference δ for an applied electric intensity E_x. Then $\delta = -\frac{1}{2}n_0^3 r_{11} d_y E_x$. This method permits assigning the correct sign to r_{11}. The dispersion of r_{11} can be studied using a birefringent compensator [844] or its modification [1303].

8.9.5.4. *Monoclinic Crystals of Class 2 (C_2). Example: Pentaeritrite*

The following method may be employed in such cases where the angle of tilt of the optical indicatrix depends upon the field **E**. Thus, in the penta-

Table 8.4. *Linear Electro-Optic Constants of Crystals (r_{ij} Values are in Units of* 10^{-8} cm/stat V)[a]

Crystal	r_{ij}
Cubic system	
Class $\bar{4}3m$ (T_d)	
1. Cuprous bromide, $CuBr$	$r_{41} = 2.5$
2. Cuprous chloride, $CuCl$	$r_{41} = 10.8$
3. Cadmium telluride, $CdTe$	$r_{41} = 20.4$
4. Eulytite, $Bi_4Si_3O_{12}$	$r_{41} = 3$
5. Gallium arsenide, $GaAs$	$r_{41} = 4.5$
6. Gallium phosphide, GaP	$r_{41} = 3.2$
7. Germanoeulytite, $Bi_4Ge_3O_{12}$	$r_{41} = 13$
8. Hexamethylenetetramine, $C_6H_{12}N_4$	$r_{41} = 2.2 - 2.7$
9. Sphalerite, ZnS	$r_{41} = 6$
10. Stilleite, $ZnSe$	$r_{41} = 6$
11. Zinc telluride, $ZnTe$	$r_{41} = 13.7$
Class 23 (T)	
1. Ammonium chloride, NH_4Cl at -196°C	$r_{41} = 4.5$
2. Ammonium cadmium langbeinite, $(NH_4)_2Cd_2(SO_4)_3$	$r_{41} = 2.1$
3. Molecular compound $2NH_4Cl \cdot 3CuCl$	$r_{41} = 2$
4. $(NH_4)_2Mn_2(SO_4)_3$	$r_{41} = 1.6$
5. $Bi_{12}GeO_{20}$	$r_{41} = 10$
6. $K_2Mg_2(SO_4)_3$	$r_{41} = 1.2$
7. $Bi_{12}SiO_{20}$	$r_{41} = 10$
8. $NaClO_3$	$r_{41} = 1.2$
9. Sodium uranyl acetate, $NaUO_2(CH_3COO)_3$	$r_{41} = 2.6$
10. Thallium cadmium langbeinite, $Tl_2Cd_2(SO_4)_3$	$r_{41} = 1.8$
11. Trenhydrobromide, $N(CH_2CH_2NH_2 \cdot HBr)_3$	$r_{41} = 4.5$
12. Trenhydrochloride, $N(CH_2CH_2NH_2 \cdot HCl)_3$	$r_{41} = 5.1$
Tetragonal system	
Class $\bar{4}2m$ (D_{2d})	
1. Ammonium dihydrogen phosphate, ADP	$r_{41} = 72.0$; $r_{63} = 26.0$
2. Ammonium dideuterium phosphate, AD*P	$r_{63} = 34.7$

continued overleaf

Table 8.4 (*continued*)

Crystal	r_{ij}
3. Ammonium dihydrogen arsenate, ADA	$r_{63} = 27.6$
4. Caesium dihydrogen arsenate, CsDA	$r_{63} = 50$
5. Caesium dideuterium arsenate, CsD*A	$r_{63} = 55.8$
6. Potassium dihydrogen phosphate, KDP	$r_{41} = 26$; $r_{63} = 32$
7. Potassium dideuterium phosphate, KD*P	$r_{41} = 26$; $r_{63} = 72$
8. Potassium dihydrogen arsenate, KDA	$r_{41} = 38$; $r_{63} = 33$
9. Potassium dideuterium arsenate, KD*A	$r_{63} = 54.6$
10. Rubidium dihydrogen phosphate, RbDP	$r_{63} = 43$
11. Rubidium dihydrogen arsenate, RbDA	$r_{63} = 39$
12. Rubidium dideuterium arsenate, RbD*A	$r_{63} = 64.2$
Class 4*mm* (C_{4v})	
1. Barium titanate, $BaTiO_3$	$r_{13} = 24$; $r_{33} = 84$
2. $K_3Li_2Nb_5O_{15}$	$r_{13} = 26.7$; $r_{33} = 23.4$
3. $KSr_2Nb_5O_{15}$	$(r_{33}n_3^3 - r_{13}n_1^3)/n_3^3 = 490$
4. Lead titanate, $PbTiO_3$	$r_{13} = 40.4$; $r_{33} = 16.7$
5. $Sr_{0.75}Ba_{0.25}Nb_2O_6$	$r_{13} = 201$; $r_{33} = 4020$ $r_{51} = 126$
6. $Sr_{0.5}Ba_{0.5}Nb_2O_6$	$(r_{33}n_3^3 - r_{13}n_1^3)/n_1^3 = 654$
7. $Sr_{0.25}Ba_{0.75}Nb_2O_6$	$(r_{33}n_3^3 - r_{13}n_1^3)/n_1^3 = 135$
Class $\bar{4}$ (S_4)	
1. Pentaerythritol, $C(CH_2OH)_4$	$r_{52} = 4.1$
Hexagonal system	
Class 6*mm* (C_{6v})	
1. Greenockite, CdS	$r_{13} = 3.3$; $r_{33} = 7.2$ $r_{51} = 11.1$
2. Wurtzite, ZnS	$r_{13} = 2.8$; $r_{33} = 5.6$
3. Zincite, ZnO	$r_{13} = 4.2$; $r_{33} = 8.4$
Class 6 (C_6)	
1. Lithium iodate	$r_{13} = 12.3$; $r_{33} = 19.2$ $r_{41} = 4.2$; $r_{51} = 9.9$
2. Lithium potassium sulfate, $LiKSO_4$	$r_{13} - r_{33} = 4.8$

Table 8.4 (*continued*)

Crystal	r_{ij}
Trigonal system	
Class $3m$ (C_{3v})	
1. Caesium nitrate, $CsNO_3$	$r_{22} = 1.3$
2. Lithium niobate, $LiNbO_3$	$r_{13} = 25.8$; $r_{33} = 92.4$ $r_{22} = 21$; $r_{42} = 84$
3. Lithium sodium sulfate, $LiNaSO_4$	$r_{22} = 0.06$
4. Lithium tantalate, $LiTaO_3$	$r_{13} = 20.7$; $r_{33} = 91.2$
5. Tourmaline, $Na(LiAl)_3Al_6(OH)_{1-3}(BO_3)_3(Si_6O_{18})$	$r_{22} = 0.9$; $r_{33} - r_{13} = 1.7$
Class 32 (D_3)	
1. Caesium tartrate, $Cs_2C_4H_4O_6$	$r_{11} = 3$
2. HgS	$r_{11} = 9.3$; $r_{41} = 4.2$
3. Potassium dithionate, $K_2S_2O_6$	$r_{11} = 0.8$
4. Strontium dithionate, $Sr_2S_2O_6 \cdot 4H_2O$	$r_{11} = 0.3$
5. Quartz, SiO_2	$r_{11} = 1.41$; $r_{41} = 0.6$
6. Selenium, Se	$r_{41} = 7.5$
Orthorhombic system	
Class 222 (D_2)	
1. Ammonium oxalate, $(NH_4)_2C_2O_4 \cdot H_2O$	$r_{41} = 680$; $r_{52} = 980$ $r_{63} = 750$
2. Rochelle salt, $KNaC_4H_4O_6 \cdot 4H_2O$	$r_{41} = 6$; $r_{52} = 5.1$
Class $mm2$ (C_{2v})	
1. $Ba_2NaNb_5O_{15}$	$r_{13} = 45$; $r_{23} = 39$; $r_{33} = 144$; $r_{42} = 276$; $r_{51} = 270$
Monoclinic system	
Class 2 (C_2)	
1. Calcium pyroniobate, $Ca_2Nb_2O_7$	$r_{22} = 1.0$; $r_{32} = 41$
2. Triglycine sulfate, $(NH_2CH_2COOH)_3H_2SO_4$	$r_{22} = 21.6$; $r_{32} = 40.8$

[a] From [1042]. In this data no distinction is made between the free and clamped r_{ij}. The sign of r_{ij} is not given. All r_{ij} values are for λ in the visible region (from 5461 to 6300 Å), where the dispersion of r_{ij} is negligible. For more exhaustive data on r_{ij} of crystals, see [83; 84; 1526].

eritrite crystals of class 2 when the field acts in the direction of the symmetry axis 2 and the light is transmitted in the same direction (i.e., for longitudinal observation), the corresponding cross section of the optical indicatrix, which in the absence of the field appears to be elliptical, is rotated through an angle θ, where

$$\tan 2\theta = \frac{2r_{52}E_y}{(B_1^0 - B_3^0) + (r_{12} - r_{32})E_y} \tag{8.87}$$

where B_1^0 and B_3^0 are reciprocal values of the square of the main refractive indices in the given section, in the absence of the field. Having determined experimentally the values of the angles θ at two different voltage values on the crystal and forming simultaneous equations for r_{52} and $(r_{12} - r_{32})$ with these measurements, it is possible to determine r_{52} and $(r_{12} - r_{32})$. This experimental method is applicable also to ADP to determine r_{41} utilizing the longitudinal effect along the X or Y axis.

The numerical values of r_{ij} for some common crystals are given in Table 8.4. Table 8.5. gives quadratic electro-optic coefficients of some cubic crystals.

Table 8.5. *Quadratic Electro-Optic Constants of Some Cubic Crystals*

No.	Crystal	Kerr constants g_{ij} in units of 10^{10} e.s. units[a]			
		g_{11}	g_{12}	$g_{11} - g_{12}$	g_{44}
1	$BaTiO_3$	1.33	−0.11	1.44	—
2	$SrTiO_3$	—	—	1.56	—
3	$KTaO_3$	—	—	1.77	1.33
4	$KTa_{0.65}Nb_{0.35}O_3$	1.5	−0.42	1.93	1.63
5	$PbMg_{1/2}Nb_{2/3}O_3$	—	—	0.17	0.09

[a] From [1042]. All the values of g_{ij} are for λ at 6328 Å. For the numerical data on the quadratic electro-optic (Kerr) coefficients of 15 alkali halide crystals of the NaCl type, namely, (Li, Na, K, Rb) (F, Cl, Br, I), see [488]. For further information on g_{ij} values of crystals, see [83; 84; 1526].

8.10. Some Points of Interest on the Use of the Pockels Effect in Crystals, and Half-Wave Voltage $V_{\lambda/2}$

a. When a *Z*-cut ADP crystal plate is placed between (i) crossed plane polarizers or (ii) left and right circular polarizers, the modulation frequency is twice that of the alternating electric field applied to the crystal plate. On the other hand, if the light incident on the same *Z*-cut ADP crystal plate is circularly polarized and observed through an analyzer, the modulation frequency of the transmitted light is equal to that of the applied alternating electric field. In the first two cases an ac voltage causing a half-wave path retardation can change the transmission of light from 0 to 100%. In the third case an ac voltage producing a peak retardation of $\lambda/4$ can give 100% modulation of the transmitted light beam.

b. In many linear electro-optic devices the longitudinal effect is used, that is, the light beam and electric field are parallel. Longitudinal effect devices are particularly useful for light beams of large cross-sectional area [840]. Cubic crystals allow the design of wide-aperture electro-optic modulators that also have transverse control.

Transverse effect devices avoid the use of transparent electrodes in the light path. In addition, the voltage required for a given retardation can be reduced by increasing the ratio of the light path length to the electrode spacing. In the longitudinal-effect devices the required voltage is independent of the dimensions of the crystal; but there is a limitation on the angular aperture for these longitudinal devices when KDP-type uniaxial crystals are used.

c. It is easily seen from symmetry considerations that a longitudinal effect free of background birefringence and optical activity is obtained only with crystals of two classes: (i) the class of $\bar{4}3m$ (T_d) of the cubic system, for example, cubic zinc sulfide, cuprous chloride, gallium arsenide, and hexamethylene-tetramine (HMTA) and (ii) the class $\bar{4}2m$ (D_{2d}) of the tetragonal system represented by KH_2PO_4 (KDP) and its isomorphs. Kaminow and Turner [613] have discussed the crystal orientation for the maximum induced electro-optic birefringence. They have considered also the cases of longitudinal and transverse observation for the electric field along the [111] direction of cubic crystals.

d. Classes $\bar{4}3m$ (T_d), $3m$ (C_{3v}), $\bar{6}$ (C_{3h}), $\bar{6}m2$ (D_{3h}), $4mm$ (C_{4v}), and $6mm$ (C_{6v}) can exhibit a transverse effect which is free of background birefringence and optical activity.

e. Similar to the figure of merit of an acousto-optic material [296],

Kaminow and Turner [613] have defined F, the figure of merit of an electro-optic material, as

$$F = n^7 r_{ij}^2 / \varkappa \tag{8.88}$$

where $\varkappa$ is the dielectric constant at the modulating frequency and n the appropriate refractive index. On this basis, we can compare the figures of merit of different electro-optic materials.

Thus, for example, Nash *et al.* [855] find for $LiNbO_3$ the quantity $n^7 r_{63}^2/\varkappa$ equal to 3.1×10^{-22}, which is approximately four times that for KDP. But in practical applications, the availability of KDP in large-sized crystals of excellent optical and dielectric quality outweighs the high figure of merit of $LiNbO_3$.

f. Another important characteristic of an electro-optic crystal specimen is its half-wave voltage $V_{\lambda/2}$, which may be defined as the voltage required on a crystal specimen of unit dimensions in order to produce a path retardation of half a wave. Thus $V_{\lambda/2} = E_{\lambda/2} d$, where $E_{\lambda/2}$ is the applied electric field strength along a thickness d of the crystal in order to produce a path retardation of half a wave; if d is 1 cm, then $V_{\lambda/2} = E_{\lambda/2}$.

$V_{\lambda/2}$ can be calculated from a knowledge of the electro-optic constants of the crystal. For example, for $LiNbO_3$ [belonging to the $3m$ (C_{3v}) class] when the electric field is along Z and light is propagated in the XY plane, it is easily seen that

$$V_{\lambda/2} = \frac{\lambda}{n_o^3 r_{33} - n_e^3 r_{13}} \tag{8.89}$$

The reverse process of determining the electro-optic constants from a knowledge of $V_{\lambda/2}$ has been described in Section 8.9.

g. Some of the essential requirements for electro-optic crystals to serve efficiently are (a) highest attainable optical homogeneity, with the minimum amount of absorption and scattering, (b) sufficient hardness to allow good machining and workability, (c) minimum anisotropy of thermal expansion and minimum thermal variation of the effective coefficients being used (such as electro-optic, piezo-optic, etc.), (d) insolubility in ordinary solvents including water, and (e) optimum dielectric properties.

h. Since the development of the laser in 1960, KDP crystals have been the most widely used electro-optic materials, primarily because samples of the required optical quality have been available. The frequency range of laser-related experiments in which linear electro-optic crystals have been used is sufficiently broad, e.g., Kaminow has used KDP rods to modulate laser beams at 9.2 GHz. It was found that the electro-optic coefficients were of the same order as measured in the megacycle range.

i. Related to the Pockels effect is the frequency doubling or frequency mixing occurring in intense optical-frequency electric fields obtained from lasers. Such effects have been first observed by Franken, Hill, Peters, and Weinreich [388] in quartz, KDP, and triglycine sulfate.

8.11. Some Technological Applications of Pockels Cells (Linear Electro-Optic Devices)

8.11.1. Use of the Electro-Optic Effect in Technology

Broadly speaking the electro-optic devices can be grouped under three categories: (1) amplitude, phase, and frequency type modulators of light; (2) light shutters and beam splitters; (3) function converters, in which a voltage varying according to a certain law is converted into the intensity of light which varies according to the same law.

Modulators of light are used in the following:

1. optical range finders or devices for measuring distance by means of light
2. for light communication (light telephone)
3. for sound recording on cine films
4. for reproducing images in color television, phototelegraphy, and electron microscopy
5. automatic polarimetric and other modulation measuring devices
6. for the modulation of induced radiation of quantum mechanical light generators (lasers)

Electro-optic shutters are used in

1. photography and high-speed cinematography
2. light telegraphy

Function converters are used in

1. tunable narrow-band interference polarization optical filters
2. devices for high-voltage measurement
3. optical elements of computer systems

8.11.2. Some Applications of Electro-Optic Devices

8.11.2.1. Optical Range Finders or Electro-Optic Devices for Measuring Distances

The electro-optic method of determining distances is essentially the same as the well-known Fizeau toothed wheel and Michelson rotating mirror methods for measuring the velocity of light; here the process of computation is reversed, that is, knowing the velocity of light one calculates the distance. The chopping of a light beam with a known frequency using either of the methods above is replaced here by modulating light using a Pockels cell. Thus, Fizeau's rotating toothed wheel or Michelson's prismatic mirror is replaced by two Pockels cells which operate synchronously; one cell transmits the light signal while the other receives it. Prior to the introduction of the Pockels cell, a Kerr cell filled with nitrobenzene was successfully employed by Karolus and Mittelstaedt [629] in 1928 and by Anderson [38] in 1941. A large number of liquid mixtures have recently been found to be more efficient than nitrobenzene with respect to their Kerr effect [723]. However, the use of Kerr cells presents a number of difficulties [145], which can be avoided if electro-optic crystal modulators are employed.

There is a limitation imposed on the frequency of modulation and the accuracy obtained earlier by Fizeau's and Michelson's methods because of the mechanical devices employed in these methods. In the electro-optic devices, however, this modulation being achieved by electronic means, the modulation frequency can be raised to 10^6 Hz and even beyond with a consequent improvement of the accuracy. The relative error in the measurement of distances is of the order of a centimeter in a kilometer. Distances up to 50 km can be measured with very intense light sources like lasers, the range being limited only by the light losses in the atmosphere.

Quartz has been used [150] in distance-measuring modulators. ADP and KDP crystals can be used as modulators up to about 10^9 Hz. At a frequency of the order of 10^9 Hz the electro-optic elements work under the piezoelectric clamped conditions, far above the acoustic modes. Hence in this case the modulation is entirely due to the primary or true electro-optic effect, and consequently the voltage requirements are far higher than those needed at the acoustic frequency, where the secondary effect is very much enhanced. Thus, at $\lambda = 5500$ Å, a voltage of 14.8 kV must be applied to an ADP crystal and 8.5 kV to a KDP crystal, while at low frequencies, 9.3 and 7.7 kV, respectively are needed for the same orientation of the crystals.

8.11.2.2. Sound Recording on Cine Film

One of the traditional methods for the recording of sound in cinematography employs a Kerr cell with nitrobenzene as the electro-optic medium. Obviously a Pockels cell is preferable to a Kerr cell.

The essential component of the optical system (Fig. 8.15) employed for recording the sound on film is a Pockels cell which consists of a Z-cut ADP crystal fitted with transparent electrodes. This unit is then placed between crossed polaroids, with a mica $\lambda/4$ plate interposed between the polarizer and the ADP crystal. A parallel beam of monochromatic light passes through the Pockels cell and is focused on the cine film; when an electric field is applied to the cell, the central dark part of the conoscopic pattern becomes transparent because it splits up into two dark spots under the applied voltage. By applying an audio-frequency voltage to the crystal the sound will be recorded on the film.

The modulating crystal must be so designed that it has no piezoelectric resonance within the audio-frequency range. The main difficulty in this device is due to the use of transparent electrodes. The thin films forming the electrodes should not absorb an appreciable percentage of light. Furthermore, when ADP crystals are used as modulating devices they must be kept in oil or sealed vessels, and be strictly maintained within a certain temperature range, in order to prevent the electrodes from getting detached from the crystal surfaces.

8.11.2.3. Measurement of High Voltage

A plate of ADP subjected to an electric field in a suitable direction develops artificial birefringence due to the Pockels effect, and from a knowledge of the applied voltage one can obtain r_{ij} for the experimental orientation of the crystal. Conversely, knowing the value of r_{ij} one can calculate the voltage applied to the ADP plate. Hence the experiment consists in determining the linear electro-optic phase difference developed in the crystal, for the given voltage, using any of the well-known methods to study elliptically polarized light.

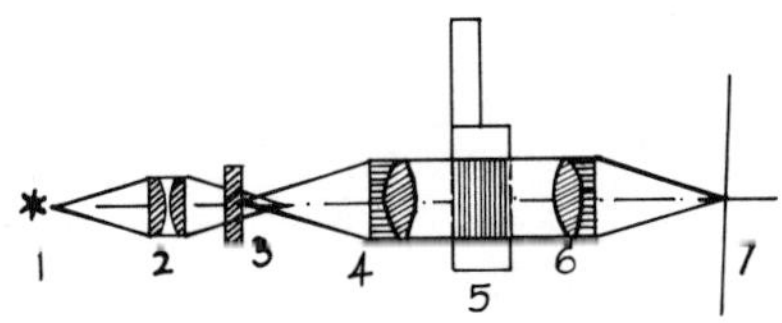

Fig. 8.15. Electro-optic sound modulator for sound recording on cine film. (1) Light source; (2) light condenser; (3) slit; (4) collimating lens; (5) $\lambda/4$ mica plate and ADP crystal between crossed polaroids; (6) focusing lens; (7) cine film. From [1417].

We shall consider two cases.

a. To utilize the transverse effect, a plate of ADP with its edges along the directions [110], [1$\bar{1}$0], [001] is subjected along the Z axis to the voltage V to be measured, and a Babinet compensator may be used to determine the phase difference. For this orientation, the phase difference $\hat{\delta}$ is given by

$$\hat{\delta} = \frac{\pi}{\lambda} r_{63} n_o^3 E = \frac{\pi}{\lambda} r_{63} n_o^3 \frac{V}{t} \tag{8.90}$$

where V is the applied voltage across the dimension t of the crystal. Hence knowing $\hat{\delta}$, one can calculate V.

b. Taking the longitudinal effect, Namba [838] developed the following method to determine V. When the electric field and observation are along the Z axis of the ADP plate (Fig. 8.16), the path difference is $\delta = r_{63} n_o^3 E$ per unit dimension along the light propagation direction [see Eq. (8.77)]. Hence for a crystal of dimension t along E, the phase difference is

$$\hat{\delta} = r_{63} n_o^3 (V/t) t (2\pi/\lambda)$$

and the path difference is

$$\delta = r_{63} n_o^3 V \tag{8.91}$$

With crossed Nicols whose vibration directions are set parallel to the X, Y axes of the ADP plate, on the application of the electric field the original circular cross section of the indicatrix becomes now elongated into an ellipse with its axes rotated by 45° with respect to the vibration direction of the crossed Nicols. As the field is increased, the intensity of light will rise from zero, reaching its maximum when the path retardation equals $\lambda/2$ or the phase difference equals π.

The intensity I of the light transmitted through the analyzer now is given by

$$I = I_0 \sin^2(\pi n_o^3 r_{63} V/\lambda) \tag{8.92}$$

where I_0 is the intensity of the incident light.

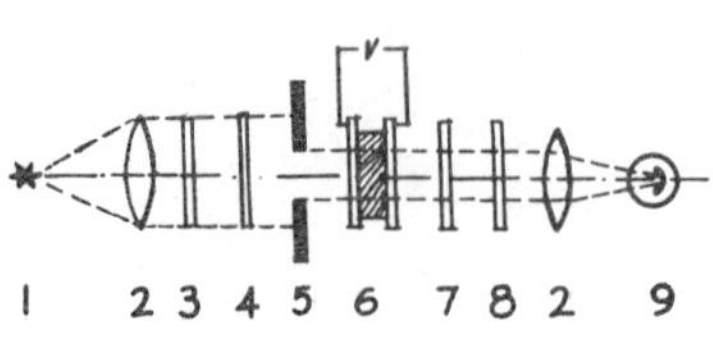

Fig. 8.16. Schematic diagram for measuring high voltages. (1) Light source; (2) converging lenses; (3) light filter; (4) polarizer; (5) diaphragm; (6) ADP plate; (7) $\lambda/4$ plate; (8) analyzer; (9) photocell. From [838].

If the analyzer is rotated through an angle ϕ, the fraction of the light transmitted is given by

$$\frac{I}{I_0} = \cos^2\left(\frac{\pi}{\lambda}\,\delta - \phi\right) \tag{8.93}$$

Therefore the intensity of the transmitted light is a function of ϕ, which itself depends upon the applied voltage; the applied voltage can therefore be measured from a knowledge of δ, the path difference, or $\hat{\delta}$, the phase difference.

I becomes zero if

$$\left(\frac{\pi\delta}{\lambda} - \phi\right) = \frac{\pi}{2}$$

or (8.94)

$$\delta = \frac{\lambda}{\pi}\left(\frac{\pi}{2} + \phi\right)$$

Hence

$$V = \frac{\delta}{r_{63}n_o^{\,3}} = \frac{\lambda}{\pi r_{63}n_o^{\,3}}\left(\frac{\pi}{2} + \phi\right) \tag{8.95}$$

Thus V can be obtained from a knowledge of the setting of the analyzer for extinction of the field of view. (See Sec. 8.9.4.3.)

8.11.2.4. Electro-Optic Shutters

A Z-cut plate of ADP placed between crossed polaroids acts as an optical shutter for light propagation if there is also an electric field along the Z axis, because the crystal becomes biaxial owing to the longitudinal Pockels effect, with neither of its new optic axes coinciding with its Z axis.

Three types of optical systems are given in Fig. 8.13 (I, II, III) for use as optical shutters which can also serve as modulators of light. Cubic crystals like ZnS and CuCl can also be used in place of crystals of ADP type.

a. The crystal is placed between crossed polaroids and a parallel or convergent light beam is transmitted. (See Fig. 8.13–I.)

b. The crystal is placed between right and left circular polarizers each comprising a polaroid and a $\lambda/4$ plate. (See Fig. 8.13–II.)

c. The crystal is placed between crossed polaroids with a $\lambda/4$ plate inserted between the crystal and analyzer. The $\lambda/4$ plate has its vibration directions at $\pm 45°$ with that of the polarizer. This forms the setup for a Sénarmont compensator. (See Fig. 8.13–III.)

In the absence of an electric field in the first two cases the field of view is dark when parallel light is used and the center is dark with convergent light, whereas in the third case, even in the absence of an electric field, the intensity of the transmitted light is not zero owing to the presence of the $\lambda/4$ plate. On the application of an electric field in the first two arrangements the intensity of the transmitted light increases to a maximum at the half-wave voltage, but in the third arrangement the intensity drops to zero or increases to a maximum at the electric field strength which causes an additional path difference of $\lambda/4$. Consequently, in order to achieve 100% modulation of light, the voltage to be applied in the third case is only half of that required for the first and second cases. This is a definite advantage.

Furthermore, in the first and second cases, the modulation frequency is twice that of the alternating electric field applied to the system, but in the third case the modulation frequency of light is the same as that of the applied electric field [149; 459].

In all the three systems the transmission is governed by the optical path difference produced along the Z axis consequent upon the application of the electric field.

The limitations on the performance of each of the systems are of importance in actual practice. Taking, for example, the second arrangement where a Z-cut ADP plate is placed between two circular polarizers (right and left handed, respectively, on either side), the intensity of the transmitted light [149] is

$$I = \frac{1}{2} I_0 \sin^2 \frac{\hat{\delta}}{2} \tag{8.96}$$

where the phase difference is

$$\hat{\delta} = \frac{2\pi d}{\lambda} \left(\frac{n_o^2 - n_e^2}{2n_o n_e} \right) \sin^2 \theta \tag{8.97}$$

θ being the inclination of the ray with the Z axis. It is seen from Eq. (8.97) that $\sin^2 \theta \propto 1/d$ at any particular λ; hence the angular field increases as the crystal becomes thinner.

Strictly speaking the phase difference is zero at zero voltage only for rays parallel to the Z axis (optic axis); but in practice it is impossible to produce an ideal parallel beam of light, and the so-called parallel beam in reality contains also rays slightly converging or diverging. These rays enter the crystal at different angles and therefore are subject to varying degrees of phase difference; on emergence from the crystal plate and the analyzer they interfere producing the well-known interference patterns. This means

Table 8.6. *Dependence of Residual Intensity on the Angular Field and Thickness of a Z-Cut ADP Plate Placed between Crossed Polaroids*

Thickness	Angular field	
	1% intensity	0.1% intensity
2.0 mm	2°20′	44′
6.3 mm	1°20′	26′

that an ADP crystal enclosed between crossed polaroids never gives complete extinction. The brightness still persisting can be reduced only through severely limiting the aperture of the field of view. The problem becomes more acute with increasing natural birefringence of the electro-optic crystal. Conversely cubic crystals like CuCl and ZnS which exhibit no natural birefringence facilitate a larger field of view.

Therefore the problem that arises now with ADP-type crystals is how best to increase the angular field without unduly increasing the background noise. The dependence of residual intensity at zero electric field on the thickness of a double refracting crystal can be appreciated if we consider the thickness versus intensity relationship of ADP crystal as an example (given above) [1417] (see Table 8.6).

On the basis of the table above, one cannot jump to the conclusion that thinning down of a crystal will appreciably solve the problem because other physical properties demand attention. For example, ADP crystals develop fracture spontaneously if the plates are thinner than about 1.2 mm. Thus the design of the electro-optic crystals will have to take all these factors into consideration.

The alternative is to explore other methods of increasing the angular field. In this connection, the following are some other methods that have been tried.

ADP, being a negative uniaxial crystal, is combined with a positive uniaxial crystal of requisite thickness so that the phase difference introduced into a ray traveling slightly inclined to the Z axis is canceled out by that introduced by the positive uniaxial plate, thereby considerably reducing the residual intensity of the interference pattern. This will greatly enhance the angular field. Namba [838] has discussed this problem thoroughly, but one should be cautioned even here, because the dispersions of the two types of crystals are not the same, and, furthermore, though the compensation is satisfactory at zero electric field, complications may arise when an electric field is applied to the ADP plate because, owing to the Pockels effect, the fast and slow rays change their directions in ADP under electric field. Therefore the canceling effect by the second plate becomes vitiated.

8.11.2.5. Tunable Electro-Optic Narrow Passband Polarizing Interference Optical Filter

The principle underlying the action of the polarizing interference optical filter is based on the phenomenon of interference of polarized light passing through a double refracting crystal, say quartz [752]. A plate of quartz of 1–2 mm thickness and with its Z axis parallel to its broad faces is placed between two crossed polaroids whose vibration directions make an angle of 45° with the Z axis of the quartz plate. A parallel beam of white light is normally incident on the quartz plate, and the transmitted light is examined by a spectroscope. The spectrum is found to consist of a number of bright and dark bands whose separation depends upon the thickness of the quartz plate—the thicker the plate the smaller the separation and the larger the number of dark bands across the spectrum. By placing in series a number of quartz plates of the same orientation, the thicknesses of which are precisely in the geometrical progression 1:2:4:8. . . , and each plate being enclosed between a pair of crossed polaroids, the resulting spectrum shows just a few narrow and widely separated bright bands. An ordinary light filter is then used to select only the desired color for transmission. By varying the temperature of the unit the wavelength of the transmitted light can be altered suitably. The separation of consecutive bright bands in any spectral region is equal to that of the transmission maxima of the thinnest plate, while the width of each bright band is that given by the thickest plate.

The tunable electro-optic polarizing interference filter consists of a unit of quartz plates and polaroids described above, together with Z-cut ADP plates as in Fig. 8.17, and it is the longitudinal electro-optic effect in ADP that is used for tuning. The path difference in a Z-cut ADP crystal is obviously independent of its thickness but varies linearly with the applied voltage. The tuning or the variation of the mean wavelength of the transmitted light is therefore achieved by controlling the electric field applied to the Z-cut ADP plates.

Here in the electro-optically tunable filter, the set of quartz plates functions as a fixed filter in order to isolate certain discrete bands, and the

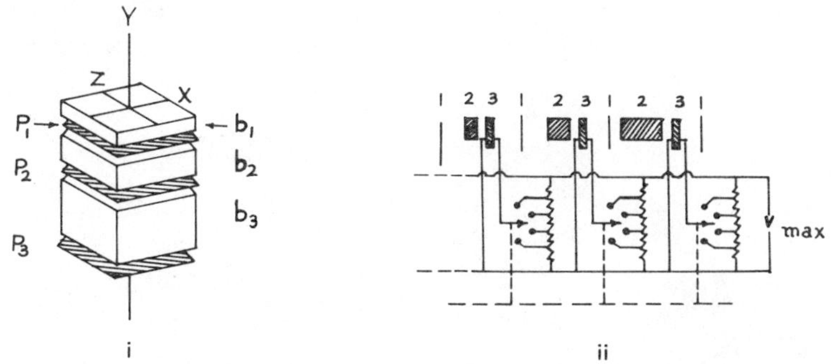

Fig. 8.17. Interference polarizing monochromator. (i) Fixed type: P_1, P_2, P_3 are polaroids, b_1, b_2, b_3 are quartz plates, and *X*, *Y*, *Z* are axes of quartz plates. (ii) Electronically controlled type: (1) polaroids; (2) quartz plates; (3) ADP plates. From [1417].

ADP plates enable one to adjust the position and bandwidth of each transmitted band. The pass bandwidth can be varied at will from a fraction of an angstrom to hundreds of angstroms.

The electro-optically tunable light filters can be used in a variety of situations, including

a. the study of the solar corona and prominences without the necessity of a total eclipse,
b. the frequency modulation of light,
c. the spectroscopic studies on very short duration processes,
d. the color reproduction, colorimetry, and color television,
e. the study of the fine structure of Raman lines.

Bibliography

1. Aas, H. G., and Erf, R. K., Application of ultrasonic standing waves to the generation of optical beam scanning. *J. Acoust. Soc. Am.* (USA) **36**, 1906–1913 (1964).
2. Abrams, R. L., and Pinnow, D. A., Acousto-optic properties of crystalline germanium. *J. Appl. Phys.* (USA) **41**, 2765–2768 (1970).
3. Achyuthan, K., and Breazeale, M. A., Photoelastic constants of ammonium dihydrogen phosphate (ADP). *J. Opt. Soc. Am.* (USA) **51**, 914–915 (1961).
4. Achyuthan, K., Ultrasonic measurement of the photoelastic constants of ADP. *J. Acoust. Soc. Am.* (USA) **33**, 856–858 (1961).
5. Achyuthan, K., and Breazeale, M. A., Measurement of sound pressure amplitude in transparent solids. *J. Acoust. Soc. Am.* (USA) **33**, 1673 (1961).
6. Achyuthan, K., "Ultrasonic methods and photoelastic constants." Ph. D. Thesis, Michigan State University, East Lansing, Michigan (USA), 1962.
7. Adams, L. H., and Williamson, E. D., The relation between birefringence and stress in various types of glass. *J. Washington Acad.* **9**, 609–623 (1919).
8. Adhav, R. S., Linear electro-optic effects in tetragonal arsenates. *Nature* (GB) **220**, 1230 (1968).
9. Adhav, R. S., Some physical properties of single crystals of normal and deuterated KDA. II: Electro-optical & dielectric properties. *J. Appl. Phys.* (USA) **39**, 4095–4098 (1968).
10. Adhav, R. S., Linear electro-optic effects in tetragonal phosphates and arsenates. *J. Opt. Soc. Am.* (USA) **59**, 414–418 (1969).
11. Adhav, R. S., Electro-optic properties of RbDA and D-RbDA. *J. Phys. D* (*Appl. Phys.*) (GB), Ser. 2, **2**, 177–182 (1969).
12. Adhav, R. S., Piezoelectric effect in tetragonal crystals. *J. Appl. Phys.* (USA) **46**, 2808 (1975).
13. Adler, R., Interaction between light and sound. *IEEE Spectrum* (USA) **4**, 42–54 (1967).
14. Adrianova, I. I., Frequency characteristics of diffraction modulators of light with ferro-electric ceramic emitters of ultrasound. *Opt. & Spectrosc.* (USA) **12**, 48–51 (1962).
15. Adrianova, I. I., Photoelastic light modulator. *Opt. & Spectrosc.* (USA) **14**, 70–74 (1963).
16. Adrianova, I. I., Nesterova, Z. V., and Popov, Yu. V., Modulation method for phase detection in an electro-optical light modulator. *Sov. J. Opt. Technol.* (USA) **36**, 662–665 (1969).
17. Adrianova, I. I., Berezhnoi, A. A., Dubenskii, K. K., and Sokolov, V. A., Electro-optic effect in zinc selenide crystals. *Sov. Phys. Solid State* (USA) **12**, 1972–1973 (1971).

18. Adrianova, I. I., Volkonskii, V. B., and Kordev, Yu. G., Diffraction of polarized light by ultrasonic waves in KDP crystals. *Opt. & Spectrosc.* (USA) **30**, 601–603 (1971).
19. Adrianova, I. I., Berezhnoi, A. A., Nesterova, Z. V., and Rusetskaya, V. S., Microwave modulation of light based on the electro-optical effect in a zinc selenide crystal. *Sov. J. Quantum Electron.* (USA) **2**, 72–73 (1972).
20. Adrianova, I. I., Berezhnoi, A. A., Nefedova, E. V., Pessmennyi, V. A., Popov, Yu. V., and Skornyakova, K. P., Electro-optical effect in lead magnesium niobate crystals of pyrochlore structure. *Opt. & Spectrosc.* (USA) **36**, 547–548 (1974).
21. Afanasev, I. I., Adrianova, L. K., Mamontov, I. Ya., and Reiterov, V. M., Photoelastic properties and residual stresses in magnesium fluoride crystals. *Sov. Phys.-Solid State* (USA) **17**, 2006–2007 (1975).
22. Aggarwal, R. R., Diffraction of light by ultrasonic waves. *Proc. Indian Acad. Sci.* **A31**, 417–426 (1950).
23. Aggarwal, K. G., and Szigeki, B., Photoelastic constants and the Claussius–Mossotti field in the alkali halides. *J. Phys. C (Solid State Physics)* (GB) **3**, 1097–1111 (1970).
24. Ahmed, M. S., and Ley, J. M., Defects in the performance of electro-optic modulators using class $\bar{4}2m$ crystals. *J. Phys. D (Applied Physics)* (GB) **4**, 1637–1641 (1971).
25. Akulenok, E. M., Danileiko, Yu. K., Khaimov-Malkov, Ya. V., Determination of piezo-optic moduli (q_{11}, q_{12}, q_{13}, q_{31}, q_{33}) in corundum single crystals by an interferometric method. *Sov. Phys.-Crystallogr.* (USA) **15**, 717–720 (1970).
26. Aleksandrov, K., and Ryzheva, T., The elastic properties of crystals. *Sov. Phys.-Crystallogr.* (USA) **6**, 228–252 (1961).
27. Aleksandrov, V. I., Kitaeva, V. E., Kozlov, I. V., Osiko, V. V., Sobolev, N. N., Tatarintsev, V. M., and Chistyi, I. L., Mandelstamm–Brillouin scattering of light and elastic properties of zirconium dioxide single crystals stabilized by yttrium. *Sov. Phys.-Solid State* (USA) **16**, 1456–1459 (1975).
28. Aleksandrovskii, A. L., Maskaev, Y. A., and Naumova, I. I., Electro-optic diffraction of light by growth domains in barium–sodium niobate crystals. *Sov. Phys.-Solid State* (USA) **17**, 2108–2110 (1975).
29. Ali, L., Frequency change of light through the Doppler effect in diffraction due to ultrasonic waves. *Helv. Phys. Acta* (Switzerland) **8**, 502–505 (1935), in German.
30. Ali, L., Evidence of frequency variation of light due to the Doppler effect in the diffraction of light by supersonic waves in fluids. *Helv. Phys. Acta* (Switzerland) **9**, 63–83 (1936), in German.
31. Almairac, R., and Vergnoux, A. M., Principal birefringences of α-sulphur in the infrared range, effects of uniaxial pressure. *C.R. Hebd. Seances Acad. Sci. B* (France) **272**, 1442–1444 (1971), in French.
32. Alphonse, G. A., Broad-band acousto-optic deflectors using sonic gratings for first order beam steering. *R. C. A. Review* (USA) **33**, 543–594 (1972).
33. Alphonse, G. A., Broad band AO deflectors: New results. *Appl. Opt.* (USA) **14**, 201–207 (1975).
34. Anastassakis, E., Performing optical measurements in solids under variable uniaxial stress and temperature: A system. *Appl. Opt.* (USA) **13**, 1971–1974 (1974).
35. Anastassakis, E., Relation between piezo-optical and thermo-optical properties of crystals. *J. Opt. Soc. Am.* (USA) **64**, 944–951 (1974).

36. Anastassakis, E., and Burstein, E., The full symmetry of the photoelastic tensor of elastic waves. *J. Phys. C (Solid State Physics)* (GB) **7**, 1374–1380 (1974).
37. Anderson, L. K., Holographic optical memory for bulk data storage. *Bell. Lab. Rec.* (USA) **46**, 318–325 (1968).
38. Anderson, W. C., Final measurements of the velocity of light. *J. Opt. Soc. Am.* (USA) **31**, 187–197 (1941).
39. Angress, I. F., Cooke, C., and Haiden, A. J., Infrared electro-optical effects in diamond. *J. Phys. C (Solid State Physics)* (GB) Ser. 2, **1**, 1769–1772 (1968).
40. Anistratov, A. T., The electro-optical properties of ferroelectrics near the phase transitions. *Acta Crystallogr. A* (Denmark) **21**, Part. 7, Suppl. A, 191 (1966).
41. Anistratov, A. T., and Aleksandrov, K. S., Conditions for independent measurement of the linear and quadratic electro-optic effects. *Sov. Phys.-Crystallogr.* (USA) **11**, 232–234 (1966).
42. Anistratov, A. T., Anomalies in the linear electro-optical effect in Rochelle salt. *Sov. Phys.-Crystallogr.* (USA) **11**, 700–702 (1966).
43. Anistratov, A. T., and Aleksandrov, K. S., Electro-optical properties of Rochelle salt. *Sov. Phys.-Crystallogr.* (USA) **12**, 392–396 (1967).
44. Anistratov, A. T., and Melnikova, A. V., Optical and electro-optical properties of crystals belonging to the $Me^{+}NH_4RO_4 \cdot nH_2O$ group. *Bull. Acad. Sci. USSR, Phys. Ser.* (USA) **39**, 144–148 (1975).
45. Anistratov, A. T., Electro-optic and photo-elastic phenomena in ferroelectric crystals. *Izv. VUZ Fiz.* (USSR) **12**, 7–22 (1975), in Russian.
46. Appel, A. V., Betz, H. T., and Pontarelli, D. A., Infrared polariscope for photoelastic measurements of semiconductors. *Appl. Opt.* (USA) **4**, 1475–1478 (1965).
47. Ardenne, M. Von, A new large surface light relay for intensity, colour or plane of polarization control. *Telegraphen Fernsprech Tech.* (Berlin) **28**, 226–231 (1939), in German.
48. Armbruster, A., Thoma, R., and Wehrle, H., Measurement of the elastic constants of $LuAsO_4$ and $LuPO_4$ by Brillouin scattering and determination of the Debye temperatures. *Phys. Status Solidi A* (Germany) **24**, K71–K73 (1974).
49. Arumov, G. P., Voromin, E. S., Illinskii, Yu. A., and Solomatin, V. S., Bragg diffraction of $\lambda = 10.6\,\mu$ radiation by acoustic waves in proustite. *Sov. J. Quantum Electron.* (USA) **4**, 397–398 (1974).
50. Ascarelli, G., Dynamic piezo-optic study of exciton and polaron state in AgBr. *Phys. Rev.* (USA) **179**, 797–815 (1969).
51. Ascoli, A., and Dorsi, M., Growth of electro-optical crystals from aqueous solutions. *Energia Nucleare* (Italy) **17**, 545–550 (1970).
52. Aspnes, D. E., and Cardona, M., Electro-optic measurements of PbS, PbSe and PbTe. *Phys. Rev.* (USA) **173**, 714–728 (1968).
53. Baak, T., Thermal coefficient of refractive index of optical glasses. *J. Opt. Soc. Am.* (USA), **59**, 851 857 (1969).
54. Babits, V. A., and Kicks, H. H., Electrical colour filters. *Electronics* (USA) **23**, 112–115 (1950).
55. Bachem, C., Hiedemann, E. A., and Asbach, H. R., Ultrasonic waves in liquids. *Z. Phys.* (Germany) **87**, 734–737 (1934), in German.
56. Bachem, C., and Hiedemann, E. A., Interference of ultrasonic waves. *Z. Phys.* (Germany) **91**, 418 421 (1934), in German.
57. Bachem, C., Hiedemann, E. A., and Asbach, H. R., New methods for direct vi-

sualisation of ultrasonic waves and for the measurement of ultrasonic velocity. *Nature* (GB) **133**, 176 (1934).

58. Badikov, V. V., Bogdanov, S. V., Godovikov, A. A., Zubrinov, I. I., and Sheloput, D. V., Proustite single crystals as a material for acousto-optic devices. *Sov. Phys. Acoust.* (USA) **17**, 250–251 (1971).
59. Badox, J., Photoelectric measurements of weak birefringences and very small rotatory powers. *J. Phys. Rad.* (France) **17**, Suppl. No. 11, 143A–149A (1956), in French.
60. Bagaev, V. S., Belousova, T. Ya., Berozashvilli, Yu. N., and Lordkipanidze, D. Sh., Electro-optical effect in gallium arsenide and cadmium telluride. *Sov. Phys. Semicond.* (USA) **3**, 1418–1420 (1970).
61. Bagdavadze, V. N., Berozashvilli, Yu. N., Gogolins, O. V., Solomko, V. E., and Tsitsishvilli, E. G., Electro-optic effect in deformed gallium arsenide. *Sov. Phys. Semicond.* (USA) **10**, 246–247 (1976).
62. Baillou, J., and Lemercier, M., Variation of the electro-optical coefficient of cubic ZnSe with wavelength ($0.5\,\mu < \lambda < 0.6\,\mu$). *C. R. Hebd. Seances Acad. Sci. B* (France) **276**, 83–85 (1973), in French.
63. Baillou, J., Lemercier, M., Rivaud, G., and Pavis, B., Electro-optical effects in ZnSe crystals. *Rev. Phys. Appl.* (France) **8**, 121–126 (1973), in French.
64. Ballman, A. A., Electro-optic effects in strontium barium niobate, in: *133rd Meeting of the Electrochemical Society, Abstracts, Boston, Massachusetts* (USA), 5–9 May, 1968, Electrochemical Society, New York (1968), p. 17.
65. Banerjee, K., Theory of photoelasticity. *Indian J. Phys.* **2**, 195–242 (1927).
66. Bansigir, K. G., and Iyengar, K. S., Stress and wavelength dependence of birefringence in cubic crystals in the visible region. *Curr. Sci.* (India) **27**, 436–437 (1958).
67. Bansigir, K. G., and Iyengar, K. S., Stress birefringence relation in some alkali halides. *Proc. Phys. Soc. London* **71**, pt. 2, 225–230 (1958).
68. Bansigir, K. G., "Photoelastic dispersion in crystals." Ph. D. Thesis, Osmania University, India, 1958.
69. Bansigir, K. G., and Iyengar, K. S., Photoelastic dispersion in barite. *J. Opt. Soc. Am.* (USA) **50**, 183 (1960).
70. Bansigir, K. G., and Iyengar, K. S., Piezo-optic birefringence in NaCl structure crystals, Part I: Theory. *Acta Crystallogr. A* (Denmark) **14**, 670–674 (1961).
71. Bansigir, K. G., and Iyengar, K. S., Piezo-optic birefringence in NaCl structure crystals, Part II: Applications. *Acta Crystallogr. A* (Denmark) **14**, 727–732 (1961).
72. Bansigir, K. G., Strain polarizability constant—A measure of ion overlap in crystals of NaCl structure. *Nature* (GB) **216**, 256–257 (1967).
73. Bansigir, K. G., Effect of λ on the sign of the photoelastic constant $(p_{11} - p_{12})$ of crystals of sodium chloride structure. *Acta Crystallogr. A* (Denmark) **23**, 505–506 (1967).
74. Banza, J. M., Carles, C., Torguet, R., and Thomson, C., Theoretical and experimental study of light deviation by strong acousto-optic interaction—I: Theory. *Acustica* (Germany) **30**, 137–146 (1974), in French.
75. Banza, J. M., Carles, C., and Torguet, R., Theoretical and experimental study of the deviation of light by strong acousto-optic interaction—II: Experiences. *Acustica* (Germany) **31**, 161–167 (1974), in French.
76. Bär, R., Refraction of light by ultrasonic waves. *Helv. Phys. Acta* (Switzerland) **6**, 570–580 (1933), in German.

77. Bär, R., and Meyer, E., Diffraction of light by ultrasonic sound waves. *Phys. Zeits.* (Germany) **34**, 393–396 (1933), in German.
78. Barkovskii, L. M., Lavrukovich, V. I., and Bobrovich, V. P., Passage of light through lithium niobate, potassium dihydrogen phosphate and quartz plates in an electrical field. *Zh. Prikl. Spektrosk.* (USSR) **16**, 1073–1078, (1972), in Russian.
79. Barnes, J. M., and Hiedemann, E. A., Determination of elastic constants of optical glasses by an ultrasonic method. *J. Acoust. Soc. Am.* (USA) **28**, 1218–1221 (1956).
80. Barriol, J., Chapelle, J., and Taurel, L., Study of Rayleigh diffusion by anisotropic crystals. *J. Chim. Phys.* (France) **51**, 331–340 (1954), in French.
81. Barron, T. H. K., and Batana, A., Pressure dependence of the optical properties of alkali halide crystals. *Philos. Mag.* (GB) **20**, 619–628 (1969).
82. Bechmann, R., Piezo-optic and electro-optic constants of zinc blende. *J. Phys. Chem. Solids* (GB) **16**, 100–101 (1960).
83. Bechmann, R., Piezo-optic and electro-optic constants of crystals, in: *Landolt-Börnstein Tables, New Series*, K. H. Hellwege, ed., Group III, Vol. 1, Springer-Verlag, New York (1966), pp. 124–149.
84. Bechmann, R., First and second order piezo-optic and electro-optic constants of crystals, in: *Landolt-Börnstein Tables, New Series*, K. H. Hellwege, ed., Group III, Vol. 2, Springer-Verlag, New York (1969), pp. 126–166.
85. Beck, M., and Granicher, H., Electro-optical investigations on single crystals of the KH_2PO_4 group. *Helv. Phys. Acta* (Switzerland) **23**, 522–524 (1950), in German.
86. Beckers, J. M., Variations in the birefringence of quartz. *Appl. Opt.* (USA), **6**, 1279 (1967).
87. Belikova, G. S., Belyaev, L. M., Golovei, M. P., Pisaverskii, Yu. V., Silverstrova, I. M., and Turskaya, T. I., Growth of rubidium and caesium biphthalate crystals and study of their optico-acoustic and non-linear optical properties. *Sov. Phys.-Crystallogr.* (USA) **19**, 351–355 (1974).
88. Belouet, C., Crystal defects and electro-optical properties of D-KDP. *Rev. Phys. Appl.* (France) **11**, 253–262 (1976).
89. Belyaev, L. M., Vlokh, O. G., Gilvarg, A. B., Dobhrzhanskii, G. F., Netesov, G. B., Shamburov, V. A., and Shuvalov, L. A., Linear electro-optical effect in crystals of hexamethylenetetramine (Urotropine) $C_6H_{12}N_4$. *Sov. Phys.-Crystallogr.* (USA) **8** 383–384 (1963).
90. Belyaev, L. M., Dobrzhanskii, G. F., Pisarevskii, Yu. V., Chernyshev, K. S., and Shaldin, Yu. V., Electro-optical properties of copper chloride and bromide crystals. *Sov. Phys.-Solid State* (USA) **6**, 2988 (1965).
91. Ben-Amar, A., and Wiener-Avnear, E., Raman studies of the electro-optic properties of CuX (X = Cl, Br, I) multicrystallite samples. *Appl. Phys. Lett.* (USA) **27**, 410–412 (1975).
92. Benckert, L., and Backstrom, C., Determination of the photoelastic constants of NaCl, NaF, KCl and KI by means of Brillouin scattering. *Phys. Rev. B* (USA) **8**, 5888–5893 (1973).
93. Benckert, L., and Backstrom, G., Elastic constants of KCl and NaCl from Brillouin scattering. *Phys. Scr.* (Sweden) **11**, 43–46 (1975).
94. Bendow, B., and Gianino, P. D., Photoelastic constants of infrared transmitting materials, in: *Laser Induced Damage in Optical Materials*. A. J. Glass and A. H. Guenther, eds., N. B. S. Special Publication No. 387, U. S. Govt. Printing Office, Washington, D. C. (1973).

95. Bendow, B., Optical properties of infrared transmitting materials. Paper presented at the Conference on Preparation and Properties of Electronic Materials, 26–29 August 1973, Las Vegas, Nevada (USA), 35 pp.

96. Bendow, B., Gianino, P. D., Tsay, Y. F., and Mitra, S. S., Pressure and stress dependence of the refractive index of transparent crystals. *Appl. Opt.* (USA) **13**, 2382–2396 (1974).

97. Bendow, B., and Gianino, P. D., Calculations of the frequency dependence of elasto-optic constants of infrared laser window materials. Air Force Cambridge Research Laboratories Physical Science Research papers, No. 608, Hanscom (Massachusetts) (USA), 1974, 111 pp.

98. Benedek, G. B., and Fritsch, K., Brillouin scattering in cubic crystals. *Phys. Rev.* (USA) **149**, 647–662 (1966).

99. Benes, F., The elasto-optic effect at elastic resonance of quartz crystals. *Czech. J. Phys.* (*Czechoslovakia*) **B19**, 1033–1043 (1969), in German.

100. Berezhnoi, A. A., Longitudinal electro-optic effect in lead magnoniobate crystals. *Sov. Phys.-Solid State* (USA) **14**, 2230–2233 (1973).

101. Berezhnoi, A. A., and Mamedov, A. M., Electro-optical effect in ferroelectric semiconductor SbSi. *Sov. Phys.-Solid State* (USA) **15**, 1023–1024 (1973).

102. Berezina, Ye. Ye., Photoelastic constants of optical glasses. *Sov. J. Opt. Tech.* (USA) **37**, 110–111 (1970).

103. Bergman, J. G., Turner, E. H., Bjorkholm, J. E., and Kurtz, S. K., Nonlinear optical iodates. *Am. Ceram. Soc. Bull.* (USA) **47**, 389 (1968), 70th Annual Meeting of the American Ceramic Society, Chicago, 20–25 April 1968.

104. Bergman, J. G., Ashkin, A., Boyd, G. D., Nash, F. R., and Kurtz, S. K., Nonlinear optical materials—metal oxides with non-bonded electrons. *Acta Crystallogr. A* (Denmark) **25**, S58 (1969).

105. Bergman, J. G., McFee, J. H., and Crane, G. R., Nonlinear optical properties of $CdHg(SCN)_4$ and $ZnHg(SCN)_4$. *Mater. Res. Bull.* (USA) **5**, 913–917 (1970).

106. Bergman, J. G., and Kurtz, S. K., Nonlinear optical materials. *Mater. Sci. Engineering* (Netherlands) **5**, 235–250 (1970).

107. Bergman, J. G., Crane, J. G., and Guggenheim, H., Linear and non-linear optical properties of ferroelectric $BaMgF_4$ and $BaZnF_4$. *J. Appl. Phys.* (USA), **46**, 4645–4646 (1975).

108. Bergmann, L., and Fues, E., Determination of elasto-optical constants from diffraction experiments. *Naturwissenschaften* (Germany) **24**, 492 (1936) in German; also see *Ann. Phys.* (Germany) **v-21**, 553–563 (1934).

109. Bergmann, L., Elastic constants of cubic crystals by ultrasonic elastogram method. *Z. Naturforsch.* (Germany) **12a**, 229–233 (1957) in German.

110. Bergmann, L., *Der Ultraschall*, 6th ed., S. Hirzel Verlag, Stuttgart and Zurich (1964), in German.

111. Berkowicz, R., and Skettrup, T., Brillouin scattering, piezobirefringence, and dispersion of photoelastic coefficients of CdS and ZnO. *Phys. Rev. B* (USA) **11**, 2316–2326 (1975).

112. Berkowicz, R., and Kietis, B. P., Optical transmission modulation and light scattering by acoustoelectric domain in Cds. *Phys. Status Solidi A* (Germany) **28**, 425–432 (1975).

113. Berlincourt, D., Jaffe, H., and Shiozawa, L. R., Electroelastic properties of the

sulphides, selenides and tellurides of zinc and cadmium. *Phys. Rev.* (USA) **129**, 1009–1017 (1963).

114. Berlincourt, D. A., Curran, D. R., and Jaffe, H., Piezoelectric and piezomagnetic materials and their function in transducers, in: *Physical Acoustics*, W. P. Mason, ed., Vol. I, Part A, Academic Press, New York and London (1964), pp. 170–270.
115. Bernal, E., Chen, G. D., and Lee, T. C., Low frequency electro-optic and dielectric constants of lithium niobate. *Phys. Lett.* (Netherlands) **21**, 259–260 (1966).
116. Berry, M. V., *The Diffraction of Light by Sound*, Academic Press, New York and London (1966).
117. Bhagavantam, S., Photoelastic effect in crystals. *Proc. Indian Acad. Sci.* **A16**, 359–365 (1942).
118. Bhagavantam, S., and Bhimasenachar, J., Elastic constants of crystals—A new method and its application to pyrites and galena. *Proc. Indian Acad. Sci.* **A20**, 298–303 (1944).
119. Bhagavantam, S., and Suryanarayana, D., Photoelastic effect in crystals—A new result for the T_h class. *Proc. Indian Acad. Sci.* **A26**, 97–109 (1947).
120. Bhagavantam, S., and Ramachandra Rao, B., Diffraction of light by very high frequency ultrasonic waves—Effect of tilting the wave front. *Proc. Indian Acad. Sci.* **A28**, 54–62 (1948).
121. Bhagavantam, S., and Suryanarayana, D., Photoelastic effect in barium nitrate. *Nature* (GB) **162**, 740 (1948).
122. Bhagavantam, S., and Suryanarayana, D., Crystal symmetries and physical properties—Application of group theory. *Acta Crystallogr.* (Denmark) **2**, 21–26 (1949).
123. Bhagavantam, S., and Suryanarayana, D., Photoelastic behaviour of ammonium alum crystals. *Acta Crystallogr.* (Denmark) **2**, 26–30 (1949).
124. Bhagavantam, S., Photoelastic properties of crystals: Some new results. *Acta Crystallogr.* (Denmark) **5**, 591–593 (1952).
125. Bhagavantam, S., Photoelastic behaviour of cubic crystals: New methods of distinguishing between different classes. *Proc. Indian Acad. Sci.* **A37**, 585–588 (1953).
126. Bhagavantam, S., and Krishna Rao, K. V., Photoelastic behaviour of cubic crystals: Displacement of principal axes caused by linear stresses. *Proc. Indian Acad. Sci.* **A37**, 589–591 (1953).
127. Bhagavantam, S., and Krishna Rao, K. V., Photoelastic effect in lead nitrate. *Nature* (GB) **172**, 415 (1953).
128. Bhagavantam, S., and Krishna Rao, K. V., Photoelastic behaviour of barium nitrate and lead nitrate crystals. *Acta Crystallogr.* (Denmark) **6**, 799–801 (1953).
129. Bhagavantam, S., and Krishna Rao, K. V., Photoelastic behaviour of some cubic crystals. *Curr. Sci.* (India) **23**, 257–258 (1954).
130. Bhagavantam, S., Elastic properties of single crystals and polycrystalline aggregates. *Proc. Indian Acad. Sci.* **A41**, 72–90 (1955).
131. Bhagavantam, S., and Krishnamurty, Y., Temperature dependence of photoelastic constants in plastics. *Proc. Indian Acad. Sci.* **A43**, 203–209 (1956).
132. Bhagavantam, S., and Krishnamurty, Y., Temperature dependence of photoelastic constants in alkali halides. *Proc. Indian Acad. Sci.* **A46**, 399–405 (1957).
133. Bhagavantam, S., Group theory and crystal properties, in: *Advanced Methods of Crystallography*, G. N. Ramachandran, ed., Academic Press, New York and London (1964), pp. 97–119.

134. Bhagavantam, S., *Crystal Symmetry and Physical Properties*, Academic Press, New York and London (1966).

135. Bhagavantam, S., and Venkatarayudu, T., *Theory of Groups and Its Application to Physical Problems*, Academic Press, New York and London (1969).

136. Bhatia, A. B., and Krishnan, K. S., Light scattering in homogeneous media regarded as reflection from appropriate thermal elastic waves. *Proc. R. Soc. A* (GB) **192A**, 181–195 (1948).

137. Bhatia, A. B., and Noble, W. J., Diffraction of light by ultrasonic waves: I—General theory, and II—Approximate expressions for the intensities and comparison with experiment. *Proc. R. Soc. A* (GB) **220A**, 356–385 (1953).

138. Biegelsen, D. K., An ultrasonic technique for measuring the absolute signs of photoelastic coefficients and its application to fused silica and cadmium molybdate. *Appl. Phys. Lett.* (USA) **22**, 221–223 (1973).

139. Biegelsen, D. K., Photoelastic tensor of silicon and the volume dependence of the average gap. *Phys. Rev. Lett.* (USA) **32**, 1196–1199 (1974).

140. Biegelsen, D. K., Frequency dependence of the photoelastic coefficients of silicon. *Phys. Rev. B* (USA) **12**, 2427–2431 (1975).

141. Biegelsen, D. K., Chen, Tu, and Zesch, J. C., Acousto-optic parameters of $Bi_2(MoO_4)_3$. *J. Appl. Phys.* (USA) **46**, 941–942 (1975).

142. Biegelsen, D. K., Zesch, J. C., and Schwab, C., Photoelasticity of the cuprous halides. *Phys. Rev. B* (USA) **14**, 3578–3582 (1976).

143. Biegelsen, D. K., and Zesch, J. C., Optical frequency dependence of the photoelastic coefficients of fused silica. *J. Appl. Phys.* (USA) **47**, 4024–4025 (1976).

144. Billings, B. H., A tunable narrow-band optical filter. *J. Opt. Soc. Am.* (USA) **37**, 738–746 (1947).

145. Billings, B. H., The electro-optic effect in uniaxial crystals of the type XH_2PO_4: I—Theoretical. *J. Opt. Soc. Am.* (USA) **39**, 797–801 (1949).

146. Billings, B. H., The electro-optic effect in uniaxial crystals of the type XH_2PO_4: II—Experimental. *J. Opt. Soc. Am.* (USA) **39**, 802–808 (1949).

147. Billings, B. H., Sage, S., and Draisin, W., A narrow passband polarization interference filter for hydrogen alpha. *Rev. Sci. Instrum.* (USA) **22**, 1009–1017 (1951).

148. Billings, B. H., A monochromatic depolarizer. *J. Opt. Soc. Am.* (USA) **41**, 966–975 (1951).

149. Billings, B. H., The electro-optic effect in uniaxial crystals of the dihydrogen phosphate (XH_2PO_4) type IV; angular field of the electro-optic shutter. *J. Opt. Soc. Am.* (USA) **42**, 12–20 (1952).

150. Billings, B. H., The electro-optic effect in crystals and its possible application to distance measurement, in: *Optics in Metrology*, P. Mollet, ed., Pergamon Press, New York (1960), pp. 119–134.

151. Biot, M., Sur une nouvelle propriété physique quacquierent les lames de verre quando elles exécutent des vibrations longitudinales. *Ann. de Chim. et de Phys.* (France) **II-13**, 151–155 (1820), in French.

152. Bird, M. J., Jackson, D. A., and Powles, J. G., Brillouin and Rayleigh scattering in the plastic crystalline phase of pivalic acid (phonon life-times, elastic constants and Pockels coefficients). *Mol. Phys.* (GB) **25**, 1051–1060 (1973).

153. Birnbaum, G., Cory, E., and Gow, K., Interferometric null method for measuring stress induced birefringence. *Appl. Opt.* (USA) **13**, 1660–1669 (1974).

154. Bjerhammar, A., Electro-optical distance measuring. *K. Tekn. Hogsk. Handl.* (Sweden) **151**, 101 pp. (1960); see also *Pure Appl. Math. Phys.* **4**, 3–99 (1960).
155. Bjerhammar, A., An electro-optical device for measuring distance. *J. Geophys. Res.* (USA) **65**, 412–417 (1960).
156. Blitz, J., *Fundamentals of Ultrasonics*, 2nd ed., Butterworths, London (1967).
157. Blocker, A., Visible sound waves. *Nature* (GB) **146**, 369–373 (1940).
158. Blocker, A., Visible sound waves. *J. Acoust. Soc. Am.* (USA) **12**, 455 (1940).
159. Bloombergen, N., *Nonlinear Optics*, W. A. Benjamin, New York (1965).
160. Blumenthal, R. H., Design of a microwave-frequency light modulator. *Proc. IEEE* (USA) **50**, 452–456 (1962).
161. Boccara, N., and Zarembontich, A., Study of the polarization of elastic waves with a velocity extremum propagating in crystals. *C. R. Hebd. Seances Acad. Sci. B* (France) **257**, 4167–4169 (1963), in French.
162. Bohaty, L., Haussühl, S., and Nohl, G., Linear electro-optic tensor of monoclinic lithium sulfate–monohydrate. *Z. Kristallogr.* (Germany) **139**, 33–38 (1974), in German.
163. Bonch-Bruevich, A. M., Passage of polarized light through a medium with stationary ultrasonic waves. *Sov. Phys.-Tech. Phys.* (USA) **1**, 428–430 (1956).
164. Bond, W. L., and Andreus, J., Photographs of the stress field around edge dislocations (in silicon). *Phys. Rev.* (USA) **101**, 1211 (1956).
165. Bond, W. L., Measurement of the refractive indices of several crystals. *J. Appl. Phys.* (USA) **36**, 1674–1677 (1965).
166. Borcherds, P. H., The scattering of light by elastic waves in cubic crystals. *Opt. Acta* (GB) **20**, 147–159 (1972).
167. Bonner, W. A., Dearborn, E. F., Geusic, J. E., Marcos, H. M., and Van Uitert, L. G., Dielectric and electro-optical properties of lead magnesium niobate. *Appl. Phys. Lett.* (USA) **10**, 163–165 (1967).
168. Bonner, W. A., Gensic, J. E., Levinstein, H. J., Singh, S., and Van Uitert, L. G., A new and stable nonlinear optical material $K_3Li_2Nb_5O_{15}$. *Appl. Phys. Lett.* (USA) **11**, 161 (1967).
169. Bonnei, W. A., and Zydrik, G. J., Growth of single crystal lead molybdate for acousto-optic applications. *J. Cryst. Growth* (Netherlands) **7**, 65–68 (1970).
170. Bonner, W. A., Singh, S., Van Uitert, L. G., and Warner, A. W., High quality tellurium dioxide for acousto-optic and non-linear application. *J. Electron. Mater.* (USA) **1**, 115–165 (1972).
171. Borgnis, F., Special directions of longitudinal wave propagation in anisotropic media. *Phys. Rev.* (USA) **98**, 1000–1005 (1955).
172. Born, M., and Bollnow, O. F., Der Aufbau der festen Mateıie. Theoretische Grundlagen, in: *Handbuch der Physik*, H. Geiger and K. Scheel, eds., Julius Springer Verlag, Berlin (1927), Vol. 24, pp. 370–465, in German.
173. Born, M., and Huang, K., The Brillouin components of the thermal scattering of light, in: *Dynamical Theory of Crystal Lattices*, Oxford University Press, London (1954), pp. 373–381.
174. Borrelli, N. F., Electro-optic effect of ferroelectric microcrystals in a glass matrix. *Appl. Phys. Lett.* (USA) **7**, 117–118 (1965).
175. Borrelli, N. F., Electro-optic effect in transparent niobate glass ceramic systems. *J. Appl. Phys.* (USA) **38**, 4243–4247 (1967).
176. Borrelli, N. F., and Miller, R. A., Determination of the individual strain optical coefficients in glass by an ultrasonic technique. *Appl. Opt.* (USA) **7**, 745–750 (1968).

177. Borrelli, N. F., and Layton, M. M., Electro-optic properties of transparent ferroelectric glass ceramic systems. *IEEE Trans. Electron Devices* (USA) **ED-16**, 511–514 (1969).
178. Borrelli, N. F., Electric field induced birefringence in glasses. *Phys. & Chem. Glasses* (GB) **12**, 93–96 (1971).
179. Bortfeld, D. P., and Meier, H., Refractive indices and electro-optic coefficients of the eulitites $Bi_4Ge_3O_{12}$ and $Bi_4Si_3O_{12}$. *J. Appl. Phys.* (USA) **43**, 5110–5111 (1972).
180. Boyd, G. D., Miller, R. C., Nassau, K., Bond, W. L., and Savage, A., $LiNbO_3$: An efficient phase matchable non-linear optical material. *Appl. Phys. Lett.* (USA) **5**, 234–236 (1964).
181. Boyd, G. D. and Kleinman, D. A., Parametric interaction of focused Gaussian light beams. *J. Appl. Phys.* (USA) **39**, 3597–3639 (1968).
182. Boyd, G. D., Buehler, E., and Storz, F. G., Linear and non-linear optical properties of $ZnGeP_2$ and CdSe. *Appl. Phys. Lett.* (USA) **18**, 301–304 (1971).
183. Bradley, C. J., and Cracknell, A. P., *The Mathematical Theory of Symmetry in Solids*, Clarendon Press, Oxford (London) (1972).
184. Bragg, W. L., The refractive indices of calcite and aragonite. *Proc. R. Soc. A* (GB) **A105**, 370–386 (1924).
185. Bragg, W. L., The influence of atomic arrangement on refractive index. *Proc. R. Soc. A* (GB) **A106**, 346–368 (1924).
186. Brauer, K. H., Feuerstake, J., Frohlich, F., and Mohr, U., Infrared photoelastic investigations on semiconductor crystals. *Krist. & Tech.* (Germany) **8**, 253–269 (1973), in German.
187. Bravais, A., D'un nouveau polariscope, et recherches sur les doubles refractions peu énergiques. *Ann. Chim. Phys.* (France) III, **43**, 129–149 (1855), in French; see also *Ann. Phys.* (Germany) II, **96**, 395–414 (1855).
188. Braybon, J. E. H., A new method of measurement of the variation with wavelength of the refractive index and absolute stress optical coefficients of amorphous solids. *Proc. Phys. Soc. London*, **B63**, 446–451 (1950).
189. Brazhe, R. A., Grigorev, M. A., and Nayanov, V. I., Efficiency of light diffraction by pure elastic wave modes. *Sov. Phys.-Solid State* (USA) **17**, 562–566 (1975).
190. Breazeale, M. A., and Hiedemann, E. A., Simple way to observe optical diffraction patterns produced by shear waves. *J. Acoust. Soc. Am.* (USA) **27**, 1220–1221 (1955).
191. Breazeale, M. A., and Ganster, W. B., Detector for measurement of ultrasonic strain amplitudes in solids. *Rev. Sci. Instrum.* (USA) **37**, 1544–1548 (1966).
192. Brewster, D., On the effects of simple pressure in producing that species of crystallization which forms two oppositely polarized images and exhibits the complementary colours by polarized light. *Philos. Trans. A* (GB) **105**, 60–64 (1815).
193. Brewster, D., On the communication of the structure of doubly refracting crystals to glass, muriate of soda, fluor spar, and other substances, by mechanical compression and dilatation. *Philos. Trans. A* (GB) **106**, 156–178 (1816).
194. Brewster, D., On the effects of compression and dilatation in altering the polarising structure of double refracting crystals. *Trans. Roy. Soc. Edinburgh* **8**, 281–286 (1818); see also *Trans. Roy. Soc. Edinburgh* **8**, 309 (1818).
195. Brewster, D., On the laws which regulate the distribution of the polarizing force in plates, tubes, and cylinders of glass that have received the polarizing structure. *Trans. Roy. Soc. Edinburgh* **8**, 353–371 (1818).

196. Brezina, B., and Havrankova, M., Crystal growth of ferroelectric Langbeinities $(NH_4)_2Cd_2(SO_4)_3$ and $TlCd_2(SO_4)_3$. *J. Crystal Growth* (Netherlands) **21**, 77–81 (1974).
197. Bridgman, P. W., Compressibility of several artificial and natural glasses. *Am. J. Sci.* (USA) **10**, 359–367 (1925).
198. Bridgman, P. W., *The Physics of High Pressure*. G. Bell and Sons, London (1949).
199. Bridoux, E., Rouvaen, J. M., Moriamez, M., Torguet, R., and Hartemann, P., Optoacoustic testing of microsound devices. *J. Appl. Phys.* (USA) **45**, 5156–5159 (1974).
200. Brienza, M. J., and De Maria, A. J., Continuosly variable ultrasonic optical delay line. *Appl. Phys. Lett.* (USA) **9**, 312 (1966).
201. Brillouin, L., Diffusion of light in transparent homogeneous solids due to vibrations of thermal origin. *Ann. Phys.* (France), Ser. 9, **17**, 88–122 (1922), in French.
202. Brillouin, L., Diffraction of light by ultrasonics, in: *Actualities Scientifiques et Industrielles No. 59*, Hermann et Cie, Paris (1933), in French.
203. Brodovich, N. A., Some aspects of the diffraction of light by ultrasonic waves. *Opt. & Spectrosc.* (USA) **33**, 411–413 (1972).
204. Brody, E. M., and Cummins, H. Z., Elasto-optic anomaly in KDP. *Phys. Rev. Lett.* (USA) **23**, 1039–1041 (1969).
205. Brugger, K., Pure modes for elastic waves in crystals. *J. Appl. Phys.* (USA) **36**, 759–768 (1965).
206. Bruton, T. M., and White, E. A. D., Low frequency linear electro-optic effect in $PbTa_2O_6$. *J. Mater. Sci.* (GB) **7**, 1233–1234 (1972).
207. Buck, P., and Haussühl, S., Piezo-optical constants of alums. *Optik* (Germany) **24**, 146–151 (1966-67), in German.
208. Buerger, M. J., *Elementary Crystallography*, John Wiley and Sons, New York (1956).
209. Buhrer, C. F., Baird, D., and Conwell, E. M., Optical frequency shifting by electro-optic effect. *Appl. Phys. Lett.* (USA) **1**, 46–49 (1962).
210. Buhrer, C. F., and Ho, L., Electro-optic effect of gallium arsenide. *Appl. Opt.* (USA) **2**, 647–648 (1963).
211. Buhrer, C. F., Bloom, L. R., and Baird, D. H., Electro-optic light modulation with cubic crystals. *Appl. Opt.* (USA) **2**, 839–846 (1963).
212. Buhrer, C. F., and Ho, L., Electro-optic effect in $(NH_4)_2Cd_2(SO_4)_3$ and (NH_4) $Mn_2(SO_4)_3$. *Appl. Opt.* (USA) **3**, 314 (1964).
213. Buhrer, C. F., Ho, L., and Zucker, J., Electro-optic effect in optically active crystals. *Appl. Opt.* (USA) **3**, 517–521 (1964).
214. Buhrer, C. F., and Ho, L., The Pockels effect of hexamethylene tetramine. *Appl. Opt.* (USA) **3**, 1500 (1964).
215. Burstein, E., Smith, P. L., and Henvis, B., The photoelastic constants of cubic crystals. *Phys. Rev.* (USA) **73**, 1262 (1948).
216. Burstein, E., and Smith, P. L., Photoelastic properties of cubic crystals. *Phys. Rev.* (USA) **74**, 229–230 (1948).
217. Burstein, E., and Smith, P. L., The photoelastic properties of diamond. *Phys. Rev.* (USA) **74**, 1880–1881 (1948).
218. Burstein, E., and Smith, P. L., The photoelastic properties of ionic crystals. *Proc. Indian Acad. Sci.* **A28**, 377–387 (1948).
219. Burstein, E., and Davisson, J. W., Smith, P. L., and Dehnel, J. E., Projection

cathode ray tubes using electro-optic crystals. *J. Opt. Soc. Am.* (USA) **41**, 288–289 (1951).

220. Burstein, E., Ito, R., Pinczuk, A., and Shaud, M., Resonance-enhanced Brillouin scattering in crystals. *J. Acoust. Soc. Am.* (USA) **49**, 1013–1025 (1971).
221. Busch, M., Toledano, J. C., and Torres, J., Brillouin scattering in the orthorhombic and tetragonal phases of gadolinium molybdate (elastic constants). *Opt. Commun.* (Netherlands) **10**, 273–276 (1974).
222. Byer, R. L., Harris, S. E., Kuizenga, D. J., Young, J. F., and Feigelson, R. S., Non-linear optical properties of $Ba_2NaNb_5O_{15}$ in the tetragonal phase. *J. Appl. Phys.* (USA) **40**, 444–445 (1969).
223. Caddes, D. E., and Wilkinson, C. D. W., Large photoelastic anisotropy of sapphire. *IEEE J. Quantum Electronics* (USA) **QE-2**, 330–331 (1966).
224. Caddes, D. E., Quate, C. F., and Wilkinson, C. D. W., Conversion of light into sound by electrostrictive mixing in solids. *Appl. Phys. Lett.* (USA) **8**, 309 (1966).
225. Cady, W. G., *Piezoelectricity*, Dover Publications, New York (1964).
226. Callan, H., Crystal symmetry and macroscopic laws. *Am. J. Phys.* (USA) **36**, 735–748 (1968).
227. Carleton, H. R., and Soref, R. A., Modulation of 10.6 μ laser radiation by ultrasonic diffraction. *Appl. Phys. Lett.* (USA) **9**, 110–112 (1966).
228. Carpenter, R. O'B., Dynamic measurement of electro-optic coefficients in ADP crystals. *Phys. Rev.* (USA) **76**, 467 (1949).
229. Carpenter, R. O'B., The electro-optic effect in uniaxial crystals of the dihydrogen phosphate type-III: Measurement of coefficients. *J. Opt. Soc. Am.* (USA) **40**, 225–229 (1950).
230. Carpenter, R. O'B., "The electro-optic and photo-elastic effects in dihydrogen phosphate crystals." Ph. D. Thesis, Harvard University, Cambridge, Massachusetts (USA), 1951.
231. Carpenter, R. O'B., Electro-optic sound-on-film modulator. *J. Acoust. Soc. Am.* (USA) **25**, 1145–1148 (1953).
232. Carpenter, R. O'B., Photoelastic constants of ammonium dihydrogen phosphate and potassium dihydrogen phosphate. *J. Opt. Soc. Am.* (USA) **44**, 425 (1954).
233. Casasent, D., and Keicher, W., Dielectric and optical properties of electron-beam-addressed KD_2PO_4. *J. Opt. Soc. Am.* (USA) **64**, 1575–1581 (1974).
234. Casella, R. C., Optical properties of $SrTiO_3$ under applied stress and electric field. *Phys. Rev.* (USA) **154**, 743–749 (1967).
235. Caton, W. M., Electro-optic light modulation in a planar zinc-blende type crystal wave guide. *J. Appl. Phys.* (USA) **46**, 260–265 (1975).
236. Cechi, L., On the theory of the Brillouin effect in anisotropic media. *C. R. Hebd. Seances Acad. Sci. B* (France) **260**, 6852–6855 (1965), in French.
237. Cechi, L., Theoretical and experimental study of Brillouin scattering in anisotropic media. *J. Phys.* (France) **26**, 649–654 (1965), in French.
238. Chandrasekharan, V., Thermal scattering of light in crystals II: Diamond. *Proc. Indian Acad. Sci.* **A32**, 379–385 (1950).
239. Chandrasekharan, V., Thermal scattering of light in birefringent crystals. *Curr. Sci.* (India) **19**, 371–372 (1950).
240. Chandrasekharan, V., Thermal scattering of light in crystals III: Theory of birefringent crystals. *Proc. Indian Acad. Sci.* **A33**, 183–198 (1951).

241. Chandrasekharan, V., Thermal scattering of light in cubic crystals. *J. Indian Inst. Sci.* **34**, 269–286 (1952).
242. Chandrasekharan, V., Thermal scattering of light in birefringent crystals—Intensities of Brillouin components. *Proc. Natl. Inst. Sci. India* **19**, 547–561 (1953).
243. Chandrasekharan, V., Exact equation for Brillouin shifts. *J. Phys.* (France) **26**, 655–658 (1965).
244. Chang, I. C., Tunable acousto-optic filter utilizing acoustic beam walk off in crystal quartz. *Appl. Phys. Lett.* (USA) **25**, 323–324 (1974).
245. Chang, I. C., Noncollinear acousto-optic filter with large angular aperture. *Appl. Phys. Lett.* (USA) **25**, 370–372 (1974).
246. Chang, I. C., Acousto-optic devices and applications. *IEEE Trans. Sonics & Ultrason.* (USA) **SU23**, 2–22 (1976).
247. Chapelle, J., and Taurel, L., Theory of diffusion of light by strongly piezoelectric crystals. *C. R. Hebd. Seances Acad. Sci.* **240** 743–745 (1955), in French.
248. Chapelle, J., and Taurel, L., Study of Rayleigh diffusion by quartz. *J. Chim. Phys.* (France) **52**, 307–313 (1955), in French.
249. Chen, C. S., Szczesniak, J. P., and Corelli, J. C., Infrared stress birefringence in KBr, KCl, LiF and ZnSe (window materials). *J. Appl. Phys.* (USA) **46**, 303–309 (1975).
250. Chen, F. S., Guesic, J. E., Kurtz, S. K., Skinner, J. G., and Wemple, S. H., The use of perovskite paraelectrics in beam deflectors and light modulators. *Proc. IEEE* (USA) **52**, 1258–1259 (1964).
251. Chen, F. S., Guesic, J. E., Kurtz, S. K., Skinner, J. G., and Wemple, S. H., Light modulation and beam deflection with potassium tantalate–niobate crystals. *J. Appl. Phys.* (USA) **37**, 388–398 (1966).
252. Cheng, J. C., Nafie, L. A., and Stephens, P. J., Polarization scrambling using a photoelastic modulator: Application to circular dichroism measurement. *J. Opt. Soc. Am.* (USA) **65**, 1031–1035 (1975).
253. Cheng, J. C., Nafie, L. A., Allen, S. D., and Braunstein, A. I., Photoelastic modulator for the 0.55–13 μm range. *Appl. Opt.* (USA) **15**, 1960–1965 (1976).
254. Chia-Lun Hu, Linear electro-optic retardation schemes for the 20 classes of linear electro-optic crystals and their applications. *J. Appl. Phys.* (USA) **38**, 3275–3284 (1967).
255. Childress, J. D., and Hambrick, C. J., Interactions between elastic waves in an isotropic solid. *Phys. Rev.* (USA) **136**, A411–A418 (1964).
256. Christmas, T. M., and Wildey, C. G., Pulse measurement of r_{63} in KDP. *Electron. Lett.* (GB) **6**, 152–153 (1970).
257. Chubachi, N., Kikuchi, Y., and Wakatsuki, N., Light diffraction by ultrasonic waves in anisotropic crystals, in: *Proceedings of the 7th International Congress on Acoustics, Budapest, Hungary*, 18–26 August 1971, Akadamiai Kiado, Budapest (1971), pp. 245–248.
258. Clark, J. A., Whole field measurements of acousto-optic refraction. *J. Sound & Vib.* (GB) **46**, 306–310 (1976).
259. Clarke, R., and Ainger, F. W., The electro-optic properties of ferroelectric KSN crystals. *Ferroelectrics* (GB) **7**, 101–102 (1974).
260. Clayson, C. H., Low-voltage light-amplitude modulation. *Electron. Lett.* (GB) **2**, 138 (1966); reply by Ley, J. M., *ibid.* p. 139.
261. Cohen, M. G., and Gordon, E. I., Electro-optic $KTa_xNb_{1-x}O_3$ (KTN) gratings

for light beam modulation and deflection. *Appl. Phys. Lett.* (USA) **5**, 181–182 (1964).

262. Cohen, M. G., and Gordon, E. I., Acoustic beam probing using optical techniques. *Bell Syst. Tech. J.* (USA) **44**, 693–721 (1965).
263. Cohen, M. G., and Gordon, E. I., Acoustic scattering of light in a Fabry–Perot resonator. *Bell Syst. Tech. J.* (USA) **45**, 945–966 (1966).
264. Cohen, M. G., A simple highly efficient acousto-optic Q-switch. 1973 *IEEE/OSA Conference on Laser Engineering and Applications Digest of Technical Papers Washington, D. C.*, 30 May–1 June 1973, New York, IEEE, New York (1973), pp. 15 and 16.
265. Coker, E. G., The optical determination of stress. *Philos. Mag.* (GB) **VI-20**, 740–751 (1910).
266. Coker, E. G., and Filon, L. N. G., *A Treatise on Photoelasticity* (revised by H. T. Jessop), Cambridge University Press, Cambridge, England (1957).
267. Cole, W. F., and Lancucki, C. J., Shaping of soft cleavable crystals *Rev. Sci. Instrum.* (USA) **36**, 1669–1670 (1965).
268. Cole, T. W., Electro-optical processing in radio astronomy. *Opt. Acta* (GB) **22**, 81–92 (1975).
269. Conrad, M. A., "Ultraviolet piezobirefringence of diamond." Ph.D. Thesis, University of Michigan, Ann Arbor, Michigan (USA) (1959).
270. Cook, B. D., and Hiedemann, E. A., Diffraction of light by ultrasonic waves of various standing wave ratios. *J. Acoust. Soc. Am.* (USA) **33**, 945–948 (1961).
271. Coquin, G. A., Griffin, J. P., and Anderson, L. K., Wide band acoustooptic deflectors using acoustic beam steering. *IEEE Trans. Sonics & Ultrasonics* (USA) **SU-17**, 33–40 (1970).
272. Coquin, G. A., Acousto-optic interactions in piezoelectric crystals. *IEEE Trans. Sonics & Ultrasonics* (USA) **SU-17**, 66 (1970).
273. Coquin, G. A., Pinnow, D. A., and Warner, A. W., Physical properties of lead molybdate relevant to acousto-optic device applications. *J. Appl. Phys.* (USA) **42**, 2162–2168 (1971).
274. Cound, V. M., Davies, P. H., Hulme, K. F., and Robertson, D., The electro-optic coefficients of silver thiogallate ($AgGaS_2$). *J. Phys. C* (*Solid State Phys.*) (GB) **3**, L83–L84 (1970).
275. Couture, L., and Mathieu, J. P., Raman effect in cubic crystals. *Ann. Phys.* (France) **3**, 521–549 (1948), in French.
276. Cuchy, Z., and Landovsky, J., Electro-optic modulators. *Krist. & Tech.* (Germany) **6**, 563–572 (1971), in English.
277. Cummins, H. Z., and Knable, N., Single side band modulation of coherent light by Bragg reflection from acoustical waves. *Proc. IEEE* (USA) **51**, 1246 (1963).
278. Curie, J., and Curie, P., Development by pressure of polar electricity in hemihedral crystals with inclined faces. *Bull. Soc. min. de France*, **3**, 90–93 (1880), in French.
279. Cutchen, J. T., and Haertling, G. H., Material and electro-optic properties of the (Pb, La) (Hf, Ti)O_3 system. *J. Am. Ceram. Soc.* (USA) **56**, 225–226 (1973).
280. Damon, R. W., Maloney, W. T., and McMahon, D. H., Interaction of light with ultrasound: Phenomena and applications, in: *Physical Acoustics*, W. P. Mason and R. N. Thurston, eds., Academic Press, New York (1970), Vol. 7, pp. 273–366.
281. Davis, T. A., and Vedam, K., Photoelastic properties of sapphire (α-Al_2O_3). *J. Appl. Phys.* (USA) **38**, 4555–4556 (1967).

282. Davis, T. A., and Vedam, K., Pressure dependence of the refractive indices of the tetragonal crystals: ADP, KDP, $CaMoO_4$, $CaWO_4$, and rutile. *J. Opt. Soc. Am.* (USA) **58**, 1446–1451 (1968).

283. Davis, T. A., and Vedam, K., Variation of refractive indices of calcite with pressure to 7 kbar. *Phys. Status Solidi* (Germany) **26**, 285–290 (1968).

284. Deas, F., On the spectra formed by the passage of polarised light through doubly refracting crystals. *Trans. Roy. Soc. Edinburgh* **26**, 177–188 (1872).

285. Debye. P., Zur Theorie der spezifischen Warmen. *Ann. Phys.* (Germany) **IV-39**, 789–839 (1912), in German.

286. Debye, P., and Sears, F. W., On the scattering of light by supersonic waves. *Proc. Natl. Acad. Sci., Washington* (USA) **18**, 409–414 (1932).

287. Debye, P., Sack, H., and Coulon, F., Diffraction of light by ultrasonic waves. *C. R. Hebd. Seances Acad. Sci.* (France) **198**, 922–924 (1934), in French.

288. Delaplace, R., Malfait, G., and Jerome, D., High hydrostatic pressure equipment up to 40 kbar for low temperature measurements 500 K $> T >$ 1.4 K. *Rev. Phys. Appl.* (France) **11**, 327–335 (1976), in French.

289. Denning, R. M., Giardini, A. A., Poindexter, E., and Slawson, C. B., Piezobirefringence in diamond—Further results. *Am. Mineral.* (USA) **40**, 556–563 (1957).

290. Denton, R. T., Kinsel, T. S., and Chen, F. S., 224 Mc/s Optical pulse code modulator. *Proc. IEEE* (USA) **54**, 1472–1473 (1966).

291. Denton, R. T., Chen, F. S., and Ballman, A. A., Lithium tantalate light modulators. *J. Appl. Phys.* (USA) **38**, 1611–1617 (1967).

292. Deviot, B., Elasto-optic constants of ADP and $NiSO_4$, $6H_2O$. *J. Phys. Rad.* (France) **16**, 162 (1955), in French.

293. DiDomenico, M., Jr., and Anderson, L. K., Broadband electro-optic travelling-wave light modulators. *Bell Syst. Tech. J.* (USA) **43**, 2621–2678 (1963).

294. Dixon, R. W., and Cohen, M. G., A new technique for measuring magnitudes of photoelastic tensors and its application to lithium niobate. *Appl. Phys. Lett.* (USA) **8**, 205–207 (1966).

295. Dixon, R. W., and Chester, A. N., An acoustic light modulator for 10.6 microns. *Appl. Phys. Lett.* (USA) **9**, 190–192 (1966).

296. Dixon, R. W., Photoelastic properties of selected materials and their relevance for applications to acoustic light modulators and scanners. *J. Appl. Phys.* (USA) **38**, 5149–5153 (1967).

297. Dixon, R. W., Acoustic diffraction of light in anisotropic media, in: Proceedings of the Symposium on Modern Optics at the Polytechnic Institute of Brooklyn, New York, 22–24 March, 1967, pp. 265–286.

298. Dixon, R. W., Acoustic diffraction of light in anisotropic media. *IEEE J. Quantum Electron.* (USA) **QE-3**, 85–93 (1967).

299. Dixon, R. W., and Mathews, H., Diffraction of light by elastic waves in yttrium iron garnet (YIG), a ferroelectric material. *Appl. Phys. Lett.* (USA) **10**, 195–197 (1967).

300. Dixon, R. W., and Gordon, E. I., Acoustic light modulator using optical heterodyne mixing. *Bell Syst. Tech. J.* (USA) **46**, 367–389 (1967).

301. Dixon, R. W., Acousto-optic interactions and devices. *IEEE Trans. Electron Devices* (USA) **ED-17**, 229–235 (1970).

302. Dneprovskii, V. S., Parygin, V. N., Kisil, I. I., Sysoev, L. A., and Solomatin, V. S.,

Electro-optical effect in wurtzite, ZnS. *Sov. Phys.-Crystallogr.* (USA) **13**, 141–142 (1968).

303. Dosner, R. D., Turner, E. H., and Kaminow, I. P., Clamped electro-optic coefficients of KDP and quartz. *Appl. Opt.* (USA) **6**, 778 (1967).
304. Dosner, R. D., and Turner, E. H., Electro-optic coefficients in calcium pyroniobate. *Appl. Opt.* (USA) **7**, 171–173 (1968).
305. Dressler, R., and Chesnes, A., Sound on film-recording using electro-optic crystal techniques. *J. Soc. Motion Pict. Televis. Eng.* (USA) **60**, 205–215 (1953).
306. Drozhzhin, A. N., Kosovskii, L. A., and Mikhailova, L. N., Acousto-optic light deflector with 50 positions. *Sov. J. Quantum Electron.* (USA) **4**, 887–889 (1975).
307. Dubenskii, K. K., Kaplyanskii, A. A., and Lozovskaya, N. G., Photoelastic properties of cubic ZnSe in the vicinity of the intrinsic absorption edge of the crystals. *Sov. Phys.-Solid State* (USA) **8**, 1644–1648 (1967).
308. Dudmik, O. F., Kopylov, Yu. L., and Kravchenko, V. B., Effective modulation of light with barium–strontium niobate single crystals. *JETP Lett.* (USA) **18**, 239–240 (1973).
309. Duguay, M. A., and Hansen, J. W., An ultra-fast light gate. *Appl. Phys. Lett.* (USA) **15**, 192–194 (1969).
310. Dumitrica, A., and Popescu, N., Broadband modulators for optical communications. *Stud. Cercet. Fiz.* (Rumania) **27**, 399–417 (1975), in Rumanian.
311. Duran, J., and Pauttier-Carrier, S., Criticism and improvement of the sensitivity of the methods of measurement of elasto-optic constants by ultrasonics. *C. R. Hebd. Seances Acad. Sci.* (France) **267**, 1308–1311 (1968), in French.
312. Duran, J., Critical study and improvement of methods of measurement of elasto-optical constants by ultrasonics. *Rev. Phys. Appl.* (France) **4**, 57–62 (1969).
313. Durelli, A. J., Stress distribution in a stressed specimen and use of spacers of lead, cardboard and paper, in: Proceedings of the 13th Photoelasticity Conference, Massachusetts Institute of Technology (USA) (1941), p. 25.
314. Eden, D. D., and Thiess, G. H., Measurement of the direct electro-optic effect in quartz at UHF. *Appl. Opt.* (USA) **2**, 868–869 (1963).
315. Eisner, J., Dielectric and Electro-optic properties of ADP crystals doped with chromate ions. *Phys. Lett. A* (Netherlands) **47A**, 375–376 (1974).
316. Eklund, H., Roos, A., and Eng, S. T., Rotation of laser beam polarisation in acousto-optic devices. *Opt. & Quantum Electron.* (GB) **7**, 73–79 (1975).
317. Emmensegger, F., Nitsche, R., and Miller, A., Crystal growth and electro-optic effect of some double sulfates with the Langbeinite structure. *J. Appl. Phys.* (USA) **39**, 3039–3043 (1968).
318. Engeler, W. E., Garfinkel, M., and Tiemann, J. J., Observation of phonon assisted indirect transitions by stress-modulated optical transmission. *Phys. Rev. Lett.* (USA) **16**, 239–243 (1966).
319. Ennos, A. E., High precision tunable retardation plate for use with visible and ultraviolet polarized light. *Opt. Acta* (GB) **10**, 105–114 (1963).
320. Ennos, A. E., and Opperman, K. W., Birefringence of natural quartz. *Appl. Opt.* (USA) **5**, 170 (1966).
321. Eppendahl, B. R., On the piezooptical properties of topaz. *Ann. Phys.* (Germany) **IV-61**, 595–632 (1920), in German.
322. Erdos, P., The determination of the components of a tensor characterizing a crystal. *Helv. Phys. Acta* (Switzerland) **37**, 493–504 (1964), in English.

323. Ermolieff, A., Brillouin scattering in ice and deuterated ice as a function of temperature. *Solid State Commun.* (USA) **17**, 1013–1016 (1975).
324. Esayan, S. Kh., Lemanov, V. V., Rez, I. S., and Shakin, O. V., Photoelastic properties of proustite. *Sov. Phys.-Solid State* (USA) **15**, 627–628 (1973).
325. Esayan, S. Kh., Laikthman, B. D., and Lemanov, V. V., Elastic and photoelastic properties of gadolinium molybdate crystals near the phase transition. *Sov. Phys.-JETP* (USA) **41**, 342–345 (1975).
326. Eschler, H., and Weidinger, F., Acousto-optic properties of dense flint glasses. *J. Appl. Phys.* (USA) **46**, 65–70 (1975).
327. Ethiraj, R., and Bansigir, K. G., Piezo-optic birefringence in NaCl structure crystals—Part III. *Acta Crystallogr. A* (Denmark) **A29**, Part 2, 157–160 (1973).
328. Ethiraj, R., Krishna Murty, V. G., and Bansigir, K. G., Piezo-optic birefringence in CsCl-structure crystals. *Acta Crystallogr. A* (Denmark) **A29**, Part 6, 636–639 (1973).
329. Ethiraj, R., and Bansigir, K. G., Piezo-optic birefringence in calcite. *J. Jiwaji University*, Gwalior (India) **1**, 76–79 (1973).
330. Ethiraj, R., "Studies on piezo-optic birefringence in some cubic crystals," Ph.D. Thesis, Osmania University (India), 1975.
331. Evans, W., The birefringent filter. *J. Opt. Soc. Am.* (USA) **39**, 229–242 (1949).
332. Ewald, P. P., "Dispersion und Doppelbrechnung von Elektronen-gittern (Kristallen)", Ph.D. Dissertation, University of Munich, Munich (Germany), 1912, in German.
333. Ewald, P. P., Zur Begrundung der Kristalloptik Teil I: Theorie der Dispersion; Teil II: Theorie der Reflection und Brechung. *Ann. Phys.* (Germany) **49**, 1–38 (1916); **49**, 117–143 (1916), in German.
334. Ewell, A. W., Artificial rotatory polarization. *Phys. Rev.* (USA) **33**, 480–491 (1911).
335. Ewell, A. W., Rotationspolarisation durch Torsion. *Phys. Zeits.* (Germany) **13**, 100–104 (1912), in German.
336. Extermann, R., and Wannier, G., Theory of diffraction of light by ultrasonic waves, I. *Helv. Phys. Acta* (Switzerland) **9**, 520–532 (1936), in French.
337. Extermann, R., Theory of diffraction of light by ultrasonic waves, II. *Helv. Phys. Acta* (Switzerland) **10**, 185–217 (1937), in French.
338. Fajans, K., and Joos, G., Molrefraktion von Ionen und Molekulen im Lichte der Atomstruktur. *Z. Kristallogr.* (Germany) **23**, 1–46 (1924), in German.
339. Falicov, L. M., *Group Theory and Its Physical Applications*, Univ. of Chicago Press, Chicago (1966), 224 pp.
340. Farnell, G. W., Elastic waves in trigonal crystals. *Canad. J. Phys.* (Canada) **39**, 65–80 (1961); also see Technical Report No. 40, The Eaton Electronics Research Laboratory, Department of Physics, McGill University (July, 1960).
341. Fay, H., Electro-optic modulation of light propagation near the optic axis in $LiNbO_3$. *J. Opt. Soc. Am.* (USA) **59**, 1399–1404 (1969).
342. Federov, F. I., The theory of elastic waves in crystals—Comparison with isotropic medium. *Sov. Phys.-Crystallogr.* (USA) **8**, 159–163 (1963).
343. Federov, F. I., Quasilongitudinal waves—Approximate theory—Crystals. *Sov. Phys.-Crystallogr.* (USA) **8**, 312–315 (1963).
344. Federov, F. I., A general approximate theory of elastic waves in crystals. *Sov. Phys.-Dokl.* (USA) **9**, 254–256 (1964).
345. Federov, F. I., and Barkovskii, L. M., On the induced optical anisotropy of transparent uniaxial crystals. *Opt. & Spectrosc.* (USA) **18**, 589–591 (1965).

346. Federov, F. I., and Barkovskii, L. M., Phenomenological theory of the linear electro-optical effect in uniaxial crystals. *Zh. Prikl. Spektrosk.* (USSR) **3**, 83–91 (1965), in Russian.

347. Federov, F. I., and Barkovskii, L. M., Theory of the linear electro-optic effect in cubic crystals. *Zh. Prikl. Spektrosk.* (USSR) **5**, 371–380 (1966), in Russian.

348. Federov, F. I., Some general properties of the nonlinear polarizability tensor of crystals. *Sov. Phys.-Crystallogr.* (USA) **12**, 233–236 (1967).

349. Feldman, A., High stress-optical birefringence in pure and degenerate *n*-type germanium. *Phys. Rev.* (USA) **150**, 748–757 (1966).

350. Feldman, A., and Horowitz, D., Dispersion of the piezobirefringence of GaAs. *J. Appl. Phys.* (USA) **39**, 5597–5599 (1968).

351. Feldman, A., Horowitz, D., and Waxler, R. M., Relative importance of electrostriction and the Kerr effect to self-focusing in optical glasses. *Appl. Phys. Lett.* (USA) **21**, 260–262 (1972).

352. Feldman, A., Horowitz, D., and Waxler, R. M., Mechanism for self-focusing in optical glasses. *IEEE J. Quantum Electron.* (USA) **QE-9**, 1054–1061 (1973).

353. Feldman, A., Horowitz, D., and Waxler, R. M., Stress-optic measurements in the infrared, in: *Proceedings of the Third Conference on High Power Infrared Laser Window Materials, Hyannis, Mass.* (USA), Nov. 1973, Vol. 1, *Optical Properties*, Carl A. Pitha and Bernard Bendow, eds., Air Force Cambridge Research Laboratories, Bedford, Massachusetts (USA) (Feb. 1974), pp. 403–409.

354. Feldman, A., Malitson, I., Horowitz, D., Waxler, R. M., and Dodge, M. J., Characterization of infrared laser window materials at the National Bureau of Standards, in: Proceedings of the Symposium on Laser Induced Damage in Optical Materials, Boulder Colorado (USA), 22–23 May 1974, edited by A. J. Glass and A. H. Guenther, National Bureau of Standards Publication No. 414, pp. 141–148.

355. Feldman, A., Horowitz, D., Waxler, R. M., Malitson, I., and Dodge, M. J., Optical Materials Characterization, Semi-Annual Technical Reports: (i) NBSIR 74-525 (July 1974), pp. 1–15; (ii) NBSIR 75-639 (1975); (iii) NBSIR 75-781 (August 1975); NBSIR 76-1010 (February 1976), Washington, D.C. (USA).

356. Feldman, A., Waxler, R. M., and Horowitz, D., Measuring photoelastic and elastic constants of transparent materials by application of static stress, in: Proceedings of the Conference on Optical Properties of Highly Transparent Solids, Waterville, New Hampshire (USA), 4–6 February 1975, pp. 1–9; also available in *Optical Properties of Highly Transparent Solids*, S. S. Mitra and B. Bendow, eds., Plenum Publishing Corporation, New York (1975), pp. 517–525.

357. Feldman, A., and McKean, W. J., Improved stressing apparatus for photoelasticity measurements. *Rev. Sci. Instrum.* (USA) **46**, 1588–1589 (1975).

358. Feldman, A., Horowitz, D., and Waxler, R. M., Piezo-optical constants in the infrared, in: Proceedings of the Fifth Infrared Laser Window Materials Conference, Las Vegas, Nevada (USA), December 1975, pp. 944–951.

359. Feldman, A., Horowitz, D., and Waxler, R. M., Photoelastic constants of infrared materials, in: *Proceedings of the Laser Induced Damage in Optical Materials Meeting*, July 1975, A. J. Glass and A. H. Guenther, eds., Natl. Bur. Standards, Washington (USA), Special Publication No. 435 (April 1976), pp. 164–169.

360. Fesefeldt, H., Influence of temperature on absorption spectra of alkali halide crystals. *Z. Phys.* (Germany) **64**, 623–628 (1930), in German.

361. Fieschi, R., Matter tensors in the crystallographic groups of Cartesian symmetry. *Physica* (Netherlands) **23**, 972–976 (1957).
362. Filippini, J. C., Kerr cell light modulator with crossed fields. *Appl. Opt.* (USA) **14**, 3007–3011 (1975).
363. Filon, L. N. G., On the variation with the wavelength of the double refraction in strained glass. *Camb. Phil. Soc. Proc.* (GB) **11**, 478–492 (1902); **12**, 55–64; 313–336 (1903–1904).
364. Filon, L. N. G., Preliminary note on a new method of measuring directly the double refraction in strained glass. *Proc. R. Soc. A* (GB) **A79**, 440–442 (1907).
365. Filon, L. N. G., On the dispersion in artificial double refraction. *Philos. Trans. A* (GB) **A207**, 263–306 (1907).
366. Filon, L. N. G., Measurements of the absolute indices of refraction in strained glass. *Proc. R. Soc. A* (GB) **A-83**, 572–579 (1910).
367. Filon, L. N. G., On the temperature variation of the photoelastic effect in strained glass. *Proc. R. Soc. A* (GB) **A89**, 587–593 (1912).
368. Filon, L. N. G., and Jessop, H. T., On the stress-optical effect in transparent solids strained beyond the elastic limit. *Philos. Trans. A* (GB) **A223**, 89–125 (1922).
369. Filon, L. N. G., and Harris, F. C., On the photoelastic dispersion of vitreous silica. *Proc. R. Soc. A* (GB) **A-130**, 410–431 (1931).
370. Flannery, M., and Marburger, J., Theory of elasto-optic coefficients in polycrystalline materials. *Appl. Phys. Lett.* (USA) **28**, 600–601 (1976).
371. Fleischmann, R., and Schopper, H., A photometric precision method for measuring absolute light phases with the aid of a phase uniform field of view. *Z. Phys.* (Germany) **131**, 225–230 (1952), in German.
372. Flinchbaugh, D. E., Acousto-optic laser modulation today. *Electro-optic Syst. Des.* (USA) **6**, 24–26 (1974).
373. Flubacher, P., Leadbetter, A. J., Morrison, J. A., and Stoicheff, B. P., Brillouin effect in solids. *J. Phys. Chem. Solids* (GB) **12**, 53–65 (1959).
374. Flynn, J. T., Kaul, R., and Reeder, T. M., Optical heterodyne detection of microwave signal from acousto-optical interactions. *Proc. IEEE* (USA) **64**, 388–389 (1976).
375. Flytzanis, C., Electro-optic effect and effective changes in III–V compounds. *C. R. Hebd. Séances Acad. Sci. B* (France) **269**, 325–328 (1969), in French.
376. Flytzanis, C., Electro-optic coefficients in III–V compounds. *Phys. Rev. Lett.* (USA) **23**, 1336–1339 (1969).
377. Flytzanis, C., Electro-optic and Raman coefficients in cubic semiconductors. *Phys. Lett. A* (Netherlands) **34**, 99–100 (1971).
378. Folland, R., Jackson, D. A., and Rajagopal, S., Light scattering organic plastic crystal norbornylene. *Mol. Phys.* (GB) **30**, 1053–1061 (1975).
379. Fornges, D., Measurement of electro-optical coefficients of crystals in the micro wave region. *Arch. Elektron. & Uebertragungstech.* (Germany) **30**, 83–90 (1976), in German.
380. Forno, C., and Jones, O. C., Hexamine electro-optic light modulators. *J. Phys. E (Sci. Instrum.)* (GB) **7**, 101–104 (1974).
381. Forno, C., Growth of large crystals of hexamine from solution. *J. Cryst. Growth* (Netherlands) **21**, 61–64 (1974).
382. Fountain, W. D., Comments on transient elasto-optic effects and *Q*-switching

performance in lithium niobate and KDP Pockels cells. *Appl. Opt.* (USA) **10**, 972–973 (1971).

383. Fowler, V. J., Buhrer, C. F., and Bloom, L. R., Electro-optic light beam deflector. *Proc. IEEE* (USA) **52**, 193–194 (1964).
384. Fox, A. J., The longitudinal quadratic electro-optic effect in potassium tantalo-niobate (KTN), in: Abstracts of the IEEE symposium on the application of ferro-electrics, Yorktown Heights, N.Y. (USA) (1971), pp. 45–46.
385. Fox, A. J., Nonlinear longitudinal KTN modulator. *Appl. Opt.* (USA) **14**, 343–352 (1975).
386. Fox, A. J., Electro-optic materials and their applications. *Electron. Equip. News* (GB) **17**, 33–36 (1975).
387. Francois, G. E., and Siegman, A. E., Effect of Gaussian beam spread on phase velocity matching in C. W. optical second harmonic generation. *Phys. Rev.* (USA) **139**, A4–A9 (1965).
388. Franken, P. A., Hill, A. E., Peters, C. W., and Weinreich, G., Generation of optical harmonics. *Phys. Rev. Lett.* (USA) **7**, 118–119 (1961).
389. Franken, P. A., and Ward, J. F., Optical harmonics and non-linear phenomena. *Rev. Modern Phys.* (USA) **35**, 23–39 (1963).
390. Franklin, R., Holder for lapping water soluble crystals. *Rev. Sci. Instrum.* (USA) **36**, 1650 (1965).
391. Fresnel, A., Résumé d'un mémoire sur la réflexion de la lumière. *Ann. Chim. Phys.* (Germany) **II-15**, 379–386 (1820), in French; see also *Ann. der Phys.* (Germany) **II-30**, 255-261 (1836), in German.
392. Fresnel, A., Note sur la double réfraction du verre comprimé. *Ann. Chim. Phys.* (France) **II-20**, 376–383 (1822), in French.
393. Fresnel, A., Über die Doppelbrechung des zusammengedruckten Glasses. *Ann. der Phys.* (Germany) **II-19**, pp. 539-545 (1830), in German.
394. Froome, K. D., and Bradsell, R. H., Distance measurement by means of a light ray modulated at a microwave frequency. *J. Phys. E* (*Sci. Instrum.*) (GB) **38**, 458–462 (1961).
395. Froome, K. D., Distance measurement by means of modulated light. *Alta Freq.* (Italy) **41**, 753–758 (1972), in English.
396. Frova, A., Electro-optical and magneto-optical effects in solids. *Nuovo Cimento Supp.* (Italy) **6**, 726–744 (1968), in Italian.
397. Fujii, Y., and Sakudo, T., Interferometric determination of quadratic electro-optic coefficients in $SrTiO_3$ crystal. *J. Appl. Phys.* (USA) **41**, 4118–4120 (1970).
398. Fujii, Y., and Hayashi, H., Acousto-optic tunable filter with controllable passband. *J. Appl. Phys.* (USA) **46**, 5046–5048 (1975).
399. Fukuda, S., Shiosaki, T., and Kawabata, A., Electro-optic effect in thallium doped selenium crystal at 10.6 μm. *Jap. J. Appl. Phys.* (Japan) **12**, 1944–1945 (1973).
400. Fukuda, S., Shiosaki, T., and Kawabata, A., Acousto-optic properties of tellurium single crystal at 10.6 μm. *Jap. J. Appl. Phys.* (Japan) **15**, 927–928 (1976).
401. Fukumoto, A., and Watanabe, A., Liquid materials and their figures of merit as acousto-optical deflector. *Jap. J. Appl. Phys.* (Japan) **9**, 662–665 (1970).
402. Fukumoto, A., Kawabuchi, M., and Hayami, H., Polarization considerations in the operation of a two-dimensional TeO_2 abnormal Bragg deflector. *Appl. Opt.* (USA) **14**, 812–813 (1975).

403. Fumi, F. G., Third-order elastic coefficients of crystals. *Phys. Rev.* (USA) **83**, 1274 (1951).
404. Fumi, F. G., Third-order elastic coefficients in trigonal and hexagonal crystals. *Phys. Rev.* (USA) **86**, 561 (1952).
405. Fumi, F. G., Physical properties of crystals: The direct inspection method. *Acta Crystallogr.* (Denmark) **5**, 44–48 (1952).
406. Fumi, F. G., The direct-inspection method in systems with a principal axis of symmetry. *Acta Crystallogr.* (Denmark) **5**, 691–694 (1952).
407. Fumi, F. G., Matter tensors in symmetrical systems. *Nuovo Cimento* (Italy) **9**, 739–756 (1952), in English.
408. Furuhata, I., Guidelines for looking for single crystal materials for opto-electronics. *Solid State Phys.* (*Japan*) **6**, 36–39 (1971), in Japanese.
409. Gainon, D. J., Linear electro-optic effect in CdS. *J. Opt. Soc. Am.* (USA) **54**, 270–271 (1964).
410. Gainon, D. J., Jaffe, H., and Sliker, T., Optical observation of elastic resonances in $NH_4H_2PO_4$. *J. Appl. Phys.* (USA) **35**, 1166–1168 (1964).
411. Galt, J. K., Mechanical properties of NaCl, KBr, KCl. *Phys. Rev.* (USA) **73**, 1460–1462 (1948).
412. Gans, P., *Vibrating Molecules—An Introduction to the Interpretation of Infrared and Raman Spectra.* Chapman and Hall Ltd., London (1971).
413. Gates, H. F., and Hiedemann, E. A., Application of two ultrasonic methods to the determination of the ratio p/q of the photoelastic constants of optical glasses. *J. Acoust. Soc. Am.* (USA) **28**, 1222–1227 (1956).
414. Gerlach, H., A new method for measuring electro-optic constants at high frequencies. *Appl. Phys.* (Germany) **1**, 279–283 (1973).
415. Gavini, A., and Cordana, M., Piezobirefringence and deformation potentials of the alkali halides. *Phys. Rev.* (USA) **177**, 1351–1357 (1969).
416. Geusic, J. E., Kurtz, S. K., Van Uitert, L. G., and Wemple, S. H., Electro-optic properties of some ABO_3 perovskites in the paraelectric phase. *Appl. Phys. Lett.* (USA) **4**, 141–143 (1964).
417. Ghafelebhashi, M., Dandekar, D. P., and Ruoff, A. L., Pressure and temperature dependence of the elastic constants of RbCl, RbBr, and RbI. *J. Appl. Phys.* (USA) **41**, 652–666 (1970).
418. Giardini, A. A., "Piezobirefringence in strontium titanate." Doctoral thesis, University of Michigan, Ann Arbor, Michigan (USA), 1956.
419. Giardini, A. A., Stress optic study of strontium titanate. *J. Opt. Soc. Am.* (USA) **47**, 726–735 (1957).
420. Giardini, A. A., Piezobirefringence in silicon. *Am. Mineral* (USA) **43**, 249–262 (1958).
421. Giardini, A. A., and Poindexter, E., Stress-optical behaviour of synthetic periclase (MgO). *J. Opt. Soc. Am.* (USA) **48**, 556–557 (1958).
422. Gibson, A. F., Non-linear optics. *Science Progress* (GB) **56**, 479–497 (1968).
423. Giordmaine, J. A., Nonlinear optics. *Phys. Today* (USA) **22**, 32–44 (1969).
424. Gires, F., and Paillette, M., Electro-optical effect in some coloured glasses. *C. R. Hebd. Séances Acad. Sci. B* (France) **261**, 4690–4692 (1965), in French.
425. Gisin, B. V., Optimal orientation of crystals in utilizing the linear electro-optical effect for light modulation. *Sov. Phys.-Crystallogr.* (USA) **16**, 118–122 (1971).

426. Gisin, B. V., Orientation of electro-optical crystals for phase modulators. *Sov. Phys.-Crystallogr.* (USA) **16**, 547–548 (1971).
427. Glurdgydze, L. N., Remenyuk, A. D., and Shmartsev, Yu. V., Free carrier piezobirefringence in GaP. *Solid State Commun.* (USA) **15**, 143–144 (1974).
428. Glurdgydze, L. N., Remenyuk, A. D., Vul, A. Ya., and Golubev, L. V., Piezo-optic phenomena due to free carriers in gallium antimonide. *Sov. Phys.-Semicond.* (USA) **8**, 1114–1116 (1975).
429. Glushkov, V. F., and Sheloput, D. V., Photoelastic properties and chemical bond type in acoustico-optical materials. *Inorg. Mater.* (USA) **9**, 1422–1425 (1973).
430. Glushkov, V. F., and Sheloput, D. V., Refraction of light by ultrasound in an anisotropic medium. *Sov. Phys.-Acoust.* (USA) **21**, 73–74 (1975).
431. Gniadek, K., and Petykiewicz, J., Application of optical methods in diffraction theory of elastic waves, in: *Progress in Optics*, E. Wolf, ed., North-Holland, Amsterdam (1971), Vol. 9, pp. 283–310.
432. Gobeli, G. W., and Kane, E. O., Dependence of optical constants of silicon on uniaxial stress. *Phys. Rev. Lett.* (USA) **15**, 142–146 (1965).
433. Godovikov, A. A., Zukrinov, I. I., Semenov, V. I., and Sheloput, D. V., Elastic and photoelastic properties of proustite. *Sov. Phys.-Acoust.* (USA) **19**, 582–583 (1974).
434. Goldstein, L. F., Thompson, J. S., Schroeder, J. B., and Slattery, J. E., Stress optic coefficients of ZnSe. *Appl. Opt.* (USA) **14**, 2432–2434 (1975).
435. Goodman, L. E., and Sutherland, J. G., Elasto-plastic stress-optical effect in silver chloride single crystals. *J. Appl. Phys.* (USA) **24**, 577–584 (1953).
436. Gorbach, S. S., Deptsova, L. A., Dobrazhanskii, G. F., Markovskii, V. Yu., and Shaskolskaya, M. P., Piezo-optical constants of single crystals of LiF, NaF, NaCl, NaI, KBr, KI, RbI. *Sov. Phys.-Crystallogr.* (USA) **14**, 627–629 (1970).
437. Gordon, E. I., and Rigden, J. D., The Fabry–Perot electro-optic modulator. *Bell Syst. Tech. J.* (USA) **42**, 155–179 (1963).
438. Gordon, E. I., and Cohen, M. G., Electro-optic diffraction grating for light beam modulation and diffraction. *IEEE J. Quantum Electron.* (USA) **QE-1**, 191–198 (1965).
439. Gordon, E. I., Figures of merit for acousto-optical deflection and modulation devices. *IEEE J. Quantum Electron* (USA) **QE-2**, 104–105 (1966).
440. Gordon, E. I., A review of acousto-optical deflection and modulation devices. *Proc. IEEE* (USA) **54**, 1391–1401 (1966).
441. Gordon, E. I., and Cohen, M. G., Acoustic modulation devices. Presented at the 1966 Conference on Electron Device Research, California Institute of Technology, Pasadena, California.
442. Goruk, W. S., Normiandin, R., Vella, P. J., and Stegeman, G. I., Optical diffraction studies of sound waves in lithium niobate. *J. Phys. D.* (GB) **9**, 999–1007 (1976).
443. Gottlieb, M., Isaacs, T. J., and Feichtner, J. D., Opto-acoustic properties of thallium phosphorous selenide—$TlPSe_4$. *J. Opt. Soc. Am.* (USA) **63**, 1329–1330 (1973).
444. Gottlieb, M., Isaacs, T. J., Feichtner, J. D., and Rowland, G. W., Acousto-optic properties of some chalcogenide crystals. *J. Appl. Phys.* (USA) **45**, 5145–5151 (1974).
445. Graf, P., Modulators and deflectors. *Alta Freq.* (Italy) **41**, 726–743 (1972), in English.

446. Grandjean, C., and Desvignes, F., Coefficient of photoelasticity of germanium and of silicon. *C. R. Hebd. Séances Acad. Sci. B* (France) **250**, 1183–1185 (1960), in French.
447. Grant, K., High frequency interruption of light. *Nature* (GB) **120**, 586 (1927).
448. Grimsditch, M. H., and Ramdas, A. K., Brillouin scattering in diamond. *Phys. Rev. B* (USA) **11**, 3139–3148 (1975).
449. Grimsditch, M. H., and Holah, G. D., Brillouin scattering and elastic moduli of silver thiogallate ($AgGaS_2$). *Phys. Rev. B* (USA) **12**, 4377–4382 (1975).
450. Grimsditch, M. H., and Ramdas, A. K., Elastic and elasto-optic constants of rutile from a Brillouin scattering study. *Phys. Rev. B* (USA) **14**, 1670–1682 (1976).
451. Grodzinski, P., Photoelastic investigations on diamond crystals, in Summarized proceedings of a conference on stress analysis, London, 1950, in: *Brit. J. Appl. Phys.* **1**, 241–251 (1950).
452. Groshans, R. G., and Mayer, W. G., Frequency modulation of a laser beam by standing ultrasonic waves. *Acustica* (Germany) **18**, 361 (1967).
453. Gross, E., Über die Änderung der Wellenge bei Lichtzerstreuung in Kristallen. *Z. Phys.* (Germany) **63**, 685–687 (1930), in German.
454. Gross, E., Scattering of light by crystals. *Comptes Rendus (Doklady) de l'Acad. des Sciences* (USSR) **18**, 93–94 (1938), in English; see also **26**, 757 (1940).
455. Gross, E. F., Abolin'sh, Ya. Ya., and Shultin, A. A., Optico-acoustic effect in crystals. *Sov. Phys. Tech. Phys.* (USA) **3**, 2070–2072 (1958).
456. Grunstra, B. R., and Perkins, H. B., A method for the measurement of optical retardation angles near 90°. *Appl. Opt.* (USA) **5**, 585–588 (1966).
457. Gualtieri, J. G., and Aucoin, T. R., Laser performance of large Nd-pentaphosphate crystals. *Appl. Phys. Lett.* (USA) **28**, 189–192 (1976).
458. Gunter, P., Electro-optical properties of $KNbO_3$. *Opt. Commun.* (Netherlands) **11**, 285–290 (1974).
459. Gunter, R. S., Electro- and magneto-optical shutter for aerial cameras. *J. Opt. Soc. Am.* (USA) **38**, 419–420 (1948).
460. Günther, N., Investigation of the effect of mechanical and electrical forces on the double refraction of quartz. *Ann. Phys.* (Germany), Ser. 5, **13**, 783–801 (1932), in German.
461. Haas, W., Johannes, R., and Cholet, P., Light beam deflection using the Kerr effect in single crystal prisms of $BaTiO_3$. *Appl. Opt.* (USA) **3**, 988–989 (1964).
462. Haertling, G. H., and Land, C. E., Hot pressed (Pb, La)(Zr · Ti)O_3 ferroelectric ceramics for electro-optic applications. *J. Am. Ceram. Soc.* (USA) **54**, 1–11 (1971).
463. Haertling, G. H., and Land, C. E., Recent improvements in the optical and electro-optical properties of PLZT ceramics. *Ferroelectrics* (GB) **3**, 269–280 (1972).
464. Hagelberg, M. P., "The diffraction of linearly polarized light by ultrasonic waves in transparent solids," Ph.D. Thesis, Michigan State University, East Lansing, Michigan (USA), 1961.
465. Hammond, V. J., and Carter, R., Visualization of mode conversion of an ultrasonic beam in fused quartz. *Nature* (GB) **182**, 790 (1958).
466. Hance, H. V., and Parks, J. K., Wide band modulations of a laser beam, using Bragg-angle diffraction by amplitude modulated ultrasonic waves. *J. Acoust. Soc. Am.* (USA) **38**, 14–23 (1965).
467. Hance, H. V., Parks, J. K., and Tsai, C. S., Optical imaging of a complex ultrasonic field by diffraction of a laser beam. *J. Appl. Phys.* (USA) **38**, 1981–1983 (1967).

468. Hargrove, L. E., Optical effects of ultrasonic waves producing phase and amplitude modulation. *J. Acoust. Soc. Am.* (USA) **34**, 1547–1552 (1962).
469. Hargrove, L. E., and Achyuthan, K., Use of light diffraction in measuring the parameter of nonlinearity of liquids and the photoelastic constants of solids. *Physical Acoustics*, W. P. Mason, ed., Academic Press, New York (1965), Vol. 2, Part B, pp. 333–369.
470. Harris, F. C., The photoelastic constants of glass as affected by high temperatures and by lapse of time. *Proc. R. Soc.* (GB) **A106**, 718–723 (1924).
471. Harris, S. E., and Siegman, A. E., A technique for optical frequency translation utilizing the quadratic electro-optic effect in cubic crystals. *Appl. Opt.* (USA) **3** 1090–1098 (1964).
472. Harris, S. E., Ammann, E. O., and Chang, I. C., Optical network using birefringent crystals, I: Synthesis of lossless networks of equal length crystals. *J. Opt. Soc. Am.* (USA) **54**, 1267–1279 (1964).
473. Harris, S. E., and Wallace, R. W., Acousto-optic tunable filter. *J. Opt. Soc. Am.* (USA) **59**, 744–747 (1969).
474. Harris, S. E., Nieh, S. T. K., and Winslow, D. K., Electromechanically tunable acousto-optic filter. *Appl. Phys. Lett.* (USA) **15**, 325–326 (1969).
475. Harris, S. E., Nieh, S. T. K., and Feigelson, R. S., $CaMoO_4$ electronically tunable optical filter. *Appl. Phys. Lett.* (USA) **17**, 223–225 (1970).
476. Hart, S., The measurement of elastic constants of four alkali halides. *J. Phys. D* (*Applied Physics*) Ser. 2 (GB) **1**, 1277–1283 (1968).
477. Hartshorne, N. H., and Stuart, A., *Crystals and the Polarising Microscope*, 4th ed., Edward Arnold (Publishers) Ltd., London (1970).
478. Harthum, N., Laser modulation with Pockels cell. *Nachrichtentech. Z.* (*NTZ*) (Germany) **27**, 167 (1974), in German.
479. Hasseem, H. Md., and Subramanian, A., A simple compensator for photoelasticity. *J. Phys. E* (*Sci. Instrum.*) (GB) **2**, 830–831 (1969).
480. Hauret, G., and Taurel, L., Study of the Brillouin effect in ADP. *J. Phys.* (France) **31**, 657–664 (1970), in French.
481. Hauret, G., Chapelle, J. P., and Taurel, L., Brillouin scattering in sodium nitrite at room temperature. *Phys. Status Solidi A* (Germany) **11**, 225–261 (1972).
482. Hauret, G., Taurel, L., and Chapelle, J. P., Intensity of some Brillouin–Rayleigh scattering in RbDP (ferroelectric transition). *Solid State Commun.* (USA) **10**, 727–729 (1972).
483. Hauser, S. M., Smith, L. S., Marlowe, D. G., and Yoder, Jr. P. R., The stressed shutter, a new moderate speed electro-optical light modulator. *Appl. Opt.* (USA) **2**, 1175–1179 (1963).
484. Haussühl, S., Elastic constants of crystals from measurements of the natural frequencies, using Schaefer–Bergmann pattern. *Naturwissenschaften* (Germany) **43**, 394–395 (1956), in German.
485. Haussühl, S., Elasticity of alums. *Fortschr. Mineral.* (Germany) **36**, 75–77 (1958), in German.
486. Haussühl, S., and Hesse, H., Piezo-optic constants of calcium formate. *Z. Kristallogr.* (Germany) **123**, 338–341 (1966), in German.
487. Haussühl, S., Piezoelectric and electrical properties of $LiIO_3$. *Phys. Status Solidi* (Germany) **29**, K.159–162 (1968), in German.
488. Haussühl, S., and Hesse, H., Quadratic electro-optical effect (Kerr effect) in

alkali halides of the NaCl type. *Phys. Status Solidi* (Germany) **30**, 209–214 (1968), in German.

489. Haussühl, S., and Uhl, K., Determination of the piezo-optic tensor components of monoclinic crystal taurine. *Z. Kristallogr.* (Germany) **128**, 418–423 (1969), in German.

490. Haussühl, S., and Siegert, H., Determination of the elasticity tensor components of $CuSO_4$, $5H_2O$, a triclinic crystal. *Z. Kristallogr.* (Germany) **129**, 142–146 (1969), in German.

491. Haussühl, S., Isotope effects in the optical, electrical, elastic and thermal properties of the Schlippe salts Na_3SbS_4, $9H_2O$ and Na_3SbS_4, $9D_2O$. *Z. Kristallogr.* (Germany) **132**, 255–265 (1970), in German.

492. Haussühl, S., and Weber, H. J., Elasto-optic constants of α-sulphur and α-iodic acid. *Z. Kristallogr.* (Germany) **132**, 266–271 (1970), in German.

493. Haussühl, S., and Walda, G., Measurement of the absolute quadratic electro-optical effects in crystals, examples LiF and alpha $TlAl(SO_4)_2 \cdot 12H_2O$. *Phys. Status Solidi A* (Germany) **5**, K163–165 (1971), in German.

494. Haussühl, S., Piezoelectric, elastic and optical properties of $(AlIO_3)_3 \cdot 2HIO_3$. *Z. Kristallogr.* (Germany) **135**, 287–293 (1972), in German.

495. Haussühl, S., Crystals with large non-linear optical effects, in: Proceedings of the 2nd National Conference on Crystal Growth, Crystallography Institute, Freiburg in Breisgau, Germany, 1972, pp. 15 and 16, in German.

496. Haussühl, S., Elastic and thermoelastic properties of 1,3,5-triphenyl-benzene derived from ultrasonic wave velocities. *Acta Crystallogr. A* (Denmark) **A30**, 106–107 (1974).

497. Haussühl, S., and Madhava, M. S., Linear electro-optical constants of orthorhombic strontium formate, strontium formate dihydrate and barium formate single crystals. *Z. Kristallogr.* (Germany) **141**, 25–30 (1975), in English.

498. Havelock, T. H., The dispersion of double refraction in relation to crystal structure. *Proc. R. Soc.* (GB) **A80**, 28–44 (1908).

499. Havlice, J., Quate, C. F., and Richardson, B., Visualization of sound beams in quartz and sapphire near one GHz, in: Proceedings of the IEEE Symposium on Sonics and Ultrasonics, Vancouver, Canada, paper 14, 1967, p. 68.

500. Hearmon, R. F. S., The elastic constants of anisotropic materials. *Rev. Mod. Phys.* (USA) **18**, 409–440 (1946).

501. Hearmon, R. F. S., The elastic constants of piezoelectric crystals. *J. Phys. D (Appl. Phys.)* (GB) **3**, 120–124 (1952).

502. Hearmon, R. F. S., The elastic constants of anisotropic materials II. *Adv. Phys.* (GB) **5**, 323–382 (1956).

503. Hearmon, R. F. S., Equations for transforming elastic and piezoelectric constants of crystals. *Acta Crystallogr.* (Denmark) **10**, Part 2, 121–124 (1957).

504. Hearmon, R. F. S., *An Introduction to Applied Anisotropic Elasticity*, Oxford University Press, London (1961).

505. Heckmann, G., Zur Gittertheorie deformierbarer Ionen. *Z. Kristallogr.* (Germany) **61**, 250–292 (1925), in German.

506. Heilmeier, G. H., The dielectric and electro-optic properties of a molecular crystal —hexamine. *Appl. Opt.* (USA) **3**, 1281–1287 (1964).

507. Henderson, D. M., and Abrams, R. L., A comparison of acousto-optic and electro-optic modulators at 10.6 μ. *Opt. Commun.* (Netherlands) **2**, 223–226 (1970).

508. Hernandez, E. N., Analysis of four optical distance measuring configurations using electro-optical polarization modulators. *Appl. Opt.* (USA) **11**, 609–612 (1972).

509. Herzfeld, K. F., On the theory of forced double refraction. *J. Opt. Soc. Am.* (USA) **17**, 26–36 (1928).

510. Herzfeld, K. F., and Lee, R., Theory of forced double refraction. *Phys. Rev.* (USA) **44**, 625–631 (1933).

511. Hiedemann, E. A., and Siefen, N., Optischer Nachweis der Oberschwingungen eines Piezoquarzes und der starken Abflachung seiner Resonanzkurve durch Flussigkeitsdampfung. *Z. Phys.* (Germany) **91**, 413–417 (1934), in German.

512. Hiedemann, E. A., Measurement of the elastic constants of isotropic transparent solids. *Nature* (GB) **136**, 337 (1935).

513. Hiedemann, E. A., and Hoesch, K. H., On the visualization of lengthwise and thickness vibrations in a vibrating quartz. *Naturwissenschaften* (Germany) **23**, 511–512 (1935), in German.

514. Hiedemann, E. A., and Hoesch, K. H., A new method of measuring the elastic constants of transparent isotropic solids. *Naturwissenschaften* (Germany) **23**, 577–578 (1935), in German.

515. Hiedemann, E. A., and Hoesch, K. H., Schubwellen als optische Gitter. *Naturwissenschaften* (Germany) **23**, 705–706 (1935), in German.

516. Hiedemann, E. A., and Hoesch, K. H., Stationary ultrasonic waves in transparent solids. *Z. Phys.* (Germany) **96**, 268–272 (1935), in German.

517. Hiedemann, E. A., Precision determination of elastic constants of transparent bodies. *Z. Phys.* (Germany) **96**, 273–276 (1935), in German.

518. Hiedemann, E. A., and Hoesch, K. H., Visualization of stationary ultrasonic waves in transparent solids. *Z. Phys.* (Germany) **98**, 141–144 (1935), in German.

519. Hiedemann, E. A., and Hoesch, K. H., Zur Frage der optischen Beugungserscheinungen an schwingenden Glaskörpern. *Naturwissenschaften* (Germany) **24**, 60 (1936), in German.

520. Hiedemann, E. A., and Osterhammel, K., Optische Untersuchungen der Richtcharakteristik von Ultraschallwellen. *Z. Phys.* (Germany) **107**, 273–282 (1937), in German.

521. Hiedemann, E. A., and Schreuer, E., Periodicities in images of ultrasonic waves. *Z. Phys.* (Germany) **107**, 463–473 (1937), in German.

522. Hiedemann, E. A., Versuche Zur Nath–Muellerschen Theorie der Beugung von Licht an schwingenden Festkörpern. *Z. Phys.* (Germany) **108**, 592–597 (1938), in German.

523. Hiedemann, E. A., *Ultraschallforschung*, De Gruyter, Berlin (1939), in German.

524. Hiedemann, E. A., and Breazeale, M. A., Secondary interference in the Fresnel zone of gratings. *J. Opt. Soc. Am.* (USA) **49**, 372–375 (1959).

525. Higginbotham, C. W., Cardona, M., and Pollok, F. H., Intrinsic piezobirefringence of GeSi and GeAs. *Phys. Rev.* (USA) **184**, 821–829 (1969).

526. Higman, B., *Applied Group-Theoretic and Matrix Methods*, Clarendon Press, Oxford, England (1955).

527. Hines, R. H., and Hollinshed, W. L., An electro-optical distance measuring instrument for surveying, in: *Proceedings of the Technical Program—Electrical and Optical Systems Design Conference, West Anaheim, California*, 1971, Industrial and Scientific Conference Management, Inc., Chicago (1971), pp. 324–328.

528. Hirano, H., Dependence of linear electro-optic effect on melt composition of $LiNbO_3$. *Jap. J. Appl. Phys.* (Japan) **8**, 399 (1969).
529. Hirano, H., Takei, H., and Koide, S., The linear electro-optic effect of $(Ba_{1-x}Sr_x)_2NaNb_5O_{15}$ Crystals. *Jap. J. Appl. Phys.* (Japan) **8**, 972–973 (1969).
530. Ho. L., and Buhrer, C. F., Electro-optic effect of gallium arsenide. *Appl. Opt.* (USA) **2**, 647–648 (1963).
531. Hoesch, K., Elastic constants of transparent isotropic solids. *Z. Phys.* (Germany) **109**, 606–624 (1938), in German.
532. Holister, G. S., The photoelastic method applied to materials research. *Appl. Mater. Res.* (GB) **1**, 149–159 (1962).
533. Holmes, C. H., Spencer, E. G., Ballman, A. A., and Lenzo, P. V., The electro-optic effect in calcium pyroniobate. *Appl. Opt.* (USA) **4**, 551–553 (1965).
534. Holshauser, D. F., Van Foerster, H., and Clark, G. L., Microwave modulation of light using the Kerr effect. *J. Opt. Soc. Am.* (USA) **51**, 1360–1365 (1961).
535. Hookabe, K., and Matsuo, Y., Novel type of cut for KDP crystals for low voltage light modulation. *Electron. Lett.* (GB) **6**, 550–551 (1970).
536. Hookabe, K., and Matsuo, Y., Novel cut of KDP-type crystals (45°Z–45°Y cut). *Mem. Inst. Sci. Ind. Res., Osaka Univ.* (Japan) **28**, 23–27 (1971).
537. Hookabe, K., and Matsuo, Y., Optimum cut of KDP type crystals for light modulation. *Electron. Commun. Jap.* (USA) **54**, 77–84 (1972).
538. Hope, L. L., Brillouin scattering in birefringent media. *Phys. Rev.* (USA) **166**, 883–892 (1968).
539. Hrbek, G., and Watson, W., A high speed laser alphanumeric generator, in: *Proceedings of the Electro-optical Systems Design Conference*, New York, September 1970, pp. 271–275.
540. Hruska, K., Interesting directions in crystals from the aspect of the Voigt, and Laval–Raman theory of the elastic properties of crystals. *Czech. J. Phys. B* (Czechoslovakia) **17**, 17–26 (1967).
541. Hsu, H., and Kavage, W., Stimulated Brillouin scattering in anisotropic media and observation of phonons. *Phys. Lett.* (USA) **15**, 207–208 (1965).
542. Huang, Y. Y., Hunt, J. L., and Stevens, J. R., Determination of elastic constants in isotropic silicate glasses by Brillouin scattering. *J. Appl. Phys.* (USA) **44**, 3589–3592 (1973).
543. Hueter, T., and Pohlman, R., An optical method for determining the absorption of ultrasonic waves in (nontransparent) materials of low acoustic impedance. *Z. Angew. Phys.* (Switzerland) **1**, 405–411 (1949), in German.
544. Hulme, K. F., Davies, P. H., and Cound, V. M., The signs of the electro-optic coefficients for lithium niobate. *J. Phys. C (Solid State Physics)* Ser. 2 (GB) **2**, 855–857 (1969).
545. Hulme, K. F., Davies, P. H., and Cound, V. M., Optimum longitudinal electro-optic effect in oblique cut lithium niobate plates. *Electron. Lett.* (GB) **5**, 171–172 (1969).
546. Hulme, K. F., Oblique-cut longitudinal electro-optic modulators. *IEEE J. Quantum Electron.* (USA) **QE-7**, 236–239 (1971).
547. Humphreys, L. B., and Maradudin, A. A., Ionic Brillouin effect. *Phys. Rev. B* (USA) **6**, 3868–3886 (1972).
548. Huntington, H. B., The elastic constants of crystals, in: *Solid State Physics*, F. Seitz and D. Turnbull, eds., Academic Press, New York and London (1958), Vol. 7, pp. 233–251.

549. Hutcheson, L. D., and Hughes, R. S., Rapid acousto-optic tuning of a dye laser. *Appl. Opt.* (USA) **13**, 1395–1398 (1974).

550. Hylleras, E., The arrangement of the atoms in the tetragonal crystals of the mercury halogen compounds Hg_2Cl_2, Hg_2Br_2, Hg_2I_2. Calculation of the optical double refraction of Hg_2Cl_2. *Z. Phys.* (Germany) **36**, 859–896 (1926), in German.

551. Hylleras, E., Bestimmung der Lage der Sauerstoffatome in Rutil und Anatas mit Hilfe der optischen Doppelbrechung. *Z. Kristallogr.* (Germany) **65**, 469–492 (1927), in German.

552. IEEE Standards on Piezoelectric Crystals, 1949. *Proc. IRE* (USA) **37**, 1378–1395 (1949).

553. Iosilevskii, Ya. A., Elastic waves in crystals of cubic and tetragonal symmetry. *Sov. Phys.-Crystallogr.* (USA) **12**, 57–62 (1967).

554. Iosilevskii, Ya. A., Covariant forms of the third and fourth rank tensors such as piezoelectric moduli and piezo-optical constants. *Phys. Status Solidi* (Germany) **61**, 767–777 (1974).

555. Ippen, E. P., Electro-optic deflection with $BaTiO_3$ prisms. *IEEE J. Quantum Electron.* (USA) **QE-2**, 152 (1966).

556. Irisova, N. A., Kozlov, G. V., Narytnik, T. N., and Chernyshev, I. M., Temperature dependence of the elasto-optical effect of single $SrTiO_3$ crystals in the submillimeter range. *Bull. Acad. Sci. USSR, Phys. Ser.* (USA) **39**, 133–135 (1975).

557. Isaacs, T. J., Gottlieb, M., and Feichtner, J. D., Optoacoustic properties of thallium phosphorus selenide, Tl_3PSe_4. *Appl. Phys. Lett.* (USA) **24**, 107–109 (1974).

558. Isaacs, T. J., Gottlieb, M., Daniel, M. R., and Feichtner, J. D., Tl_3VS_4 as an acousto-optic and surface wave material. *J. Electron. Mater.* (USA) **4**, 67–75 (1975).

559. Ishida, A., and Sumi, M., Acousto-optical measurements of acoustic wave generation in piezoelectrics coupled to a Gunn oscillator. *Appl. Phys. Lett.* (USA) **18**, 252–253 (1971).

560. Isyanova, E. D., Kamach, Yu. E., Kozlovskiy, E. N., and Ochinnikov, V. M., Dispersion properties of controllable electro-optic elements. *Sov. J. Opt. Technol.* (USA) **37**, 149 (1970).

561. Ivanov, N. R., and Shivalov, L. A., Rotation of the optical indicatrix of triglycine sulfate crystals by an E_y electric field. *Sov. Phys.-Crystallogr.* (USA) **11**, 648–651 (1966).

562. Ivanov, I. V., Morozov, N. A., and Yudin, Yu. M., Absolute temperature stabilization of electro-optical crystals. *Sov. Phys.-Tech. Phys.* (USA) **18**, 977–978 (1974).

563. Ivanov, N. R., and Koniak, C., Optical activity and the electro-optical effect in crystals of the langbeinite $Cd_2(NH_4)_2(SO_4)_3$. *Sov. Phys.-Crystallogr.* (USA) **19**, 755–756 (1975).

564. Iwasaki, H., Toyoda, H., Niizeki, N., and Kubota, H., Temperature and optical frequency dependence of the d.c. electro-optic constant r_{22}^T of $LiNbO_3$. *Jap. J. Appl. Phys.* (Japan) **6**, 1419–1422 (1967).

565. Iyengar, K. S., "Photoelastic properties of solids as studied by ultrasonic methods." Ph. D. Thesis, Osmania University (India), 1953.

566. Iyengar, K. S., Photoelastic dispersion in cubic crystals. *Nature* (GB) **176**, 1119–1120 (1955).

567. Iyengar, K. S., Refraction of light in solids by ultrasonic waves. *Proc. Indian Acad. Sci.* **A41**, 25–29 (1955).

568. Izrailenko, A. N., Orlov, R. Yu., and Koptsik, V. A., Ammonium oxalate—a new optically non-linear material. *Sov. Phys.-Crystallogr.* (USA) **13**, 136–137 (1968).
569. Jacob, P. J., and Narasimhamurty, T. S., Photoelastic properties of barium fluoride. *Curr. Sci.* (India) **31**, 328 (1962).
570. Jacquerod, A., The Kerr effect in potassium dihydrogen phosphate and potassium dihydrogen arsenate. Quoted in *Helv. Phys. Acta* (Switzerland) **15**, 324–325 (1942).
571. Jaffe, H., Electro-optic effect in crystals of the primary phosphate. *Phys. Rev.* (USA) **73**, 95 (1948).
572. Jaffe, H., and Smith, C. S., Elastic constants of ADP, and Laval theory. *Phys. Rev.* (USA) **121**, 1604–1607 (1961).
573. Jagodzinski, H., Kristallographie–Kristallsymmetrie und physikalische Eigenschaften homogener Körper, in "Kristallphysik". *Handbuch der Physik*, S. Flugge, ed., Springer-Verlag, Berlin (1955), PH Vol. VII/1, pp. 40–59, in German.
574. Jahn, H. A., Elastic constants of anisotropic solids. *Z. Kristallogr.* (Germany) **98**, 191–200 (1937), in English.
575. Jahn, H. A., Note on the Bhagavantam–Suryanarayana method of enumerating the physical constants of crystals. *Acta Crystallogr.* (Denmark) **2**, 30–33 (1949).
576. James, B. W., Optimum values of the elastic constants of single crystals with particular reference to $CaMoO_4$. *J. Appl. Phys.* (USA) **45**, 3201–3202 (1974).
577. Jannetaz, E., Optische Erscheinungen in komprimierten Alaunen. *Z. Kristallogr.* (Germany) **4**, 421 (1880), in German.
578. Jantx, W., Wettling, W., and Sandercock, J. R., Determination of magnetic and elastic properties of $FeBO_3$ by light scattering. *J. Phys. C* (GB) **9**, 2220–2240 (1976).
579. Jeffreys, H., *Cartesian Tensors*, Cambridge Univ. Press, London (1931).
580. Jelatis, J. G., An improved light modulator utilizing the crystalline electro-optic effect. *J. Opt. Soc. Am.* (USA) **43**, 335 (1953).
581. Jeppesen, M. A., Some optical, thermo-optical and piezo-optical properties of synthetic sapphire. *J. Opt. Soc. Am.* (USA) **48**, 629–632 (1958).
582. Jerrard, H. G., Transmission of light through birefringent and optically-active media—the Poincaré sphere. *J. Opt. Soc. Am.* (USA) **44**, 634–640 (1954).
583. Jessop, H. T., On the Tardy and Sénarmont methods of measuring fractional relative retardations. *Brit. J. Appl. Phys.* (GB) **4**, 138–140 (1953).
584. Jessop, H. T., A tilting stage method for three-dimensional photoelastic investigations. *Brit. J. Appl. Phys.* (GB) **8**, 30–32 (1957).
585. Jog, E. S., Dispersion of the stress optical coefficient of vitreous silica. *J. Indian Inst. Sci.* **A39**, 93–100 (1957).
586. Jog, E. S., Dispersion of piezo-optic constants of vitreous silica. *J. Indian Inst. Sci.* **A39**, 101–107 (1957).
587. Jog, E. S., and Krishnan, R. S., Dispersion of the photoelastic constants of fused silica. *Nature* (GB) **179**, 540–541 (1957).
588. Johannsen, A., *Manual of Petrographic Methods*, McGraw-Hill Book Co., New York (1914), p. 343.
589. Johnson, C. J., Some dielectric and electro-optic properties of $BaTiO_3$ single crystals. *Appl. Phys. Lett.* (USA) **7**, 221–223 (1965).
590. Johnson, C. J., Electro-optic effect in CdTe at 23.35 and 27.95 microns. *Proc. IEEE* (USA) **56**, 1719–1722 (1968).
591. Johnston, A. R., The strain-free electro-optic effect in single crystal barium titanate. *Appl. Phys. Lett.* (USA) **7**, 195–198 (1965).

592. Johnston, A. R., and Weingart, J. M., Determination of the low-frequency linear electro-optic effect in tetragonal $BaTiO_3$. *J. Opt. Soc. Am.* (USA) **55**, 828–834 (1965).

593. Johnston, A. R., and Nakamura, T., Determination of the low frequency electro-optic coefficients of $NaNO_2$. *J. Appl. Phys.* (USA) **40**, 3656–3658 (1969).

594. Johnston, A. R., Dispersion of the electro-optic effect in $BaTiO_3$, in: Program of the 1971 spring meeting of the Optical Society of America, Tucson, Arizona (USA), 5–8 April 1971, p. 38 (abstract only).

595. Johnston, A. R., Dispersion of electro-optic effect in $BaTiO_3$. *J. Appl. Phys.* (USA) **42**, 3501–3507 (1971).

596. Johnston, W. D., and Kaminow, I. P., Contributions to optical nonlinearity in GaAs as determined from Raman scattering efficiencies. *Phys. Rev.* (USA) **188**, 1209–1211 (1969).

597. Jones, O. C., and Forno, C., Linear electro-optic effect in hexamine. *Opt. Acta.* (GB) **17**, 559–562 (1970).

598. Juretschke, H. J., *Crystal Physics—Macroscopic Physics of Anisotropic Solids*, W. A. Benjamin, Inc., Reading, Massachusetts (1974).

599. Kachalov, O. V., Dobrzhanskii, G. F., and Shpilko, I. A., Mandelstam–Brillouin scattering in α-HIO_3 crystals. *Sov. Phys.-Crystallogr.* (USA) **16**, 678–680 (1971).

600. Kachalov, O. V., Intensity of Mandelstam–Brillouin components in calcite. *Sov. Phys.-JETP* (USA) **34**, 719–722 (1971).

601. Kachalov, O. V., and Shpilko, I. A., Determination of the elastic and photoelastic constants of an $NaNO_3$ single crystal on the basis of the Mandelstam–Brillouin scattering spectra. *Sov. Phys.-JETP* (USA) **35**, 957–959 (1972).

602. Kaiser, W., and Zurek, R., Brillouin and critical light scattering in $SrTiO_3$ crystals. *Phys. Lett.* (Netherlands) **23**, 668–670 (1966).

603. Kalymnios, D., Optimisation of electro-optical crystal modulators. *Electron. Lett.* (GB) **6**, 804–805 (1970).

604. Kalymnios, D., Linear electro-optic (Pockels) effect in *m*-nitroaniline, $C_6H_4(NO_2)$-NH_2. *J. Phys. D* (*Appl. Phys.*) (GB) **5**, 667–669 (1972).

605. Kamach, Yu. A., Kozlovskiy, Ye. N., Ovchinnikov, V. M., and Sirotin, Yu. I., Crystal deflectors with variable angle of reflection. *Sov. J. Opt. Technol.* (USA) **37**, 167–172 (1970).

606. Kamach, Yu. A., Kozlovskii, Ye. N., Ovchinnikov, V. M., and Sirotin, Yu. I., Design of electro-optic deflectors with a controlled angle of internal reflection. *Sov. J. Quantum Electron.* (USA) **4**, 899–904 (1975).

607. Kaminow, I. P., Microwave modulation of the electro-optic effect in KH_2PO_4. *Phys. Rev. Lett.* (USA) **6**, 528–530 (1961).

608. Kaminow, I. P., Kompfner, R., and Louisell, W. H., Improvements in light modulators of the travelling wave type. *IEEE Trans. Microwave Theory & Tech.* (USA) **MTT-10**, 311–313 (1962).

609. Kaminow, I. P., Splitting up of Fabry–Perot rings by microwave modulation of light. *Appl. Phys. Lett.* (USA) **2**, 41–42 (1963).

610. Kaminow, I. P., Strain effects in electro-optic light modulators. *Appl. & Opt.* (USA) **3**, 511–515 (1964).

611. Kaminow, I. P., Microwave dielectric properties of $NH_4H_2PO_4$, KH_2AsO_4 and partially deuterated KH_2PO_4. *Phys. Rev.* (USA) **138**, A1539–A1543 (1965).

612. Kaminow, I. P., Barium titanate light phase modulator. *Appl. Phys. Lett.* (USA) **7**, 123–125 (1965); **8**, 54 (1965).

613. Kaminow, I. P., and Turner, E. H., Electro-optic light modulators. *Proc. IEEE* (USA) **54**, 1374–1390 (1966).

614. Kaminow, I. P., Barium titanate light modulator, II. *Appl. Phys. Lett.* (USA) **8**, 305–306 (1966).

615. Kaminow, I. P., and Johnston, W. D., Jr., Quantitative determination of source of the electro-optic effect in $LiNbO_3$ and $LiTaO_3$. *Phys. Rev.* (USA) **160**, 519–522 (1967).

616. Kaminow, I. P., and Sharpless, W. M., Performance of $LiTaO_3$ and $LiNbO_3$, light modulators at 4 GHz. *Appl. Opt.* (USA) **6**, 351–352 (1967).

617. Kaminow, I. P., Electro-optic materials, in: *Ferroelectricity*, E. F. Weller, ed., Elsevier, New York (1967).

618. Kaminow, I. P., Measurements of the electro-optic effect in CdS, ZnTe and GaAs at 10.6 μ. *IEEE J. Quantum Electron.* (USA) **QE-4**, 23–26 (1968).

619. Kaminow, I. P., Electro-optic materials. *J. Opt. Soc. Am.* (USA) **61**, 666 (1971), in: Program of the 1971 spring meeting of the Optical Society of America, Tucson, Arizona (USA), 5–8 April 1971.

620. Kaminow, I. P., and Turner, E. H., Linear electro-optical materials, in: *Handbook of Lasers*, R. J. Pressley, ed., Chemical Rubber Co., Cleveland, Ohio (1971), pp. 447–459.

621. Kaminow, I. P., and Turner, E. H., Temperature dependence of Raman scattering and the electro-optic properties of CuCl. *Phys. Rev. B* (USA) **5**, 1564–1578 (1972).

622. Kaminow, I. P., *An Introduction to Electrooptic Devices*. Academic Press, New York (1974).

623. Kaplan, H., Shaham, J., and Low, W., Brillouin scattering in KBr. *Phys. Lett.* (Netherlands) **31A**, 201–202 (1970).

624. Kaplan, M. S., Krasovitskaya, I. M., Shumskov, I. V., and Sumin, V. I., Temperature dependence of the electron polarizability of alkali halide crystals. *Sov. Phys.-Solid State* (USA) **17**, 1673–1674 (1975).

625. Kaplan, M. S., Nagornaya, L. L., Shumskov, I. V., Stroilova, D. L., and Sumin, V. I., Temperature dependence of the photoelastic constants of caesium iodide crystals. *Sov. Phys.-Solid State* (USA) **17**, 1999 (1975).

626. Kaplyanskii, A. A., and Lozovskaya, N. G., Anomalous dispersion in deformation induced double refraction, and the connection between photoelasticity and piezo-spectroscopic phenomena in crystals. *Sov. Phys.-Dokl.* (USA) **10**, 613–616 (1966).

627. Kaplyanskii, A. A., Malkin, B. Z., Medredev, V. N., and Skvortsov, A. P., Quadratic electro-optical effect in crystals having the fluorite structure. *Sov. Phys.-Solid State* (USA) **16**, 221–224 (1974).

628. Kara, R., Mathelu, J. P., and Poulet, H. J., Piezo-optic and electro-optic properties of zincblende. *J. Phys. Radium* **15**, 60–61 (1954), in French.

629. Karolus, Von A., and Mittelstaedt, O., Die Bestimmung der Lichtgeschwindigkeit unter Verwendung des elektrooptischen Kerr-Effektes. *Physikalische Zeitschrift* (Germany) **29**, 698–702 (1928), in German.

630. Kato, Y., and Stoicheff, B. P., Absolute intensity measurements of Brillouin spectra of liquid and solid krypton, and determination of the elasto-optic constants. *Phys. Rev. B* (USA) **11**, 3984–3989 (1975).

631. Kaufman, I., and Strawn, R. S., Electronically variable optical beam deflection and focussing with PLZT ceramics. *Electron. Lett.* (GB) **12**, 122–123 (1976).

632. Kazaryan, R. A., Vartanyan, E. S., and Manuchayan, R. G., Adjustable electro-optical ultra high frequency light modulator. *Instrum. Exper. Tech.* (USA) **76**, 1573–1574 (1969).

633. Keller, O., Acousto-electric dispersion of the Brillouin scattering cross section in CdS. *Solid State Commun.* (USA) **18**, 1227–1229 (1976).

634. Kelley, R. L., Pockels effect in zinc-blende structure ionic crystals. *Phys. Rev.* (USA) **151**, 721–726 (1966).

635. Kemp, J. C., Piezo-optical birefringence modulators: New use for a long known effect. *J. Opt. Soc. Am.* (USA) **59**, 950–954 (1969).

636. Kerr, J., A new relation between electricity and light: dielectrified media birefringent. *Philos. Mag.* (GB), 4th Series, **50**, 337–348 (1875); also papers in subsequent issues.

637. Kerr, J., On the birefringent action of strained glass. *Philos. Mag.* (GB) **V-26**, 321–342 (1888).

638. Khaschina, M. V., Raff, V. S., Zamkov, V. A., Apparatus for measuring Kerr constants. *Instrum. Exp. Tech.* (USA) **18**, No. 2, Part 2, 580–582 (1975).

639. Kidani, Y., Photoelastic constant of rock-salt crystals. *Proc. Phys. Math. Soc.* (Japan) **21**, 457–460 (1939).

640. Kiefer, J. E., and Yariv, A., Electro-optic characteristics of CdTe at 3.39 and 10.6 μ. *Appl. Phys. Lett.* (USA) **15**, 26–27 (1969).

641. Kiefer, J. E., Electro-optic characteristics of CdTe at 3.39 and 10.6 μ. *IEEE J. Quantum Electron.* (USA) **QE-6**, 7 (1970).

642. Kielich, S., Nonlinear optical and electro-optical properties of dielectrics and ferroelectrics. *Ferroelectrics* (GB) **4**, 257–282 (1973).

643. Kikuchi, Y., Chubachi, N., and Yasuda, S., Experiments on Bragg diffraction of He–Ne laser beam by means of ultrasonic waves. *Rec. Elec. Commun. Eng. Conversazione Tohoku Univ.* (Japan) **40**, 60–64 (1971), in Japanese.

644. Kimura, M., Nanamatsu, S., Kawamura, T., and Matsushita, S., Ferro-electric, electro-optic and piezoelectric properties of Nd_2TiO_7 single crystals. *Jap. J. Appl. Phys.* (Japan) **13**, 1473–1474 (1974).

645. Kinase, W., Uemura, Y., and Kikuchi, M., Correction of dipole field due to lattice deformation of a perovskite-type crystal. *J. Phys. Chem. Solids* (GB) **30**, 441–447 (1969).

646. Kinase, W., and Mori, T., Photoelastic effect of alkali halide crystals. *J. Phys. Soc.* (Japan) **31**, 426–430 (1971).

647. Kirk, J. L., and Vedam, K., Piezo-optic behaviour of forsterite, Mg_2SiO_4. *J. Phys. Chem. Solids* (GB) **33**, 1251–1255 (1972).

648. Kitahiro, I., Yano, T., and Watanabe, A., Electro-optic properties of $Sr_4KLiNb_{10}O_{30}$ type ferroelectrics. *Jap. J. Appl. Phys.* (Japan) **8**, 807 (1969).

649. Klein, W. R., Cook, B. D., and Mayer, W. G., Light diffraction by ultrasonic gratings. *Acustica* (Germany) **15**, 67–74 (1965).

650. Klein, W. R., Tipnis, C. B., and Hiedemann, E. A., Experimental study of Fraunhofer light diffraction. *J. Acoust. Soc. Am.* (USA) **38**, 229–233 (1965).

651. Klein, W. R., Theoretical efficiency of Bragg devices. *Proc. IEEE* (USA) **54**, 803–804 (1966).

652. Klein, W. R., and Cook, B. D., Unified approach to ultrasonic light diffraction. *IEEE Trans. Sonics & Ultrasonics* (USA) **SU-14**, 123–134 (1967).
653. Kludzin, V. V., Photoelastic constants of $LiNbO_3$ crystals. *Sov. Phys.-Solid State* (USA) **13**, 540–541 (1971).
654. Kludzin, V. V., Characteristics of light intensity variation in diffraction order as a function of angle of incidence of a light beam on the surface of an ultrasound light modulator. *Izv. VUZ Radiofiz.* (USSR) **17**, 261–264 (1974), in Russian.
655. Knox, R. S., and Gold, A., *Symmetry in the Solid State*, W. A. Benjamin, Inc., New York (1964).
656. Koelsch, A. C., A high speed, low voltage light modulator. *Inst. Radio Engrs., Convention Rec.* (USA) **5**, 169–177 (1957).
657. Koerber, G. G., *Properties of Solids*. Prentice-Hall, Inc., Englewood Cliffs, New Jersey (1962).
658. Koestser, H., Measurement of r_{63} for ADP up to electric breakdown. *Electron Lett.* (GB) **3**, 52–54 (1967).
659. Kogelnik, H., and Li, T., Laser beams and resonator. *Proc. IEEE* (USA) **54**, 1312–1329 (1966).
660. Kojima, C., Shikama, T., Kuninobu, S., Kawabata, A., and Tanaka, T., The electro-optic effect in cubic ZnSe at 10.6 μ. *Jap. J. Appl. Phys.* (Japan) **8**, 1361–1362 (1969).
661. Kojima, S., Ohi, K., and Nakamura, T., Temperature dependence of electro-optic coefficient and spontaneous birefringence of ferroelectric gadolinium molybdate. *J. Phys. Soc. Jap.* (Japan) **41**, 162–166 (1976).
662. Kolb, J., and Loeber, A. P., The study of a sound field by means of optical refraction effects. *J. Acoust. Soc. Am.* (USA) **26**, 249–251 (1954).
663. Kolsky, H., *Stress Waves in Solids*, Oxford University Press, London (1953).
664. Konak, C., Use of the electro-optic modulators for the measurements of birefringence. *Czech. J. Phys. B* (Czechoslovakia) **B24**, 355–365 (1974).
665. Konak, C., and Fousek, J., Birefringent and electro-optical properties of improper ferroelectric $Cd_2(NH_4)_2(SO_4)_3$. *Ferroelectrics* (GB) **6**, 235–240 (1974).
666. Koptsik, V. A., and Sirotin, Yu. I., The symmetry of piezoelectric and elastic tensors and the symmetry of the physical properties of crystals. *Sov. Phys.-Crystallogr.* (USA) **6**, 612–614 (1962).
667. Korolos, A., The physical basis of the measurement of distance by electro-optical technique. *Ber. Akad. Wissenschaften Abhand* (Germany) **92** (1958), in German.
668. Korpel, A., Adler, R., and Alpiner, B., Direct observation of optically induced generation and amplification of sound. *Appl. Phys. Lett.* (USA) **5**, 86 (1964).
669. Korpel, A., Adler, R., Desmares, P., and Smith, T. M., An ultrasonic light deflection system. *IEEE J. Quantum Electron.* (USA) **QE-1**, 60–61 (1965).
670. Korpel, A., Adler, R., Desmares, P., and Watson, W., A television display using acoustic deflection and modulation of coherent light. *Appl. Opt.* (USA) **5**, 1667 1675 (1966); *Proc. IEEE* (USA) **54**, 1429–1437 (1966), joint issue.
671. Korpel, A., Visualization of the cross-section of a sound beam by Bragg diffraction of light. *Appl. Phys. Lett.* (USA) **9**, 425–427 (1966).
672. Korpel, A., Acoustic imaging and holography. *IEEE Spectrum* (USA) **5**, 45–52 (1968).
673. Korpel, A., "Optical imaging of ultrasonic fields by acoustic Bragg diffraction." Ph.D. Thesis, University of Delft, Holland, 1969.

674. Korpel, A., Spatial imaging of sound fields by Bragg diffraction of light. *International J. Nondestructive Testing* (GB) **1**, 337–351 (1970).

675. Korpel, A., Acousto-optics, in: *Applied Solid State Science, Advances in Materials and Device Research*, R. Wolfe, ed., Academic Press, New York (1972), Vol. 3, pp. 72–180.

676. Korpel, A., Principles of acousto-optics, in: *Proceedings of the Society of Photo-optical Instrumentation Engineers Seminar on Electro-optics Principles and Applications, Boston, Massachusetts*, 30 April–1 May 1973, Society of Photo-optical Instrumentation Engineers, Redondo Beach, California (1973), pp. 3–8.

677. Krainik, N. N., Gokhberg, L. S., and Mylnikova, I. E., Electro-optical effect in $PbZn_{1/3}Nb_{2/3}O_3$ crystals. *Sov. Phys.-Solid State* (USA) **12**, 1885–1888 (1971).

678. Krainik, N. N., Kamzina, L. S., Sakharov, D. G., Volik, B. A., Pismennyi, V. A., and Skomyakova, K. P., Characteristic features of the electro-optic effect in lead magnoniobate subjected to strong alternating electric fields in the region of broad phase transitions. *Sov. Phys.-Solid State* (USA) **17**, 122–124 (1975).

679. Krasser, W., Buchenau, V., and Haussühl, S., Brillouin scattering in cubic potassium cyanide. *Solid State Commun.* (USA) **18**, 287–289 (1976).

680. Krause, J. T., Kurkjian, C. R., Pinnow, D. A., and Sigety, E. A., Low acoustic loss chalcogenide glasses—A new category of materials for acoustic and acousto-optic applications. *Appl. Phys. Lett.* (USA) **17**, 367–368 (1970).

681. Krishna Murty, V. G., "Temperature dependence of the photo-elastic behaviour of crystals." Ph.D. Thesis, Osmania University, India, 1964.

682. Krishnan, R. S., The reciprocity theorem in colloid optics and its generalisation. *Proc. Indian Acad. Sci.* **7A**, 21–34 (1938).

683. Krishnan, R. S., The scattering of light in diamond and its Raman spectrum. *Proc. Indian Acad. Sci.* **26A**, 399–418 (1947).

684. Krishnan, R. S., and Chandrasekharan, V., Thermal scattering of light in crystals, Part I: Quartz. *Proc. Indian Acad. Sci.* **31A**, 427–434 (1950).

685. Krishnan, R. S., Elastic constants of crystals from light scattering measurements. *Proc. Indian Acad. Sci.* **41A**, 91–97 (1955).

686. Krishnan, R. S., Jog, E. S., and Srinivasan, R., Stress optical coefficients of different specimens of vitreous silica. *Nature* (GB) **181**, 692–693 (1958).

687. Krishnan, R. S., Chandrasekharan, V., and Rajagopal, E. S., *Elastic and Photo-elastic Constants of Crystals*, Indian Institute of Science Golden Jubilee Research volume, Bangalore, India (1959), pp. 150–162.

688. Krishnan, R. S., *Progress in Crystal Physics*, Interscience Publishers, Inc., New York (1960).

689. Krishna Rao, K. V., and Kalpagam, V., Photoelastic constants of thallium alum. *Curr. Sci.* (India) **24**, 228 (1955).

690. Krishna Rao, K. V., and Krishna Murty, V. G., Variation with temperature of the photoelastic constants of sodium chloride. *Nature* (GB) **190**, 429–430 (1961).

691. Krishna Rao, K. V., Determination of photoelastic constants in the presence of tilt of the axes. *Indian J. Phys.* **35**, 341–345 (1961).

692. Krishna Rao, K. V., and Krishna Murty, V. G., Photoelastic constants of magnesium oxide. *Acta Crystallogr.* (Denmark) **17**, 788 (1964).

693. Krishna Rao, K. V., and Krishna Murty, V. G., Temperature dependence of the photoelastic properties of crystals. I. *Proc. Indian Acad. Sci.* **A64**, 24–35 (1966).

694. Krishna Rao, K. V., and Krishna Murty, V. G., Temperature variation of photoelastic constants of NH_4Cl. *Indian J. Phys.* **41**, 150–151 (1967).
695. Krishna Rao, K. V., and Krishna Murty, V. G., Temperature dependence of the piezo-optic behaviour of the nitrates of barium, lead and strontium. *Curr. Sci.* (India) **42**, 87–88 (1973).
696. Krishna Rao, K. V., and Krishna Murty, V. G., Effect of temperature on the piezo-optic dispersion in some alkali halide crystals. *Curr. Sci.* (India) **42**, 310–311 (1973).
697. Kroll, N. M., Excitation of hypersonic vibrations by means of photoelastic coupling of high-intensity light waves to elastic waves. *J. Appl. Phys.* (USA) **36**, 34–43 (1965).
698. Kundt, A., Über des optisches Verhalten des Quarzes im elektrischen Felds. *Ann. Phys. Chem.* (Germany) **18**, 228–233 (1883), in German.
699. Kuppers, H., Diffraction of light by transverse ultrasonic waves in crystals. *Acustica* (Germany) **16**, 365–367 (1965–1966).
700. Kuppers, H., Diffraction of light by ultrasonic waves in anisotropic solids. *Phys. Status Solidi* (Germany) **37**, K59–K61 (1970).
701. Kurtz, S. K., and Robinson, F. N. H., A physical model of the electro-optic effect. *Appl. Phys. Lett.* (USA) **10**, 62–65 (1967).
702. Kurtz, S. K., and Perry, T. T., A powder technique for the evaluation of non-linear optical materials. *J. Appl. Phys.* (USA) **39**, 3798–3813 (1968).
703. Kurtz, S. K., Perry, T. T., and Bergman, Jr., J. G., α-iodic acid: a solution-grown crystal for non-linear optical studies and applications. *Appl. Phys. Lett.* (USA) **12**, 186–188 (1968).
704. Kurtzing, A. J., Faraday rotation in birefringent crystals. *J. Appl. Phys.* (USA) **42**, 3494–3498 (1971).
705. Kusters, J. A., Wilson, D. A., and Hammond, D. L., Optimum crystal orientation for acoustically tuned optical filters. *J. Opt. Soc. Am.* (USA) **64**, 434–440 (1974).
706. Lafitte, F., Pauliquem, J., and Segard, N., Photoelastic properties of quartz at resonance—Synchronization of a pulsed laser. *Ann. Soc. Sci. Brusselles I* (Belgium) **83**, 146–149 (1969).
707. Lagarde, A., and Jacquesson, R., Measurement of the mechanical and photoelastic characteristics of a plastic material submitted to sinusoidal forces at low frequencies (less than 50 C/s). *C. R. Hebd. Seances Acad. Sci.* (France) **250**, 969–971 (1960), in French.
708. Lagarde, A., Application of the method of harmonic excitation to the measurement of the mechanical and photoelastic characteristics of a plastic material. *C. R. Hebd. Seances Acad. Sci.* (France) **250**, 2796–2798 (1960), in French.
709. Lagarde, A., and Oheix, P., On adaptations of Sénarmont's method (in photoelasticity). *C. R. Hebd. Seances Acad. Sci. B* (France) **281**, 1–4 (1975), in French.
710. Laiho, R., and Korpela, A., Stress-optical dispersion in RbBr, RbI, CsCl and CsI. *Ann. Acad. Sci. Fennicae A VI* (Finland) **272**, 1–9 (1968).
711. LaMacchia, J. T., Optical memories—A progress report. *Laser Focus* (USA) **6**, 35–39 (February 1970).
712. Land, C. E., and Thacher, P. D., Ferroelectric ceramic electro-optic materials and devices. *Proc. IEEE* (USA) **57**, 751–768 (1969).
713. Land, C. E., and Haertling, G. H., Electro-optic ceramics—A review of materials and applications. *Am. Ceram. Soc. Bull.* (USA) **49**, 411 (1970).

714. Landheer, D., Jackson, H. E., Mclaren, R. A., and Stoicheff, B. P., Elastic constants of krypton single crystals determined by Brillouin scattering. *Phys. Rev. B* (USA) **13**, 888–895 (1976).
715. Langer, D. W., Temperature and pressure dependence of the index of refraction of CdS. *J. Appl. Phys.* (USA) **37**, 3530–3532 (1966).
716. Langevin, M. P., and Chilowsky, M. C., Procédés et appareils pour la production des signaux sous-marins dirigés et pour la localisation à distance d'obstacles sous-marins. French Patent No. 502913 (1918), in French.
717. Langevin, M. P., Sound transmitters for under-water signalling. British Patent No. 145691 (1920).
718. Larionov, A. L., and Malkin, B. Z., Electro-optical constants of ionic crystals. *Sov. Phys.-Solid State* (USA) **16**, 21–24 (1974).
719. Laval, J., Physique cristalline—Élasticité des cristaux. *C. R. Hebd. Séances Acad. Sci.* (France) **232**, 1947–1948 (1951), in French.
720. Lean, E. G. H., Quate, C. F., and Shaw, H. J., Continuous deflection of laser beams. *Appl. Phys. Lett.* (USA) **10**, 48–57 (1967).
721. Lee, R. W., Linear electro-optic (Pockels) effect in hexamethylenetetramine: Influence of crystal strain. *J. Opt. Soc. Am.* (USA) **59**, 1574–1580 (1969).
722. Lee, R. W., The linear electro-optic (Pockels) effect in cubic crystals—wide angle light intensity modulation. *Appl. Opt.* (USA) **8**, 1385–1392 (1969).
723. Lee, S. M., and Hauser, S. M., Kerr constant evaluation of organic liquids and solutions. *Rev. Sci. Instrum.* (USA) **35**, 1679–1680 (1964).
724. Leibssle, H., Die piezo-optischen Eigenschaften einiger Alkalihalogenide des Steinsalz-Typus. *Z. Kristallogr.* (Germany) **114**, 457–467 (1960), in German.
725. Lelong, P., New method for measurement of stationary dynamic birefringence. *C. R. Hebd. Seances Acad. Sci. B* (France) **277**, 199–202 (1973), in French.
726. Lemanov, V. V., Shakin, O. V., and Smolenskii, G. A., Scattering of light by hypersonic waves in lithium niobate crystals. *Sov. Phys.-Solid State* (USA) **13**, 426–428 (1971).
727. Lemanov, V. V., Acousto-optical study of ferroelectric crystals. *Ferro-electrics* (GB) **7**, 11–16 (1974).
728. Lenzo, P. V., Spencer, E. G., and Nassau, K., Electro-optic coefficients in single-domain ferro-electric lithium niobate. *J. Opt. Soc. Am.* (USA) **56**, 633–635 (1966).
729. Lenzo, P. V., Spencer, E. G., and Ballman, A. A., Optical activity and electro-optic effect in bismuth germanium oxide ($Bi_{12}GeO_{20}$). *Appl. Opt.* (USA) **5** 1688–1689 (1966).
730. Lenzo, P. V., Turner, E. H., Spencer, E. G., and Ballman, A. A., Electro-optic coefficients and elastic wave propagation in single-domain ferro-electric lithium tantalate. *Appl. Phys. Lett.* (USA) **8**, 81–82 (1966).
731. Lenzo, P. V., Ballman, A. A., and Spencer, E. G., Electro-optic coefficients of ferroelectric strontium barium niobate. *Appl. Phys. Lett.* (USA) **11**, 23–24 (1967).
732. Lenzo, P. V., Spencer, E. G., and Ballman, A. A., Temperature dependence of the electro-optic effect in strontium barium niobate. *IEEE J. Quantum Electron.* (USA) **QE-4**, 331 (1968).
733. Leontowitsch, M., and Mandelstamm, S., Molecular scattering of light in solids: Part V. *Z. Phys.* (Germany) **75**, 350–351 (1932); see also *Phys. Z. Sowjet* **1**, 317 (1931), in German.
734. Leroy, O., On the exact system of difference-differential equations of Raman–Nath

for the diffraction of light by supersonic waves. *Simon Stevin* (Netherlands) **44**, 103–109 (1971).

735. Levi, L., D'Alessio, J. T., and Otheguy, J. L., Pockels effect electro-optic modulators. *Acta Cient.* (Argentina) **3**, 37–42 (1970), in Spanish.
736. Ley, J. M., Low-voltage light-amplitude modulation. *Electron. Lett.* (GB) **2**, 12–13 (1966).
737. Librecht, F. M., and Francois, C. H., Optimum cut for a $LiNbO_3$ transverse light modulator. *IEEE J. Quantum Electron.* (USA) **QE-7**, 374–376 (1971).
738. Lipnick, R., Reich, A., and Schoen, G. A., Nonmechanical scanning of light using acoustic waves. *Proc. IEEE* (USA) **52**, 853–854 (1964).
739. Lippmann, G., Principle de la conservation de l'électricité. *Ann. Chim. Phys.* Ser. 5 (France) **24**, 145–178 (1881), in French.
740. Litvin, B. N., Shaldin, Yu. V., and Pitovranova, I. E., Synthesis and electro-optical properties of single crystals of Si-sellenite. *Sov. Phys.-Crystallogr.* (USA) **13**, 971–973 (1969).
741. Liu, H., Schock, R. N., and Anderson, D. L., Temperature dependence of single crystal spinel ($MgAl_2O_4$) elastic constants from 293 K to 423 K measured by light–sound scattering in the Raman–Nath region. *Geophys. J. R. Astron. Soc.* (GB) **42**, 217–250 (1975).
742. Loeber, A. P., and Hiedemann, E. A., A simple method for photographing the sound pressure distribution in a stationary ultrasonic wave. *J. Acoust. Soc. Am.* (USA) **26**, 257 (1954).
743. Loeber, A. P., and Hiedemann, E. A., Investigation of stationary ultrasonic waves by light refraction. *J. Acoust. Soc. Am.* (USA) **28**, 27–35 (1956).
744. Lomova, L. G., Sonin, A. S., and Regulskaya, T. A., Spontaneous electro-optic effect in triglycine sulphate single crystals. *Sov. Phys.-Crystallogr.* (USA) **13**, 68–70 (1968).
745. Lotspeich, J. F., Electro-optic light beam deflection. *IEEE Spectrum* (USA) **5**, 45–52 (1968).
746. Lucas, R., and Biquard, P., Optical properties of solids and liquids under ultrasonic vibrations. *J. Phys. Radium* 7th Series (Paris) **3**, 464–477 (1932), in French.
747. Lucas, R., and Biquard, P., New optical properties produced in liquids by high frequency sound waves. *C. R. Hebd. Séances Acad. Sci.* (France) **194**, 2132–2134; **195**, 121–123 (1932), in French.
748. Lucas, R., Sur les phénomènes de mirage optique dus aux ondes élastiques. *C. R. Hebd. Séances Acad. Sci.* (France) **199**, 1107–1108 (1934), in French.
749. Luukkala, M., and Merilainen, P., Image scanning by acousto-electrooptic interaction. *Electron Lett.* (GB) **10**, 80–81 (1974).
750. Luspin, Y., and Hauret, G., Study of the Brillouin effect in ammonium sulphate at room temperature (elastic constants). *C. R. Hebd. Séances Acad. Sci. B* (France) **274**, 995–997 (1972), in French.
751. Lynch, R. T., Dillon, Jr., J. F., and Van Uitert, L. G., Stress birefringence in ferrimagnetic garnets. *J. Appl. Phys.* (USA) **44**, 225–229 (1973).
752. Lyot, B., Optical apparatus with wide field using interference of polarized light. *C. R. Hebd. Séances Acad. Sci.* (France) **197**, 1593–1595 (1933), in French.
753. Mace de Lepinay, J., Recherches sur la double réfraction accidentalle. *C. R. Hebd. Séances Acad. Sci. B* (France) **84**, 1024–1026 (1877); **86**, 326–328 (1878), in French.

754. Mace de Lepinay, J., Recherches expérimentales sur la double réfraction accidentalle. *Ann. Chim. Phys.* (France) **v-19**, 5–90 (1880), in French.
755. Mach, E., Über die temporare Doppelbrechung der Körper durch einseitigen Druck. *Ann. Phys. II Leipzig* **146**, 313–316 (1872), in German.
756. Macheleidt, G., The influence of piezoelectric coupling on material constants determining Brillouin scattering. *Acta Crystallogr.* (Denmark) **A31**, 801–803 (1975).
757. Magdich, L. N., Moiya, A. M., and Pankratov, V. M., Electro-optical light modulator with a low controlling voltage. *Opt. & Spectrosc.* (USA) **26**, 248–249 (1969).
758. Mahajan, V. N., and Gaskill, J. D., Diffraction of light by two orthogonal sound waves. *J. Appl. Phys.* (USA) **45**, 1518–1520 (1974).
759. Mahajan, V. N., and Gaskill, J. D., Diffraction of light by sound waves according to the vector wave equation. *J. Opt. Soc. Am.* (USA) **64**, 400–401 (1974).
760. Mahajan, V. N., and Gaskill, J. D., Doppler interpretation of the frequency shifts of light diffracted by sound waves. *J. Appl. Phys.* (USA) **45**, 2799–2800 (1974).
761. Mahajan, V. N., and Gaskill, J. D., Bragg diffraction of light by a standing sound wave. *Opt. Acta* (GB) **21**, 893–902 (1974).
762. Mahajan, V. N., Diffraction of light by sound waves of arbitrary standing wave ratio. *J. Appl. Phys.* (USA) **46**, 3707–3709 (1975).
763. Malitson, I. H., Dodge, M. J., Waxler, R. M., and Bower, W. S., Refractive properties of $CaMO_3$, in: Program of the 1971 Annual Meeting of the Optical Society of America, Ottawa, Canada, 5–8 October 1971, p. 10.
764. Maloney, W. T., and Carleton, H. R., Light diffraction by transverse ultrasonic waves in hexagonal crystals. *IEEE Trans. Sonics & Ultrason.* (USA) **SU-14**, 135–139 (1967).
765. Mandarino, J. A., "Some optical and stress optical properties of synthetic ruby." Ph.D. Thesis, University of Michigan, Ann Arbor, Michigan (USA), 1957.
766. Manenkov, A. A., Prokhorov, A. M., and Ritus, A. I., Rayleigh and Brillouin light scattering in optical glasses. Hypersonic velocity, elasto-optical constants and extinction coefficients. *Sov. Phys.-Solid State* (USA) **17**, 706–709 (1975).
767. Mandelstamm, L., Landsberg, G., and Leontowitsch, M., Über die Theorie der molekularen Lichtzerstreuung in Kristallen (klassische Theorie). *Z. Phys.* (Germany) **60**, 334–344 (1930), in German.
768. Maradudin, A. A., and Burstein, E., Relation between photoelasticity, electrostriction and first order Raman effect in crystals of the diamond structure. *Phys. Rev.* (USA) **164**, 1081–1099 (1967).
769. Maradudin, A. A., and Burstein, E., A lattice theory of the electrooptic effects in semi-conductors. Report No. TR-20 (AD-679591), Pennsylvania University, Philadelphia, Pennsylvania (USA), December 1968, Contract No. NONR-551 (51), 11 pp.
770. Marck, P., and Taurel, L., Molecular interpretation of the elasto-optical constants of NaCl. *J. Phys. Radium* (France) **14**, 8–13 (1953), in French.
771. Marculescu, L., and Hauret, G., Study of the Brillouin effect at room temperature in $LiNbO_3$. *C. R. Hebd. Séances Acad. Sci. B* (France) **276**, 555–558 (1973), in French.
772. Marecek, V., Dobiosova, R., and Novak, J., Preparation of non-wedge-shaped KDP and DKD crystals. *Krist. & Tech.* (Germany) **4**, 39–44 (1969), in English.

773. Marie, G., and Donjon, J., Single crystal ferroelectrics and their application in light valve display devices. *Proc. IEEE* (USA) **61**, 942–958 (1973).
774. Marie, J., and Mathieu, J. P., Modifications caused in the Raman spectrum of a quartz crystal by compression along a binary axis. *C. R. Hebd. Seances Acad. Sci.* (France) **223**, 147 (1946), in French.
775. Maris, H. B., Photoelastic properties of transparent cubic crystals. *J. Opt. Soc. Am.* (USA) **15**, 194–200 (1927).
776. Markham, M. F., Measurement of elastic constants by the ultrasonic pulse method. *Brit. J. Appl. Phys. Suppl.* **6** (*Physics of Nondestructive Testing*) S56–S63 (1957).
777. Martynov, V. G., Aleksandrov, K. S., and Anistratov, A. T., Investigation of the photoelasticity of ferroelectrics by the diffraction of light on ultrasound. *Sov. Phys.-Solid State* (USA) **15**, 1950–1952 (1974).
778. Martynov, V. G., Aleksandrov, K. S., Anistratov, A. T., and Polkhovskaya, T. M., Photoelasticity of gadolinium molybdate. *Sov. Phys.-Crystallogr.* (USA) **20**, 753–754 (1975).
779. Marugin, A. M., and Ochinnikov, V. M., Influence of piezo-optic effects on the transmission of electro-optic shutters. *J. Opt. Technol.* (USA) **37**, 147–148 (1970).
780. Marx, J., Use of the piezoelectric gauge for internal friction measurements. *Rev. Sci. Instrum.* (USA) **22**, 503–509 (1951).
781. Mason, W. P., The elastic, piezoelectric, and dielectric constants of ADP and KDP. *Phys. Rev.* (USA) **69**, 173–194 (1946).
782. Mason, W. P., Optical properties, and the electrooptic and photoelastic effects in crystals expressed in tensor form. *Bell. Syst. Tech. J.* (USA) **29**, 161–188 (1950).
783. Mason, W. P., *Piezoelectric Crystals and Their Application to Ultrasonics*. D. Van Nostrand, Princeton, New Jersey (1950).
784. Mason, W. P., Tensor properties, in: *Physical Acoustics and the Properties of Solids*, D. Van Nostrand, New York (1958), pp. 382–392.
785. Mason, W. P., *Crystal Physics of Interaction Processes*, Academic Press, New York and London (1966).
786. Maxwell, J., On the equilibrium of elastic solids. *Trans. R. Soc. Edinburgh* **20**, Part I, 87–120 (1853).
787. Mayburg, S., Effect of pressure on the low frequency dielectric constant ionic crystals. *Phys. Rev.* (USA) **79**, 375–382 (1950).
788. Maydan, D., Acousto-optical pulse modulators. *IEEE J. Quantum Electron.* (USA) **QE-6**, 15–24 (1970).
789. Mayer, W. G., and Hiedemann, E. A., Optical methods for the ultrasonic determination of elastic constants of sapphire. *J. Acoust. Soc. Am.* (USA) **30**, 756–760 (1958).
790. Mayer, W. G., and Hiedemann, E. A. Ultrasonic determination of elastic constants and structural irregularities in transparent single crystals (measurements in sapphire). *Acta Crystallogr.* (Denmark) **12**, 1–6 (1959).
791. Mayer, W. G., and Hiedemann, E. A.. Light diffraction by progressive ultrasonic waves in plexiglass. *Acustica* (Germany) **10**, 251–252 (1960).
792. Mayer, W. G., Light diffraction by ultrasonic waves for oblique incidence. *J. Acoust. Soc. Am.* (USA) **36**, 779–781 (1964).
793. Mayer, W. G., and Hiedemann, E. A., Evaluation of two methods for the determination of the elastic constants of trigonal crystals. *Proc. Indian Acad. Sci.* **A60**, 147–152 (1964).

794. Mayer, W. G., Cook, B. D., and Klein, W. R., Light diffraction by ultrasonic gratings. *Acustica* (Germany) **15**, 67–74 (1965).
795. McLaren, R. A., Kiefte, H., Landheer, D., and Stoicheff, B. P., *Elastic constants of neon single crystals determined by Brillouin scattering. Phys. Rev. B* (USA) **11**, 1705–1717 (1975).
796. McQuaid, R. W., Electro-optic properties of zinc selenide. *Proc. IRE* (USA) **50**, 2484–2485 (1962); see also Correction to Electro-optic properties of zinc selenide. *Proc. IEEE* (USA) **51**, 470 (1963).
797. McQuaid, R. W., The Pockels effect of hexamethylenetetramine. *Appl. Opt.* (USA) **2**, 320–321 (1963).
798. McQuaid, R. W., Cubic piezoelectric crystals for electro-optic modulation, in: Proceedings of the 1963 National Aerospace Electronics Conference (USA), pp. 282–286.
799. McQuaid, R. W., and Watkins, M. C., Laser modulation with hexamethylene–tetramine. *J. Opt. Soc. Am.* (USA) **53**, 1339 (1963).
800. McSkimin, H. J., Notes and references for the measurements of elastic moduli by means of ultrasonic waves. *J. Acoust. Soc. Am.* (USA) **33**, 606–615 (1961).
801. McSkimin, H. J., Ultrasonic methods for measuring the mechanical properties of liquids and solids, in: *Physical Acoustics*, W. P. Mason, ed., Academic Press, New York and London (1964), Vol. 1, Part A, pp. 271–334.
802. McVoy, K. W., Symmetry groups in physics. *Rev. Mod. Phys.* (USA) **37**, 84–104 (1965).
803. Meisner, L. B., Electro-optic effect in the ionic model of barium titanate single crystals. *Sov. Phys.-Solid State* (USA) **10**, 741–742 (1968).
804. Meisner, L. B., and Sonin, A. S., Quadratic electro-optic effect in strontium titanate single crystals. *Sov. Phys.-Solid State* (USA) **10**, 998–999 (1968).
805. Meisner, L. B., and Rez, I. S., Calculation of the electro-optical constants in "ionic" crystals. *Sov. Phys.-Crystallogr.* (USA) **18**, 288–290 (1973).
806. Melngailis, J., Maradudin, A. A., and Seeger, A., Diffraction of light by ultra-sound in anharmonic crystals. *Phys. Rev.* (USA) **131**, 1972–1975 (1963).
807. Merkulov, L. G., and Yakovlev, L. A., Propagation and reflection of ultrasonic beams in crystals. *Sov. Phys.-Acoust.* (USA) **8**, 72–77 (1962).
808. Mertens, R., On the theory of diffraction of light by supersonic waves. *Simon Stevin, Wis-Naturkundig Tijdschrift* (Holland) **27**, 212–230 (1949/1950); **28**, 1–12 (1950–1951).
809. Mertens, R., Diffraction of light by standing supersonic waves. General theory. *Simon Stevin, Wis-Naturkundig Tijdschrift* (Holland) **28**, 164–180 (1950-1951).
810. Michael, A. J., Intensity method of stress-optical measurements. *J. Opt. Soc. Am.* (USA) **58**, 889–894 (1968).
811. Milek, J. T., and Welles, S. J., Linear electro-optic modulator materials. Electronic properties Information Center, Hughes Aircraft Co., Culver City, California, 1970.
812. Miller, R. C., Optical second harmonic generation in piezoelectric crystals. *Appl. Phys. Lett.* (USA) **5**, 17–19 (1964).
813. Miller, R. C., Boyd, G. D., and Savage, A., Nonlinear optical interactions in $LiNbO_3$ without double refraction. *Appl. Phys. Lett.* (USA) **6**, 77–79 (1965).
814. Minard, J., Experimental study by a new method of the absolute retardation in electric birefringence of liquids. Electrostriction. *C. R. Hebd. Séances Acad. Sci.* (France) **250**, 694–696 (1960), in French.

815. Minck, R. W., Terhune, R. W., and Wang, C. M., Nonlinear optics. *Proc. IEEE* (USA) **54**, 1357–1374 (1966).
816. Minkoff, J., Bennett, W. R., Lambert, L. B., Arm, M., and Bernstein, S., The operation of wide band fused-silica Debye–Sears light modulators, in: *Proceedings of the Symposium on Modern Optics at the Polytechnic Institute of Brooklyn* (*New York*), 22–24 March 1967, Polytechnic Press, Brooklyn, New York (1967), pp. 703–715.
817. Minkoff, J. B., Bennett, W. R., Lambert, L. B., and Arm, M., Transfer characteristics and frequency response of wide-band amorphous-solid Debye–Sears light modulators. *J. Appl. Phys.* (USA) **39**, 2024–2029 (1968).
818. Miyakawa, K., New derivation of elastic equations for trigonal holoaxial crystals. *Phys. Rev.* (USA) **107**, 677–682 (1957).
819. Modine, F. A., Major, R. W., and Sonder, E., High frequency polarization modulation method for measuring birefringence, *Appl. Opt.* (USA) **14**, 757–760 (1975).
820. Modine. F. A., and Major, R. W., High frequency polarization modulation method for measuring optical rotation. *Appl. Opt.* (USA) **14**, 761–764 (1975).
821. Moigno, F., and Soleil, H., Note sur un nouveau caractère distinctif entre les cristaux et un axe, positifs et négatifs. *C. R. Hebd. Séances Acad. Sci.* (France) **30**, 361–362 (1850), in French.
822. Morky, P., Electro-optical effect in $NH_4H_2PO_4$. *Czech. J. Physics* (Czechoslovakia) **2**, 245 (1943), in Russian.
823. Morris, C. E., Jamieson, J. C., and Yarger, F. L., Ultrasonic measurements at elevated pressure (9 GPa) to determine the Poisson's ratio and other elastic moduli of NaCl and NaF. *J. Appl. Phys.* (USA) **47**, 3979–3986 (1976).
824. Mueller, H., Theory of photoelasticity in amorphous solids. *Phys.* (USA) **6**, 179–184 (1935).
825. Mueller, H., Properties of Rochelle salt. *Phys. Rev.* (USA) **47**, 175–191 (1935).
826. Mueller, H., Theory of the photoelastic effect of cubic crystals. *Phys. Rev.* (USA) **47**, 947–957 (1935).
827. Mueller, H., The intensity and polarization of the light diffracted by supersonic waves in solids. *Phys. Rev.* (USA) **52**, 223–229 (1937).
828. Mueller, H., The theory of photoelasticity. *J. Am. Ceram. Soc.* (USA) **21**, 27–33 (1938).
829. Mueller, H., Determination of elasto-optical constants with supersonic waves. *Z. Kristallogr.* (Germany) **A99**, 122–141 (1938), in English.
830. Mueller, H., On the theory of scattering of light. *Proc. R. Soc. London* **A166**, 425–449 (1938).
831. Mueller, H., Properties of Rochelle salt, IV. *Phys. Rev.* (USA) **58**, 805–811 (1940).
832. Murnaghan, F. D., *The Theory of Group Representations*, The Johns Hopkins Press, Baltimore (1938), pp. 336–346.
833. Myers, M. B., and Vedam, K., Effect of pressure on the optical rotatory power and dispersion of crystals. *J. Opt. Soc. Am.* (USA) **55**, 1180–1181 (1965).
834. Myers, M. B., and Vedam, K., Effect of pressure on the optical rotatory power and dispersion of α-quartz. *J. Opt. Soc. Am.* (USA) **56**, 1741–1745 (1966).
835. Myers, M. B., and Vedam, K., Effect of pressure on the optical rotatory power and dispersion of crystalline sodium chlorate. *J. Opt. Soc. Am.* (USA) **57**, 1146–1148 (1967).

836. Myers, R. A., and Pershan, P. S., Light modulation experiments at 16GC/sec. *J. Appl. Phys.* (USA) **36**, 22–28 (1965).
837. Nagai, K., and Ishikawa, T., Stress-optical coefficient of poly-(dimethyl siloxane) networks. *J. Chem. Phys.* (USA) **45**, 3128–3140 (1966).
838. Namba, S., High voltage measurement by ADP crystal plates. *Rev. Sci. Instrum.* (USA) **27**, 336 (1956).
839. Namba, S., Electro-optic effect in zincblende. *J. Appl. Phys.* (Japan) **28**, 482–487 (1959).
840. Namba, S., Electro-optical effect of zincblende. *J. Opt. Soc. Am.* (USA) **51**, 76–79 (1961).
841. Narasimhamurty, T. S., Photoelastic behaviour of ammonium chloride. *Curr. Sci.* (India) **23**, 149–150 (1954).
842. Narasimhamurty, T. S., Photoelastic studies of optically active crystals—sodium chlorate and sodium bromate. *Proc. Indian Acad. Sci.* **40A**, 167–175 (1954).
843. Narasimhamurty, T. S., "Photoelasticity of crystals: Optical and Ultrasonic Methods," Ph.D. Thesis, Osmania University, India, 1955.
844. Narasimhamurty, T. S., and Ziauddin, M., Birefringent compensator for studying very small changes in double refraction. *J. Opt. Soc. Am.* (USA) **51**, 574–578 (1961).
845. Narasimhamurty, T. S., Ultrasonic methods of determining elasto-optic constants of uniaxial and biaxial crystals. *Acta Crystallogr.* (Denmark) **14**, 1176–1179 (1961).
846. Narasimhamurty, T. S., Photoelastic behaviour of rubidium halides. *J. Opt. Soc. Am.* (USA) **51**, 914 (1961).
847. Narasimhamurty, T. S., and Jeelani, G., Interference spectroscope method for studying the absolute photoelastic constants of glasses and their variation with wavelength. *J. Osmania University Science* (India) **2**, 49–56 (1965).
848. Narasimhamurty, T. S., Photoelastic behaviour of sodium nitrate. *Nature* (GB) **211**, 401–402 (1966).
849. Narasimhamurty, T. S., and Veerabhadra Rao, K., A new technique to study the dispersion of natural birefringence in uniaxial and biaxial crystals. *Indian J. Pure & Appl. Phys.* **6**, 367–370 (1968).
850. Narasimhamurty, T. S., Photoelastic constants of α-quartz, *J. Opt. Soc. Am.* (USA) **59**, 682–686 (1969).
851. Narasimhamurty, T. S., Photoelastic behaviour of Rochelle salt. *Phys. Rev.* (USA) **186**, 945–948 (1969).
852. Narasimhamurty, T. S., Kerr-electro-optic effect in crystals, in: *Dr. Bhagavantam's Commemoration Volume* (S. Balakrishna, M. Krishna Murty, and B. Ramachandra Rao, eds.), The Bangalore Printing and Publishing Co. Ltd., Bangalore, India, 1969, pp. 170–172.
853. Narasimhamurty, T. S., Interference spectroscope: Some further applications. *J. Opt. Soc. Am.*, Bulletin on topical meeting at Massachusetts Institute of Technology, Cambridge, Massachusetts, 1970, p. 6.
854. Narasimhamurty, T. S., Veerabhadra Rao, K., and Pettersen, H. E., Photoelastic constants of ADP. *J. Materials Sci.* (GB) **8**, 577–580 (1973).
855. Nash, F. R., Bergman, J. G., Boyd, G. D., and Turner, E. H., Optical nonlinearities in $LiNbO_3$. *J. Appl. Phys.* (USA) **40**, 5201–5206 (1969).
856. Nasyrov, A. N., Akhmedzhanov, F. R., and Akhmedzhanov, M. M., Acoustic

and acousto-optic properties of magnesium fluoride crystals. *Sov. Phys.-Solid State* (USA) **17**, 2379–2380 (1975).

857. Nath, G., and Haussühl, S., Large nonlinear optical coefficient and phase matched second harmonic generation in $LiIO_3$. *Appl. Phys. Lett.* (USA) **14**, 154–156 (1969).
858. Nath, G., and Haussühl, S., Strong second harmonic generation of a ruby laser in lithium iodate. *Phys. Lett.* (Netherlands) **29a**, 91–92 (1969).
859. Nath, N. S. N., Diffraction of light by high frequency sound waves—Generalised theory. *Proc. Indian Acad. Sci.* **4A**, 222–242 (1936).
860. Nath, N. S. N., The visibility of ultrasonic waves and its periodic variations. *Proc. Indian Acad. Sci.* **4A**, 262–274 (1936).
861. Nath, N. S. N., and Mueller, H., Diffraction of light by supersonic waves in solids. *Nature* (GB) **141**, 37 (1938).
862. Nekrasov, M. M., Belyakov, B. M., and Rodionov, M. K., Refractive index and electro-optical effect in impure KCl crystals. *Sov. Phys.-Crystallogr.* (USA) **13**, 270–271 (1968).
863. Nelson, D. F., and Reinhart, F. K., Light modulation by the electro-optic effect in reverse-biased GaP *p–n* junctions. *Appl. Phys. Lett.* (USA) **5**, 148–150 (1964).
864. Nelson, D. F., The electro-optic effect as a probe for studying the optical waveguide properties of *p–n* junctions, in: *Electronic Structure in Solids, Cheria, Crete*, 30 June–14 July 1968, Plenum, London (1969), pp. 133–143.
865. Nelson, D. F., and Turner, E. H., Electro-optic and piezo-electric coefficients and refractive index of gallium phosphide. *J. Appl. Phys.* (USA) **39**, 3337–3343 (1968).
866. Nelson, D. F., and Lax, M., New symmetry foı acousto-optic scattering. *Phys. Rev. Lett.* (USA) **24**, 378–380 (1970).
867. Nelson, D. F., and Lax, M., Theory of the photoelastic interaction. *Phys. Rev. B* (USA) **B-3**, 2778–2794 (1971); also Erratum, **B-4**, 3779 (1971).
868. Nelson, D. F., Lazay, P. D., and Lax, M., Brillouin scattering in anisotropic media —calcite. *Phys. Rev. B* (USA) **6**, 3109–3120 (1972).
869. Nelson, D. F., and Lax, M., Higher order photo-elastic interactions—Theory and experiment, in: *Photo-elastic Effect and Its Applications—Proceedings of the Conference at Ottigneis, Belgium*, 9–14 September 1973, Springer-Verlag, Berlin (1975), pp. 351–375.
870. Nelson, D. F., General solution for the electro-optic effect. *J. Opt. Soc. Am.* (USA) **65**, 1144–1151 (1975).
871. Neumann, F. E., Über Gesetze der Doppelbrechung des Lichtes in comprimierten oder ungleichformig Erwamten unkrystallischen Körpern. *Abh. Akad. Wiss. Berlin*, Part II, pp. 1–254 (1841); see also *Ann. der Phys. Chem.* (Germany) **II-54**, 449–476 (1841), in German.
872. New, B. M., Versatile electro-optic alignment system for field applications. *Appl. Opt.* (USA) **13**, 937–941 (1974).
873. Nieh, S. T. K., and Harris, S. E., Aperture-bandwidth characteristics of acousto-optic filter. *J. Opt. Soc. Am.* (USA) **62**, 672–776 (1972).
874. Nikitenko, V. I., and Martynenko, G. P., Certain photoelastic properties of gallium arsenide and silicon. *Sov. Phys.-Solid State* (USA) **7**, 494–496 (1965).
875. Nissle, T. R., and Babcock, C. L., Stress optical coefficient as related to glass composition. *J. Am. Cerem. Soc.* (USA) **56**, 596–598 (1973).
876. Nitsche, R., Crystal growth and electro-optic effect of bismuth germanate, $Bi(GeO_4)_3$. *J. Appl. Phys.* (USA) **36**, 2358–2360 (1965).

877. Nitsche, R., and Wild, P., Crystal growth and electro-optic effect of copper-tantalum-selenide Cu_3TaSe_4. *J. Appl. Phys.* (USA) **38**, 5413–5414 (1967).

878. Nomoto, O., Über eine neue Sichtbarmachungsmethode stehender Ultraschallwellen in Flüssigkeiten. *Proc. Phys. Math. Soc.* (Japan) **18**, 402–424 (1936); also see **19**, 337–365 (1937), in German.

879. Nomoto, O., Intensitätsverteilung des Lichtes in den auf fertschreiftenden Ultraschallwellen in Flüssigkeiten erzeugten Beugungspektren. *Proc. Phys. Math. Soc.* (Japan) **22**, 314–320 (1940), in German.

880. Nomoto, O., Studien über die Beigung von Licht an Ultraschallwellen. *Proc. Phys. Math. Soc.* (Japan) **24**, 380–400 (1942), in German.

881. Nomoto, O., and Torikai, Y., Diffraction of light by ultrasound: Method of successive diffraction computation. *J. Acoust. Soc.* (Japan) **26**, 406–417 (1970), in English.

882. Nomoto, O., Diffraction of light by ultrasound—Extension of the Brillouin theory. *Japan J. Appl. Phys.* (Japan) **10**, 611–622 (1971).

883. Nomoto, O., Diffraction of light by ultrasound—Calculation of spectral intensities for the case of Bragg incidences of various orders using the theory of Brillouin. *J. Phys.* (France) **33**, No. 11-12, Suppl. PC6, 189–195 (1972), in French.

884. Nomura, S., Arima, H., and Kojima, F., Quadratic electro-optic effect in the system $Pb(Zn_{1/3}Nb_{2/3})O_3$–$PbTiO_3$. *Jap. J. Appl. Phys.* (Japan) **12**, 531–535 (1973).

885. Nye, J. F., *Physical Properties of Crystals*, 2nd ed., Oxford University Press, London (1964); Clarendon Press, Oxford, England (1967).

886. Ogawa, T., Piezo-optic frequency analyzer. *J. Acoust. Soc. Am.* (USA) **30**, 46–47 (1958).

887. Ogawa, T., Piezo-optic ADP crystal resonator. *J. Appl. Phys.* (Japan) **28**, 321–326 (1959).

888. Ohm, E. A., A linear optical modulator with high FM sensitivity. *Appl. Opt.* (USA) **6**, 1233–1235 (1967).

889. Ohmachi, Y., and Uchida, N., Temperature dependence of elastic, dielectric and piezoelectric constants in TeO_2 single crystals. *J. Appl. Phys.* (USA) **41**, 2307–2311 (1970).

890. Ohmachi, Y., and Uchida, N., Acoustic and acousto-optic properties of Pb_2MoO_5 single crystal. *J. Appl. Phys.* (USA) **42**, 521–524 (1971).

891. Ohmachi, Y., and Uchida, N., Vitreous As_2Se_3—Investigation of acousto-optical properties and application to infrared modulator. *J. Appl. Phys.* (USA) **43**, 1709–1712 (1972).

892. Ohmachi, Y., and Uchida, N., Acousto-optic property of single crystal $5PbO \cdot 3GeO_2$. *J. Appl. Phys.* (USA) **43**, 3583–3584 (1972).

893. Ohmachi, Y., Uchida, N., Niizeki, N., Acoustic wave propagation in TeO_2 single crystal. *J. Acoust. Soc. Am.* (USA) **51**, 164–168 (1972).

894. Ohmachi, Y., and Uchida, N., Acoustic and acousto-optical properties of TeO_2 single crystal. *Rev. Ele. Commun. Lab.* (Japan) **20**, 529–541 (1972).

895. Ohmachi, Y., and Uchida, N., Investigation of acousto-optical properties of vitreous As_2Se_3 and its application to infrared modulator. *Electron. Commun. Labs. Tech. J.* (Japan) **21**, 181–185 (1972), in Japanese.

896. Okada, M., and Ieiri, S., Extinction ratio of electro-optic light modulator. *IEEE J. Quantum Electron.* (USA) **QE-6**, 522–523 (1970).

897. Okolicsanyl, F., The wave-slot, an optical television system. *Wireless Eng.* **14**, 527–536 (1937).
898. Onuki, K., Uchida, N., and Saku, T., Interferometric method for measuring electro-optic coefficients in crystals. *J. Opt. Soc. Am.* (USA) **62**, 1030–1032 (1972).
899. Ott, J. H., and Sliker, T. R., Linear electro-optic effects in KH_2PO_4 and its isomorphs. *J. Opt. Soc. Am.* (USA) **54**, 1442–1444 (1964).
900. Owen, J. R., and White, E. A. D., Second harmonic generation in organic crystals. *J. Mater. Sci.* (GB) **11**, 2165–2169 (1976).
901. Pakhnev, A. V., Gorbach, S. S., and Shaskol'skaya, M. P., Temperature dependence of elasto-optical constants p_{11} and p_{12} of NaCl, KCl, KBr and KI crystals. *Sov. Phys.-Crystallogr.* (USA) **18**, 687 (1974).
902. Pakhnev, A. V., Shaskol'skaya, M. P., and Gorbach, S. S., On the temperature dependence of the elasto-optic effect in some alkali halide crystals. *Sov. Phys.-Crystallogr.* (USA) **20**, 648–649 (1975).
903. Pakhnev, A. V., Shaskol'skaya, M. P., and Gorbach, S. S., Calculation of the components of the elasto-optic effect in alkali halide crystals. *Izv. VUZ Fiz.* (USSR) **12**, 28–34 (1975), in Russian.
904. Parker, Jr., J. H., Kelley, F., and Bolef, D. I., An ultrasonic-optical determination of the third order elastic constant c_{111} for NaCl single crystals. *Appl. Phys. Lett.* (USA) **5**, 7–9 (1964).
905. Parthasarathy, S., Diffraction of light by ultrasonic waves—Part I. *Proc. Indian Acad. Sci.* **A3**, 442–447 (1936).
906. Parthasarathy, S., Diffraction of light by ultrasonic waves—Part II: Reflection and transmission phenomena. *Proc. Indian Acad. Sci.* **A3**, 594–606 (1936).
907. Parthasarathy, S., and Nath, N. S. N., The visibility of ultrasonic waves in liquids. *Proc. Indian Acad. Sci.* **A4**, 555–561 (1936).
908. Parthasarathy, S., Visibility of ultrasonic waves in liquids. *Curr. Sci.* (India) **5**, 136 (1936).
909. Parygin, V. N., and Timershanova, R. S., Direction of the strongest electro-optic effect in crystals of different classes. *Sov. J. Quantum Electron.* (USA) **4**, 835–838 (1975).
910. Pataj, K., Turska, I., and Wardzynski, W., Birefringence in gallium arsenide induced by uniaxial stress. *Bull. Acad. Pol. Sci., Ser. Sci. Tech.* (Poland) **17**, 723–729 (1969).
911. Pedinoff, M., and Seguin, H., Direct measurement of infrared photoelastic constants of silicon. *IEEE J. Quantum Electron.* (USA) **QE-3**, 31–32 (1967).
912. Pepperi, R., Digital laser beam deflection and its application. *Phys. Bl.* (Germany) **29**, 352–361 (1973), in German.
913. Perfilova, V. E., and Sonin, A. S., Experimental detection of the Kerr effect in non-centro symmetric crystals. *Sov. Phys.-Crystallogr.* (USA) **10**, 349–350 (1965).
914. Perfilova, V. E., Sonin, A. S., and Lomova, L. G., Change in the optical properties of crystals when placed in electric fields (Kerr effect). *Sov. Phys.-Crystallogr.* (USA) **10**, 588–591 (1966).
915. Perfilova, V. E., and Sonin, A. S., The electro-optic properties of single crystals of barium titanate. *Sov. Phys.-Solid State* (USA) **8**, 82–84 (1966).
916. Perfilova, V. E., Quadratic electro-optical effect in single crystals ADP. *Sov. Phys.-Crystallogr.* (USA) **12**, 621–622 (1968).

917. Peters, C. G., Refractive index of glass at high temperatures. *Bureau of Standards, Sci. Papers* (USA), No. 521, 635–659 (1926).
918. Peters, C. J., Gigacycle band width coherent light travelling wave phase modulator. *Proc. IEEE* (USA) **51**, 147–153 (1963).
919. Peters, C. J., Gigacycle bandwidth coherent light travelling wave amplitude modulator. *Proc. IEEE* (USA) **53**, 455–460 (1965).
920. Peterson, G. E., Ballman, A. A., Lenzo, P. V., and Bridenbaugh, P. M., Electro-optic properties of $LiNbO_3$. *Appl. Phys. Lett.* (USA) **5**, 62–64 (1964).
921. Pettersen, H. E., Determination of the ratio p/q for glasses by Mueller's method C; a modification due to Klein—Experimental results. Ultrasonics Laboratory Memo No. 26, Michigan State University, East Lansing, Michigan (USA), 1964.
922. Pettersen, H. E., "The strain-optical constants of some sodium halides," Doctoral Dissertation, Michigan State University, East Lansing, Michigan (USA), 1968.
923. Pettersen, H. E., Cook, B. D., and Hiedemann, E. A., Ultrasonic determination of the ratios of photoelastic constants, in: *Physics of the Solid State, Commemoration Volume to Professor S. Bhagavantam*, S. Balakrishna, M. Krishnamurthy, and B. Ramachandra Rao, eds., Academic Press, London and New York (1969), 389–399.
924. Pettersen, H. E., New ultrasonic technique for the determination of the ratios of strain-optical constants. *J. Acoust. Soc. Am.* (USA) **48**, 1093–1097 (1970).
925. Pettersen, H. E., Ratios of strain-optical constants of alkali halides by an ultrasonic technique. *J. Opt. Soc. Am.* (USA) **63**, 1243–1245 (1973).
926. Pettersen, H. E., and Narasimhamurty, T. S., Determination of the signs of the ratios of strain-optical constants measured ultrasonically, *J. Acoust. Soc. Am.* (USA) **54**, 1635–1638 (1973).
927. Phariseau, P., On the diffraction of light by progressive supersonic waves. *Proc. Indian Acad. Sci.* **A-44**, 165–170 (1956).
928. Phariseau, P., Diffraction of light by a three-dimensional system of ultrasonics. *Physica* (Netherlands) **24**, 985–995 (1958).
929. Phariseau, P., The diffraction of light by ultrasonic waves. *Simon Stevin, Wis-Natuurkundig Tijdschrift* (Holland) **33**, 72–93 (1959).
930. Phatak, S. D., Srivastava, R. C., and Subbarao, E. C., Elastic constants of orthorhombic $KNbO_3$ by X-ray diffuse scattering. *Acta Crystallogr. A* (Denmark) **28**, Part 3, 227–231 (1972).
931. Phillips, F. C., *An Introduction to Crystallography*, 4th ed., ELBS and Longman Group Ltd., London (1971).
932. Piltch, M. S., Cantrell, C. D., and Sze, R. C., Infrared second harmonic generation in nonbirefringent cadmium telluride. *J. Appl. Phys.* (USA) **47**, 3514–3517 (1976).
933. Pine, A. S., Brillouin scattering in semi-conductors, in: *Light Scattering in Solids*, M. Cardona, ed., Springer-Verlag, Berlin (1975), pp. 253–273.
934. Pinnow, D. A., and Dixon, R. W., α-Iodic acid: A solution grown crystal with a high figure of merit for acousto-optic device applications. *Appl. Phys. Lett.* (USA) **13**, 156–158 (1968).
935. Pinnow, D. A., Van Uitert, L. G., Warner, A. W., and Bonner, W. A., Lead molybdate—A melt-grown crystal with a high figure of merit for acousto-optic device applications. *Appl. Phys. Lett.* (USA) **15**, 83–86 (1969).
936. Pinnow, D. A., and Dixon, R. W., α-Iodic acid—A solution grown crystal with a high figure of merit for acousto-optic device applications. *IEEE Trans. Sonics Ultrasonics* (USA) **SU-16**, No. 1, 25 (1969).

937. Pinnow, D. A., and Williamson, S. R., Acousto-optic light deflection—the design and operation of a simple *X–Y* deflection system, Presented at the spring meeting of the Optical Society of America, San Diego, California, April 1969.
938. Pinnow, D. A., Guidelines for the selection of acousto-optic materials. *IEEE J. Quantum Electron.* (USA) **QE-6**, 223–238 (1970).
939. Pinnow, D. A., Acousto-optic light deflection—Design considerations for first order beam steering transducers. *IEEE Trans. Sonics & Ultrasonics* (USA) **SU-18**, 209–214 (1971).
940. Pinnow, D. A., Elasto-optical materials, in: *Handbook of Lasers*, R. J. Pressley, ed., The Chemical Rubber Co., Cleveland, Ohio (1971).
941. Pisarevskii, Yu., Tregubov, G. A., and Shaldin, Yu. V., The electro-optical properties of ADP, KDP and $N_4(CH_2)_6$ crystals in UHF fields. *Sov. Phys.-Solid State* (USA) **7**, 530–531 (1965).
942. Pisarevskii, Yu., Tregubov, G. A., and Shaldin, Yu. V., Measurement of electro-optical coefficients in microwave fields. *Instrum. Exper. Tech.* (USA), No. 5, 1180–1182 (1965).
943. Pishoraty, P. Rama, On the visibility of ultrasonic waves. *Proc. Indian Acad. Sci.* **A-4**, 27–36 (1936).
944. Plancke-Schuyten, G., and Mertens, R., The diffraction of light by standing supersonic waves—Normal incidence of light—II: Exact solution of Raman–Nath equations. *Simon Steven* (Belgium) **47**, 17–31 (1973).
945. Plechata, R., Crystal and stress optics—Contribution to the mutual relationship of. *Acta Tech.* (Prague) **1**, 394–412 (1956).
946. Plechata, R., A geometric interpretation of double refraction in photoelasticity. *Acta Tech.* (Prague) **2**, 230–261 (1957).
947. Pockels, F., Über den Einfluss elastischer Deformationen, speciell einseitigen Druckes, auf das optische Verhalten kristallinischer Körper. *Ann. Phys.* (Germany) **III-37**, 144–172 (1889), in German.
948. Pockels, F., Über den Einfluss elastischer Deformationen, speciell einseitigen Druckes, auf das optische Verhalten krystallinischer Körper—Beobachtungen an Quarz. *Ann. Phys.* (Germany) **III-37**, 269–305 (1889), in German.
949. Pockels, F., Über den Einfluss elastischer Deformationen, speciell einseitigen Druckes, auf das optische Verhalten kristallinischer Körper. *Ann. Phys.* (Germany) **III-37**, 372–394 (1889), in German.
950. Pockels, F., Über die durch einseitigen Druck hervorgerufene Doppelbrechung regulärer Kristalle, speciell von Steinsalz und Sylvin. *Ann. Phys.* (Germany) **III-39**, 440–469 (1890), in German.
951. Pockels, F., Über die Änderung des optischen Verhaltens von Alaun und Beryll durch einseitigen Druck. *Neues Jahrb. Miner.* **8**, 217–268 (1892), in German.
952. Pockels, F., Über die Änderung des optischen Verhaltens von Alaun und Beryll durch einseitigen Druck. *Z. Kristallogr.* (Germany) **23**, 639–642 (1892), in German.
953. Pockels, F., Über den Einfluss des elektrostatischen Feldes auf das optische Verhalten piezoelektrischer Kristalle. *Abh. Gött.* (Germany) **39**, 1–204 (1894), in German.
954. Pockels, F., Über die Änderung des optischen Verhaltens verschiedener Gläser durch elastische Deformation. *Ann. Phys.* (Germany) **IV-7**, 745–771 (1902); also see **IV-9**, 220–223 (1902); **IV-11**, 651–653 (1902); both in German.

955. Pockels, F., Über die Änderung der Lichtfortpflanzung in Kalkspat durch Deformationen. *Ann. Phys.* (Germany) **IV**(11), 726–753 (1903), in German.
956. Pockels, F., *Lehrbuch der Kristalloptik*, B. G. Teubner, Leipzig, Germany (1906), in German.
957. Pockels, F., Bemerkungen zur Theorie der durch äussere Einwirkung erzeugten Doppelbrechung. *Phys. Z.* (Germany) **9**, 314–317 (1908), in German.
958. Podmaniczky, A., Some properties of TeO_2 light deflectors with small interaction length. *Opt. Commun.* (Netherlands) **16**, 161–165 (1976).
959. Pohl, D. W., and Schwarz, S. E., Thermo- and elasto-optic parameters of NaF and their implications for light scattering from second sound. *Phys. Rev. B* (USA) **7**, 2735–2739 (1973).
960. Poincaré, H., *Théorie Mathématique de la Lumière*, Gauthier-Villars, Paris (1892), Vol. 2, Chap. 12, in French.
961. Poindexter, E., "Piezobirefringence in diamond." Doctoral thesis, University of Michigan, Ann Arbor (USA), 1955.
962. Poindexter, E., Piezobirefringence in diamond. *Am. Mineral.* (USA) **40**, 1032–1054 (1955).
963. Polyakova, A. L., Elastic non-linearity in stimulated Mandelshtam–Brillouin scattering. *JETP Lett.* (USA) **7**, 57–60 (1968).
964. Popov, S. G., Acoustic diffraction of light in a α-iodic acid (α-HIO_3) crystal. *C. R. Acad. Bulg. Sci.* (Bulgaria) **26**, 1319–1322 (1973), in Russian.
965. Popova, A. S., and Rips, L. B., The true electro-optic effect in crystals of ADP. *Sov. Phys. Crystallogr.* (USA) **10**, 347–348 (1965).
966. Porreca, F., On the propagation of narrow light beams in liquids travelled by ultrasound. *Nuovo Cimento* (Italy) **6**, 274–281 (1952).
967. Poulet, H., The piezo-optic and electro-optic properties of zinc blende (ZnS). *J. Phys. Rad.* (France) **16**, 237–238 (1955), in French.
968. Prasad, P. L., and Singh, S., Calculation of electro-optical parameters: a new approach. *Chem. Phys. Lett.* (Netherlands) **31**, 535–538 (1975).
969. Primak, W., Delbecq, C. J., and Yuster, P. H., Photoelastic observations of the expansion of alkali halides on irradiation. *Phys. Rev.* (USA) **98**, 1708–1715 (1955); see also **102**, 1688–1689 (1956).
970. Primak, W., Radiation induced changes of dimensions and dispersion of LiF. *Phys. Rev.* (USA) **112**, 1075–1088 (1958).
971. Primak, W., and Post, D., Photoelastic constants of vitreous silica, and its elastic coefficient of refractive index. *J. Appl. Phys.* (USA) **30**, 779–788 (1959).
972. Primak, W., and Luthra, J., Radiation induced expansion and increase in refractive index of MgO; evidence for the F-centre. *Phys. Rev.* (USA) **150**, 551–561 (1966).
973. Primak, W., and Anderson, T. T., Secondary (electro) photoelastic effect in lithium niobate. *J. Appl. Phys.* (USA) **41**, 4745–4746 (1970).
974. Prince, E., and Wooster, W. A., Determination of elastic constants of crystals from diffuse reflection of X-rays—III: Diamond. *Acta Crystallogr.* (Denmark) **6**, 450–459 (1953).
975. Pursey, H., Merran, P. A., and Musgrave, M. J., Measurement of the Pockels effect in KDP at 9G c/s. *Brit. J. Appl. Phys.* **18**, 285–291 (1967).
976. Pursey, H., and Newman, R. J., Measurements of the Pockels effect in quartz at 9 GHz. *Brit. J. Appl. Phys.* (*J. Phys. D*) *Ser. 2* (GB) **1**, 707–710 (1968).
977. Pursey, H., Page, P. A., and Musgrave, M. J., On the dispersion of optical and

electro-optical coefficients in zinc telluride. *J. Phys. C (Solid State Phys.), Ser. 2* (GB) **2**, 1085–1091 (1969).

978. Pursey, H., and Page, P. A., Dispersion of optical and electro-optical coefficients in semiconductors. *J. Phys. C (Solid State Phys.)* (GB) **3**, 431–437 (1970).

979. Quate, C. F., Wilkinson, C. D., and Winslow, D. K., Interaction of light and microwave sound. *Proc. IEEE* (USA) **53**, 1604–1623 (1965).

980. Rabin, H., Non-linear optics and defect solid state. *Radiat. Eff.* (GB) **4**, 167–171 (1970): International Conference on nonmetallic crystals, New Delhi (India), January 1969.

981. Radhakrishnan, T., The dispersion formulae of the alkali halides. *Proc. Indian Acad. Sci.* (India) **A27**, 165–170 (1948).

982. Radhakrishnan, T., Temperature variation of the refractive index of lithium fluoride. *Proc. Indian Acad. Sci.* (India) **A31**, 224–228 (1950).

983. Radhakrishnan, T., The refraction of magnesium oxide and its temperature variation. *Proc. Indian Acad. Sci.* (India) **A31**, 229–233 (1950).

984. Rahman, A., and Iyengar, K. S., Dispersion of the strain-optical ratios in cubic crystals. *Acta Crystallogr.* (Denmark) **20**, Part I, 144–145 (1966).

985. Rahman, A., and Iyengar, K. S., Piezo-optic birefringence sign reversal and isotropy in cubic crystals. *Phys. Lett.* (Netherlands) **25A**, 478–479 (1967).

986. Rahman, A., "Ultraviolet study of piezo-optic behaviour in crystals." Ph.D. Thesis, Osmania University, Hyderabad, India, 1969.

987. Rahman, A., Piezo-optic dispersion in the nitrates of lead and strontium in the ultraviolet, in *Bhagavantam Volume* (Dr. S. Bhagavantam's commemoration volume), Balakrishna, S., Krishnamurty, M., and Ramachandra Rao, B., eds., The Bangalore Printing and Publishing Co. Ltd., Bangalore, India (1969), pp. 173–178.

988. Rahman, A., and Iyengar, K. S., Dispersion of piezo-optic constants of some alkali halides in the ultraviolet region—I: Determination of absolute values. *Acta Crystallogr. A* (Denmark) **26A**, 128–133 (1970).

989. Rahman, A., and Iyengar, K. S., Dispersion of piezo-optic constants of some alkali halides in the ultraviolet region—II: Deformation potential constants of the valence band. *Acta Crystallogr. A* (Denmark) **26A**, 359–363 (1970).

990. Rajagopal, E. S., Orientational misalignments in the determination of elastic constants of cubic crystals. *J. Sci. Ind. Res.* (India) **20B**, 50–52 (1961).

991. Rajagopalan, S. R., and Ramaseshan, S., Rotating elliptic analysers for the automatic analysis of polarised light: I. *Proc. Indian Acad. Sci.* (India) **60A**, 297–312, 379–395 (1964).

992. Rajaram, N., and Ramaseshan, S., Dispersion of the birefringence of quartz. *J. Opt. Soc. Am.* (USA) **11**, 1531 1532 (1971).

993. Ramachandran, G. N., Photoelastic constants of diamond. *Proc. Indian Acad. Sci.* **A-25**, 208–219 (1947).

994. Ramachandran, G. N., Thermo-optic behaviour of solids—III: Fluorspar. *Proc. Indian Acad. Sci.* (India) **A25**, 286–295 (1947).

995. Ramachandran, G. N., Thermo-optic behaviour of solids—V: Alkali halides. *Proc. Indian Acad. Sci.* (India) **A25**, 481–497 (1947).

996. Ramachandran, G. N., Photoelastic constants of diamond—Corrections. *Proc. Indian Acad. Sci.* (India) **A27**, 421 (1948).

997. Ramachandran, G. N., Photoelasticity of diamond. *Proc. Indian Acad. Sci.* (India) **A32**, 171–173 (1950).
998. Ramachandran, G. N., and Chandrasekharan, V., Photoelastic constants of sodium chlorate. *Proc. Indian Acad. Sci.* **A33**, 199–215 (1951).
999. Ramachandran, G. N., and Wooster, W. A., Determination of elastic constants of crystals from diffuse reflexions of X-rays—I: Theory of the method. *Acta Crystallogr.* (Denmark) **4**, 335–344 (1951).
1000. Ramachandran, G. N., and Wooster, W. A., Determination of elastic constants of crystals from diffuse reflexions of X-rays—II: Application to some cubic crystals. *Acta Crystallogr.* (Denmark) **4**, 431–440 (1951).
1001. Ramachandran, G. N., and Ramaseshan, S., Magneto-optic rotation in birefringent media—Application of the Poincaré sphere. *J. Opt. Soc. Am.* (USA) **42**, 49–56 (1952).
1002. Ramachandran, G. N., and Ramaseshan, S., Electro-optics, in: *Handbuch der Physik, Crystal Optics—Diffraction*, S. Flugge, ed., Springer-Verlag, Berlin (1961), PH Vol. XXV-1, pp. 191–198.
1003. Ramachandran, G. N., and Ramaseshan, S., Photoelasticity, in: *Handbuch der Physik, Crystal Optics—Diffraction*, S. Flugge, ed., Springer-Verlag, Berlin (1961), PH Vol. XXV-1, pp. 202–217.
1004. Ramachandran, V., Palmer, A. W., Scibor-Rylski, M., and Widdis, F. C., A novel electro-optic diffraction deflector. *J. Phys. D* (*Appl. Phys.*) (GB) **8**, L115–L117 (1975).
1005. Ramachandran, V., and Palmer, A. W., An improved diffraction-type electro-optic modulator deflector. *J. Phys. D* (*Appl. Phys.*) (GB) **8**, L199–L201 (1975).
1006. Ramachandran, V., and Palmer, A. W., Comparison of deflection and modulation characteristics of symmetrically and asymmetrically electroded electro-optic devices. *J. Phys. D* (*Appl. Phys.*) (GB) **9**, 1387–1390 (1976).
1007. Raman, C. V., and Nath, N. S. N., The diffraction of light by high frequency sound waves, Part I. *Proc. Indian Acad. Sci.* (India) **2A**, 406–412 (1935).
1008. Raman, C. V., and Nath, N. S. N., The diffraction of light by sound waves of high frequency, Part II. *Proc. Indian Acad. Sci.* **2A**, 413–420 (1935).
1009. Raman, C. V., and Nath, N. S. N., The diffraction of light by high frequency sound waves, Part III: Doppler effect and coherence phenomena. *Proc. Indian Acad. Sci.* **3A**, 75–84 (1936).
1010. Raman, C. V., and Nath, N. S. N., The diffraction of light by high frequency sound waves, Part IV: Generalized theory. *Proc. Indian Acad. Sci.* (India) **3A**, 119–125 (1936).
1011. Raman, C. V., and Nath, N. S. N., The diffraction of light by high frequency sound waves, Part V: General considerations—Oblique incidence and amplitude changes. *Proc. Indian Acad. Sci.* (India) **3A**, 459–465 (1936).
1012. Raman, C. V., and Venkateswaran, C. S., Optical observations of the Debye heat waves in crystals (gypsum). *Nature* (GB) **142**, 250 (1938).
1013. Raman, C. V., and Venkataraman, K. S., Determination of the adiabatic piezo-optic coefficient of liquids. *Proc. R. Soc. A* (GB) **171A**, 137–147 (1939).
1014. Raman, C. V., and Viswanathan, K. S., On the theory of the elasticity of crystals. *Proc. Indian Acad. Sci.* (India) **A42**, 51–70 (1955).
1015. Ramanadham, M., Magnetic birefringence in liquids of the aliphatic series. *Indian J. Phys.* **4**, 15–38 (1929–30).

1016. Ramaseshan, S., and Chandrasekharan, V., Faraday effect and birefringence. *Curr. Sci.* (India) **20**, 150–151 (1951).

1017. Ramaseshan, S., Faraday effect and birefringence I. *Proc. Indian Acad. Sci.* (India) **34A**, 32–40 (1951).

1018. Ramaseshan, S., Faraday effect and birefringence II—Corundum. *Proc. Indian Acad. Sci.* (India) **34A**, 97–100 (1951).

1019. Ramaseshan, S., A magneto-optic method for the determination of piezo-optic coefficients. *Proc. Indian Acad. Sci.* (India) **A40**, 184–188 (1954).

1020. Ramaseshan, S., An analyser for elliptically polarized light. *J. Indian Inst. Sci.* (India) **37A**, 195–199 (1955).

1021. Ramaseshan, S., and Sivaramakrishnan, V., Stress-optic dispersion in glasses. *Curr. Sci.* (India) **25**, 246–248 (1956).

1022. Ramaseshan, S., and Ranganath, G. S., Piezo-optic phenomena, in: *Physics of the Solid State, Commemoration Volume to Professor S. Bhagavantam*, Balakrishna, S., Krishnamurty, M., and Ramachandra Rao, B., eds., Academic Press, London and New York (1969), pp. 427–470.

1023. Ramaseshan, S., Ranganath, G. S., Ramesh, T. G., and Rajagoplan, S. R., Piezo-absorption coefficients of weakly and strongly absorbing crystals. *Mater. Res. Bull* (USA) **4**, 535–544 (1969).

1024. Ramavataram, K., Ultrasonic diffraction patterns in some optical glasses. *J. Opt. Soc. Am.* (USA) **45**, 749–750 (1955).

1025. Ranganath, G. S., and Ramaseshan, S., The evaluation of some piezo-rotatory coefficients of α-quartz. *Curr. Sci.* (India) **38**, 303–304 (1969).

1026. Ranganath, G. S., and Ramaseshan, S., Piezo-rotatory coefficients, and crystal symmetry. *J. Opt. Soc. Am.* (USA) **59**, 1229–1232 (1969).

1027. Ranganath, G. S., and Ramaseshan, S., Piezo-rotatory coefficients and stress-induced optical activity. *Proc. Indian Acad. Sci.* (India) **70A**, 275–291 (1969).

1028. Ranganath, G. S., Rajagopalan, S. R., and Ramaseshan, S., Photoelastic behaviour of ionic crystals. International Conference on Non-metallic Crystals, New Delhi, India, January, 1969, in: *Radiat. Eff.* (GB) **4**, Nos. 3–4, 313–318 (July, 1970).

1029. Ranganath, G. S., Optics of polycrystalline minerals. *Curr. Sci.* (India) **41**, 386–391 (1972).

1030. Ranganath, G. S., and Ramaseshan, S., Photoelasticity in polycrystalline aggregates. *Pramana* (India) **1**, 78–87 (1973).

1031. Rao, B. R., and Murty, J. S., Diffraction of light by weak ultrasonic fields. *Z. Phys.* (Germany) **152**, 440–447 (1958).

1032. Rasetti, F., Incoherent scattered radiation in diatomic molecules. *Phys. Rev.* (USA) **34**, 367–371 (1929).

1033. Rau, R. Y. N., "Theory of the linear electro-optic effect in KDP." Thesis, New Mexico State University, USA, 1967, 56 pp.

1034. Rayleigh, Lord, On the light from the sky, its polarization and colour. *Philos. Mag.* (GB) **41**, 107–120, 274–279 (1871).

1035. Rayleigh, Lord, On the scattering of light by small particles. *Philos. Mag.* (GB) **41**, 447–454 (1871).

1036. Rayleigh, Lord, On the transmission of light through an atmosphere containing small particles in suspension, and on the origin of the blue of the sky. *Philos. Mag.* (GB) **47**, 375–384 (1899).

1037. Rayleigh, Lord, Does motion through the ether cause double refraction? *Philos. Mag.* (GB) **4**, 678–683 (1902).
1038. Rayleigh, Lord, *The Theory of Sound*, Dover Publications, New York (1945), Vol. II, pp. 89–96.
1039. Recker, K., Wallrafen, F., and Haussühl, S., Single crystal growth and optical, elastic and piezoelectric properties of polar magnesium barium fluoride. *J. Crystal Growth* (Netherlands) **26**, 97–100 (1974).
1040. Reeb, O., Über eine Methode zur Messung der Dispersion der künstlichen Doppelbrechung. *Ann. Phys.* (Germany) **IV-71**, 427–456 (1923), in German.
1041. Reintjes, J., and Schulz, M. B., Photoelastic constants of selected ultrasonic delay-line crystals. *J. Appl. Phys.* (USA) **39**, 5254–5258 (1968).
1042. Rez, I. S., Crystals with non-linear polarizability, *Sov. Phys.-Usp.* (USA) **10**, 759–782 (1968).
1043. Rez, I. S., Ferroelectric materials for optical applications. *Bull. Acad. Sci. USSR Phys. Ser.* (USA) **33**, 266–272 (1969).
1044. Rez, I. S., Meisner, L. B., Safonov, A. I., Baryshev, S. A., Plotinskaya, T. A., and Klyuev, V. P., Optical properties of single crystals of the selenite type. *Sov. Phys.-Crystallogr.* (USA) **15**, 1020–1022 (1971).
1045. Reza, A., and Babonas, G., Effect of directional stress on the birefringence of CdSe. *Sov. Phys.-Solid State* (USA) **16**, 909–911 (1974).
1046. Reza, A., Babonas, G., and Silcika, A., Quasi-cubic state and piezobirefringence of CdS and CdSe. *Phys. Status Solidi B* (Germany) **72**, 421–429 (1975).
1047. Reznikov, B. A., Sirotin, Yu. I., and Voropaeva, N. E., Determination of the temperature dependence of the piezo-optical and elasto-optical coefficients of crystals. *Phys. Status Solidi* (Germany) **33**, 633–640 (1969).
1048. Rice, R. R., and Burkhart, G. H., Single crystal modelocks YAG and doubles frequency. Reported in *Laser Focus* (USA) **11**, 89 (1975).
1049. Rigrod, W. W., and Kaminow, I. P., Wide band microwave light modulation. *Proc. IEEE* (USA) **51**, 137–140 (1963).
1050. Riley, W. A., and Klein, W. R., Piezo-optic coefficients of liquids. *J. Acoust. Soc. Am.* (USA) **42**, 1258–1261 (1967).
1051. Riskaer, S., Determination of shear deformation potentials from the free-carrier piezo-birefringence in germanium and silicon. *Phys. Rev.* (USA) **152**, 845–849 (1966).
1052. Riskaer, S., and Balslev, I., Interband piezo-birefringence in germanium. *Phys. Lett.* (Netherlands) **21**, 16–18 (1966).
1053. Rissmann, S., and Vosahlo, H., Light modulation and image production by means of electro-optical crystals. *Jenaer Jahrbuch* (Germany) **I**, 228–264 (1960).
1054. Rivoallan, L., and Favre, F., KRS5—A new acousto-optical material of high performance. *Opt. Commun.* (Netherlands) **8**, 404–405 (1973), in French.
1055. Rivoallan, L., and Favre, F., Optical elastic and elasto-optical properties of KRS6. *Opt. Commun.* (Netherlands) **11**, 296–298 (1974), in French.
1056. Robert, A., The application of Poincaré sphere to photoelasticity. *Int. J. Solids & Struct.* (GB) **6**, 423–432 (1970).
1057. Robertson, G., An intensity method for photoelastic birefringence measurements. *Brit. J. Appl. Phys.* (GB) **16**, 207–218 (1965).
1058. Robinovitch, M. M., and Rouzeyre, M., Photoelectric arrangement of great sen-

sibility for the measurement of the Kerr electro-optic effect. *C. R. Hebd. Séances Acad. Sci.* (France) **247**, 441–442 (1958), in French.

1059. Robinson, D. M., The supersonic light control and its application to television with special reference to the Scophony television receiver. *Proc. IRE* (USA) **27**, 483–486 (1939).

1060. Robinson, F. N. H., Nonlinear optical coefficients. *Bell Syst. Tech. J.* (USA) **46**, 913–956 (1967).

1061. Robinson, F. N. H., Electronic contribution to the electro-optic effect in ammonium chloride. *J. Phys. C.* (*Solid State Physics*) (GB), Ser. 2 **1**, 286–289 (1968).

1062. Rodgers, K. F., The Pockels effect in hexamine. *Appl. Opt.* (USA) **8**, 2369–2370 (1969).

1063. Roland, G. W., Gottlieb, M., and Feichtner, J. D., Optoacoustic properties of hallium arsenic sulphide, Tl_2AsS_4. *Appl. Phys. Lett.* (USA) **21**, 52–54 (1972)

1064. Ronnekleiv, A., Shaw, H. J., and Souquet, J., Grating acoustic scanners. *Appl. Phys. Let*'. (USA) **28**, 361–362 (1976).

1065. Röntgen, W. C., Über die durch elektrische Kräfte erzeugte Änderung der Doppelbrechung des Quarzes. *Ann. Phys. Chem.* (Germany) **18**, 213–228 (1883); also see 534–551 (1883), in German.

1066. Röntgen, W. C., Bemerkung zu der Abhandlung des Hrn. A. Kundt: Über des optisches Verhalten des Quarzes im elektrischen Felds. *Ann. Phys. Chem.* (Germany) **19**, 319–323 (1883), in German.

1067. Rosenthal, A. H., Application of ultrasonic light modulation to signal recording, display, analysis and communication. *IRE Trans. Ultrasonic Engineering* (USA) **UE-8**, 1–5 (1961).

1068. Rosner, R. D., Turner, E. H., and Kaminow, I. P., Clamped electro-optical coefficients of KDP and quartz. *Appl. Opt.* (USA) **6**, 778 (1967).

1069. Rosner, R. D., and Turner, E. H., Electro-optic coefficients in calcium pyroniobate. *Appl. Opt.* (USA) **7**, 171–173 (1968).

1070. Rozenberg, L. D., Survey of methods used for the visualization of ultrasonic fields. *Sov. Phys.–Acoust.* (USA) **1**, 105–116 (1955).

1071. Rupp, W. J., Conventional optical polishing techniques. *Opt. Acta* (GB) **18**, 1–16 (1971).

1072. Ruscio, J. T., A coherent light modulator. *IEEE J. Quantum Electron.* (USA) **QE-1**, 182–183 (1965).

1073. Rytov, S. M., Lichtbeugung an Ultraschallwellen. *Phys. Z. Sowjetunion* **8**, 626–643 (1935), in German.

1074. Rytov, S. M., Diffraction of light by ultrasonic waves. *C.R.* (*Doklady*) *Acad. Sci.* (USSR) **2**, 229–233 (1936), in French.

1075. Rytov, S. M., Diffraction of light by ultrasonic waves. *C. R.* (*Doklady*) *Acad. Sci.* (USSR) **3**, 151–156 (1936), in French.

1076. Sailer, E., Fouskova, A., Konak, C., and Unruh, H. G., Brillouin scattering in $(NH_4)_2Cd_2(SO_4)_3$ and $Tl_2Cd_2(SO_4)_3$. *Phys. Status Solidi A* (Germany) **29**, k73–78 (1975).

1077. Sakane, T., A new method of measuring the photoelastic constants. *Jap. J. Appl. Phys.* (*Japan*) **13**, 739–740 (1974).

1078. Sakane, T., and Kashiwage, H., Simple method for the measurement of photoelastic constants and its application to Vycor glass. *Jap. J. Appl. Phys.* (Japan) **13**, 1992–1995 (1974).

1079. Salje, E., Electro-optical studies on monoclinic potassium iodate, in *Discussion Group of the Crystal Science Section of the German Mineralogical Society* (abstract only), Karlsruhe, Germany, 1971, Akademische Verlagsgesellschaft, Frankfurt (1971), p. 20, in German.
1080. Sanders, H. F., Intensity measurements in the diffraction of light by ultrasonic waves. *Canad. J. Res.* (Canada) **A14**, 158–171 (1936).
1081. Sapriel, J., Cinnabar (α HgS), a promising acousti-optical material, *Appl. Phys. Lett.* (USA) **19**, 533–535 (1971).
1082. Sapriel, J., Rivoallan, L., and Ribet, J. L., Acoustic and acousto-optical properties of α-sulphur. *J. Phys.* (France) **33**, No. 11–12, suppl., C6.150–155 (1972), in French.
1083. Sasaki, H., Tsubouchi, K., Chubachi, N., and Mikoshiba, N., Photoelastic effect in piezoelectric semiconductor: ZnO. *J. Appl. Phys.* (USA) **47**, 2046–2049 (1976).
1084. Savinkov, A. I., A high temperature chamber for optical investigations (crystals and glasses). *Instrum. Exp. Tech.* (USA) **18**, No. 1, Part 2, 226–227 (1975).
1085. Savur, S. R., On the stress optical coefficients for direct tension and pressure in the case of glass. *Philos. Mag.* (GB) **VI-50**, 453–463 (1925).
1086. Schaefer, C., and Bergmann, L., New diffraction effect with oscillating crystals, *Naturwissenschaften* (Germany) **22**, 685–690 (1934), in German.
1087. Schaefer, C., and Bergmann, L., Zur Frage der optischen Beugungserscheinungen an schwingenden Glaskörpern. *Naturwissenschaften* (Germany) **23**, 799–800 (1935), in German.
1088. Schaefer, C., Bergmann, L., Fues, L., and Ludloff, H., Diffraction effects with oscillating crystals. *Preuss. Akad. Wiss. Berlin* (Germany) **14**, 222–239 (1953), in German.
1089. Schaefer, C., Bergmann, L., and Goehlich, H. J., Die Bestimmung der elastischen Konstanten optischer Glaser mittels der Licht Beugung an Ultraschall-wellen. *Glastech. Ber.* (Germany) **15**, 447–449 (1937), in German.
1090. Schaefer, C., and Bergmann, L., Measurement of elastic constants of optical glass from diffraction patterns of glass cubes vibrating at high frequencies. *Ann. Phys.* (Germany) **3**, 72–76 (1948), in German.
1091. Schaefer, C., and Nassenstein, H., Determination of the photoelastic constants p and q of optical glasses. *Z. Naturforsch.* (Germany) **8a**, 90–95 (1953), in German.
1092. Schaefer, C., and Dransfeld, K., The determination of the ratio p/q of the photoelastic constants of optical glasses by a dynamic method. *Z. Naturforsch.* (Germany) **8a**, 96–99 (1953), in German.
1093. Schaefer, B., Duppy, C., and Saucier, H., Photoelastic study of dislocations in coloured LiF. *Phys. Status Solidi* (Germany) **9**, 753–765 (1965), in French.
1094. Schatter, W., Experiments on light diffraction by ultrasound in quartz. *Z. Phys.* (Germany) **198**, 409–413 (1967), in German.
1095. Schieber, M., Davies, T. J., Schnepple, N. F., Randtke, P. T., and Carlston, R. C., Crystal growth and electro-optical characterization of bismuth tri-iodide. *J. Appl. Phys.* (USA) **45**, 5371–5372 (1974).
1096. Schmidt-Tiedemann, K. J., Stress-optical constants of germanium. *J. Appl. Phys.* (USA) **22**, 2058–2059 (1961).
1097. Schmidt-Tiedemann, K. J., Optical double refraction caused by free carriers in semiconductors. *Z. Naturforsch.* (Germany) **16A**, 639 (1961), in German.
1098. Schmidt, E., Precision of ellipsometric measurements. *J. Opt. Soc. Am.* (USA) **60**, 490–494 (1970).

1099. Schmidt, E. D. D., and Vedam, K., Variation of refractive indices of CaF_2, BaF_2, and β-PbF_2 with pressure to 7 kbar. *J. Phys. Chem. Solids* (GB) **27**, 1563–1566 (1966).

1100. Schmidt, E. D. D., Kirk, J. L., and Vedam, K., Variation of the refractive index of diamond with hydrostatic pressure to 7 kilobars. *Am. Mineral.* (USA) **53**, 1404–1406 (1968); also see *Am. Mineral.* (USA) **54**, 569 (1969).

1101. Schmidt, H., and Schwarzmuller, J., Review of ferroelectric materials usable for passive electro-optic alphanumeric display devices. *Ferroelectrics* (GB) **10**, 283–293 (1976).

1102. Schneider, W. C., and Vedam, K., Piezo-optic properties of amorphous selenium at a wavelength of 1.15 μ. *J. Opt. Soc. Am.* (USA) **60**, 800–804 (1970).

1103. Schramm, C., Electro-optical effect in zinc blende. *Ann. Phys.* (Germany) **V-25**, 309–336 (1936).

1104. Schreuer, E., Precision measurements of the velocity of ultrasonic waves in different liquids. *Akust. Zh.* (USSR) **4**, 215–230 (1939), in German.

1105. Schueler, D. G., and Smith, W. D., Self-limiting electro-optic variable density filters. *1973 WESCON Technical Papers. Western Electronic Show and Convention, San Francisco, California*, 11–14 September 1973, Western Electronic Show & Convention, Los Angeles (1973), Vol. 17, pp. 30/4-6.

1106. Schulz, M. B., Polarization of light Bragg diffracted by sound on optically isotropic solids. *IEEE J. Quantum Electron.* (USA) **QE-4**, 232–234 (1968).

1107. Schwab, C. S., and Robino, P., Photoelastic properties of cuprous halides. *Opt. Commun.* (Netherlands) **4**, 304–306 (1971).

1108. Schumate, M. S., Interferometric determination of the principal refractive indices of barium titanate single crystals. *Appl. Phys. Lett.* (USA) **5**, 178–179 (1964).

1109. Sealer, D. A., and Hsu, H., Stimulated Brillouin scattering as a parametric interaction. *IEEE J. Quantum Electron.* (USA) **QE-1**, 116–124 (1965).

1110. Shakin, O. V., Bryzhina, M. F., and Lemanov, V. V., Photoelastic constant of alkaline earth fluorides. *Sov. Phys.-Solid State* (USA) **13**, 3141–3142 (1972).

1111. Shaldin, Yu. V., Pisarevsku, Yu. V., and Melnikov, Yu. S., Measurement of the electro-optical effect in crystals. *Zh. Priklad. Spektrosk.* (USSR) **3**, 463–467 (1965), in Russian.

1112. Shaldin, Yu. V., Lobachev, A. N., and Dem'yanets, L. N., Nonlinear optical properties of $LiWO_4$ crystals. *Sov. Phys.-Solid State* (USA) **11**, 1395–1396 (1969).

1113. Shaldin, Yu. V., The measurement of the primary linear electro-optical effect in ammonium dihydrogen phosphate crystals. *Sov. Phys.-Crystallogr.* (USA) **15**, 146–147 (1970).

1114. Shaldin, Yu., V., and Belogurov, D. V., Induced double refraction study of the distribution of macroscopic electric fields within semiconductors. *Sov. Phys.-Semicond.* (USA) **6**, 807–808 (1972).

1115. Shaldin, Yu. V., Smirnow, S. L., Belyoev, L. M., and Shpilko, I. A., Nonlinear optical properties of paratellurite (TeO_2) crystals. *Sov. Phys.-Crystallogr.* (USA) **18**, 358–359 (1973).

1116. Shaldin, Yu. V., and Belogurov, D. A., Anisotropy of a nonlinear refractive index in acentric crystals: linear electro-optic effect. *Opt. & Spectrosc.* (USA) **35**, 403–407 (1973).

1117. Shaldin, Yu. V., Belogurov, D. A., Belyaev, L. M., and Chikladze, O. A., Low frequency electro-optical properties of lithium iodate crystals. *Sov. J. Quantum Electron.* (USA) **4**, 283–285 (1974).

1118. Shamburov, V. A., and Blokh, O. G., Electro-optical effect in crystals and its applications. *Radio Eng. & Electron. Phys.* (USA) **9**, 408–415 (1964).

1119. Shamburov, V. A., and Kharikonova, I. V., Transverse electro-optical effect in $4/m$ class crystals. *Zh. Priklad. Spektrosk.* (USSR) **9**, 637–643 (1968), in Russian.

1120. Shamburov, V. A., Kuznetsov, V. A., Lobachev, A. N., Kharikonova, I. V., and Soshnikov, B. G., Electro-optical properties of cubic ZnS crystals produced by a hydrothermal method. *Sov. Phys.-Crystallogr.* (USA) **15**, 252–255 (1970).

1121. Shandarov, S. M., Investigation of the Schaefer–Bergmann diffraction in a lithium niobate crystal at hypersonic frequency. *Sov. Phys.-Solid State* (USA) **15**, 1722–1724 (1974).

1122. Sharma, H. P., Shanker, J., and Verma, M. P., Analysis of the photoelastic effect in ionic crystals. *J. Phys. & Chem. Solids* (GB) **37**, 1077–1079 (1976).

1123. Shaskolskaya, M. P., Pakhnev, A. V., Gorbach, S. S., and Malikova, N. M., Influence of impurities on the photoelastic constants of potassium chloride crystals. *Sov. Phys.-Crystallogr.* (USA) **18**, 817 (1974).

1124. Shileikar, A. Yu., Cardona, M., and Pollak, F. H., Intrinsic piezobirefringence of AlSb. *Solid State Commun.* (USA) **7**, 1113–1117 (1969).

1125. Shimizu, H., Tsukamoto, M., Ishibashi, Y., and Takagi, Y., Photoelastic constants of $NaNO_2$. *J. Phys. Soc. Jap.* (Japan) **38**, 195–201 (1975).

1126. Shimomura, O., and Furuya, N., Electro-optical characteristics of DKDP single crystals and their relation with Curie points. *Electr. Eng. Jap.* (USA) **93**, 1–7 (1973).

1127. Shimomura, O., Furuya, N., and Fujii, Y., Optimum cut of KDP-type crystals for optical modulation considering crystal dimensions. *Electron. Commun. Jap.* (USA) **57**, 105–113 (1974).

1128. Shintani, R., Okuda, S., and Soezima, Y., New methods for precision measurements of birefringence using Babinet Compensator. *Technol. Rep. Osaka Univ.* (Japan) **16**, 199–204 (1966).

1129. Shtyrkova, A., Internal stress and refractive indices of quartz irradiated by fast neutrons. *Sov. Phys. Crystallogr.* (USA) **14**, 624–626 (1970).

1130. Shull, H. E., and Vedam, K., Nonlinear variation of the induced birefringence of vitreous silica with uniaxial stress to 7 kbar. *J. Appl. Phys.* (USA) **43**, 3724–3728 (1972).

1131. Sibaiya, L., Scattering of light in a Rochelle salt crystal. *Proc. Indian Acad. Sci.* **8A**, 393–396 (1938).

1132. Sidnenko, E. V., and Zheludev, I. S., Measurement of the electromechanical and electro-optical coefficients of single crystal ADP by means of a Twyman Interferometer. *Sov. Phys. Crystallogr.* (USA) **12**, 397–399 (1967).

1133. Siegman, A. E., Quate, C. F., Bjorkholm, J., and Francois, G., Frequency translation of an He–Ne laser's output frequency by acoustic output coupling inside the resonant cavity. *Appl. Phys. Lett.* (USA) **5**, 1 (1964).

1134. Silverstein, L., and Sucher, M., Determination of the Pockels electro-optic coefficient in ADP at 5.5 GHz. *Electron. Lett.* (GB) **2**, 437–438 (1966).

1135. Silvestrova, I. M., Barta, Ch., Dobrzhanskii, G. F., Balyaev, L. M., and Pisarevskii, Yu. V., Acousto-optical properties of calomel crystals Hg_2Cl_2. *Sov. Phys.-Crystallogr.* (USA) **20**, 649–651 (1975).

1136. Simonder, F., Michard, F., and Torguet, R., Use of optical heterodyning for local measurement of the phase velocity of elastic waves. *Opt. Commun.* (Netherlands) **16**, 411–416 (1976), in French.

1137. Sinyakov, E. V., Golovyanko, A. A., and Savehenko, V. G., Electro-optical effect in monocrystals of solid solutions $BaTiO_3$-$ZnTiO_3$. *Ukr. Fiz. Zh.* (USSR) **16**, 1371–1374 (1971), in Russian.

1138. Sinyakov, E. V., Savehenko, V. G., and Golovyanko, A. A., Quadratic electro-optic effect in single crystals of $BaTiO_3$–$NiTiO_3$ solid solutions. *Sov. Phys.-Crystallogr.* (USA) **19**, 400–401 (1974).

1139. Sirota, N. N., and Korina, R. V., Electro-optical properties of triglycine selenate. *Krist. & Tech.* (Germany) **9**, 55–61 (1974), in English.

1140. Sittig, E. K., Design, technology and limitations of elasto-optic light deflection, in: Proceedings of the Electro-optical Systems Design Conference, New York, September 1970, pp. 681–690.

1141. Sittig, E. K., Elasto-optic light modulation and deflection, in: *Progress in Optics*, Wolf, E., ed., North-Holland Publishing Co., Amsterdam (1972), Vol. X, pp. 231–288.

1142. Skinner, J. G., Comment on light beam deflectors. *Appl. Opt.* (USA) **3**, 1504 (1964).

1143. Slater, J. C., Interaction of waves in crystals. *Rev. Mod. Phys.* (USA) **30**, 197–222 (1958).

1144. Slawson, C. G., and Denning, R. M., Stress and double refraction in diamond. *Am. Mineral.* (USA) **40**, 1135–1139 (1955).

1145. Sliker, T. R., Linear electro-optic effects in class 32, 6, $3m$, and $\bar{4}3m$ crystals. *J. Opt. Soc. Am.* (USA) **54**, 1348–1351 (1964).

1146. Sliker, T. R., and Jost, J. M., Linear electro-optic effect and refractive indices of cubic ZnTe. *J. Opt. Soc. Am.* (USA) **56**, 130–131 (1966).

1147. Smakula, P. H., and Claspy, P. C., The electro-optic effect in $LiNbO_3$ and KTN. *Trans. Metall. Soc. AIME* (USA) **239**, 421–424 (1967).

1148. Smirnov, A. I., Electro-optical effect in cubic crystals. *Izv. Vuz. Fiz.* (USSR) **9**, 127–129 (1970), in Russian.

1149. Smith, A. W., and Burns, G., Optical properties and switching in $Gd_2(MoO_4)_3$. *Phys. Lett.* (Netherlands) **28A**, 501–502 (1969).

1150. Smith, A. W., Burns, G., and Scott, B. A., Nonlinear optical properties of potassium–lithium niobates. *J. Appl. Phys.* (USA) **42**, 684–686 (1971).

1151. Smith, R. G., Use of the acousto-optic light deflector as an optical isolator. *IEEE J. Quantum Electron.* (USA) **QE-9**, 545–546 (1973).

1152. Smith, T., and Korpel, A., Measurement of light–sound interaction efficiencies in solids. *IEEE J. Quantum Electron.* (USA) **QE-1**, 283–284 (1965).

1153. Smolenskii, G. A., Krainik, N. N., Berezhnov, A. A., Electro-optical effect in $PbMg_{1/3}Nb_{2/3}O_3$ crystals. *Sov. Phys. Solid State* (USA) **10**, 365–369 (1968).

1154. Smolenskii, G. A., Krainik, N. N., Berzhnoi, A. A., and Mylnikova, I. E., Electro-optical properties of $PbZn_{1/3}Nb_{2/3}O_3$, $PbMg_{1/3}Ta_{2/3}O_3$ and $PbNi_{1/3}Nb_{2/3}O_3$ ferroelectric crystals. *Sov. Phys.-Solid State* (USA) **10**, 2105–2108 (1968).

1155. Smolenskii, G. A., Berezhnoi, A. A., Pisarev, R. V., Prokhorova, S. D., Krainik, N. N., and Mylnikova, I. E., Anomalous dispersion of the electro-optical effect in ferroelectric $PbNi_{1/3}Nb_{2/3}O_3$. *Sov. Phys.-Solid State* (USA) **11**, 913–916 (1969).

1156. Smolenskii, G. A., Kamzina, L. S., and Krainik, N. N., Orientation processes and elasto-optical effect in lead magnoniobate—a ferroelectric with a diffuse transition. *Bull. Acad. Sci. USSR, Phys. Ser.* (USA) **39**, 141–143 (1975), in Russian.

1157. Smushkov, I. V., Kaplan, M. S., and Sumin, V. I., Piezo-optic constants of KCl and KBr crystals as a function of temperature. *Sov. Phys.-Solid State* (USA) **12**, 1540–1542 (1971).

1158. Sokolov, V. I., Kushev, D. B., and Subashiev, V. K., Proposed interference method for determining the signs of the electro-optical coefficients. *Sov. Phys.-Crystallogr.* (USA) **18**, 200–201 (1973).

1159. Sonin, A. S., Vasilevskaya, A. S., and Strukov, B. A., Electro-optical properties of KDP and deuterated KDP crystals in the region of their phase transitions. *Sov. Phys.-Solid State* (USA) **8**, 2758–2760 (1967).

1160. Sonin, A. S., and Vasilevskaya, A. S., Electro-optical properties of triglycine sulphate: (iv) Deuterated triglycine sulphate. *Sov. Phys.-Crystallogr.* (USA) **12**, 386–391 (1967).

1161. Sonin, A. S., and Vasilevskaya, A. S., Electro-optical properties of triglycine sulphate. *Sov. Phys.-Solid State* (USA) **9**, 1536–1539 (1968).

1162. Sonin, A. S., and Apkaryants, P. A., Electro-optical properties of potassium sodium niobate single crystals. *Sov. Phys.-Solid State* (USA) **10**, 234–235 (1968).

1163. Sonin, A. S., and Perfilova, V. E., Electro-optical effect in strontium titanate crystals. *Sov. Phys.-Solid State* (USA) **10**, 242–243 (1968).

1164. Soref, R. A., and McMahon, D. H., Optical design of Wollaston prism digital light deflectors. *Appl. Opt.* (USA) **5**, 425–434 (1966).

1165. Sorok, V. V., Theory of light diffraction by sound waves in anisotropic media. *Sov. Phys.-Acoust.* (USA) **19**, 560–564 (1974).

1166. Soule, H. V., *Electro-optical Photography at Low Illumination Levels*. John Wiley & Sons, New Yoιk (1968), 392 pp; see "Modular considerations," pp. 182–184, and "Laser beam writing capabilities," p. 186.

1167. Spangenberg, K., and Haussühl, S., Elastic constants of alkali halides of the rocksalt type. *Z. Kristallogr.* (Germany) **109**, 422–437 (1957), in German.

1168. Spariel, J., Cinnabar (α-HgS), a promising acousto-optical material. *Appl. Phys. Lett.* (USA) **19**, 533–535 (1971).

1169. Spencer, E. G., Lenzo, P. V., and Nassau, K., Ferroelectric behaviour of lithium niobate. *Appl. Phys. Lett.* (USA) **7**, 67–69 (1965).

1170. Spencer, E. G., Lenzo, P. V., and Nassau, K., Optical interactions with elastic waves in lithium niobate. *IEEE J. Quantum Electron.* (USA) **QE-2**, 69–70 (1966).

1171. Spencer, E. G., Lenzo, P. V., and Ballman, A. A., Dielectric materials—electro-optic, elasto-optic, and ultrasonic device applications. *Proc. IEEE* (USA) **55**, 2074–2108 (1967).

1172. Spinner, S., and Waxler, R. M., Relations between refractive index and density of glasses resulting from annealing compared with corresponding relation resulting from compression. *Appl. Opt.* (USA) **5**, 1887–1889 (1966).

1173. Srinath, L. S., and Sarma, A. V. S. S. S. R., Effects of stress induced optical activity in photoelasticity. *J. Phys. D (Appl. Phys.)* (GB) **5**, 883–895 (1972).

1174. Srinivasan, R., Dispersion of stress-optic coefficient of NaCl and potassium halides. *Nature* (GB) **182**, 1660–1661 (1958).

1175. Srinivasan, R., Dispersion of stress-optic coefficient of alkali halides. *Z. Phys.* (Germany) **155**, 281–289 (1959).

1176. Srinivasan, R., Dispersion of the stress-optic coefficient of LiF. *Naturwissenschaften* (Germany) **48**, 96 (1961).
1177. Stadnik, B., Crystal modulators of type ADP and KDP. *Acta Tech. CSAV* (Czechoslovakia) **12**, 236–239 (1967).
1178. Stadnik, V., New method for studying the linear electro-optic effect. *Opt. & Spectrosc.* (USA) **32**, 308–309 (1972).
1179. Stafsudd, O. M., Haak, F. A., and Radisavljevic, K., Electro-optic constant of cadmium telluride. *Appl. Opt.* (USA) **6**, 1276–1277 (1967).
1180. Staudte, J. H., and Cook, B. D., Visualization of quasilongitudinal and quasi-transverse elastic waves. *J. Acoust. Soc. Am.* (USA) **41**, 1547–1548 (1967).
1181. Stepanov, A. V., and Bobrikov, V. P., Dependence of the optical limit of elasticity in the system (111):(011) on temperature in rocksalt crystals. *Sov. Phys.–Tech. Phys.* (USA) **1**, 777–778 (1957).
1182. Stevenson, J. L., The linear electro-optic coefficients of meta-nitroaniline. *J. Phys. D (Appl. Phys.)* (GB) **6**, 13–16 (1973).
1183. Stone, S. M., A microwave electro-optic modulator which overcomes transit time limitation. *Proc. IEEE* (USA) **52**, 409–410 (1964).
1183a. Subbaramaiya, D. S., Diffraction of light by ripples on liquid surfaces—Part I. *Proc. Indian Acad. Sci.* **6A**, 333–339 (1937).
1184. Sueta, T., Matsushima, T., Nishimoto, T., and Makimoto, T., Modulation of 10.6 micron laser radiation by CuCl. *Proc. IEEE* (USA) **58**, 1378–1379 (1970).
1185. Sugie, M., and Tada, K., A theory on the photoelastic coefficients and unclamped values of the linear electro-optic coefficients. *Jap. J. Appl. Phys.* (Japan) **15**, 257–264 (1976).
1186. Sugie, M., and Tada, K., Measurements of the linear electro-optic coefficients and analysis of the nonlinear susceptibilities in cubic GaAs and hexagonal CdS. *Jap. J. Appl. Phys.* (Japan) **15**, 421–430 (1976).
1187. Suryanarayana, D., "Studies in some aspects of crystal physics: (i) Crystal symmetry and physical properties, (ii) Photoelastic effect in cubic crystals and (iii) Ultrasonic studies in some piezoelectric crystals." D. Sc. Thesis, Andhra University (India) (1948).
1188. Suvorov, V. S., and Sonin, A. S., Nonlinear optical materials. *Sov. Phys.-Crystallogr.* (USA) **11**, 711–723 (1966).
1189. Swanson, D., Brunel, L. C., and Dows, D. A., Brillouin scattering by molecular crystals: s-trichlorobenzene. *J. Chem. Phys.* (USA) **63**, 3863–3869 (1975).
1190. Swartz, K. D., and Granato, A. V., Experimental test of the Laval–Raman theory of elasticity. *J. Acoust. Soc. Am.* (USA) **38**, 824–826 (1965).
1191. Sypek, D. W., Modulators move ahead on two fronts. *Opt. & Spectrosc.* (USA) **6**, 22–23 (1972).
1192. Szivessy, G., The linear electro-optic effect in crystals—Pockels effect, in: *Handbuch der Physik*, Geiger, H., and Scheel, K., eds., Vol. 21, *Light and Matter*, Julius Springer, Berlin (1929), pp. 790–802, in German.
1193. Szivessy, G., Mechanical double refraction—Photoelastic effect in crystals, in: *Handbuch der Physik*, Geiger, H., and Scheel, K., eds., Vol. 21, *Light and Matter*, Julius Springer, Berlin (1929), pp. 832–866, in German.
1194. Tada, K., Aoki, M., Washiyama, M., and Murai, T., Linear electro-optic effect in ZnTe. *J. Fac. Eng. Univ. Tokyo A* (Japan), *Ser. A*, No. 8, 34–35 (1970), in Japanese.

1195. Tada, K., and Aoki, M., Linear electro-optic properties of ZnTe at 10.6 microns. *Jap. J. Appl. Phys.* (Japan) **10**, 998–1001 (1971).
1196. Tada, K., Murai, T., Aoki, M., Muto, K., and Awazu, K., Electro-optic light beam deflection with $Sr_{0.75}Ba_{0.25}Nb_2O_6$ prism. *Jap. J. Appl. Phys.* (Japan) **11**, 1622–1627 (1972).
1197. Tada, K., and Sugie, M., Methods for measuring the linear electro-optic coefficients in the infrared region with high accuracy. *Oyo Buturi* (Japan) **45**, 260–265 (1976).
1198. Takasaki, H., Polarization interferometer. *Appl. Opt.* (USA) **8**, 2344–2351 (1969).
1199. Talbot, H. F., Facts relating to optical science (periodicity of grating visibility). *Philos. Mag.* (GB) **9**, 401–407 (1836).
1200. Tamm, Ig., Quantum theory of molecular light scattering in solids. *Z. Phys.* (Germany) **60**, 345–363 (1930), in German.
1201. Tanaka, M., Yamada, M., and Hamaguchi, C., Brillouin scattering in GaSe. *J. Phys. Soc. Jap.* (Japan) **38**, 1708–1714 (1975).
1202. Tashiro, M., The effects of the polarization of the constituent ions on the photoelastic birefringence of glass. *J. Soc. Glass Technol.* (UK) **40**, 353T–362T (1956).
1203. Taylor, G. W., and Miller, A., Feasibility of electro-optic devices utilizing ferroelectric bismuth titanate. *Proc. IEEE* (USA) **58**, 1220–1229 (1970).
1204. Teich, M. C., and Kaplan, T., Electro-optic effect in trigonal selenium at 10.6 μ. *IEEE J. Quantum Electron.* (USA) **QE-2**, 702–703 (1966).
1205. Teleschewskii, W. I., The acousto-optical raster as a grating for length measurements. *Feingeraete Tech.* (Germany) **23**, 65–69 (1974), in German.
1206. Tell, B., Worlock, J. M., and Martin, R. J., Enhancement of elasto-optic constants in the neighbourhood of a band gap in ZnO and CdS. *Appl. Phys. Lett.* (USA) **6**, 123–124 (1965).
1207. Thacher, P. D., and Land, C. E., Ferroelectric electro-optic ceramics with reduced scattering. *IEEE Trans. Electron Devices* (USA) **ED-16**, 515–521 (1969).
1208. Thacher, P. D., A linear electro-optic effect in ferro-electric ceramics PLZT 12/40/60. *Ferroelectrics* (GB) **3**, 147–156 (1972).
1209. Theimer, O., The first order Raman effect in crystals, particularly in diamond. *Proc. Phys. Soc.* (GB) **64A**, 1012–1030 (1951).
1210. Theimer, O., On the relations between the photoelastic properties and the Raman effect in crystals. *Proc. Phys. Soc.* (GB) **65A**, 38–45 (1952).
1211. Theimer, O., Thermal scattering of light by ionic crystals, *J. Phys. Chem. Solids* (GB) **29**, 1077–1079 (1968).
1212. Theophanous, N. G., A $Gd_2(MoO_4)_3$ longitudinal electro-optic modulator at 6328 Å. *IEEE J. Quantum Electron.* (USA) **QE-12**, 507–510 (1976).
1213. Tien, P. K., Parametric amplification and frequency mixing in propagating circuits. *J. Appl. Phys.* (USA) **29**, 1347–1357 (1958).
1214. Timan, B. L., and Minkov, B. I., Manifestation of the elastic nonlinearity in diffraction of light by ultrasonic waves in crystals. *Sov. Phys.-Solid State* (USA) **13**, 2573–2575 (1972).
1215. Trevelyan, B., The electro-optical effect in proustite. *Opto-Electron.* (GB) **1**, 9–12 (1969).
1216. Tsai, C. S., and Hance, H. V., Optical imaging of the cross-section of a microwave acoustic beam in rutile by Bragg diffraction of a laser beam. *J. Acoust. Soc. Am.* (USA) **42**, 1345–1347 (1967).

1217. Tsai, C. S., and Yao, S. K., Bragg diffraction by standing ultrasonic waves with application to optical demultiplexing. *J. Appl. Phys.* (USA) **43**, 5081–5084 (1972).
1218. Tsay, Y. F., Mitra, S. S., Bendow, B., and Gianino, P. D., Theory of photoelasticity of semiconductor crystals. Report No. AFCRL-TR-74-0636, Rhode Island University, Kingston, Rhode Island (USA), 1974, 7 pp.
1219. Tsay, Y. F., Bendow, B., and Mitra, S. S., Theory of Brillouin effect—calculation of elasto-optic constants of diamond and zinc blende semiconductors. *Proceedings of the Third International Conference on Light Scattering in Solids, Campinas, Brazil*, 25–30 July 1975, Flammarion Sci., Paris (1976), pp. 442–446.
1220. Tsi-Ze, Ny., Experimental study of the deformations and of the changes in the optical properties of quartz under the influence of an electric field. *C. R. Hebd. Séances Acad. Sci.* (France) **185**, 195–197 (1927); also see *J. Phys. Rad.* (France) **9**, 13–37 (1928); both in French.
1221. Tubs, M. R., Dispersion effects in interference methods for the measurement of refractive index. *J. Phys. Chem. Solids* (GB) **30**, 2323–2325 (1969).
1222. Tujisawa, Y., Measurement of electro-optic coefficient r_{63} of ADP crystal at microwave frequencies. *J. Appl. Phys.* (Japan) **27**, 657–659 (1958).
1223. Tukhtasunov, I. T., Shamburov, V. A., and Zheludev, I. S., Electro-optical effect in triglycine sulphate crystals at room temperature. *Sov. Phys.-Crystallogr.* (USA) **12**, 252–255 (1967).
1224. Tukhtasunov, I. T., and Shamburov, V. A., Rotation effect on the optic axis of a Rochelle salt crystal under the influence of an electric field. *Sov. Phys.-Crystallogr.* (USA) **14**, 800–802 (1970).
1225. Turner, E. H., and Kaminow, I. P., Electro-optic effect in gallium arsenide. *J. Opt. Soc. Am.* (USA) **53**, 523 (1963).
1226. Turner, E. H., High frequency electro-optic coefficients of lithium niobate. *Appl. Phys. Lett.* (USA) **8**, 303–304 (1966).
1227. Turner, E. H., Electro-optic coefficients of some crystals at 3.39 μ, in: Program of the Annual Meeting of the Optical Society of America (USA), paper No. ThA13, October 1966.
1228. Turner, E. H., Linear electro-optic effect in HgS (cinnabar) at 0.63 and 3.39 microns. *IEEE J. Quantum Electron.* (USA) **QE-3**, 695–696 (1967).
1229. Turner, E. H., Kaminow, I. P., and Kolb, E. D., Electro-optic effect in trigonal selenium at 1.15 microns. *IEEE J. Quantum Electron.* (USA) **QE-4**, 234 (1968).
1230. Turner, E. H., Nash, F. R., and Bridenbaugh, P. M., Dependence of linear electro-optic effect and dielectric constant on melt composition in lithium niobate. *J. Appl. Phys.* (USA) **41**, 5278–5281 (1970).
1231. Turner, E. H., Buehler, E., and Kasper, H., Electro-optic behaviour and dielectric constants of $ZnGeP_2$ and $CuGeS_2$. *Phys. Rev. B* (USA) **9**, 558–561 (1974).
1232. Twyman, F., and Perry, J. W., The determination of Poisson's ratio and of the absolute stress variation of refractive index. *Proc. Phys. Soc. London* (GB) **34**, 151–154 (1922).
1233. Twyman, F., and Perry, W., On the refractive index changes in optical glass occasioned by chilling and tempering. *J. Soc. Glass Tech.* (UK) **7**, 199–207 (1923).
1234. Uchida, N., Elasto-optic coefficients of liquids determined by ultrasonic light diffraction method. *Jap. J. Appl. Phys.* (Japan) **7**, 1259–1266 (1968).
1235. Uchida, N., Direct measurement of photoelastic coefficients by ultrasonic light diffraction technique. *Jap. J. Appl. Phys.* (Japan) **8**, 329–333 (1969).

1236. Uchida, N., and Iwasaki, H., Two dimensional acousto-optical deflector. *Jap. J. Appl. Phys. Lett.* (Japan) **8**, 811 (1969).

1237. Uchida, N., and Ohmachi, Y., Elastic and photoelastic properties of TeO_2 single crystal. *J. Appl. Phys.* (USA) **40**, 4692–4695 (1969).

1238. Uchida, N., and Ohmachi, Y., Acousto-optical light deflector using TeO_2 single crystal. *Jap. J. Appl. Phys.* (Japan) **9**, 155–156 (1970).

1239. Uchida, N., Optical properties of single crystal paratellurite (TeO_2). *Phys. Rev. B* (USA) **4**, 3736–3745 (1971).

1240. Uchida, N., Schaefer–Bergmann diffraction pattern due to the abnormal Bragg reflection in birefringent media. *IEEE J. Quantum Electron.* (USA) **QE-7**, 160–165 (1971).

1241. Uchida, N., and Saito, S., Elastic and photoelastic constants of ZnS. *J. Appl. Phys.* (USA) **43**, 971–976 (1972).

1242. Uchida, N., Saku, T., Iwasaki, H., and Onuki, K., Electro-optic properties of ferroelectric $5PbO \cdot 3GeO_2$ single crystal. *J. Appl. Phys.* (USA) **43**, 4933–4936 (1972).

1243. Uchida, N., and Niizeki, N., Acousto-optic deflection materials and techniques. *Proc. IEEE* (USA) **61**, 1073–1092 (1973).

1244. Uchida, N., and Saito, S., Acousto-optic tunable filter using TeO_2. *Proc. IEEE* (USA) **62**, 1281–1282 (1974).

1245. Uchida, N., Niizeki, N., Saito, S., and Miyazawa, S., Design and performance of acousto-optic light deflectors. *Oyo Buturi* (Japan) **43**, 171–177 (1974).

1246. Vacher, R., Boyer, L., and Danyach, L., Measurement of elastic constants of $NaClO_3$ by Brillouin scattering. *C. R. Hebd. Séances Acad. Sci. B* (France) **270**, 1430–1433 (1970), in French.

1247. Vacher, R., and Boyer, L., Brillouin scattering—a tool for the measurement of elastic and photoelastic constants. *Phys. Rev. B* (USA) **6**, 639–673 (1972).

1248. Vacher, R., Boissier, M., and Boyer, L., Brillouin scattering measurement of the algebraic values of the photoelastic constants in a birefringent crystal. *Phys. Status Solidi A* (Germany) **18**, 523–529 (1973).

1249. Valasek, J., Properties of Rochelle salt related to the piezoelectric effect. *Phys. Rev.* (USA) **20**, 639–664 (1922).

1250. Van Cittert, P. H., Theory of diffraction of light by ultrasonic waves. *Physica* (Netherlands) **4**, 590–594 (1937).

1251. Van Raalte, John A., Linear electro-optic effect in ferroelectric KTN. *J. Opt. Soc. Am.* (USA) **57**, 671–673 (1967).

1252. Van Uitert, L. G., Singh, S., Levinstein, H. J., Geusic, J. E., and Bonner, W. A., A new and stable nonlinear optical material. *Appl. Phys. Lett.* (USA) **11**, 161–163 (1967).

1253. Van Uitert, L. G., Rubin, J. J., Grodkiewicz, W. H., and Bonner, W. A., Some characteristics of Ba, Sr, Na niobates. *Mater. Res. Bull.* (USA) **4**, 63–73 (1969).

1254. Van der Ziel, J. P., Electro-optic amplitude modulation of laser-generated second harmonics in KH_2PO_4 (KDP). *Appl. Phys. Lett.* (USA) **5**, 27–29 (1964).

1255. Vasileva, R. V., Dobrynin, B. M., and Shingarkina, V. A., Electro-optical system of comparing pulsed sources of light. *Zh. Prikl. Spektrosk.* (USSR) **12**, 142–144 (1970), in Russian.

1256. Vasilevskaya, A. S., Sonin, A. S., and Slepkov, I. A., Electro-optic properties of triglycine sulphate II. *Sov. Phys.-Crystallogr.* (USA) **11**, 640–643 (1966).

1257. Vasilevskaya, A. S., The electro-optical properties of crystals of KDP type. *Sov. Phys.-Crystallogr.* (USA) **11**, 644–647 (1966).
1258. Vasilevskaya, A. S., and Sonin, A. S., Electro-optical and elasto-optical properties of deuterated ADP crystals. *Sov. Phys.-Solid State* (USA) **8**, 2756–2757 (1967).
1259. Vasilevskaya, A. S., and Sonin, A. S., Electro-optical properties of triglycine sulphate III. *Sov. Phys.-Crystallogr.* (USA) **12**, 248–251 (1967).
1260. Vasilevskaya, A. S., Sonin, A. S., and Rez, I. S., Electro-optical and elasto-optical properties of alkali metal dihydrogen arsenates. *Sov. Phys.-Solid State* (USA) **9**, 86–87 (1967).
1261. Vasilevskaya, A. S., Kollobskaya, M. F., Lomova, L. G., Popova, V. P., Regulskaya, T. A., Rez, I. S., Sobesskii, Yu. P., Sonin, A. S., and Suvarov, V. S., Some physical properties of single crystal rubidium dihydrogen phosphate. *Sov. Phys.-Crystallogr.* (USA) **12**, 383–385 (1967).
1262. Vasilevskaya, A. S., Volkova, E. N., Koptsik, V. A., Rashkovich, L. N., Regulskaya, T. A., Rez, I. S., Sonin, A. S., and Suvarov, V. S., Nonlinear electro-optical properties of single crystals of deuterated ADP. *Sov. Phys.-Crystallogr.* (USA) **12**, 446–447 (1967).
1263. Vasilevskaya, A. S., and Sonin, A. S., The contribution of the quadratic effect to the electro-optical properties of ADP crystals. *Sov. Phys.-Crystallogr.* (USA) **13**, 272 (1968).
1264. Vasilevskaya, A. S., Kuznetsova, L. I., Rez, I. S., and Sonin, A. S., A new crystal with a large electro-optical effect. *Sov. Phys.-Solid State* (USA) **10**, 733–734 (1968).
1265. Vasilevskaya, A. S., and Sonin, A. S., On the phenomenological description of electro-optical and elasto-optical effects in crystals. *Opt. & Spectrosc.* (USA) **24**, 511–513 (1968).
1266. Vasilevskaya, A. S., Popova, V. P., Rez, I. S., and Sonin, A. S., Primary electro-optic effect in trenchloride crystals. *Opt. & Spectrosc.* (USA) **25**, 332–333 (1969).
1267. Vasilevskaya, A. S., Ganeev, I. G., Rez, I. S., and Sonin, A. S., Electro-optical properties of thallium–cadmium langbeinite crystals. *Sov. Phys.-Crystallogr.* (USA) **14**, 421–422 (1969).
1268. Vasilevskaya, A. S., and Sonin, A. S., Electro-optical and elasto-optical properties of crystals in the KDP group and their connection with structure. *Sov. Phys.-Crystallogr.* (USA) **14**, 611–613 (1970).
1269. Vasilevskaya, A. S., and Sonin, A. S., Relationship between the dielectric and electro-optical properties of ferroelectric KDP type crystals in the paraelectric phase. *Sov. Phys.-Solid State* (USA) **13**, 1299–1304 (1971).
1270. Vasilevskaya, A. S., and Sonin, A. S., Electro-optical properties of caesium nitrate and potassium tartrate. *Opt. & Spectrosc.* (USA) **31**, 164 (1971).
1271. Vassel, M., and Conwell, E. M., Electro-optic effect in NH_4Cl. *Phys. Rev.* (USA) **140**, A2110–A2117 (1965).
1272. Vasudevan, T. N., and Michael, A. J., Intensity method for stress optical measurements—Stress optical dispersion of single crystal NaF. *J. Opt. Soc. Am.* (USA) **62**, 344–347 (1972).
1273. Vasudevan, T. N., and Krishnan, R. S., Dispersion of the stress-optic coefficient in glasses. *J. Phys. D (Applied Physics)* (GB) **5**, 2283–2287 (1972).
1274. Vasudevan, T. N., and Krishnan, R. S., Stress-optical dispersion in rubidium chloride and rubidium bromide crystals. *Phys. Lett. A* (Netherlands) **43A**, 461–462 (1973).

1275. Vazquez, F., Singh, R. S., and Gonzalo, J. A., Elastic and elasto-optic constants of ammonium perchlorate. *J. Phys. Chem. Solids* (GB) **37**, 451–455 (1976).
1276. Vedam, K., A new method of measuring the elastic constants of solids. *Curr. Sci.* (India) **19**, 205–206 (1950).
1277. Vedam, K., The elastic and photoelastic constants of fused quartz. *Phys. Rev.* (USA) **78**, 472–473 (1950).
1278. Vedam, K., Elastic and photoelastic properties of some optical glasses. *Proc. Indian Acad. Sci.* (India) **A31**, 450–458 (1950).
1279. Vedam, K., Photoelastic properties of barite. *Proc. Indian Acad. Sci.* (India) **A34**, 161–172 (1951).
1280. Vedam, K., and Ramachandran, G. N., Photoelastic constants of sodium chlorate from ultrasonic diffraction. *Proc. Indian Acad. Sci.* (India) **A34**, 240–244 (1951).
1281. Vedam, K., and Ramaseshan, S., Photoelastic effect, in: *Progress in Crystal Physics*, Krishnan, R. S., ed., Interscience Publishers, Inc., New York (1960), pp. 102–138.
1282. Vedam, K., and Schmidt, E. D. D., Effect of hydrostatic pressure on the refractive index of LiF. *Solid State Commun.* (USA) **3**, 373–375 (1965).
1283. Vedam, K., and Schmidt, E. D. D., Piezo-optic behaviour of rubidium chloride up to transition point. *J. Mater. Sci.* (GB) **1**, 301–302 (1966).
1284. Vedam, K., and Schmidt, E. D. D., Variation of refractive index of MgO with pressure to 7 kbar. *Phys. Rev.* (USA) **146**, 548–554 (1966).
1285. Vedam, K., and Schmidt, E. D. D., Nonlinear piezo-optic behaviour of sphalerite (β-ZnS). *Phys. Rev.* (USA) **150**, 766–767 (1966).
1286. Vedam, K., Schmidt, E. D. D., and Roy, R., Non-linear variation of refractive index of vitreous silica with pressure to 7 kbars. *J. Am. Ceram. Soc.* (USA) **49**, 531–535 (1966).
1287. Vedam, K., and Schmidt, E. D. D., Variation of refractive indices of CaF_2, BaF_2 and PbF_2 with pressure to 7 kbar. *J. Phys. Chem. Solids* (GB) **27**, 1563–1566 (1966).
1288. Vedam, K., Schmidt, E. D., Myers, M. B., and Roy, R., *Precision Optical Measurements at High Pressures*. ASME Publication No. 66-WA/PT-3, published by the American Society of Mechanical Engineers, United Engineering Center, 345 East 47th Street, New York (1966), 7 pp.
1289. Vedam, K., and Srinivasan, R., Nonlinear piezo-optics. *Acta Crystallogr. A* (Denmark) **22**, 630–634 (1967).
1290. Vedam, K., and Davis, T. A., Nonlinear variation of refractive indices of α-quartz with pressure. *J. Opt. Soc. Am.* (USA) **57**, 1140–1145 (1967).
1291. Vedam, K., and Davis, T. A., Pressure dependence of the refractive indices of the tetragonal crystals, ADP, KDP, $CaMoO_4$, $CaWO_4$ and rutile. *J. Opt. Soc. Am.* (USA) **58**, 1446–1451 (1968).
1292. Vedam, K., and Davis, T. A., Pressure and temperature variation of the optical rotatory power of α-quartz. *J. Opt. Soc. Am.* (USA) **58**, 1451–1455 (1968).
1293. Vedam, K., and Davis, T. A., Piezo- and thermo-optic behaviour of $LiNbO_3$. *Appl. Phys. Lett.* (USA) **12**, 138–140 (1968).
1294. Vedam, K., Optical interferometry at high pressures. Proceedings of the symposium on accurate characterization of the high pressure environment held at the National Bureau of Standards, Gaithersburg, Maryland (USA), October 14–18, 1968, pp. 39–44.
1295. Vedam, K., and Davis, T. A., Pressure dependence of the refractive indices of the

hexagonal crystals beryl, α-CdS, α-ZnS and ZnO. *Phys. Rev.* (USA) **181**, 1196–1201 (1969).

1296. Vedam, K., Schmidt, E. D. D., Kirk, J. L., and Schneider, W. C., Variation of refractive index of KBr and KI with pressure up to 14 kbars. *Mater. Res. Bull.* (USA) **4**, 573–579 (1969).

1297. Vedam, K., and Schneider, W. C., Variation of the refractive index of boric oxide glasses with hydrostatic pressure to 7 kbar. *J. Appl. Phys.* (USA) **43**, 3623–3627 (1972).

1298. Vedam, K., Piezo-optic properties of solids at high pressure. Proceedings of the fourth international conference on high pressure, Kyoto (Japan), 1974, pp. 339–343.

1299. Vedam, K., Kirk, J. L., and Achar, B. N. N., Piezo- and thermo-optic behaviour of spinel ($MgAl_2O_4$). *J. Solid State Chem.* (USA) **12**, 213–218 (1975).

1300. Vedam, K., and Hennessey, P., Piezo- and thermo-optical properties of $Bi_{12}GeO_{20}$. II. Refractive index. *J. Opt. Soc. Am.* (USA) **65**, 442–445 (1975).

1301. Vedam, K., Schmidt, E. D. D., and Schneider, W. C., Dispersion of the elasto-optic constants of potassium halides, in: *Optical Properties of Highly Transparent Solids*, S. S. Mitra and B. Bendow, eds., Plenum Press, New York and London (1975), pp. 169–177.

1302. Veerabhadra Rao, K., and Narasimhamurty, T. S., Photoelastic constants of CaF_2 and BaF_2. *J. Phys. Chem. Solids* (GB) **31**, 876–878 (1970).

1303. Veerabhadra Rao, K., and Narasimhamurty, T. S., A new technique to study the photoelastic dispersion in cubic and noncubic crystals. *Appl. Opt.* (USA) **9**, 155–159 (1970).

1304. Veerabhadra Rao, K., and Narasimhamurty, T. S., Photoelastic dispersion in α-quartz. *Indian J. Pure & Appl. Phys.* (India) **9**, 844–846 (1971).

1305. Veerabhadra Rao, K., and Narasimhamurty, T. S., Pockels effect in quartz. *Optica Acta* (GB) **19**, 319–325 (1972).

1306. Veerabhadra Rao, K., "Photoelastic and linear electro-optic dispersion in some non-cubic crystals." Ph.D. thesis, Osmania University, Hyderabad (India), 1972.

1307. Veerabhadra Rao, K., and Narasimhamurty, T. S., Photoelastic behaviour of KDP. *J. Mater. Sci.* (GB) **10**, 1019–1021 (1975).

1308. Veerabhadra Rao, K., and Narasimhamurty, T. S., Linear electro-optic behaviour of ADP and its dispersion. *Opt. Quantum Electron.* (GB) **8**, 55–57 (1976).

1309. Venkataraman, R., The anisotropy of elastic wave propagation in crystals. *Proc. Indian Acad. Sci.* (India) **A55**, 311–323 (1962).

1310. Venturini, E. L., Spencer, E. G., and Ballman, A. A., Elasto-optic properties of $Bi_{12}GeO_{20}$, $Bi_{12}SiO_{20}$, and $Sr_xBa_{1-x}Nb_2O_6$. *J. Appl. Phys.* (USA) **40**, 1622–1624 (1969).

1311. Verlan, E. M., and Tsyashchenko, Yu. P., Linear electro-optical effect in doped alkali halide crystals. *Ukr. Fiz. Zh.* (USSR) **20**, 1388–1390 (1975), in Russian.

1312. Vetter, R., The pressure coefficient of the infrared refractive index of silicon. *Phys. Status Solidi A* (Germany) **8**, 443–447 (1971).

1313. Viswanathan, K. S., and Venkataraman, R., Vibration directions of elastic waves in cubic crystals. *J. Sci. Indust. Res.* (India) **20B**, 502–504 (1961).

1314. Vittoz, B., Accidental birefringence and absolute variation of the indices in quartz. *Helv. Phys. Acta* (Switzerland) **27**, 502–505 (1954), in French.

1315. Vittoz, B., New method of measuring directly the piezo-optic effect application to quartz I. *Rev. Opt.* (France) **35**, 253–291 (1956), in French.

1316. Vittoz, B., New method of measuring directly the piezo-optic effect—Application to quartz—II. *Rev. Opt.* (France) **35**, 468–480 (1956), in French.
1317. Vlokh, O. G., and Zheludev, I. S., Changes in the optical parameters of crystals caused by electric fields. *Sov. Phys.-Crystallogr.* (USA) **5**, 368–380 (1960).
1318. Vlokh, O. G., Zheludev, I. S., and Shamburov, V. A., The electro-optic effect in crystals of pentaerythritol $C(CH_2OH)_4$. *Sov. Phys.-Crystallogr.* (USA) **8**, 37–40 (1963).
1319. Vlokh, O. G., and Lutsiv-Shumskii, L. F., Spontaneous electro-optic effect in triglycine sulphate crystals. *Ukr. Fiz. Zh.* (USSR) **11**, 1006–1010 (1966), in Ukrainian.
1320. Vlokh, O. G., Dispersion characteristics of the linear electro-optic effect in quartz crystals. *Ukr. Fiz. Zh.* (USSR) **11**, 1184–1190 (1966), in Ukrainian.
1321. Vlokh, O. G., and Lutsiv-Shumskii, L. F., Temperature and dispersion dependence of the electro-optic coefficient r_{63} in ADP. *Sov. Phys.-Crystallogr.* (USA) **12**, 380–382 (1967).
1322. Vlokh, O. G., and Lutsiv-Shumskii, L. F., Dispersion and temperature dependence of electro-optical and piezo-optical coefficients r_{63} and π^d_{66} for KDP and ADP crystals. *Ukr. Fiz. Zh.* (USSR) **14**, 323–327 (1969), in Russian.
1323. Vlokh, O. G., Lutsiv-Shumskii, L. F., and Lynbikskii, T. T., Temperature and wavelength dependence of r_{41} electro-optical coefficient of ADP and KDP crystals. *Bull. Acad. Sci. USSR Phys. Ser.* (USA) **33**, 260–262 (1969).
1324. Vlokh, O. G., and Lutsiv-Shumskii, L. F., Photoelastic effect in 45° *X*-cut crystals of KDP and ADP. *Sov. Phys.-Solid State* (USA) **12**, 256–257 (1970).
1325. Vlokh, O. G., Kutnyi, I. V., and Nesterenko, V. Ya., Electro-optical effect in crystals of iodic acid. *Ukr. Fiz. Zh.* (USSR) **16**, 327–328 (1971), in Russian.
1326. Vlokh, O. G., Lutsiv-Shumskii, L. F., and Pylypyshin, B. P., The electro-optical and piezo-optical properties of KH_2PO_4 and $NH_4H_2PO_4$ crystals at high temperatures. *Sov. Phys.-Crystallogr.* (USA) **16**, 717–718 (1972).
1327. Vlokh, O. G., Klimov, I. M., Lobskii, M. I., Ovchinnikov, V. M., and Pirogov, Yu. A., Electro-optical properties of quartz crystals in static fields. *Zh. Prikl. Spektrosk.* (USSR) **16**, 1079–1083 (1972), in Russian.
1328. Vlokh, O. G., and Laźko, L. A., Electro-optical effect in L-rhamnose crystals. *Sov. Phys.-Crystallogr.* (USA) **19**, 245–246 (1974).
1329. Vlokh, O. G., and Laźko, L. A., Gyrotropy and the electro-optical effect in α-HIO_3 crystals. *Ukr. Fiz. Zh.* (USSR) **20**, 510–513 (1975), in Russian.
1330. Vlokh, O. G., Krochuk, A. S., Lutsiv-Shumskii, L. F., and Smishko, E. V., Refractive indices and electro-optical coefficient r_{41} of crystals of $K(D_xH_{1-x})_2PO_4$. *Bull. Acad. Sci. USSR, Phys. Ser.* (USA) **39**, 75–77 (1975).
1331. Vlokh, O. G., Characteristic features of electro-optical effect in crystals of pentaerythritol. *Ukr. Fiz. Zh.* (USSR) **20**, 921–923 (1975), in Russian.
1332. Vlokh, O. G., Lutsiv-Shumskii, L. F., Stasyuk, I. V., and Popel, A. M., Transverse electro-optical effect in $KH_{2(1-x)}D_{2x}PO_4$ crystals. *Ukr. Fiz. Zh.* (USSR) **20**, 1380–1382 (1975), in Russian.
1333. Vlokh, O. G., Valichko, I. A., and Laźko, L. A., Characteristic features of the electro-optical effects in $LiIO_3$ crystals. *Sov. Phys.-Crystallogr.* (USA) **20**, 263–264 (1975).
1334. Voigt, W., *Lehrbuch der Kristallphysik*, B. G. Teubner, Leipzig and Berlin (1928).

1335. Voigt, W., Measurement of experimentally produced optical phase difference in natural crystals. *Optik* (Germany) **14**, 28–33 (1957), in German.
1336. Volkova, E. N., Zamkov, V. A., and Nalhandov, L. V., Precision measurements of the absolute value of Kerr constants. *Opt. & Spectrosc.* (USA) **30**, 300–303 (1971).
1337. Volkova, E. N., Dianova, V. A., Zuev, A. L., Izrail-enko, A. N., Lipatov, A. S., Parygin, V. N., Rashkovich, L. N., and Chirkov, L. E., Electro-optical and piezoelectric properties of alpha-HIO_3 crystals. *Sov. Phys.-Crystallogr.* (USA) **16**, 284–286 (1971).
1338. Volkova, E. N., Berezhnoi, B. M., Izrail-enko, A. N., Mishchenko, A. V., and Rashkovich, L. N., Electro-optical and optical properties of partially deuterated crystals of rubidium dihydrogen phosphate. *Bull. Acad. Sci. USSR, Phys. Ser.* (USA) **35**, 1690–1693 (1971).
1339. Volkova, E. N., and Velichko, I. A., Electro-optical properties of potassium dihydrogen phosphate crystals having different degrees of deuteration. *Sov. Phys.-Crystallogr.* (USA) **18**, 256–257 (1973).
1340. Von Hundelshausen, U., Electro-optic effect and dielectric properties of cadmium–mercury–thiocyanate crystals. *Phys. Lett. A* (Netherlands) **34**, 405–406 (1971).
1341. Voronov, V. V., Kuźminov, Yu. S., Osiko, V. V., Prokhorov, A. M., Shumskaya, L. S., and Shipulo, G. P., Barium–sodium–potassium niobate, a promising material for nonlinear optics. *Sov. Phys.-Dokl* (USA) **19**, 691–692 (1975).
1342. Wachi, Y., High speed camera provided with ADP shutter. *Jap. J. Appl. Phys.* (Japan) **27**, 670–674 (1958).
1343. Wade, G., and Giglio, D. A., Physical concepts in ultrasonic light diffraction. Report No. HDL-TR-1518 (AD. 712975) Harry Diamond Laboratories, Washington, D.C. (USA), 1970, 31 pp.
1344. Wakatsuki, N., Chubachi, N., and Kikuchi, Y., Bragg condition of light diffraction by ultrasonic waves in anisotropic crystals. *Jap. J. Appl. Phys.* (Japan) **13**, 1754–1764 (1974).
1345. Wakita, K., Umeno, M., Takagi, K., and Miki, S., Brillouin scattering of thermal phonons in CdS. *J. Phys. Soc.* (Japan) **35**, 149–156 (1973).
1346. Walder, J., and Tang, C. L., Photoelastic amplification and the generation of hypersound by the stimulated Brillouin process. *Phys. Rev. Lett.* (USA) **19**, 623–626 (1967).
1347. Walsh, T. E., Gallium arsenide electro-optic modulators. *RCA Rev.* (USA) **27**, 323–335 (1966).
1348. Walters, W. L., Electro-optic effect in reverse-biased GaAs p–n junctions. *J. Appl. Phys.* (USA) **37**, 916 (1966).
1349. Wang, S., Interferometric measurements of dielectric properties of solids. *J. Appl. Phys.* (USA) **39**, 5810–5811 (1968).
1350. Ward, J. F., and Franken, P. A., Structure of nonlinear optical phenomena in KDP. *Phys. Rev.* (USA) **133**, A183–190 (1964).
1351. Ward, J. F., and New, G. H. C., Optical rectification in ADP, KDP, and quartz. *Proc. R. Soc.* (GB) **A299**, 238–263 (1967).
1352. Wardzyoski, W., Birefringence induced by uniaxial stress in cadmium telluride single crystals. *J. Phys. C (Solid State Phys.)* (GB) **3**, 1251–1263 (1970).
1353. Warner, A. W., Onoe, M., and Coquin, G. A., Determination of elastic and piezoelectric constants for crystals in class (3*m*). *J. Acoust. Soc. Am.* (USA) **42**, 1223–1231 (1967).

1354. Warner, A. W., Pinnow, D. A., Bergman, J. G., and Crane, G. R., Piezoelectric and photoelastic properties of lithium iodate. *J. Acoust. Soc. Am.* (USA) **47**, 791–794 (1970).

1355. Warner, A. W., White, D. L., and Bonner, W. A., Acousto-optic light deflectors using optical activity in paratellurite. *J. Appl. Phys.* (USA) **43**, 4489–4495 (1972).

1356. Warner, A. W., Sittig, E. K., and Bonner, W. A., TeO_2 acousto-optic light deflector with 1000 spot capacity and 40 sec retrace line. *IEEE Trans. Sonics & Ultrason.* (USA) **SU-19**, 412 (1972).

1357. Warner, A. W., and Pinnow, D. A., Miniature acousto-optic modulators for optical communications. *J. Quantum Electron.* (USA) **QE-9**, 1155–1157 (1973).

1358. Warner, J., Robertson, D. S., and Parfit, H. T., Electro-optic effect of sodium uranyl acetate. *Phys. Lett.* (Netherlands) **19**, 479–480 (1965).

1359. Warner, J., The electro-optic effect in proustite (Ag_3AsS_3). *J. Phys. D* (*Appl. Phys.*) (GB), *Ser.* 2 **1**, 949–950 (1968).

1360. Watanabe, A., Nabeta, Y., and Yano, T., Acousto-optical crystals from $Pb_5(GeO_4)(VO_4)_2$. *Oyo. Buturi* (Japan) **40**, 1140–1144 (1971), in Japanese.

1361. Waterman, P. C., Orientation dependence on elastic waves in single crystals. *Phys. Rev.* (USA) **113**, 1240–1253 (1959).

1362. Watkins, M. C., Microwave modulation of light with ADP. Presented at 18th Annual National Electronics Conference, Chicago, Illinois (USA), October 8, 1962.

1363. Waxler, R. M., and Napolitano, A., Relative stress-optical coefficients of some National Bureau of Standards optical glasses. *J. Res. Natl. Bur. Stand. A* (USA) **59A**, 121–125 (1957).

1364. Waxler, R. M., and Weir, C. E., Effect of hydrostatic pressure on the refractive indices of some solids. *J. Res. Natl. Bur. Stand. A* (USA) **69A**, 325–333 (1965).

1365. Waxler, R. M., and Farabaugh, E. N., Photoelastic constants of ruby. *J. Res. Natl. Bur. Stand. A* (USA) **74A**, 215–220 (1970).

1366. Waxler, R. M., Piezo-optic and thermo-optic properties of solid laser materials. *Natl. Bur. Stand. Tech. News Bull.* (USA) **54**, 177 (1970).

1367. Waxler, R. M., Cleek, G. W., Malitson, I. H., Dodge, M. J., and Hahn, T. A., Optical and mechanical properties of some neodymium doped laser glasses. *J. Res. Natl. Bur. Stand. A* (USA) **75A**, 163–174 (1971).

1368. Waxler, R. M., Laser glass composition and the possibility of eliminating electrostrictive effects. *IEEE J. Quantum Electron.* (USA) **QE7**, 166–167 (1971).

1369. Waxler, R. M., and Cleek, G. W., The effect of temperature and pressure on the refractive index of some oxide glasses. *J. Res. Natl. Bur. Stand. A* (USA) **77A**, 755–763 (1973).

1370. Weil, R., and Sun, M. J., Piezo-optic constants of CdTe. *Proceedings of the International Symposium on Cadmium Telluride, a Material for Gamma Ray Detectors*, Centre de Recherches Nucléaires, Strassbourg, France, 19/6 pp. (1971).

1371. Weil, R., and Halido, D., Resonant piezo-electro-optic light modulation. *J. Appl. Phys.* (USA) **45**, 2258–2265 (1974).

1372. Weller, J. F., Milton, A. F., and Giallorenzi, T. G., Experimental characterization of acousto-optical deflection in multimode optical wave guides. *J. Appl. Phys.* (USA) **46**, 974–976 (1975).

1373. Wemple, S. H., Di Domenico, Jr., M., and Camlibel, I., Relationship between linear and quadratic electro-optic coefficients in $LiNbO_3$, $LiTaO_3$, and other oxygen octahedra ferroelectrics based on direct measurement of spontaneous polarization. *Appl. Phys. Lett.* (USA) **12**, 209–211 (1968).

1374. Wemple, S. H., and Di Domenico, Jr., M., Oxygen-octahedra ferroelectrics, I and II—Electro-optical and non-linear optical device applications. *J. Appl. Phys.* (USA) **40**, 720–734 and 735–752 (1969).

1375. Wemple, S. H., and Di Domenico, Jr., M., Theory of the elasto-optic effect in non-metallic crystals, *Phys. Rev.* (USA) **1**, 193–202 (1970).

1376. Wemple, S. H., and Di Domenico, Jr., M., Electro-optical and nonlinear optical properties of crystals, in: *Applied Solid State Science: Advances in Materials and Device Research*, Wolfe, R., ed., Academic Press, New York and London (1972), Vol. 3, pp. 263–383.

1377. Werner, A. J., Methods in high precision refractometry of optical glasses. *Appl. Opt.* (USA) **7**, 837–843 (1968).

1378. Wertheim, G., Mémoire sur la polarization chromatique produite par le verre comprimé. *C. R. Hebd. Séances Acad. Sci. B* (France) **32**, 289–292 (1851); **33**, 576–579 (1851), in French; see also *Ann. Phys.* (Germany) **II86**, 321–325 (1852), in German.

1379. Wertheim, G., Notes sur la double réfraction artificiallement produite dans des cristaux du système régulier. *C. R. Hebd. Séances Acad. Sci. B* (France) **33**, 498–500 (1852), in French.

1380. Wertheim, G., Mémoire sur la double réfraction temporairement produite dans les corps isotropes, et sur la relation entre l'élasticité mécanique et l'élasticité optique. *Ann. Chim. Phys.* (France) **III40**, 156–221 (1854), in French.

1381. Wertheim, G., On the double refraction temporarily produced in isotropic bodies. *Philos. Mag.* (GB) **IV8**, 241–261, 342–357 (1854).

1382. West, C. D., and Makas, A. S., Technical crystals with abnormally large stress birefringence. *J. Chem. Phys.* (USA) **16**, 427 (1948).

1383. West, C. D., and Makas, A. S., Some photoelastic constants of crystals—$(NH_4)H_2PO_4$ (ADP), and KH_2PO_4 (KDP). *Am. Mineral.* (USA) **35**, 130 (1950).

1384. West, C. D., Electro-optic and related properties of crystals with the zinc blende structure. *J. Opt. Soc. Am.* (USA) **43**, 335 (1953).

1385. West, C. D., and Carpenter, R. O'B., Magneto-optics and electro-optics, in: *American Institute of Physics Handbook*, Condon, E. U., and Odishaw, H., eds., McGraw-Hill, New York (1957), Sec. 6, pp. 6.94–6.97.

1386. White, E. D., Recent advances in crystal growing techniques. *Brit. J. Appl. Phys.* (GB) **16**, 1415–1428 (1965).

1387. White, J. M., Heidrich, P. F., and Lean, E. G., Thin film acousto-optic interaction in $LiNbO_3$. *Electron. Lett.* (GB) **10**, 510–511 (1974).

1388. Whitman, R., Korpel, A., and Lotsoff, S., Application of acoustic Bragg diffraction to optical processing techniques, in Proceedings of the symposium on modern optics at the Polytechnic Institute of Brooklyn, March 22–24, 1967, pp. 243–256.

1389. Wigner, E., On the characteristic elastic vibrations of symmetrical systems. *Nachr. Akad. Wiss. Göttingen II Math.-Phys. Kl.* (Germany) 133–146 (1930), in German.

1390. Willard, G. W., Criteria for normal and abnormal ultrasonic light diffraction effects. *J. Acoust. Soc. Am.* (USA) **21**, 101–108 (1949).

1391. Wilson, E. B., Decius, J. C., and Cross, P. C., *Molecular Vibrations—The Theory of Infrared and Raman Vibrational Spectra*, McGraw-Hill Book Co. Inc., New York (1955).

1392. Winkelmann, A., Über einige Erscheinungen die bei der Beugung des Lichtes durch das Gitter auftreten. *Ann. Phys.* (Germany) **27**, 905–954 (1908), in German.

1393. Wohlers, M. R., and Lieb, K. G., Control of laser radiation with birefringent crystals: The microwave circuit viewpoint. *J. Appl. Phys.* (USA) **35**, 2311–2312 (1964).

1394. Wolkerstofer, D. C., Raman–Nath diffraction in organic crystals. *Appl. Phys. Lett.* (USA) **14**, 323–325 (1969).

1395. Wondratschek, H., Derivation of the electro-optic effect in crystals. *Z. Kristallogr.* (Germany) **106**, 467–473 (1955), in German.

1396. Wood, E. A., *Crystals and Light—An Introduction to Optical Crystallography*, D. Van Nostrand Co., Inc., Princeton, New Jersey (USA) (1964).

1397. Wood, R. W., *Physical Optics*, The Macmillan Co., New York (USA) (1962).

1398. Wooster, W. A., *A Text-book on Crystal Physics*, Cambridge University Press, Cambridge (GB) (1949).

1399. Wooster, W. A., *Tensors and Group Theory for the Physical Properties of Crystals*, Clarendon Press, Oxford (London) (1973).

1400. Yamada, T., Okuda, S., Azuma, K., and Kataoka, T., Piezo-optical coefficients in alkali halide solid solution crystals. *Technol. Rep. Osaka Univ.* (Japan) **24**, 443–450 (1974).

1401. Yamamoto, K., Misawa, K., Shimizu, H., and Abe, K., Resonant Brillouin scattering by acousto-electrically amplified phonons in CdSe. *J. Phys. Chem. Solids* (GB) **37**, 181–182 (1976).

1402. Yanagi, T., Temperature dependence of principal refractive indices of orthorhombic KNO_3. *J. Math. Phys. Cambridge* (USA) **7**, 1056–1058 (1968).

1403. Yano, T., and Watanabe, A., Electro-optic properties of $Sr_4NaLiNb_{10}O_{30}$ type ferroelectrics. *Jap. J. Appl. Phys.* (Japan) **9**, 1008 (1970).

1404. Yano, T., Fukumoto, A., and Watanabe, A., Tellurite glass, a new acousto-optic material. *J. Appl. Phys.* (USA) **42**, 3674–3676 (1971).

1405. Yano, T., Nabeta, Y., and Watanabe, A., A new crystal $Pb_5(GeO_4)(VO_4)_2$ for acousto-optic device applications. *Appl. Phys. Lett.* (USA) **18**, 570–571 (1971).

1406. Yano, T., and Watanabe, A., Acousto-optic figure of merit of TeO_2 for circularly polarized light. *J. Appl. Phys.* (USA) **45**, 1243–1245 (1974).

1407. Yariv, A., and Gordon, J. P., The laser. *Proc. IEEE* (USA) **51**, 4–29 (1963).

1408. Yariv, A., Mead, C. A., and Parker, J. V., GaAs as an electro-optic modulator at 10.6 microns. *IEEE J. Quantum Electron.* (USA) **QE-2**, 243–245 (1966).

1409. Yasuda, A., Funahashi, A., and Takeda, S., Stabilization of optical interferometry for steady measurements. *Jap. J. Appl. Phys.* (Japan) **8**, 963–964 (1969).

1410. Yazaki, T., Kanatani, K., and Sakamoto, S., Longitudinal electro-optic effect in oblique-cut strontium barium niobate plates. *Jap. J. Appl. Phys.* (Japan) **10**, 522–523 (1971).

1411. Yu, P. Y., Cardona, M., and Pollak, F. H., Intrinsic piezo-birefringence in GaSb, InAs, and InSb. *Phys. Rev. B* (USA) **3**, 340–346 (1971).

1412. Yu, P. Y., and Cardona, M., Intrinsic piezobirefringence of several semiconducting chalcogenides. *J. Phys. Chem. Solids* (GB) **34**, 29–56 (1973).

1413. Zharikov, V. I., Light modulators with circularly polarized modulating field. *Sov. Phys.-Crystallogr.* (USA) **12**, 655–656 (1967).

1414. Zhashkov, A. A., Pankratov, V. M., Moiya, A. M., and Somov, V. G., Longitudinal electro-optical effect in oblique sections of lithium niobate. *Kvantovaya Elektron.* (USSR) **1**, 147–148 (1971), in Russian.

1415. Zheludev, I. S., Point groups of crystal symmetry and their interpretation. *Sov. Phys.-Crystallogr.* (USA) **2**, 718–722 (1957).
1416. Zheludev, I. S., and Fotchenkov, A. A., Electrostriction of linear dielectrics. *Sov. Phys.-Crystallogr.* (USA) **3**, 312–318 (1958).
1417. Zheludev, I. S., and Vlokh, O. G., The electro-optical effect in crystals. *Sov. Phys.-Crystallogr.* (USA) **3**, 647–660 (1959).
1418. Zheludev, I. S., and Lelekov, V. S., The inverse piezoelectric effect in *Z*-cut crystals of guanidine aluminium sulphate (GAS). *Sov. Phys.-Crystallogr.* (USA) **7**, 369–370 (1962).
1419. Zheludev, I. S., Electro-optical effects in crystals. *Sov. Phys.-Usp.* (USA) **9**, 97–116 (1966).
1420. Zhitnikov, R. A., and Stepanov, A. V., Photoelastic effect in crystals of the cubic system under planar compression. *Sov. Phys.* (USA) **1**, 762 (1957); also see **1**, 769 (1957).
1421. Ziauddin, M., and Narasimhamurty, T. S., Effective thickness of two birefringent crystals for the study of natural birefringence. *Rev. Sci. Instrum.* (USA) **35**, 303–305 (1964).
1422. Ziauddin, M., and Narasimhamurty, T. S., Studies on the dispersion of photoelasticity in non-cubic crystals using birefringent compensator. *Indian J. Pure & Appl. Phys.* (India) **4**, 309–311 (1966).
1423. Ziauddin, M., "Photoelastic dispersion in cubic and noncubic crystals." Ph.D. thesis, Osmania University (India), 1966.
1424. Ziauddin, M., and Narasimhamurty, T. S., Photoelastic constants of ADP. *J. Opt. Soc. Am.* (USA) **57**, 1392–1393 (1967).
1425. Zook, J. D., and Casselman, T. N., Electro-optic effects in paraelectric perovskites. *Phys. Rev. Lett.* (USA) **17**, 960–962 (1966).
1426. Zook, J. D., Chen, D., and Otto, G. N., Temperature dependence and model of the electro-optic effect in $LiNbO_3$. *Appl. Phys. Lett.* (USA) **11**, 159–161 (1967).
1427. Zukrinov, I. I., Semenov, V. I., and Sheloput, D. V., Elastic and Photoelastic properties of proustite. *Sov. Phys.-Solid State* (USA) **15**, 1921–1922 (1974).
1428. Zwicker, B., and Scherrer, P., Electro-optical behaviour of KH_2PO_4 and KD_2PO_4 crystals. *Helv. Phys. Acta* (Switzerland) **16**, 214–216 (1943).
1429. Zwicker, B., and Scherrer, P., Electro-optical properties of the signette-electric crystals KH_2PO_4 and KD_2PO_4. *Helv. Phys. Acta* (Switzerland) **17**, 346–373 (1944).
1430. Zwikker, C., *Physical Properties of Solid Materials*. Pergamon Press, London and New York (1954).

Addendum. The following references bring the bibliography up to the end of the year 1979. The journal abbreviations used here are in accordance with the standard forms given in Volume 82, Part I, Author Index, pages LJ-1 to LJ-37 (January–June, 1979) of *Physics Abstracts–Science Abstracts Series A*.

1431. Advenko, K. I., Sapozhniko, V. K., Semenov, V. I., and Sheloput, D. V., Photoelastic constant of KRS-6 monocrystals. *Avtometriya* (USSR), No. 5, 79–83 (1977), in Russian.
1432. Afanasev, I. I., Berezhnoi, A. A., Bushneva, T. S., and Prokofev, S. V., Growing

crystals of lead magnoniobate and magnotantalate. *Sov. J. Opt. Tech.* (USA) **44**, 613–615 (1977).

1433. Akimov, S. V., Stolpakova, T. M., Dudnik, E. F., and Sinyakov, E. V., Photoelastic properties of $NaBi(MoO_4)_2$. *Sov. Phys.-Solid State* (USA) **19**, 1660–1661 (1977).

1434. Akimov, S. V., Dudnik. E. F., Stolpakova, T. M., Dovchenko, G. V., Acousto-optic characteristics of $LiBi(MoO_4)_2$. *Sov. Phys.-Solid State* (USA) **20**, 547–548 (1978).

1435. Aksenov, E. T., Esepkina, N. A., and Shcherbakov, A. S., High frequency crystal acousto-optic deflectors. *Sov. Phys.-Tech. Phys.* (USA), **22**, 213–215 (1977).

1436. Aleksandrov, K. S., Anistratov, A. T., Zemkov, A. V., and Rez, I. S., Acousto-optic and photoelastic properties of KH_2PO_4 type crystals. *Sov. Phys.-Solid State* (USA) **19**, 1090–1091 (1977).

1437. Aleksandrov, K. S., Anistratov, A. T., and Beznosikov, B. V.: Acousto-optic properties of mixed chlorides $ABCl_3$. *Avtometriya* (USSR), No. 1, 50–53 (1978), in Russian.

1438. Anistratov, A. T., Martynov, V. G., Shabanova, L. A., and Rez, I. S., Elastic properties and photo-elasticity of caesium dihydrogen arsenate near ferroelectric phase transition. *Sov. Phys.-Solid State* (USA) **21**, 783–787 (1979).

1439. Apletalin, V. N., Meriakri, V. V., and Shirokov, V. A., Apparatus for investigating electro-optical properties of crystals at submillimeter wavelengths. *Instrum. & Exp. Tech.* (USA) **20**, 182–183 (1977).

1440. Arabidze, A. A., and Chikvaidze, B. Ya., Dispersion of the electro-optic effect in barium titanate crystals. *Sov. Phys.-Solid State* (USA) **21**, 722–723 (1979).

1441. Avakyants, L. P., and Kiselev, D. F., Elasticity and photoelasticity of $LiTaO_3$ near its ferroelectric phase transition. *Sov. Phys.-Solid State* (USA), **20**, 354–355 (1978).

1442. Avdienko, K. I., Bogdanov, S. V., Kidyarov, B. I., Semenov, V. I. and Sheloput, D. V., Optical, acoustic and piezoelectric properties of α-$LiIO_3$ crystals. *Bull. Acad. Sci.* (USSR) *Phys. Ser.* (USA) **41**, 40–45 (1977).

1443. Bach, H., Haussühl, H., and Weber, H. J., Quadratic electro-optic tensor of orthorhombic calcium formate. *Z. Kristallogr.* (Germany) **145**, 28–34 (1977).

1444. Bagdavadze, V. N., Berozashvili, Yu. N., Gogolin, O. G., Dundua, A. V., Solomko, V. E., and Tsitsishvili, E. G., Electro-optic effect in noncentrosymmetric cubic semiconductor crystals with allowance for spatial dispersion. *Sov. Phys.-Semicond.* (USA) **12**, 1115–1117 (1978).

1445. Baldassare, L., Cingolani, A., and Sibilano, M., The linear electro-optical effect in $CdS_{0.75}Se_{0.25}$. *Opt. Commun.* (Netherlands) **26**, 72–74 (1978).

1446. Baryshnikova, L. F., and Lyamov, V. E., Excitation of an acoustic harmonic and its energy flux in piezoelectric crystals. *Sov. Phys.-Solid State* (USA) **20**, 640–643 (1978).

1447. Bechthold, P. S., and Haussühl, S., Nonlinear optical properties of orthorhombic barium formate and magnesium barium fluoride. *Appl. Phys.* (Germany) **14**, 403–410 (1977).

1448. Berdowski, J., Opilski, A., and Szuber, J., Studies on acousto-optic properties of some piezoelectric crystals. *J. Tech. Phys.* (Poland) **18**, 211–217 (1977).

1449. Berezhnoi, A. A., Popov, Yu. V., Sherstneva, T. N., and Solntsev, V. M., Effect of twinning of zinc selenide crystals on their electro-optical properties. *Opt. & Spectrosc.* (USA) **43**, 286–288 (1977).

1450. Berezhnoi, A. A., Dubovik, M. F., Popov, Yu. V., and Sherstneva, T. N., Special features of the electro-optical properties of strontium barium niobate crystals. *Opt. & Spectrosc.* (USA) **46**, 55–58 (1979).

1451. Berezhnoi, A. A., Special features of the anisotropy of the electro-optical effect in crystals of symmetry class 42m. *Opt. & Spectrosc.* (USA) **46**, 285–287 (1979).

1452. Berezina, E. E., Levenberg, V. A., and Lunter, S. G., Photoelastic effect and polarizability of glasses. *Sov. J. Glass Phys. & Chem.* (USA) **3**, 563–569 (1977).

1453. Blumich, B. P., and Germer, R., Electro-optic shutters with PLZT ceramics for infrared applications. *J. Phys. E (Scientific Instruments)* (GB) **12**, 770–774 (1979).

1454. Bounds, P. J., and Munn, R. W., Dielectric theory of strained molecular crystals: Elasto-optic coefficients of orthorhombic sulphur. *Chem. Phys.* (Netherlands) **39**, 165–173 (1979).

1455. Brown, D. C., Jacobs, S. D., Abate, J. A., Lewis, O., and Rinefierd, J., Figures of merit and correlations of physical and optical properties in laser glasses, in: *Laser-Induced Damage in Optical Materials, Boulder, Colorado 4–6 October 1977* National Bureau of Standards, Washington, D.C. (1977), pp. 416–433.

1456. Bugnet, P., Baillou, J., Daunay, J., Effect of birefringence due to native defects on the electro-optical effect of single crystal zinc selenide. *Rev. Phys. Appl.* (France) **14**, 259–263 (1979), in French.

1457. Bystrov, V. P., Electro-optic materials in submillimeter range. *Bull. Acad. Sci.* (USSR) *Phys. Ser.* (USA) **41**, 23–29; 1977).

1458. Calligaris, F., Ciuti, P., and Gabrielli, I., Synthesis of formulae describing the temporal light modulation by ultrasound. *Acustica* (Germany) **42**, 176–178 (1979).

1459. Canal, F., Grimsditch, M., and Cardona, M., Piezobirefringence in GaP and InP. *Solid State Commun.* (USA) **29**, 523–526 (1979).

1460. Chandrasekhar, M., Grimsditch, M. H., and Cardona, M., Piezobirefringence above the fundamental edge in Si. *Phys. Rev. B* (USA) **18**, 4301–4311 (1978).

1461. Chang, I. C., Tunable acousto-optic optic filters: An overview. *Opt. Eng.* (USA) **16**, 455–460 (1977).

1462. Cheo, P. K., Integrated optical devices and applications. *J. Phys. E (Scientific Instruments)* (GB) **12**, 2–12 (1979).

1463. Chudakov, V. S., Investigation of induced birefringence with a rotating polarization element. *Instrum. & Exp. Tech.* (USA) **20** (1), pt. 2, 241–244 (1977).

1464. Cingolani, A., Ferrara, M., Lugara, M., and Levy, F., Pockels effect in gallium selenide. *Solid State Commun.* (USA) **29**, 677–679 (1979).

1465. Cook, W. R., Electro-optic coefficients (including selected indices of refraction of crystals), in: *Landolt–Börnstein Tables—New Series*, K.-H. Hellwege and A. M. Hellwege, eds., Group III, Vol. 11, Springer-Verlag, Heidelberg/New York (1979), pp. 552–651.

1466. Cook, W. R., Hearmon, R. F. S., and Nelson, D. F., Piezo optic and elasto-optic, and electro-optic coefficients of crystals—Bibliography, in: *Landolt–Börnstein Tables —New Series* K.H. Hellwege and A. M. Hellwege, eds., Group III, Vol. 11, Springer-Verlag, Heidelberg/New York (1979), pp. 651–670.

1467. Ctyroky, J., On the light diffraction by acoustic waves in an anisotropic medium. *Acta Tech. CSAV* (Czechoslovakia), **22**, 161–169 (1977).

1468. Diaz-Gongora, A., and Luty, F., Kerr effect and birefringence from paraelectric CN^- defects in nine alkali halides. *Phys. Status Solidi B* (Germany), **86**, 127–136 (1978).

1469. Drăgănescu, V., Maris, Z., Valculescu, V. G., and Tomulescu, R., Method for measuring elasto-optical constants of solids. *Appl. Phys.* (Germany) **20**, 251–253 (1979).
1470. Dudnik, E. F., and Akimov, S. V., Acousto-optical properties of single crystals of lead orthophosphate and its solid solutions. *Sov. Phys. Solid State* (USA) **20**, 1626–1627 (1978).
1471. Eggleton, R. C., and Keuwez, J. J., The zero-order Raman–Nath method for visualizing the longitudinal character of acoustic beams. *IEEE Trans. Sonics Ultrason.* (USA) **SU-25**, 203–206 (1978).
1472. Ethiraj, R., Piezo-optic birefringence in mixed crystals of KCl and KBr, *Phys. Status Solidi A* (Germany) **45**, 83–88 (1978).
1473. Ethiraj, R., Krishna Murty, V. G., and Bansigir, K. G., Piezo-optic birefringence in CsCl type crystals. *Acta Crystallogr. A* (Denmark) **34**, 318–321 (1978).
1474. Ewald, P. P., A review of my papers on crystal optics 1912–1968. *Acta Crystallogr. A* (Denmark) **35**, 1–9 (1979).
1475. Feichtner, J. D., Gottlieb, M., and Conroy, J. J., Tl_3AsSe_3 noncollinear acousto-optic filter operation at 10 μm. *Appl. Phys. Lett.* (USA) **34**, 1–3 (1979).
1476. Feichtner, J. D., Gottlieb, M., and Conroy, J. J., Tunable acousto-optic filters and their application to spectroscopy, in: *Proceedings of the Society of Photo-optical Instrumentation Engineers, Vol. 82, Unconventional Spectroscopy*, San Diego, California (1976), pp. 106–118.
1477. Feldman, A., Measuring the optical properties of infrared laser windows. *Electro-optic Syst. Des.* (USA) **8**, 36–40 (1976).
1478. Feldman, A., Horowitz, D., and Waxler, R. M., Photoelastic constants of potassium chloride at 10.6 μm (laser window materials). *Appl. Opt.* (USA) **16**, 2925–2930 (1977).
1479. Feldman, A., Waxler, R. M., and Horowitz, D., Photoelastic constants of germanium. *J. Appl. Phys.* (USA) **49**, 2589–2590 (1978).
1480. Feldman, A., Measurement of the photoelastic constants of optical materials. *Opt. Eng.* (USA) **17**, 453–462 (1978).
1481. Fick, S. E., and Palmer, C. H., 3-Dimensional real-time measurements of ultrasonic waves in transparent solids. *Appl. Opt.* (USA) **17**, 2686–2689 (1978).
1482. Fukuda, S., Shiosaki, T., and Kawabata, A., Acousto-optic properties of tellurium at 10.6 μm. *J. Appl. Phys.* (USA) **50**, 3899–3905 (1979).
1483. Galstyan, G. T., and Lomova, L. G., The electro-optic properties of triglycine sulphate single crystals doped with α-aniline (ATGS). *Izv. Akad. Nauk Arm. SSR. Fiz.* (USSR) **13**, 384–386 (1978), in Russian.
1484. Geidur, S. A., Pikhtin, A. N., Prokopenko, V. T., and Yaskov, A. D., Dispersion of induced birefringence in gallium arsenide and gallium phosphide crystals. *Opt. & Spectrosc.* (USA) **46**, 399–402 (1979).
1485. Glushkov, V. F., Efficiency of elastico-optic interaction in materials. *Inorg. Mater.* (USA) **12**, 614–617 (1976).
1486. Goyal, S. C., Prakash, R., Bakshi, B. S., and Sharma, O. P., Theory of the photoelastic constants of solids with ZnS structure. *Solid State Commun.* (USA) **29**, 731–733 (1979).
1487. Grimsditch, M. H., Anastassakis, E., and Cardona, M., Piezobirefringence in diamond. *Phys. Rev. B* (USA) **19**, 3240–3243 (1979).
1488. Grishmanovskii, A. N., Lemanov, V. V., and Sattikulov, M., Acousto-optic inter-

action in lead molybdate and paratellurite crystals at high optical intensities. *Sov. Tech. Phys. Lett.* (USA) **4**, 283–285 (1978).

1489. Gulyaev, Yu. V., Proklov, V. V., and Shkerdin, G. N., Diffraction of light by sound in solids. *Sov. Phys. Usp.* (USA) **21**, 29–57 (1978).

1490. Gulyaev, Yu. V., and Shkerdin, G. N., On the theory of nonlinear photoelasticity of solids. *Zh. Eksp. & Teor. Fiz.* (USSR) **77**, 1396–1406 (1979).

1491. Halbout, J. M., Donaldson, W., Tang, C. L., and Blit, S., Urea: A convenient and efficient nonlinear organic crystal. *J. Opt. Soc. Am.* (USA) **69**, 1464 (1979).

1492. Hall, J. L., Layer, H. P., and Deslattes, R. D., An acousto-optic frequency and intensity control system for continuous wave lasers. *IEEE J. Quantum Electron.* (USA) **QE 13**, 45–46 (1977).

1493. Haran, M. E., Visualization and measurement of ultrasonic wavefronts. *Proc. IEEE* (USA), **67**, 454–466 (1979).

1494. Haussühl, S. and Liebertz, J., Elastic, thermoelastic, piezoelectric, linear electro-optic, dielectric and pyroelectric properties of trigonal calcium vandate single crystals. *Z. Kristallogr.* (Germany) **148**, 87–93 (1978).

1495. Haussühl, S., and Preu, P., Determination of third-order elastic constants from stress-shifted resonance frequencies, observed by diffraction of light. Examples: aluminium alums of K, NH_4, Cs and CH_3NH_3. *Acta Crystallogr. A* (Denmark) **34**, 442–446 (1978).

1496. Haussühl, S., Piezoelectric, electro-optical dielectric, elastic and thermoelastic properties of hexagonal $Cs_2S_2O_6$, $LiClO_4 \cdot 3H_2O$, $LiClO_4 \cdot 3D_2O$ and $Ba(NO_2)_2 \cdot H_2O$. *Acta Crystallogr. A* (Denmark) **34**, 547–550 (1978).

1497. Hearmon, R. F. S., and Nelson, D. F., Piezo-optic and elasto-optic coefficients of crystals, in: *Landolt–Börnstein Tables, New Series*, K.-H. Hellewege and A. M. Hellwege, eds., Group III, vol. 11, Springer-Verlag, Heidelberg/New York (1979), pp. 505–551.

1498. Heiman, D., Hamilton, D. S., and Hellwarth, R. W., Brillouin scattering measurements on optical glasses. *Phys. Rev. B* (USA) **19**, 6583–6592 (1979).

1499. Hon, D. T., Electro-optic and piezo-optic tuning of second order nonlinear processes in crystals. *J. Appl. Phys.* (USA) **49**, 396–402 (1978).

1500. Ivanov, N. R., Benderskii, S. Ya., and Sluvalov, L. A., Electro-optic hysteresis and critical behaviour of the electro-optic properties of crystals belonging to triglycine sulphate family. *Sov. Phys. Crystallogr.* (USA) **22**, 64–69 (1977).

1501. Jamroz, W., Karniewicz, J., and Stachowiak, J., Nonlinear electro-optic effects in KDP and DKDP crystals. *Sov. J. Quantum Electron.* (USA) **9**, 803–805 (1979).

1502. Johnson, R. V., Temporal response of the acousto-optic modulator in the high scattering efficiency regime. *Appl. Opt.* (USA) **18**, 903–907 (1979).

1503. Joiner, R. E., Marburger, J. H., and Steier, W. H., Critical orientations for eliminating stress-induced depolarization in crystalline windows and rods, in: *Laser-Induced Damage in Optical Materials*, Boulder, Colorado *4–6 October 1977*, National Bureau of Standards, Washington, D.C. (1977), pp. 89–95.

1504. Joiner, R. E., Marburger, J. H., and Steier, W. H., Elimination of stress-induced birefringence effects in single crystal high power laser windows. *Appl. Phys. Lett.* (USA) **30**, 485–486 (1977).

1505. Kaplan, M. S., Smushkov, I. V., and Sumin, V. I., Temperature dependence of the photoelastic constants of silver chloride single crystals. *Sov. Phys. Solid State* (USA) **18**, 1837–1838 (1976).

1506. Kartalopoulos, S. V., A rapid optical method for the determination of stress-optical constants of optically isotropic and anisotropic materials. *J. Phys.* (*Applied Physics*) (GB) **9**, 2545–2553 (1976).
1507. Khatkevich, A. G., Diffraction and propagation of ultrasonic beams in single crystals. *Sov. Phys. Acoust.* (USA) **24**, 58–61 (1978).
1508. Kinase, W., Theoretical treatment of electro-optical effect. *Bull. Acad. Sci. USSR Phys. Ser.* (USA) **41**, 17–22 ;1977).
1509. Kleszczewski, Z., Properties of some acousto-optical materials. *Arch. Acoust.* (Poland) **3**, 175–184 (1978).
1510. Kleszczewski, Z., Application of laser light diffraction on ultra- and hypersonic waves to studies of acoustic and acousto-optic properties of crystals. *Postepy Fiz.* (Poland) **29**, 209–223 (1978), in Polish.
1511. Kludzin, V. V., Kulakov, S. V., and Razzhivin, B. P., Collinear diffraction of light by transverse acoustic waves. *Sov. Phys. Solid State* (USA) **18**, 1653–1655 (1976).
1512. Kobayashi, J., Uesu, Y., and Hosokawa, J., Electro-optical properties of some improper ferroelectrics. *Bull. Acad. Sci. USSR Phys. Ser.* (USA) **41**, 3–10 (1977).
1513. Kojima, S., and Nakamura, T., Electro-optical properties of gadolinium molybdate. *Phys. Rev. B* (USA) **18**, 453–458 (1978).
1514. Konak, C., Budik, L., and Novak, L., Electro-optical properties of NH_4Cl single crystals near their order–disorder phase transitions. *Phys. Status Solidi A* (Germany) **45**, K63–K66 (1978).
1515. Kondilenko, I. I., Korotkov, P. A., Grib, B. N., and Felinskii, G. S., Electro-optical deflection of light beams. *Kvantovaya Elektron., Kiev* (USSR), No. 14, 45–59 (1978), in Russian.
1516. Kopeika, N. S., A curriculum in electro-optics for electrical engineers. *Proc. IEEE* (USA) **66**, 893–901 (1978).
1517. Kotani, H., Namba, S., and Kawabe, M., Electro-optic Bragg-deflection modulators in corrugated waveguides. *IEEE J. Quantum Electron.* (USA) **QE-15**, 270–272 (1979).
1518. Krasser, W., Janik, B., Erhardt, K. D., and Haussühl, S., Brillouin scattering in the cubic phase of rubidium cyanide. *Solid State Commun.* (USA) **30**, 33–36 (1979).
1519. Kubo, H., and Nagata, R., Considerations of propagation of light waves in photoelastic materials and crystals. *Optik* (Germany) **52**, 37–47 (1978/79).
1520. Kucharczyk, W., and Stachowiak, J., Electro-optic properties of DKDP monocrystals in the paraelectric phase. *Zesz. Nauk. Politech. Lodz. Fiz.* (Poland), No. 5, 5–14 (1979), in Polish.
1521. Kukkonen, A., Laiho, R., and Levola, T., Brillouin scattering and ultrasonic measurements of phonons in K_2CuF_4. *Semicond. & Insul.* (GB) **5**, 77–83 (1979).
1522. Kumar, A., and Sharma, T. P., Nonlinear susceptibility of GaAs. *Indian J. Pure & Appl. Phys.* (India) **17**, 110–111 (1979).
1523. Kumar, G. S., Ethiraj, R., and Krishna Murty, V. G., Sign reversal of piezo-optic birefringence in mixed crystals of KCl–KBr. *Acta Crystallogr. A* (Denmark) **35** 857–858 (1979).
1524. Kushibiki, J., Chubachi, N., and Mikoshiva, N., Absolute signs of photoelastic constants of ZnO. *J. Appl. Phys.* (USA) **49**, 4961–4962 (1978).
1525. Kuzoukova, T. A., and Nilov, E. V., Investigation of acoustic phenomena in a Pockels cell using KDP and DKDP crystals modulated at high frequencies. *Opt. & Spectrosc.* (USA) **42**, 392–395 (1977).

1526. Landolt–Börnstein Tables–New Series (1979): Elastic, piezoelectric, pyroelectric, piezo-optic, electro-optic constants, and nonlinear dielectric susceptibilities of crystals, in Group III, Vol. 11, revised, updated, and extended edition of Vols. III/1 and III/2, K.-H. Hellwege and A. M. Hellwege, eds. Springer-Verlag, Heidelberg/New York (1979), pp. 495–670.

1527. Leitz, H., On the accuracy and range of electro-optical distance measurement in the case of the Zeiss ELDI-2. *Z. Vermessungswes.* (Germany) **102**, 152–156 (1977), in German.

1528. Luty, F., and Diaz-Gongora, A., Electrobirefringence from paraelectric defects in cubic crystals. *Phys. Status Solidi B* (Germany) **85**, 693–706 (1978).

1529. Madvaliev, V. M., Sandler, Yu. M., Serikov, V. I., and Serdobolskaya, O. Yu., Behaviour of some photoelastic constants of TGS and KH_2PO_4 crystals near the phase transition point. *Sov. Phys.-Solid State* (USA) **19**, 524–525 (1977).

1530. Maeda, M., and Ikada, T., Elastic properties of Rochelle salt–ammonium Rochelle salt mixed crystals. *J. Phys. Soc. Jpn.* (Japan) **45**, 162–168 (1978).

1531. Mahajan, V. N., Theory of acousto-optic interaction, standing and travelling sound waves. *Wave Electron.* (Netherlands) **2**, 4052–4055 (1976).

1532. Mamedov, A. M., Nonlinear optical properties of SbSi. *Sov. Phys.-Solid State* (USA) **19**, 488–490 (1977).

1533. Mastikhin, V. M., Bagdanov, S. V., Darvoid, T. I., Lisitskii, I. S., Sapozhnikov, V. K., and Sheloput, D. V., Acousto-optical characteristics of KRS-5 single crystals. *Sov. J. Opt. Technol.* (USA) **44**, 480–481 (1977).

1534. Mayer, W. G., "Determination of the elastic constants of sapphire by different ultrasonic methods." Ph. D. Thesis, Michigan State University, East Lansing, Michigan (USA), 1958.

1535. Mclellan, E. J., and Figueira, J. F., Ultrafast Pockels cells for the infrared. *Rev. Sci. Instrum.* (USA) **50**, 1213–1217 (1979).

1536. Melnikova, C. V., Aleksandrov, K. S., Anistratov, A. T., and Beznosikov, B. V., Birefringence and structural phase transitions in $ABCl_3$ type crystals. *Sov. Phys.-Solid State* (USA) **19**, 18–20 (1977).

1537. Mertens, R., On the symmetry of diffraction pattern obtained by the diffraction of light by ultrasonic waves. *Acustica* (Germany) **28**, 182–185 (1973).

1538. Mertens, R., and Leroy, O., Diffraction of intense laserlight by an ultrasonic wave in the case of large ultrasonic wavelengths, in: *Ultrason. Int. Brighton, England, 28–30 June, 1977*, IPC Sci. Technol. Press, Guildford, England (1977), pp. 416–421.

1539. Meseguer, F., Piezo-optical properties of alkali halides and their temperature dependence. *Phys. Status Solidi B* (Germany) **89**, 281–288 (1978).

1540. Minkov, B. I., and Timan, B. L., Light diffraction by ultrasound in anharmonic crystals. *Sov. Phys. Acoust.* (USA) **18**, 406–407 (1973).

1541. Moharam, M. G., and Young, L., Criterion for Bragg and Raman–Nath diffraction regimes. *Appl. Opt.* (USA) **17**, 1757–1759 (1978).

1542. Mohler, E., and Thomas, B., Electro-optic properties of CuCl below the absorption edge. *Phys. Status Solidi B* (Germany) **79**, 509–517 (1977).

1543. Munn, R. W., Intensity of Brillouin scattering in molecular crystals. *Chem. Phys.* (Netherlands) **39**, 303–307 (1979).

1544. Namada, M., Yamamoto, K., and Abe, K., Elastic and photoelastic constants of ZnTe measured by Brillouin scattering. *J. Phys. D (Applied Physics)* (GB) **10**, 1309–1313 (1977).

1545. Nelson, D. F., and Lazay, P. D., The dispersive asymmetry of the elasto-optical tensor. *Phys. Rev. B* (USA) **16**, 4659–4665 (1977).

1546. Nelson, D. F., *Electric, Optic and Acoustic Interaction in Dielectrics*. Wiley, Chichester, England (1979), 539 pp.

1547. Nelson, D. F., Piezo-optic and electro-optic constants of crystals—Introduction, in: *Landolt–Börnstein Tables—New Series*, K.-H. Hellwege and A. M. Hellwege, eds., Group III, Vol. 11, Springer-Verlag, Heidelberg/New York (1979), pp. 495–505.

1548. Ohtsuka, Y., Tanone, A., and Itoh, K., New visualization method of progressive acoustic wavefronts using a Köster prism. *Opt. Commun.* (Netherlands) **31**, 9–12 (1979).

1549. Ohtsuka, Y., Optical coherence modulation by two ultrasonic waves. *J. Phys.* (Canada) **57**, 1420–1425 (1979).

1550. Onaka, R., and Ito, H., Pockels effect of KDP and ADP in the ultraviolet region. *J. Phys. Soc. Jpn.* (Japan) **41**, 1303–1309 (1976).

1551. Onaka, R., and Ito, H., Electro-optical effect and vacuum ultraviolet absorption in KDP and ADP. *Bull. Acad. Sci. USSR Phys. Ser.* (USA) **41**, 59–63 (1977).

1552. Pandya, T. P., and Saxena, A. K., Measurement of gradient index profiles by Babinet fringe analysis. *Rev. Sci. Instrum.* (USA) **50**, 369–373 (1979).

1553. Pawlewicz, W. T., Mann, I. B., Lowdermilk, W. H., and Milam, D., Laser–damage–resistant transparent conductive indium tin oxide coatings. *Appl. Phys. Lett.* (USA) **34**, 196–198 (1979).

1554. Pelant, I., and Sladky, P., The ratio of photoelastic constants p_{11}/p_{12} of AgCl. *Phys. Status Solidi A* (Germany) **42**, K57–K59 (1977).

1555. Petrov, M. P., Miridinov, S. V., Stepanov, S. I., and Kulikov, V. V., Light diffraction and nonlinear image processing in electro-optic $Bi_{12}SiO_{20}$ crystals. *Opt. Commun.* (Netherlands) **31**, 301–305 (1979).

1556. Pinnow, D. A., Abrams, R. L., Lotspeich, J. F., Henderson, D. M., Plant, T. K., Stephens, R. R., and Walker, C. M., An electro-optic tunable filter. *Appl. Phys. Lett.* (USA) **34**, 391–393 (1979).

1557. Pique, J. P., Dolino, G., and Vallade, M., Optical observations and electro-optic effect of NH_4Cl near its order–disorder phase transition. *Ferroelectrics* (GB) **21**, 349–350 (1978).

1558. Pizzo, R., Laser-beam modulation: Electro-optic or acousto-optic? *Electro-optical Systems Design* (USA) **11**, 37–41 (1979).

1559. Plancke-Schyuten, G., and Mertens, R., The diffraction of light by standing supersonic waves at the normal incidence of light. I. Approximate solutions of Raman–Nath equations. *Simon Steven* (Belgium) **46**, 105–115 (1973).

1560. Prakash, R., Kumar, V., and Goyal, S. C., Photoelastic constants of cuprous halides. *Phys. Lett. A* (Netherlands) **71**, 243–244 (1979).

1561. Proklov, V. V., Antonov, S. N., and Mash, M. Y., High stability acousto-optic devices using bulk acoustic waves in TeO_2. *Electron. Lett.* (GB) **14**, 535–536 (1978).

1562. Rand, S. C., and Stoicheff, B. P., Brillouin spectrum of solid methane and determination of elastic and elasto-optic properties. *J. Opt. Soc. Am.* (USA) **68**, p. 1402 (1978).

1563. Rand, S. C., Rao, B. S., Enright, G. D., and Stoicheff, B. P., Pockels coefficients for rare-gas crystals. *Phys. Rev. B* (USA) **19**, 4205–4209 (1979).

1564. Ray, B. K., and Nigam, G. D., Quick enumeration of independent tensor components of physical properties of crystals in trigonal and hexagonal point groups. *Proc. Indian Natl. Sci. Acad. Part A* (India) **A45**, 387–391 (1979).

1565. Reibold, R., Molkenstruck, W., and Swamy, K. M., Experimental study of the integrated optical effect of ultrasonic fields. *Acustica* (Germany) **43**, 253–259 (1979).

1566. Rouvaen, J. M., Ghazaleh, M. G., Bridoux, E., and Torguet, R., On a general treatment of acousto-optic interactions in linear anisotropic crystals. *J. Appl. Phys.* (USA) **50**, 5472–5477 (1979).

1567. Salje, E., and Bismayer, U., The electro-optic effect in NH_4IO_3. *Opt. Commun.* (Netherlands) **20**, 303–304 (1977).

1568. Sapriel, J., and Vacher, R., Photoelastic tensor components of $Gd_2(MoO_4)_3$. *J. Appl. Phys.* (USA) **48**, 1191–1194 (1977).

1569. Sapriel, J., *Acousto-Optics*, John Wiley and Sons, Chichester, New York, Brisbane, and Toronto (1979), 126 pp.

1570. Senenov, V. E., and Sheloput, D. V., Acousto-optical properties of dense and ultradense flint glasses. *Avtometriya* (USSR), No. 2, 93–95 (1979), in Russian.

1571. Shaldin, Yu. V., and Semenchinskii, S. G., Electro-optic properties of KH_2PO_4 and RbH_2PO_4 crystals near phase transition. *Sov. Phys. Solid State* (USA) **19**, 709–711 (1977).

1572. Shanker, J., Bakshi, P. S., and Sharma, O. P., Theory of the photoelasticity of cuprous halides. *Solid State Commun.* (USA) **24**, 217–219 (1977).

1573. Shanker, J., Sharma, O. P., and Sharma, J. C., Volume dependence of the refractive indices of rutile crystal under hydrostatic pressure. *Solid State Commun.* (USA) **22**, 355–357 (1977).

1574. Shanker, J., Sharma, H. P., Sharma, O. P., and Sharma, J. C., Theory of the photoelastic effect in ZnO, ZnS and CdS crysatls. *Solid State Commun.* (USA) **22**, 401–404 (1977).

1575. Shanker, J., Sharma, O. P., and Lashkari, A. N. G., An empirical analysis of the photoelastic effect in XF_2 crystals. *Phil. Mag.* (GB) **35**, 1671–1673 (1977).

1576. Shanker, J., Lashkari, A. K. G., and Sharma, O. P., Analysis of volume dependence of refractive index in ionic and partially covalent crystals. *Indian J. Pure & Appl. Phys.* **16**, 578–581 (1978).

1577. Shanker, J., Lashkari, A. K. G., and Gupta, V. P., Effect of three-body interactions on the electronic polarizabilities and photoelastic behaviour of alkali halides. *Phys. Status Solidi B* (Germany) **91**, 263–268 (1979).

1578. Sharma, H. P., Shanker, J., and Verma, M. P., Effect of Lundqvist's many-body potential on the photoelastic behaviour of ionic crystals. *Phys. Rev. B* (USA) **15**, 4100–4102 (1977).

1579. Sharma, H. P., Shanker, J., and Verma, M. P., Evaluation of strain polarizability parameter in alkali halides. *J. Phys. & Chem. Solids* (GB) **38**, 255–256 (1977).

1580. Shchavlev, O. S., Molev, M. I., and Elina, N. N., The effect of alkali metal oxides on the optical stress coefficient of silicate and phosphate glasses. *Sov. J. Opt. Technol.* (USA) **43**, 538–540 (1976).

1581. Sheloput, T. A., Sheloput, D. V., and Glushkov, V. F., Photoelastic properties of some chalcogenide glasses. *Sov. J. Opt. Technol.* (USA) **44**, 418–419 (1977).

1582. Shibata, N., Jinguji, K., Kawachi, M., and Edahiro, T., Nondestructive structure measurement of optical fibre preforms with photoelastic effect. *Jpn. J. Appl. Phys.* (Japan) **18**, 1267–1273 (1979).

1583. Shih, C. C., and Yariv, A., Theoretical model for the linear electro-optical effect of crystals. *J. Opt. Soc. Am.* (USA) **69**, 1476 (1979).

1584. Shkerdin, G. N., Proklov, V. V., and Gulyaev, Yu. V., Theory of acousto-optic phenomena at high light intensities. *Sov. Phys. Solid State* (USA) **19**, 243–247 (1977).

1585. Shkerdin, G. N., Gulyaev, Yu. V., and Proklov, V. V., Theory of the diffraction of light by sound in doped semi-conductors. *Sov. Phys. Semicond.* (USA), **12**, 535–539 (1978).

1586. Shkerdin, G. N., and Gulyaev, Yu. V., Theory of the diffraction of light by sound in impurity-containing crystals. *Sov. Phys.-Solid State* (USA) **20**, 548–549 (1978).

1587. Singh, G. P., and Basu, B. K., Propagation of longitudinal ultrasonic waves in RbH_2PO_4 near its ferroelectric phase transition temperature. *Pramana* (India) **11**, 119–134 (1978).

1588. Sinha, N. K., Dispersion of birefringence in glasses and its measurement with the Babinet compensator using white light. *Phys. & Chem. Glasses* (GB) **18**, 65–69 (1977).

1589. Sinha, N. K., Normalised dispersion of birefringence of quartz and stress optical coefficient of fused silica and plate glass. *Phys. & Chem. Glasses* (GB) **19**, 69–77 (1978).

1590. Smit, J., On the signs of the elasto-optic coefficients. *J. Appl. Phys.* (USA) **49**, 2935- 2936 (1978).

1591. Smit, J., On the photoelastic shear constants of cubic ionic crystals. *J. Appl. Phys.* (USA) **50**, 3628–3629 (1979).

1592. Smith, A. M., Automated birefringence measurement system. *J. Phys. E* (*Scientific Instruments*) (GB) **12**, 927–930 (1979).

1593. Strelets, L. I., and Samartsev, V. V., Spectrometer for studying acoustical paramagnetic resonance by the Bragg light diffraction method. *Prib. & Tekh. Eksp.* (USSR) **6**, 102–104 (1978), in Russian.

1594. Tada, K., Kikuchi, K., and Sato, K., Dispersion of photoelastic coefficient in ZnSe. *Jpn. J. Appl. Phys.* (Japan) **16**, 757–760 (1977).

1595. Tanaka, K., Measurements of refractive index dispersion by optical wave guiding. *Appl. Phys. Lett.* (USA) **34**, 672–674 (1979).

1596. Tereknev, V. E., Diffraction modulators using ultrasonic standing waves in $LiNbO_3$. *Opt. & Spectrosc.* (USA) **42**, 193–196 (1977).

1597. Theocaris, P. S., and Loakimidis, N. I., On the determination of stress-optical constants by the method of reflected caustics. *J. Phys. D* (*Applied Physics*) (GB) **12**, 497–504 (1979).

1598. Thomas, P. J., Rand, S. C., and Stoicheff, B. P., Elastic constants of parahydrogen determined by Brillouin scattering. *Can. J. Phys.* (Canada) **56**, 1494–1525 (1978).

1599. Uozato, H., and Nagata, R., Modified Babinet compensator by using interference colour method for the measurement of small retardation. *Jpn. J. Appl. Phys.* (Japan) **16**, 2275–2276 (1977).

1600. Vaivod, P. A., Voronov, V. V., Ivleva, L. I., Kuz'minov, Yu. S., Makarova, O. B., and Polozkov, N. M., Dielectric and electro-optic properties of ferro-electric $Ba_{0.54}Sr_{0.46}Nb_2O_6$ doped with Y, La and Tm. *Sov. Phys.-Solid State* (USA) **19**, 1852–1853 (1977).

1601. Veerabhadra Rao, K., and Narasimhamurty, T. S., Linear electro-optic dispersion in KDP. *J. Phys. C* (*Solid State Phys.*), (GB) **11**, 2343–2347 (1978).

1602. Vlokh, O. G., Lutsiv-Shumskii, L. F., and Pirogov, Yu. A., Photoelasticity and electro-optic effects in $KH_{2(1-x)}D_{2x}PO_4$ crystals at forbidden deformation. *Ukr. Fiz. Zh.* (USSR) **23**, 146–148 (1978), in Russian.

1603. Volkov, V. V., Egorova, G. A., Lonskii, E. S., Potapov, E. V., and Rakov, A. V., Propagation of light in lithium niobate crystals along direction close to the optic axis. *Sov. Phys. Crystallogr.* (USA) **23**, 257–260 (1978).

1604. Wakatsuki, N., Chubachi, N., and Kikuchi, Y., Light diffraction by ultrasonic waves in optically active crystals. *Electron. & Commun. Jpn.* (USA) **58**, 117–125 (1975).

1605. Wasa, K., Yamada, M., and Hamaguchi, C., Brillouin scattering in ZnSe. *Technol. Rep. Osaka Univ.* (Japan) **28**, No. 1430–1458, 447–455 (1978), in Japanese.

1606. Waxler, R. M., Horowitz, D., and Feldman, A., Precision interferometer for measuring photoelastic constants. *Appl. Opt.* (USA) **16**, 20–22 (1977).

1607. Waxler, R. M., Horowitz, D., and Feldman, A., Optical and physical parameters of Plexiglas 55 and Lexan. *Appl. Opt.* (USA) **18**, 101–104 (1979).

1608. Weber, E., Electro-optical materials and their applications. *Elettron. Oggi* (Italy), No. 10, 1285–1293 (1976), in Italian.

1609. Weber, H. J., and Haussühl, S., Point symmetry, electrogyration and quadratic electro-optic effect of cubic $Sr(NO_3)_2$, $Ba(NO_3)_2$ and $Pb(NO_3)_2$. *Z. Kristallogr.* (Germany) **146**, 303–307 (1977).

1610. Weber, H. J., Piezo-optic and quadratic electro-optic effect in cubic potassium cyanide. *Z. Kristallogr.* (Germany) **146**, 309–314 (1977).

1611. Weber, H. J., Temperature dependence of quadratic electro-optic and electrostrictive effects in paraelectric phase of methylammonium alums. *Z. Kristallogr.* (Germany) **149**, 1–15 (1979).

1612. White, F. E., and Teas, D. C. (eds.), References to contemporary physics on acoustics—Acousto-optical effects, optical diffraction, ultrasonic processing, acoustic visualization, and holography. *J. Acoust. Soc. Am.* (USA) **65** (Suppl. 2), R54–R58 (1979).

1613. Wong, C. F., Birefringence measurement using a photoelastic modulator. *Appl. Opt.* (USA) **18**, 3996–3999 (1979).

1614. Xu, Z-Yi, Li, Y-Jin, and Li, Y-Yuan, Review of the research work on α-lithium iodate single crystals. *Wuli* (China) **8**, 9–19 (1979), in Chinese.

1615. Yamamoto, R., and Ishihara, S., Organic electro-optic materials. *Natl. Tech. Rep.* (Japan) **22**, 826–834 (1976), in Japanese.

1616. Yong, Z., and Deo-Fan, Z., The electro-optic, pyroelectric, dielectric and piezoelectric properties of the single crystal $Sr_4NaLiNb_{10}O_{30}$. *Acta Phys. Sin.* (China) **28**, 234–239 (1979), in Chinese.

1617. Yoshino, K., Yamaguchi, M., and Inuishi, Y., Dependence of optical damage in lithium niobate on temperature, pressure and applied electric field. *Technol. Rep. Osaka Univ.*, (Japan) **29**, No. 1459–1491, pp. 93–104 (1979).

1618. Zumsteg, F. C., Bierlein, J. D., and Giev, T. E., $K_xRb_{1-x}TiOPO_4$—A new nonlinear optical material. *J. Appl. Phys.* (USA) **47**, 4980–4985 (1976).

Author Index

This index includes only those authors whose work is cited in the text; it does not include all names given in the Bibliography.

The numbers in parentheses are pages on which an author's work is referred to, but on which his name is not explicitly given.

Subject Index